Lecture Notes in Civil Engineering

Volume 100

Lecture Notes in Civil Engineering (LNCE) publishes the latest developments in Civil Engineering - quickly, informally and in top quality. Though original research reported in proceedings and post-proceedings represents the core of LNCE, edited volumes of exceptionally high quality and interest may also be considered for publication. Volumes published in LNCE embrace all aspects and subfields of, as well as new challenges in, Civil Engineering. Topics in the series include:

- Construction and Structural Mechanics
- Building Materials
- Concrete, Steel and Timber Structures
- Geotechnical Engineering
- Earthquake Engineering
- Coastal Engineering
- Ocean and Offshore Engineering; Ships and Floating Structures
- Hydraulics, Hydrology and Water Resources Engineering
- Environmental Engineering and Sustainability
- Structural Health and Monitoring
- Surveying and Geographical Information Systems
- Indoor Environments
- Transportation and Traffic
- Risk Analysis
- Safety and Security

To submit a proposal or request further information, please contact the appropriate Springer Editor:

- Mr. Pierpaolo Riva at pierpaolo.riva@springer.com (Europe and Americas);
- Ms. Swati Meherishi at swati.meherishi@springer.com (Asia - except China, and Australia, New Zealand);
- Dr. Mengchu Huang at mengchu.huang@springer.com (China).

All books in the series now indexed by Scopus and EI Compendex database!

More information about this series at http://www.springer.com/series/15087

Zinoviy Blikharskyy
Editor

Proceedings of EcoComfort 2020

Springer

Editor
Zinoviy Blikharskyy
Lviv Polytechnic National University
Lviv, Ukraine

ISSN 2366-2557 ISSN 2366-2565 (electronic)
Lecture Notes in Civil Engineering
ISBN 978-3-030-57339-3 ISBN 978-3-030-57340-9 (eBook)
https://doi.org/10.1007/978-3-030-57340-9

This Springer imprint is published by the registered company Springer Nature Switzerland AG
The registered company address is: Gewerbestrasse 11, 6330 Cham, Switzerland

Preface

Renewable energy sources, heat and gas supply, ventilation and water supply of the buildings, as well as building constructions and construction technologies are the areas where the great improvement has been made in recent years.

After successful international conference in the field of the theoretical and practical issues of implementation of energy-saving technologies in building's life-support systems that took place in 2016, Lviv, Ukraine, Lviv Polytechnic National University has organized and held the II International Scientific Conference «EcoComfort and Current Issues of Civil Engineering» on 16–18 September 2020 in Lviv.

The Lecture Notes in Civil Engineering (LNCE) contain the latest advances, innovations and applications in the field effective methods of calculation, resource-saving technologies and progressive materials in civil and environmental engineering, as presented by leading international researchers and engineers at the II International Scientific Conference «EcoComfort and Current Issues of Civil Engineering», held in Lviv, Ukraine, on 16–18 September 2020. It covers highly diverse topics, including ecological and energy-saving technology, renewable energy sources, heat, gas and water supply, ventilation and air conditioning, microclimate provision systems, modern technology in water purification and treatment, protection of water ecosystems, as well as modern architectural shaping and structural solutions, aspects of structural behaviour and modelling of advanced methods of analysis, experimental tests and numerical simulations, innovative materials and products. The contributions, which were selected by means of a thorough international peer-review process, highlight numerous exciting ideas that will stimulate new research directions and promote multidisciplinary collaboration among different specialists.

In particular, this book introduces a number of new advances and includes interdisciplinary research, theoretical and experimental studies that advance achievements in civil engineering and energy-saving technologies in life-support systems of buildings. These advances have significant implications in a number of research areas such as renewable and clean energy, waste and wastewater management, enhanced sustainability, sustainable materials and industrial ecology, and building automation, sustainable construction and infrastructures.

The 62 papers have been carefully reviewed and selected from over 100. A total of over 150 researchers are represented in this book.

I express my sincere gratitude to the members of the Organizing Committee, to the members of the Scientific Committee and, in particular, to all the authors and participants for their substantial and valuable contributions.

I express my sincere gratitude to Dr. Dieter Merkle, Vice President of Applied Sciences, and Mr. Pierpaolo Riva, Editor of Engineering and Applied Sciences at Springer, for their great support in preparing the final book.

We hope that the reader will find the book inspirational for further research and professional life.

September 2020 Zinovy Blikharskyy

Organization

Editorial Board/Scientific Committee

Stepan Shapoval	Lviv, Ukraine
Petro Kholod	Lviv, Ukraine
İlker Tekin	Karabuk, Turkey
Todor Donchev	Kingston, Great Britain
Hsein Kew	Kingston, Great Britain

Contents

About Editor-in-Chief

Prof. Zinoviy Blikharskyy is Professor of civil engineering and Director of the Institute of Civil Engineering and Building Systems of the Lviv Polytechnic National University, Lviv, Ukraine. He has more than 35 years of teaching, research and administrative experience in higher education institutions and universities in Ukraine and abroad. He obtained his Ph.D. and Dr. Sc. degrees in the field of building constructions in Lviv Polytechnic National University. He is the winner of the State Prize for "Restoration of the Palace of Bandinelli", an award of the President of Ukraine. His scientific interests are corrosion of reinforced concrete structures taking into account load action; features of restoration of bearing capacity, operational suitability, strengthening of reinforced concrete structures under the action of loading; use of non-metal reinforcement for strengthening of normal and inclined sections of reinforced concrete structures; and reconstruction and restoration of buildings and constructions of historical and modern building. He has published more than 160 scientific papers, has supervised ten candidates' theses and was Co-founder of seven international conferences "Current Issues of Civil and Environmental Engineering and Architecture" (2007–2019 years). Also, he is Active Member of the Academy of Construction of Ukraine and Editor-in-Chief of the Journal "Theory and Practice of Construction" since 2007.

The Lecture Notes in Civil Engineering (LNCE) have as its principal aim the fostering of professional exchanges between scientists and practitioners who are interested in the field of construction and environmental engineering. Topics include ecological and energy-saving technology, renewable energy sources, heat, gas and water supply, microclimate provision systems, innovative materials and products, modern technology in water purification and treatment, protection of water ecosystems, modern architectural shaping and structural solutions and optimization.

The book includes high-quality research articles accepted for publication on the basis of thorough peer reviews. Each scientific work of the conference was reviewed by at least two members of the Scientific Committee. Their efforts have contributed to the high quality of final works, and therefore their reviewing activities are acknowledged and appreciated.

Experimental Research of Strength Characteristics of Steel Fiber Reinforced Concrete Gutters and Modeling of Their Work Using the Finite Element Method

Oleksandr Andriichuk[1(✉)], Ivan Yasiuk[1], Serhii Uzhehov[1], and Oleksandr Palyvoda[2]

[1] Lutsk National Technical University, Lutsk 43018, Ukraine
aleklutsk@gmail.com
[2] Kryvyi Rih National University, Kryvyi Rih 50027, Ukraine

Abstract. Steel fiber reinforced concrete and structures based on it have increased crack resistance, toughness and elasticity, abrasion resistance, service life and are less sensitive to vibration and shock effects than similar structures with typical reinforcement. These elements at short-term and repeated loads three series of experimental researches have been conducted. Testing of prototypes (gutters) is performed by applying a central vertical load to the metal traverse beam acting on the gutter as evenly distributed. The results of experimental research of strength characteristics the steel fiber concrete, reinforced concrete and steel fiber reinforced concrete in gutters under the action of single and repeated loads are given. The results of the simulation of gutters using the finite element method also presented. Increasing the percentage of reinforcement with steel fibers gives an increase in carrying capacity for SFRC gutters at repeated low-cycle loads. Cracks in RC and SFRRC samples were with direct nature, while in SFRC samples they were with net nature.

Keywords: Reinforced concrete · Steel fiber reinforced concrete · SFRC · Steel fiber rebar reinforced concrete · Steel fiber · Drainage · Gutter · Load-bearing capacity · Strength · Crack resistance · Modeling · Finite elements

1 First Section

Steel fiber reinforced concrete is an effective material for manufacturing many new and strengthening existing building structures [1–4].

Steel fiber reinforced concrete (SFRC) and structures based on it have increased crack resistance, toughness and elasticity, abrasion resistance, service life and are less sensitive to vibration and shock effects than similar structures with typical reinforcement [5–7].

Efficiency of SFRC application in building structures is achieved due to reduction of labor costs for reinforcement works, combination of technological operations for preparation, reinforcement, laying and tamping of SFRC mixture, reduction of costs for various types of current repair and, accordingly, prolongation of constructions life [8–10].

Z. Blikharskyy (Ed.): EcoComfort 2020, LNCE 100, pp. 1–8, 2021.
https://doi.org/10.1007/978-3-030-57340-9_1

The increased crack resistance and rigidity of steel fiber reinforced concrete in comparison with classical reinforced concrete allow to use it for the manufacture of gutters for highway drainage, which can also be used in melioration systems [11].

To investigate the features of strength and deformation characteristics of the stress-strain state of steel fiber reinforced concrete gutters and work of these elements at short-term and repeated loads three series of experimental researches have been conducted. For the planned research, it was made the samples of concrete, reinforced concrete, steel fiber rebar reinforced concrete and steel fiber reinforced concrete. Detailed design solutions, manufacturing techniques of experimental gutters are presented in the article [12]. In the first series the work of gutters made of concrete, reinforced concrete (RC), steel fiber rebar reinforced concrete (SFRRC) and steel fiber reinforced concrete (SFRC) under the action of short-term disposable loads was studied. In the second series the work of gutters made of reinforced concrete, combined reinforced concrete, and SFRC under the action of short-term repeated loads with the level of loading $\eta = 0.6$ of the destructive value was investigated. In the third series the work of SFRC gutters with a percentage of reinforcement $\mu = 1\%$; 2%; 3% at single and repeated loads with levels $\eta = 0{,}3$; $0{,}5$; $0{,}7$ was investigated.

Testing of prototypes (gutters) is performed by applying a central vertical load to the metal traverse beam acting on the gutter as evenly distributed. During the test, the lower part of the gutter is supported by a rigid base. For this purpose the hydraulic press PSU-125 is used (Fig. 1).

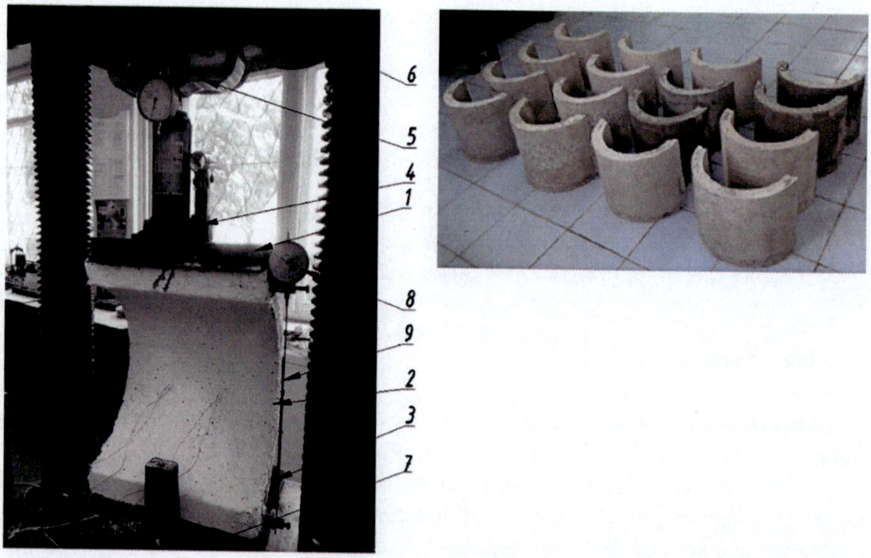

Fig. 1. General view of the drainage gutter research: 1 - metal traverse beam; 2 - experimental drainage gutter; 3 - fixed base; 4 - hydraulic jack; 5 - dynamometer; 6 - top plate of PSU-125; 7 - bottom plate of PSU-125; 8 - displacement detection sensor; 9 - displacement bar.

To improve the accuracy of measuring the acting force a dynamometer is used, which makes it possible to measure loads with an accuracy of 50 N. Load was supplied by the hydraulic jack in steps of 8–12% of the destructive force, determined by the theoretical calculation. The detailed technique of experimental gutters research is presented in works [13].

First series of research was done on gutters: 1C-1, 1C-2, 1C-3 without reinforcement; 1RC-1, 1RC-2, 1RC-3 with steel rebar frames reinforcement $\rho = 2\%$; 1SFRC-1, 1SFRC-2 and 1SFRC-3 made of SFRC with percentage of steel fiber reinforcement $\mu = 2\%$ and 1SFRRC-1, 1SFRRC-2 and 1SFRRC-3 with steel rebar frames reinforcement $\rho = 1\%$ and steel fiber reinforcement $\mu = 1\%$ (total reinforcement percentage 2%) according to the plan of the experiment [12]. Results on crack resistance obtained during the study of samples of the first series under short-term disposable loads are presented in Fig. 2.

Graphs of the average cross section displacements of the research elements of the first series are presented in Fig. 3.

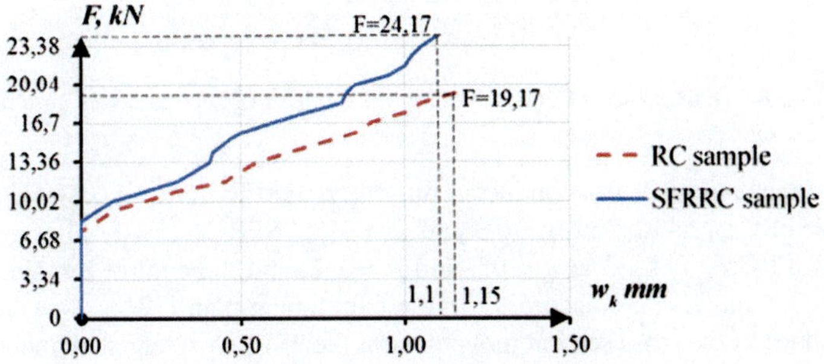

Fig. 2. Dependence of crack opening width on load.

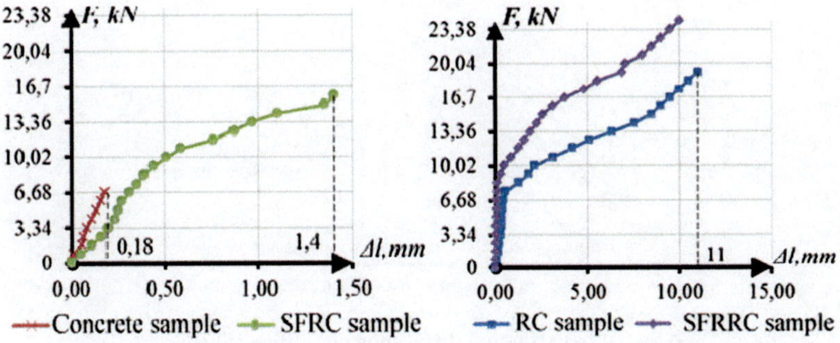

Fig. 3. Average cross section movements of research elements.

Second series of research was conducted on gutters: 2RCr-1, 2RCr-2 and 2RCr-3 with steel rebar frames reinforcement $\rho = 2\%$; 2SFRRCr-1, 2SFRRCr-2 and 2SFRRCr-3 with steel rebar frames reinforcement $\rho = 1\%$ and steel fiber reinforcement $\mu = 1\%$, and the total 2% according to the plan of the experiment [12].

Results on crack resistance obtained during the study of samples of the second series at short-term repeated loads are presented in Fig. 4.

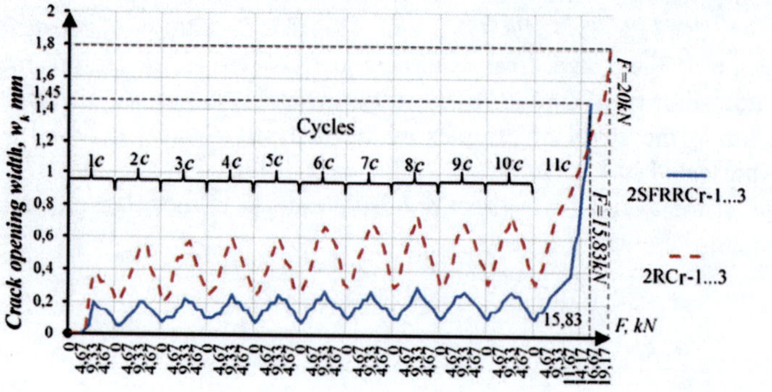

Fig. 4. Dependence of crack opening on the load of second series gutters

Third series research was conducted on gutters: $3SFRCr_{+1;+1}$–1…3 at $\eta = 0.7$ and $\mu = 3\%$; $3SFRCr_{+1;-1}$–1…3 at $\eta = 0.3$ and $\mu = 3\%$; $3SFRCr_{-1;+1}$–1…3 at $\eta = 0.7$ and $\mu = 1\%$; $3SFRCr_{-1;-1}$–1…3 at $\eta = 0.3$ and $\mu = 1\%$, and the control sample $3SFRCr_{0;0}$–1 at $\eta = 0.5$ and $\mu = 2\%$ according to the experiment plan [12].

The values of the cross section movement of the 3SFRCr research elements at 1 and 10 cycles were averaged and presented in Fig. 5 and Fig. 6.

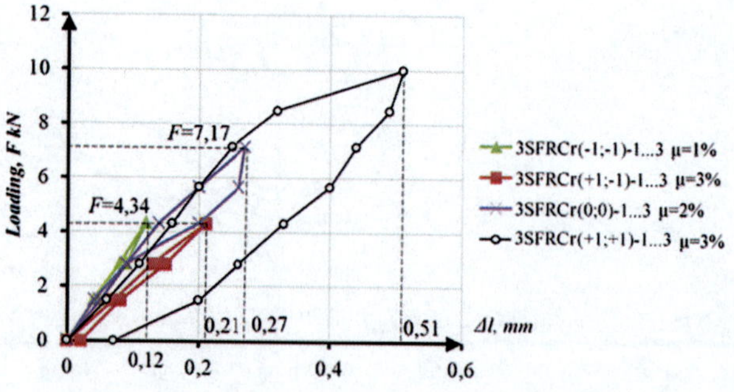

Fig. 5. Average cross-sectional displacements of the 3SFRCr on 1 cycle

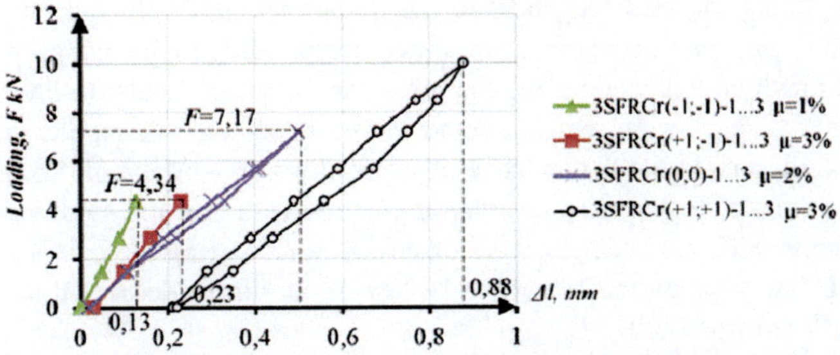

Fig. 6. Average cross-sectional displacements of the 3SFRCr on 10 cycle

According to the purpose of the work, the modeling of SFRC gutter by the finite element method (FEM) was carried out, numerical values of stresses, bending moments and displacements arising in SFRC gutters under the action of known loads on them were determined. During the research of SFRC gutters work it was applied physically nonlinear final volume elements № 236 (universal spatial 8-angle isoparametric finite elements (FE).

For this purpose, the gutter was simulated in the form of a semi-tube with dimensions in accordance with experimental models (Fig. 7), and its triangulation to the finite elements was performed under the following conditions:

- the tray wall is divided into four layers by its thickness (at $t_w = 40$ mm thickness of one layer $t_l = 10$ mm). Within one layer we accept that there is one finite element with thickness $t = 10$ mm respectively;
- the length of a half-circle along the axial radius of the gutter element is $l_c = 54$ cm. We divide it into FE with the length of the sides by $l_{fe} = 10$ mm. We have 54 finite elements along the axial radius.
- gutter length $l = 300$ mm. According to the gutter length, we divide it into FE with the condition that the length of each FE $l = 10$ mm (i.e. the FE has a cubic geometric shape). In the length of the gutter element there are 30 finite elements.

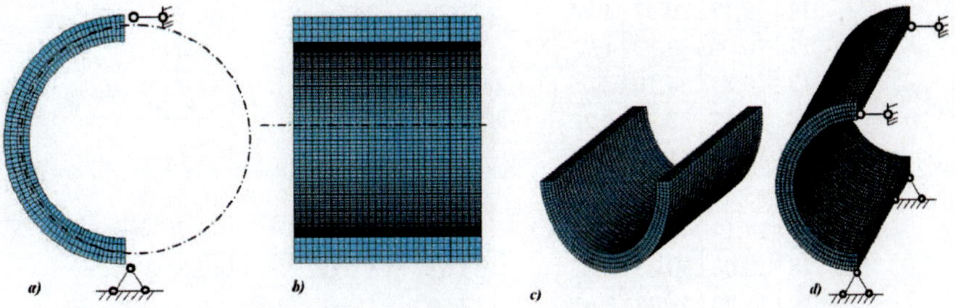

Fig. 7. Simulation of gutters: *a* - YOZ projection (front view); *b* - XOY projection (side view); *c, d* - isometric projections (position during operation and during testing)

After modeling the gutter element in the shape of a half-pipe and dividing it into finite elements, the mechanical characteristics corresponding to the real properties of the materials from which it is made using the phenomenon of physical non-linearity are set.

SFRC as a material from which a gutter in the shape of a half-pipe is made was set using a graph, describes the dependence of stress-strain (σ - ε). Evenly distributed load is set using a simple step-by-step calculation method with uniform steps (10 steps with 300 iterations in each of these steps are taken). Before starting the calculation process for the SFRC half-pipe gutter we set the 14 law of non-linear deformation ("piecewise linear law of deformation").

By using the "LIRA-SAPR" software we simulated and calculated a SFRC half-pipe gutter with parameters corresponding to the natural samples. The values of moments in the middle of the cross-section obtained as a result of this calculation are presented in Table 1. The values of certain bending moments and stresses in trays are also presented in Table 1.

Table 1. Values of bending moments and stresses in the SFRC gutter (1SFRC-1…3).

№	Efforts, *F*		Theoretically calculated in lateral cross section		LIRA-SAPR, *σ, MPa*	Error, %
	kN	*kN/m*	Moment, $(M, kN \cdot m)$	Tension (σ, MPa)		
1	0,83	2,77	0,16	1,97	2,00	1,52
2	1,67	5,57	0,32	3,97	4,00	0,75
3	2,50	8,33	0,48	5,94	7,00	17,84
4	3,33	11,10	0,63	7,91	9,00	13,78
5	4,17	13,90	0,79	9,90	11,00	11,11
6	5,00	16,67	0,95	11,88	13,00	9,42
7	5,83	19,43	1,11	13,85	16,00	15,52
8	6,67	22,23	1,27	15,84	18,00	13,63
9	7,50	25,00	1,43	17,81	20,00	12,29
10	8,33	27,77	1,58	19,78	22,00	11,22
11	9,17	30,57	1,74	21,78	24,00	10,19
12	10,00	33,33	1,90	23,75	26,00	9,47
13	10,83	36,10	2,06	25,72	29,00	12,75
14	11,67	38,90	2,22	27,72	30,00	8,22
15	12,50	41,67	2,38	29,69	33,00	11,14
16	13,33	44,43	2,53	31,66	35,00	10,54
17	14,17	47,23	2,69	33,65	37,00	9,95
18	15,00	50,00	2,85	35,63	40,00	12,26
19	15,83	52,77	3,01	37,60	41,00	9,04

2 Conclusions

At repeated loads which level does not exceed 70% of destructive ones, SFRC gutter works elastic.

Increasing the percentage of reinforcement with steel fibers with $\mu = 1\%$ to $\mu = 2\%$ gives an increase in carrying capacity (average) for SFRC gutters at repeated low-cycle loads up to 40%, and with $\mu = 1\%$ to $\mu = 3\%$ gives an increase in carrying capacity up to 58%.

Cracks in the test samples occurred in the area with maximum bending moment. Cracks in RC and SFRRC samples were with direct nature, while in SFRC samples they were with net nature.

The error of bending stress values obtained by using of "LIRA-SAPR" software and theoretical calculation is mainly within the range of 8…13% with arithmetic mean value of error $\bar{X} = 10.56$ and average square deviation $\sigma = 3.96$. Variation coefficient value $\upsilon = 37\%$.

References

1. Hameed, A.A., Mohannad, H.A.: Influence of steel fiber on the shear strength of a concrete beam. Civ. Eng. J. **4**(7), 1501–1509 (2018)
2. Babych, E.M., Andriichuk, O.V.: Strength of elements with annular cross sections made of steel-fiber-reinforced concrete under one-time loads. Mater. Sci. **52**, 509–513 (2017)
3. Babych, Y.M., Andriichuk, O.V., Kysliuk, D.Ya., Savitskiy, V.V., Ninichuk, M.V.: Results of experimental research of deformability and crack-resistance of two span continuous reinforced concrete beams with combined reinforcement. IOP Conf. Ser. Mater. Sci. Eng. – MSE **708**(1), 012043 (2019). 1–8
4. Dvorkin, L., Dvorkin, O., Ribakov, Y.: A method for optimal design of steel fiber reinforced concrete composition. Mater. Des. **32**(6), 3254–3262 (2011)
5. Kinash, R., Bilozir, V.: Deformational calculation method of bearing capability of fiber-concrete steel bending elements. Czasopismo Techniczne **8-A**(7), 49–58 (2014)
6. Shmyh, R., Bilozir, V., Vvsochenko, A., Bilozir. V.: Carrying capacity of bending concrete elements reinforced by fibro and stripes taken from used PET bottles. In: International Scientific and Practical Conference World science, ROST, vol. 1, no. 2, pp. 88–93 (2018)
7. Soetens, T., Matthys, S.: Different methods to model the post-cracking behaviour of hooked-end steel fibre reinforced concrete. Constr. Build. Mater. **73**, 458–471 (2014)
8. Oliveira, M., Ramos, E., Oliveira, D., Neto, B.: Analysis of influence of concrete element format and properties steel fibers on flexural toughness. Matéria (Rio J.) **23**(3), 1–14 (2018)
9. Sameh, Y.: Effect of steel fibers and gfrp sheet on the behavior of lightweight concrete specimens using waste lightweight sand bricks. Int. J. Eng. Res. Tech. **7**(03), 69–75 (2018)
10. Islam, M.M., Dhar, A., Patowary, F., Asif, J.H., Rahman, S., Das, S.S., Das, S., Chowdhury, M.A., Siddique, A.: Experimental investigation and finite element analysis on P-M interaction diagram of RC square columns made of steel fiber reinforced concrete (SFRC). In: Joint Conference on Advances in Bridge Engineering-III, IABSE-JSCE, pp. 192–200. Dhaka, Bangladesh (2015)
11. Vegera, P., Vashkevych, R., Blikharskyy, Z.: Fracture toughness of RC beams with different shear span. In: MATEC Web of Conferences, vol. 174, p. 02021 (2018)

12. Andriichuk, O., Babich, V., Yasyuk, I., Uzhehov, S.: The influence of repeated loading on work of the steel fiber concrete drainage trays and pipes on the roads. In: MATEC Web of Conferences, vol. 116, no. 02001, pp. 1–9 (2017)
13. Andriichuk, O., Yasyuk, I.: Metodyka eksperymentalnoho doslidzhennya dyspersno-armovanykh prydorozhnikh lotkiv vodovidvedennya. Visnyk Odeskoi natsionalnoi akademii budivnytstva ta arkhitektury: zbirnyk naukovykh prats **58**, 11–18 (2015)

Influence of Nanoparticles on the Processes of Heat Accumulation During Material Phase Transformations

Ievgen Antypov[1], Valery Gorobets[1(✉)], Yurii Bohdan[2], and Viktor Trokhaniak[1]

[1] National University of Life and Environmental Sciences of Ukraine, Kiev 03041, Ukraine
gorobetsv@ukr.net
[2] Kherson State Maritime Academy, Kherson 73000, Ukraine

Abstract. Heat accumulators are widely used to store and further utilize heat energy from solar collectors, heat pumps and other renewable energy sources. In this case, materials with phase or chemical transformations may be used. This makes it possible to significantly increase the accumulation of heat energy per unit mass of the heat storage material compared to the batteries on solid and liquid materials. To increase the efficiency of materials with phase transformations, fillers are used in the form of nanoparticles having a high coefficient of thermal conductivity. This paper presents the results of numerical modeling and experimental study of the influence of nanoparticles on the processes of heat energy accumulation in paraffin. The influence of geometric dimensions, concentration and thermophysical characteristics of nanomaterials on the processes of heat energy accumulation during paraffin phase transformations is investigated. Thermophysical properties of composite materials were investigated by optical spectroscopy. In experimental and numerical studies, the dynamics of melting processes of accumulating material with nanoparticles near cylindrical heat sources have been studied. A comparative analysis of heat-accumulating materials in the presence and absence of nanoparticles was carried out.

Keywords: Heat transfer · Phase transformations · Heat-accumulating material · Nanoparticles · Numerical modeling · Optical spectroscopy methods

1 Introduction

The limitation of traditional fuel and energy resources (gaseous, liquid and solid fuels), as well as the negative impact of their combustion products on the environment, testify to the necessity of creation and practical using of combined power supply systems for consumers, including alternative one's sources of energy.

The problem of the transition from traditional to alternative sources of power supply for industrial consumers and housing or communal services has been researched in [1–3]. It is noted that the most universal for autonomous energy supply systems, in terms of energy potential and the possibility of its public use is solar power. However, as is known, the use of this energy is complicated by the stochastic nature of its receipt,

Z. Blikharskyy (Ed.): EcoComfort 2020, LNCE 100, pp. 9–17, 2021.
https://doi.org/10.1007/978-3-030-57340-9_2

which results in the need to ensure the continuity of these systems. The latter can be achieved both by incorporating traditional power sources into them and using different types of energy accumulators. Therefore, from all considered variants of schemes of power supply of consumers, the most interesting is the variant of complex energy supply of the consumer both from the external electrical network, and from the sources of solar energy with the possibility of accumulation of thermal and electrical energy.

Of all types of existing designs of heat accumulators, heat storage accumulators with phase or chemical transformations of accumulating material are the most promising. They allow to provide high density of accumulated energy and stable temperature at thc cxit from the heat accumulator.

A number of authors [4, 5] analyzed the different methods of heat accumulation, which showed that one of the promising directions is the use of heat accumulators of periodic action, based on phase or chemical transformations of accumulating material. In accumulators of this type there per unit of mass and a stable temperature of the coolant at the outlet of the heat accumulator. Detailed information on the processes of heat and mass transfer occurring during phase transformations of the accumulating material is given in sources [4–14]. The analysis of the conducted researches shows that in order to increase the efficiency of heat accumulators, it is necessary to intensify the processes of heat and mass transfer during "charge" and "discharge" of these accumulators, as well as to reduce heat losses when storing thermal energy [6–11]. One of the most important directions for improving efficiency of the heat accumulators based on the principle of phase transformation is to improve the thermal storage and conductivity properties of the heat-accumulating materials. This in turn achieved by using of nanoparticles [15–17]. But, notwithstanding this, a number of questions regarding the thermophysical properties, concentration, type and size of nanoparticles in the composite phase change material (PCM) remain insufficient investigated. Hence one can important to carry out numerical and experimental studies of heat accumulation processes in PCM contains a fillers in the form of nanoparticles with a high coefficient of thermal conductivity.

2 Analysis and Methodology

The system of complex energy supply of consumers from the energy of traditional and alternative sources with the use of electric and thermal accumulators is proposed. It is based on the use of solar energy and/or electrical power with the possibility of accumulation of its excess in accumulators of electric and thermal energy of the improved design. The application of this system is envisaged in places where the solar energy converters are the main source, and the electrical network is auxiliary. Figure 1 shows the functional diagram of the proposed system of integrated energy supply of consumers.

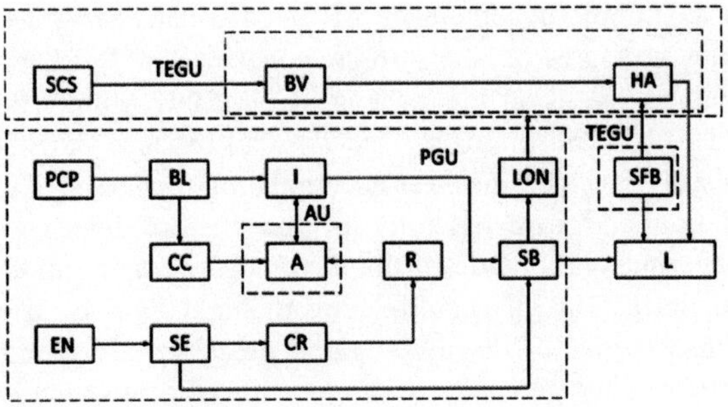

Fig. 1. Block diagram of the system of integrated energy supply of consumers with energy accumulators.

The system consists of three main units: PGU – power generation unit, AU – accumulator unit and TEGU – thermal energy generation unit, which includes: photovoltaic converter panels (PCP) that convert solar energy into electrical energy for load needs (L), the electrical network (EN), the connection of which occurs in periods when the energy generated by the PCP is not sufficient for the needs of L, at the conditions of full discharge of the accumulator unit (AU). The transformation of the voltage of the PCP from a constant to a variable is carried out using the inverter (I), the reverse – rectifier (R). When increasing the level of solar activity, in order to avoid recharge of accumulators, the system provides a controller of charge (CC) from photovoltaic converter panels and a charge regulator (CR) from the electrical network. To switch off/on the system from/to the electrical network in the system provides a switching element (SE). Transmission of energy from generating sources to the consumer is carried out through distribution equipment (DE). For the accumulation of electric energy in the system, it is proposed to use two groups of electrochemical storage accumulators. In this case, each accumulator unit is connected alternately to the voltage rectifier of the electrical network or to a photovoltaic converter panel for charging them and to the voltage inverter – for power supply to consumers in such a way that one accumulator unit always has a maximum charge (≈100%) and was ready for the backup power of consumers. In a separate block, a group of consumers is allocated (circulating pumps, heat storage backup heat sources), which constitute the load of their own needs (LON) of the proposed system.

The system also includes an automatically controlled ballast load (BL) designed to dispose of the potential excess power generated by the PCP. As BL uses, in addition to the accumulator unit, the system of additional heating of the heat exchange surface of the heat accumulator (HA) of the improved design. The connection of the latter allows both the intensification of the accumulation process and the accumulation of surplus generated electricity in it, thereby turning it into heat. But, under conditions of high cloudy, in the transition and heating periods there is no excess energy, then the system's operation goes into its consumption, not generation. In this case, the heat energy deficit is covered by the inclusion of the solid-fuel boiler (SFB) system, and the electric - from electrical network. Another one source of heat energy in TEGU can be solar collector system (SCS) which connected with HA by buffer vessel (BV).

With a view to improving efficiency of the thermal energy generation unit, an analysis of existing structures of heat storage devices [18–22] was carried out. Owing to conducted analysis by using the package COMSOL Multiphysics 3.5a., a new axonometric model of the accumulator (see Fig. 2a) proposed for the storage of heat based on the phase transition of heat-accumulating material [23, 24]. Moreover, numerical simulation and investigation of processes of hydrodynamics and heat transfer in the experimental module of the developed heat accumulator (see Fig. 2b) during phase transformations of accumulating material have been carried out. As a result, the basic regularities of the investigated processes are found and recommendations on the choice of the geometry of the location of heat sources inside volume of the heat accumulator are given in source [24].

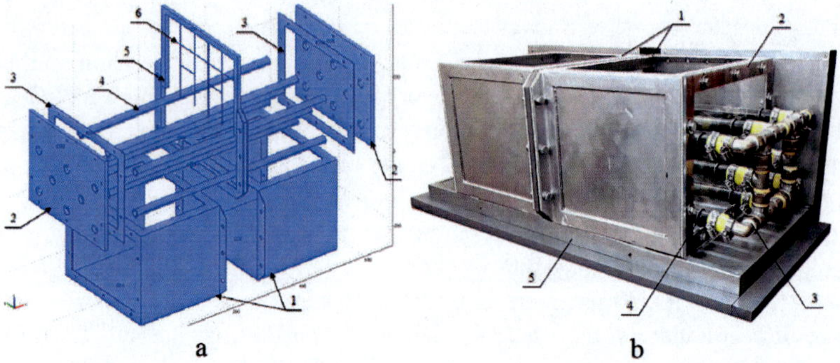

Fig. 2. Axonometric view of the model of heat accumulator (a): 1 – frame; 2 – end cap; 3 – sealing; 4 – tube bundle; 5 – movable frame; 6 – sensors grid, and general view of the experimental module (b): 1 – collapsible case; 2 – end cap; 3 – distribution collector; 4 – tube bundle; 5 – thermal insulation panel

In the proposed design of the developed heat accumulator, the coolant moves in the system of heating tubes which can emit or absorb the heat energy that is accumulated or taken from the accumulating material. As the heat-accumulating material chose Paraffin T3 with a phase transition temperature $T_f = 54–56$ °C which is wide spread applied PCM with nontoxic and noncorrosive properties. Nevertheless, utilization of it limited to its low conductivity. One of the perspective method to enhance the thermal conductivity of paraffin based PCM is uses of high conductivity nanoparticles.

Researches of the influence of micro- and nanoparticles of metals on the intensity of processes of phase changes of pure and composite paraffin was carried out on a developed experimental installation (see Fig. 3). It consists of two identical experimental modules, one of which is to be imbued with pure paraffin T3, and the second one is to be filled with the same material and nanoparticles with a high coefficient of thermal conductivity. As the example of that material, micro- and nanoparticles of copper are proposed to be used. It must be italicized that the plant is additionally equipped with a cylindrical heat energy source having a power of 1.5 kW, a laboratory autotransformer, a gauging set and equipment for temperature measuring purposes using thermo-couples placed on the surface of the heat energy source and in the volume of paraffine.

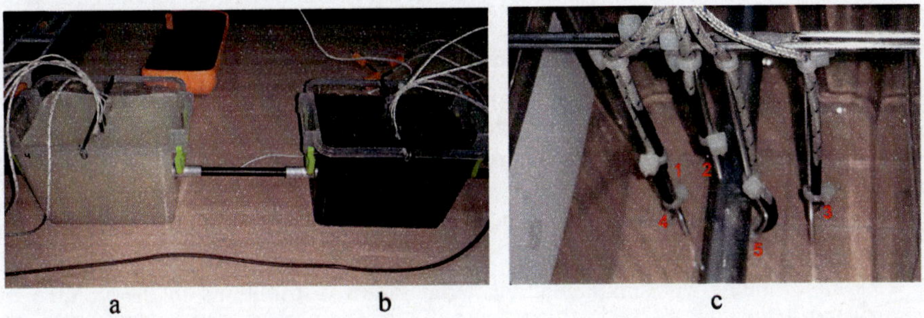

Fig. 3. General view of the experimental installation: (a) experimental modules with pure paraffin T3, (b) experimental modules with composition of paraffin T3 and nanoparticles with a high coefficient of thermal conductivity, (c) arrangement of thermocouples on a "temperature grid" installed into experimental module

For the measuring temperature procedures, carried out on the surface of a cylindrical heat source, the K-type thermocouples are used. Furthermore, in the control process of temperature fields directly in the PCM volume, 5-way thermocouples are involved creating the so-called "temperature grid" (Fig. 3c). This "grid" was executed in the form of a frame with fixed thermocouples on it and placed inside the modules. This allowed us to trace the dynamics of temperature fields in the volume of PCM of the experimental modules.

To determine the influence of temperature and impurities on the structural transformations of PCM experimentally, a spectroscopy technique (Raman spectroscopy) is utilised, which is a non-destructive method for the reflection of crystallization processes for the analysis of thermodynamic properties. The thought is particularly being given to using pure paraffin as an initial material.

Computer numerical simulation was performed for pure paraffin T3 and paraffin with metal nanoparticles. The mathematical model of heat and mass transfer processes in the phase transformations of the accumulation material includes the Navier-Stokes equation and the convective heat transfer equation with the use of the package of applications COMSOL Multiphysics 3.5a.

3 Results and Discussion

Experimental studies have shown that the distribution of metallic micro- and nanoparticles in the volume of the main substance depends on the size of the particles. In composite materials having fractions of nanoparticles of 0.2 mm or larger, the particles were noticed to be settled on the bottom of the chamber after paraffin melting.

As a result, this issue initiated their alleviation effect on the processes of heat accumulation in accumulators. Therefore, in this work, experimental studies were carried out on metallic nanoparticles that did not have a sedimentation effect.

For studying the melting processes and the dynamics of temperature distribution in PCM, experimentally based studies (Fig. 2 and 3) were successfully performed. They were carried out with more precision and certainty laboratory conditions for pure paraffin and paraffin with copper filler with a fraction size of 0.07–0.1 mm (volume ratio 20:1). As a result of the performed experimental studies of transfer processes during phase transformations of pure and composite paraffin, dependences of temperature indices of HAM on timc were obtained. The data analysis process of the obtained dependences indicates an increase in temperature throughout the volume in the presence of copper fraction compared with pure paraffin, which averaged 4–6%. This fact indicates an increase in the thermal conductivity of composite paraffin with copper nanoparticles.

Result of numerical simulation, the temperature distributions in the volume of accumulative material for pure paraffin and paraffin with copper filler can be visually presented in Fig. 4. As shown by the analysis of the obtained distributions, the averaged temperature indices in the composite PCM as being 6–9% higher than those in pure paraffin, which correlates with the data obtained in experimental studies.

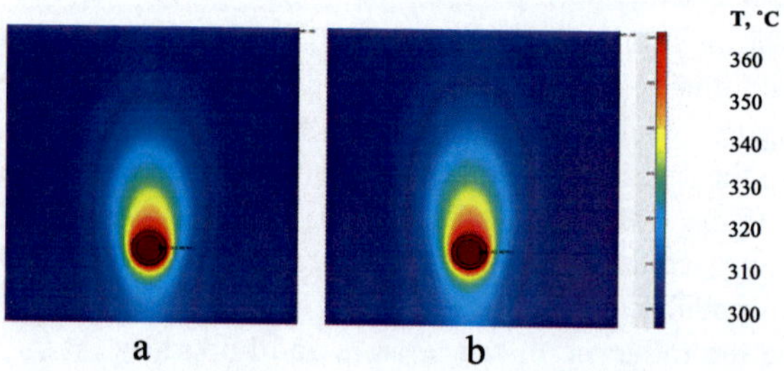

Fig. 4. Temperature distributions in volume of PCM: a - pure paraffin; b - paraffin with copper nanoparticles

In addition to numerical and experimental investigation of the processes of heat accumulation mentioned earlier, proposed heat accumulator application analysis for the system of integrated energy supply of consumers from the energy of traditional and alternative sources with the use of electric and heat accumulators. The number of photovoltaic converters (at a base power of one module of 100 W) is calculated, which is necessary to cover the load of consumers from 500 to 10000 W depending on the period of the year. Average duration of work lasting during the day 8–12 h. The results of calculations are presented in Fig. 5.

The probable duration of uninterrupted power supply of consumers is determined. For example, from two 12-V groups of accumulators, at an active power of consumers from 500 to 2000 W and the temperature of the electrolyte accumulator: +20… −20 °C. The nominal capacity of one accumulator belonging to each group is 100 A·h. The time of continuous supply of consumers (Fig. 6a), capacity consumption and the "discharge" of the heat accumulator (Fig. 6b) are determined. These values are calculated depending on the load capacity and the number of m heat accumulators in the group from 2 to 4 pcs. In addition, the magnitude of the power "discharge" is determined depending on the temperature of the electrolyte and the degree of "charge" of the heat accumulator.

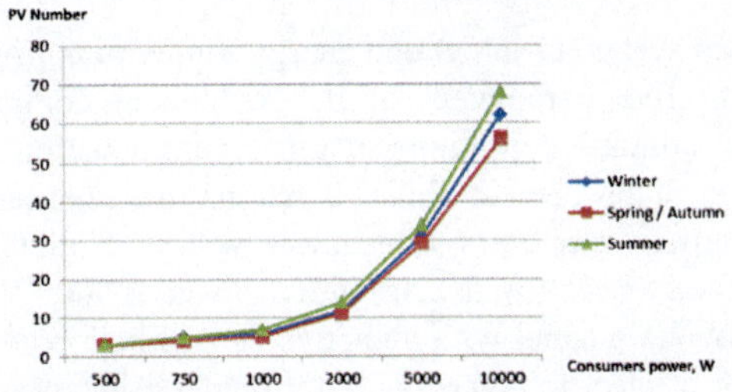

Fig. 5. The ratio of the number of PV depending on the power of consumers of el. Energy

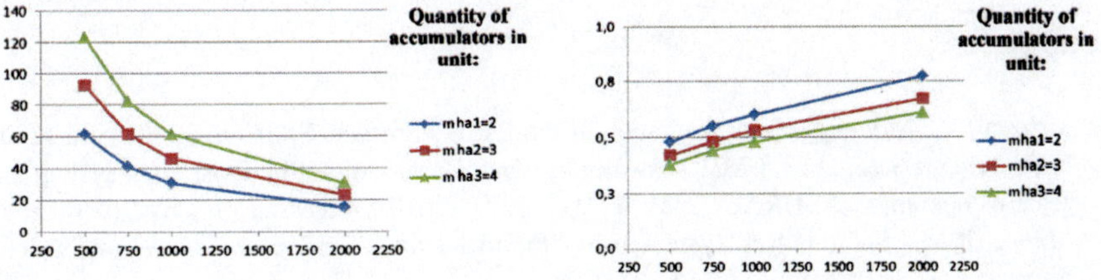

a b

Fig. 6. The ratio of the duration of the discharge (a) and the coefficient of "discharge" (b) of the accumulator unit depending on the power of consumers

It has been established that in order to ensure efficient operation of autonomous power supply of consumers with a nominal electric power of 2000 W for 24 h and a peak (up to 2 h) with a power of 5000 W, the system should consist of a photovoltaic panel with an active area of heliopoles of 7,7–8,9 m^2 and 12 V accumulator of electric power with a nominal capacity of 315–365 A·h, and with a thermal power of 700 W load - from a solar collector area of 4.5–5.8 m^2 and one accumulator of heat of phase transition with a power of 8 kW.

4 Conclusion

1. The main variants of construction of existing systems of combined power supply of consumers, among other things, from alternative sources of energy are analyzed. The low efficiency of using their power and the high cost of the unit of energy received is revealed. As a result, the system of complex energy supply of consumers with the use of alternative sources and combined energy accumulators was proposed. It is effective for use in various climatic zones.
2. Experimental and numerical investigations have proved the feasibility of using metallic micro and nanoparticles of high thermal conductivity to intensify the process of phase changes in pure paraffin and increase the efficiency of heat accumulators.
3. For the proposed system of integrated energy supply of consumers, initial data on the choice of effective parameters of its component composition is obtained, depending on the climatic conditions of the accommodation and capacity of the domestic consumer. It has been established that in order to ensure efficient operation of autonomous power supply of consumers with a nominal electric power of 2000 W for 24 h and a peak (up to 2 h) with a power of 5000 W, the system should consist of a photovoltaic panel with an active area of heliopoles of 7,7–8,9 m^2 and 12 V accumulator of electric power with a nominal capacity of 315–365 A·h, and with a thermal power of 700 W load from a solar collector area of 4.5–5.8 m^2 and one accumulator of heat of phase transition with a power of 8 kW.

References

1. Badescu, V.: Modeling Solar Radiation at the Earth's Surface. Springer, Heidelberg (2008)
2. Todorovic, M.S., et al.: 3.5 MW seawater heat pump assisted multipurpose solar system's 25 years of operation. ASHRAE Trans. **116**(1), 227+ (2010). Accessed 21 May 2020
3. Abbott, R.M.: Solar power system using thermal storage and cascaded thermal electric converters. U.S. Patent No. 6,313,391 (2001)
4. Trp, A.: An experimental and numerical investigation of heat transfer during technical grade paraffin melting and solidification in a shell-and-tube latent thermal energy storage unit. Solar Energy, **79**(6), 648–660 (2005)
5. Suganya, G., Bapu, B.R.R.: Experimental studies on performance of latent heat thermal energy storage unit integrated with solar water heater. Int. J. Chem. Sci. **14**(2), 1165–1171 (2016)
6. Zhang, Q., Huo, Y., Rao, Z.: Numerical study on solid–liquid phase change in paraffin as phase change material for battery thermal management. Sci. Bull. **61**(5), 391–400 (2016)
7. Nasieka, I., et al.: An analysis of the specificity of defects embedded into (1 0 0) and (1 1 1) faceted CVD diamond microcrystals grown on Si and Mo substrates by using E/H field discharge. J. Cryst. Growth **491**, 103–110 (2018)
8. da Cunha, J.P., Eames, P.: Thermal energy storage for low and medium temperature applications using phase change materials – a review. Appl. Energy **177**, 227–238 (2016)
9. Liu, L., Su, D., Tang, Y., Fang, G.: Thermal conductivity enhancement of phase change materials for thermal energy storage: a review. Renew. Sustain. Energy Rev. **62**, 305–317 (2016)

10. Fan, L., Khodadadi, J.M.: Thermal conductivity enhancement of phase change materials for thermal energy storage: a review. Renew. Sustain. Energy Rev. **15**(1), 24–46 (2011)
11. Agyenim, F., Hewitt, N., Eames, P., Smyth, M.: A review of materials, heat transfer and phase change problem formulation for latent heat thermal energy storage systems (LHTESS). Renew. Sustain. Energy Rev. **14**(2), 615–628 (2010)
12. Jegadheeswaran, S., Sanjay, D.: Pohekar: performance enhancement in latent heat thermal storage system: a review. Renew. Sustain. Energy Rev. **13**(9), 2225–2244 (2009)
13. Kuboth, S., König-Haagen, A., Brüggemann, D.: Numerical analysis of shell-and-tube type latent thermal energy storage performance with different arrangements of circular fins. Energies, **10**(3), 274 (2017)
14. Huo, Y., Rao, Z.: Lattice boltzmann simulation for solid-liquid phase change phenomenon of phase change material under constant heat flux. Int. J. Heat Mass Transf. **86**, 197–206 (2015)
15. Nabeel, S.D., et. al.: Experimental and numerical investigation of melting of phase change material/nanoparticle suspensions in a square container subjected to a constant heat flux. Int. J. Heat Mass Transfer, **66**, 672–683 (2013)
16. Kaviarasu, C., Prakash, D.: Review on change material with nanoparticle in engineering application. J. Eng. Sci. Technol. Rev. **9**(4), 26–36 (2016)
17. Said, M.A., Hamdy, H.: Effect of using nanoparticles on the performance of thermal energy storage of phase change material coupled with air-conditioning unit. Energy Convers. Manag. **171**(1), 903–916 (2018)
18. Sharma, A., Tyagi, V.V., Chen C.R., Buddhi, D.: Review on thermal energy storage with phase change materials and applications. Renew. Sustain. Energy Rev. **13**(2), 318–345 (2009)
19. Gorobets, V., Treputnev, V.: Heat transfer and the motion of the interphase boundary, when a heat-accumulating material is melted near a horizontal heat source with section finning. Teplofiz. Vys. Temp. **33**(4), 588–593 (1995)
20. Chen, S.L., Hsiao, M.J.: Heat pipe circuit type thermal battery. U.S. Patent No. 6,220,337 (2001)
21. Naumov, A.L., Serov, S.F., Efremov, V.V., Degtyarev, N.S.: Heat accumulator. RU Patent No. 2,436,020 (2011)
22. Thermal Batteries: All about storing solar heat (2019). Accessed 28 Feb
23. Antypov, I.: Numerical study of heat transfer processes in low-temperature heat accumulator in phase transformations accumulate material. Scientific Herald of the National University of Life and Environmental Sciences of Ukraine, vol. 224, pp. 208–213 (2015)
24. Gorobets, V., Antypov, I., Trokhaniak, V., Bohdan, Y.: Experimental and numerical studies of heat and mass transfer in low-temperature heat accumulator with phase transformations of accumulating material. In: MATEC Web of Conferences, vol. 240 (2018)

Experimental Research of Strength Characteristics of Continuous Reinforced Concrete Beams with Combined Reinforcement, and Modelling Their Work by the Finite Element Method

Evgen Babych[1], Oleksandr Andriichuk[2], Mykola Ninichuk[2(✉)], and Dmytro Kysliuk[2]

[1] National University of Water and Environmental Engineering, Rivne 33028, Ukraine
[2] Lutsk National Technical University, Lutsk 43018, Ukraine
matgen@ukr.net

Abstract. The article presents an analysis of experimental research of work of continuous combined-reinforced concrete beams. The authors conducted experimental studies of two-span reinforced concrete beams in which classical rebar frames and dispersed reinforcement of concrete with steel fibers were combined. To investigate the effect of such reinforcement on the stress-strain state of the beams, load-bearing capacity and deformation characteristics, test specimens were made with three different variants of the steel fiber volume distribution, but with the same relative percentage of reinforcement. A numerical experiment was also performed in which the operation of the experimental beams was modeled by the finite element method. The results of this experiment, their analysis and comparison with laboratory data are presented. It is established that the dispersed reinforcement by steel fibers of the stretched zones of the beam reduces the stresses in the rod reinforcement, and allows to slowing down the development of plastic deformations in the reinforcement during the redis-tribution of forces in the inseparable beam.

Keywords: Concrete · Reinforcing · Fibers · Stress-strain state · Load-bearing capacity · Finite element method

1 Introduction

Static indeterminate reinforced concrete structures, mainly continuous, multi-span beams are widely used in modern construction. In particular, they are components of the structures of industrial and civil buildings, overpasses, and bridge structures. The choice of continuous load-bearing elements is due to the peculiarities of their work, manufacturing technology, which is manifested in the most rational use of constituent materials, concrete and reinforcement.

In the design and manufacture of reinforced concrete structures, it is relevant to use dispersed reinforcement—steel fibers, as well as combined reinforcement (the combination of classical reinforcement with dispersed one). A significant advantage of this

Z. Blikharskyy (Ed.): EcoComfort 2020, LNCE 100, pp. 18–25, 2021.
https://doi.org/10.1007/978-3-030-57340-9_3

composite material—steel fibrous reinforced concrete (SFRC), is its high specific strength per unit mass. SFRC and structures based on it, have increased crack resistance, viscosity and elasticity, abrasion resistance, service life and are less sensitive to vibration and shock effects than similar structures with typical reinforcement [1–5]. Efficiency of application of SFRC in building structures can be achieved by reducing labor costs for reinforcement work, combining technological operations for the preparation, reinforcement, laying and sealing of SFRC mixture [6, 7].

It should be noted that the effect of the dispersed reinforcement on the stress-strain state, deflections, crack strength of statically indeterminate structures is poorly understood, so the results of the experimental studies considered in the article are relevant at the present time.

2 Methodology of Research

To determine the effect of the steel fiber reinforcement of concrete continuous beams on their deformation characteristics, three test specimens were tested. They were continuous two span reinforced concrete beams with a length of 300 cm with a cross-section size of 10 × 16 cm and with spans of 140 cm.

The main reinforcement was made in the form of two flat rebar frames, with the primary reinforcement in spans and over the support with Ø10 mm bars (selected with allowance for the redistribution of effort) and transverse reinforcement with Ø4 mm steel bars. The additional, dispersed reinforcement was performed by using steel fibers, in such a way that the specimens had different filling by the fibers volume of the beam, but with the same relative percentage of reinforcement, equal to $\mu = 1\%$ [6]. Thus, the beam B-1 was reinforced with fiber throughout its volume, B-2 – in stretched zones, B-3 – to the height of the two minimum concrete cover zones (Fig. 1). For the disperse reinforcement were used crimped shape steel fibers, 50 mm in length and 1 mm in diameter. The design of prototypes and the technology of their manufacture are described in detail in the article [8].

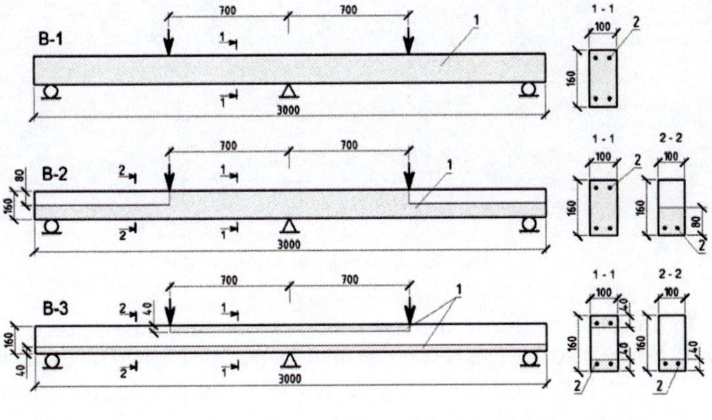

Fig. 1. Zones of reinforcing of beams by steel fibrous: 1—zone of steel fibrous reinforcement; 2—main reinforcing bars.

The experimentally established mechanical characteristics of the cement-sand matrix and reinforced concrete are given in Table 1 and Table 2.

Table 1. Cement-sand matrix properties.

Characteristic	Value
Compressive cube strength f_c, MPa	33.0
Compressive prism strength $f_{ck,prism}$, MPa	22.6
Tensile strength f_{ct}, MPa	0.5

Table 2. Steel fibrous-reinforced concrete properties.

Characteristic	Value
Compressive cube strength f_c, MPa	34.0
Compressive prism strength $f_{ck,prism}$, MPa	23.2
Tensile strength f_{ct}, MPa	1.7

A special flexural testing frame was designed and manufactured for the prototype beam testing (Fig. 2). The flat transverse bending of two-span beams in the testing device was created by means of a hydraulic jack and a steel beam, which transfers two identical (symmetrically arranged to the average support at a distance of 700 mm) actions (controlled by mechanical force gauge) from the hydraulic jack to the prototype beam.

Fig. 2. Bending test setup.

The test samples were loaded in stages. After every effort, the data were recorded and the beams were visually inspected. Detailed methods of experimental research are described in article [8]. The load was accepted as fracture when the strain of reinforcement or concrete reached the limit values.

The mechanical characteristics of the reinforcing bars were also previously determined - three bars with a length of 50 cm each, were tested in the UIM-50 bursting test machine. Tensile forces were applied in stages, after which technological endurance was made to take readings on the devices.

Deformations of each rod were measured with two strain gauges of Gugenberger with 20 mm length of base and with a value of division of 0.001 mm. During the tests, the yield strength and ultimate strength of the reinforcing bars were recorded. Tests of the rods showed that their mechanical characteristics exceed the normative values for reinforcement class A500C but do not correspond to similar indicators of reinforcement class A600C. Thus, for bars with a diameter of 10 mm, the following mechanical characteristics were obtained: strength limit σ_{ud} = 754.8 MPa, yield strength σ_y = 615.1 MPa; modulus of elasticity E_s = 195997.8 MPa, the maximum recorded deformations of the reinforcement, which match the stress σ_y, $\varepsilon_{s0} = 314.2 \times 10^{-5}$.

3 Results of Experimental Research

In the course of the research of specimens to destruction, the characteristic features of their work were observed. The fracture load for the prototype beam testing was: for B-1 F_u = 56 kN, for B-2 F_u = 54 kN, for B-3 F_u = 46 kN. So all beams collapsed by the flexural failures due to the achievement of the maximum strains of the reinforcement in spans and above the middle support, as well as the maximum strain of concrete in compressed areas. Beams B-1 and B-2 had almost the same bearing capacity, the fracture load of the beam B-2 was only 4% less than the beam B-1, which was reinforced with fibers throughout its volume. Beam B-3, which was reinforced with fibers only at the height of the two minimum concrete cover zones, collapsed during the loading, which was less than 18% of that of beams B-1, which indicates a significant effect of dispersive reinforcement throughout the volume on the bearing capacity of the reinforced concrete elements.

The deflections in all beams were developing practically in proportion to the increase of the external load to the level F = 40 kN, after which their development became more intense. Thus, the deflections of the beam B-3 were much larger than those of the other beams, and amounted to 5.08 mm at the destruction of F_u = 46 kN before fracture. The deflections of beams B-1 and B-2 at this level of loading were Δ = 3.75 mm and Δ = 3.8 mm, which is 26.2% and 25.2% less than B-3. The maximum deflection was Δ = 6.41 mm at F_u = 56 kN for beam B-1, and Δ = 4 mm at F_u = 54 kN for beam B-2 (Fig. 3).

With the help of strain gauges, the values of deformations in the armature were obtained at each load level. Thus, for the beam B-1 at the first appearance of cracks, the values of the deformation in the stretched reinforcement was in the spans $\varepsilon_{s,sp} = 47.65 \times 10^{-5}$, and $\varepsilon_{s,sup} = 70{,}21 \times 10^{-5}$ above the middle support which corresponds to the reinforcement stress $\sigma_{s,sp}$ = 125.1 MPa, and $\sigma_{s,sup}$ = 86.2 Mpa. When the beam was loaded to F = 32 kN, the deformations of the reinforcement increased to $\varepsilon_{s,sp} = 112.54 \times 10^{-5}$ and $\varepsilon_{s,sup} = 184.9 \times 10^{-5}$, that is compared to the first degree of load, they have more than doubled.

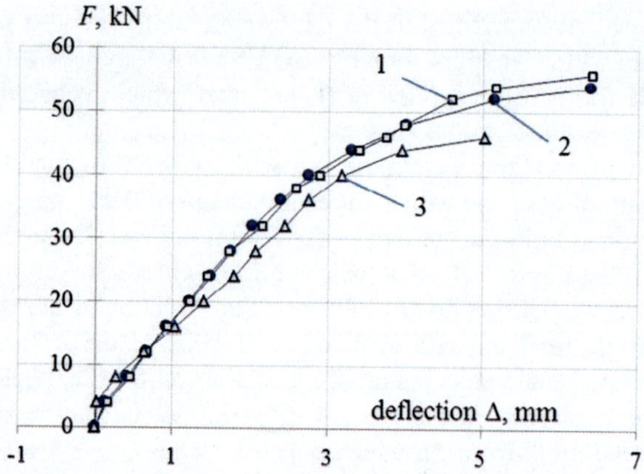

Fig. 3. Change of deflection at loading of beams:1 – Beam B-1; 2 – Beam B-2; 3 – Beam B-3.

At a comparative load levels, for three beams, $F = 40$ kN, the deformation of the reinforcement in the support section of the beam B-1, was $\varepsilon_{s,sup} = 233.7 \times 10^{-5}$, and in the span sections $\varepsilon_{s,sp} = 137.9 \times 10^{-5}$, which corresponds to the reinforcement stress $\sigma_{s,sup} = 462.3$ MPa, and $\sigma_{s,sp} = 325.4$ MPa. At the load level F = 48 kN, the deformation of the reinforcement in the reference section exceeded the limit $\varepsilon_{s0} = 314.2 \times 10^{-5}$ and $\varepsilon_{s,sup} = 371.43 \times 10^{-5}$, which indicates the formation of a plastic hinge above the support due to the stress in the reinforcement yield strength.

During the testing, the beam B-2, which was reinforced with steel fibers only in the stretched areas, the deformation of the reinforcement at the appearance of the first cracks was in the spans $\varepsilon_{s,sp} = 48.56 \times 10^{-5}$, and $\varepsilon_{s,sup} = 50.23 \times 10^{-5}$ above the middle support. At the subsequent loading deformations developed proportionally and at value of the concentrated force in span F = 40 kN, values of reinforcement deformations reached to $\varepsilon_{s,sp} = 177.02 \times 10^{-5}$, and $\varepsilon_{s,sup} = 239.19 \times 10^{-5}$ on 28.2% and 2.3%, respectively, more than the beams B-1. Reaching the rod reinforcement in the support section of the yield strength was observed at a load level $F = 46$ kN.

In the beam B-3, which was reinforced with steel fibers twice the height of the protective layer of concrete, the first appearance of cracks was recorded, as in the beam B-2, at the load level $F = 12$ kN. Deformations of the reinforcement in the span and on the support were respectively $\varepsilon_{s,sp} = 50.57 \times 10^{-5}$, and $\varepsilon_{s,sup} = 47.26 \times 10^{-5}$, that is as in the beam B-2, were almost the same.

At the loading of the beam to the comparative level $F = 40$ kN, the deformations of the reinforcement increased to the values $\varepsilon_{s,sp} = 224.58 \times 10^{-5}$, and $\varepsilon_{s,sup} = 87.03 \times 10^{-5}$ which is 62.7% and by 22.8%, correspondingly, more than the beams B-1 with the same amount of concentrated force in the span. Already at the next load level at $F = 44$ kN, the stresses in the reinforcement above the middle support reached the limit values and a plastic hinge was formed, and almost immediately after the load $F = 46$ kN, the beam was completely destroyed due to reaching the limit deformations of the reinforcement in spans and compressed concrete over the central support.

4 Finite Element Method Modelling

For the modelling work of the beams by the finite element method, the software package LIRA SAPR was used. The beams were modelled by using three-dimensional elements measuring 2 × 2 × 2 cm. The longitudinal and transverse reinforcement was set by separate rod elements.

After modelling the beams, mechanical characteristics that corresponded to the real properties of the materials and took into account their physical nonlinearity were set. To specify the stiffness of the elements, the parameters of experimentally obtained stress-strain curves of concrete and reinforcement were used.

As a result of nonlinear calculation using LIRA SAPR, the values of forces and displacements in the elements of the beams, which are displayed in the graphic form of isofields, were obtained. Also, the values of the forces in the rod elements, which were simulated beam supports, can be taken as the values of the support reactions.

The isofields of the *Nx* stresses obtained as a result of the calculation in the LIRA PC correspond to the stresses in the concrete in height σ_{cx}. Their values are highest in the zones of maximum bending moment and, accordingly, the process of crack formation is most active here. The data of *N* forces in the reinforcing bars are also obtained, which allow us to determine the actual stress in the reinforcement. For destructive loading, the load at which values of stresses in the compressed zones of concrete and the stretched armature reach limit values is accepted.

For the model of the beam B-1, the set value of the destructive load actually corresponded to the real one - F_u = 56 kN. The stresses in the concrete in the zones of compression and tension, at this level of the load, reached its limit values, which were f_{ck} = 4.4 MPa and f_{cftd} = 1.8 MPa, respectively (Fig. 4).

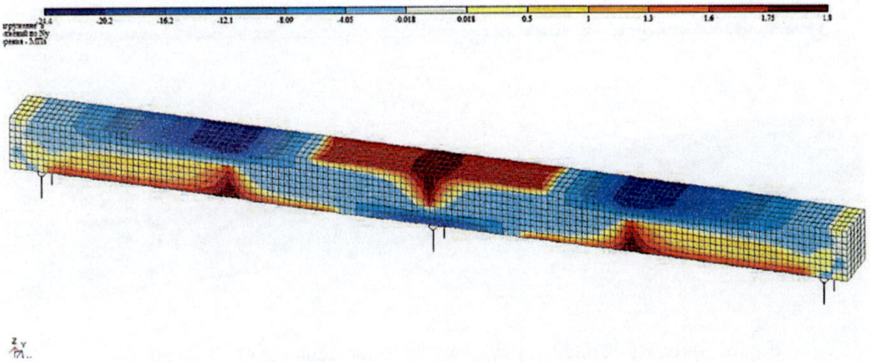

Fig. 4. Intensity isofields of σ_{cx} – stress in the concrete of the beam B-1.

The values of the forces in the reinforcing bars at the magnitude of the force in the span F_u = 56 kN were Ns = 36.9 kN, which corresponds to the stress σ_s = 470 MPa. The stress limits that correspond to the yield strength for the reinforcement simulated in the numerical experiment are σ_y = 434 MPa. This indicates that the stresses in the reinforcing bars above the span have also reached their limits. Accordingly, it can be argued that plastic hinges were formed above the support and in the spans - as a result, the load-bearing capacity of the beam is exhausted.

Also in the process of calculation the values of vertical displacements of the beam elements were obtained. However, these values were much smaller than experimentally obtained, which can be argued by the fact, that the finite element method does not allow to fully and correctly model the process of crack formation in concrete structures.

For the beam B-2, the value of the destructive load was also F_u = 56 kN, which is slightly higher than in the real experiment. The value of stress reinforcement of support was σ_s = 464 MPa > σ_y = 434 MPa (Fig. 5). Maximum deflection f = 5.9 mm.

The value of F_u for the beam B-3 was F_u = 48 kN, which also shows good agreement with the experiment. The values of the stress in reinforcing bars above the support were σ_s = 464 MPa > σ_y = 434 MPa (Fig. 6). The maximum deflection was f = 5.9 mm.

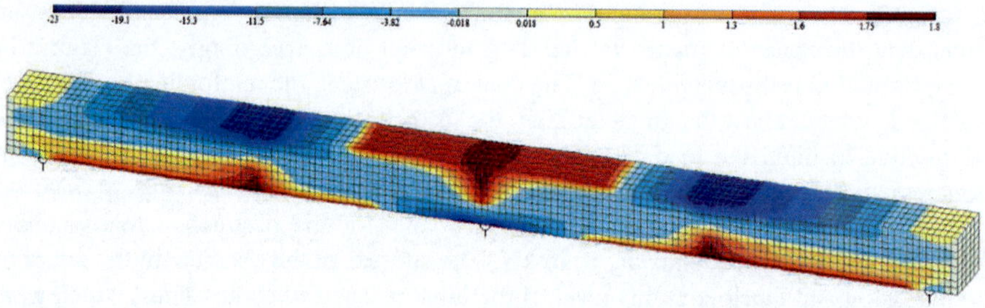

Fig. 5. Intensity isofields of σ_{cx} – stress in the concrete of the beam B-2.

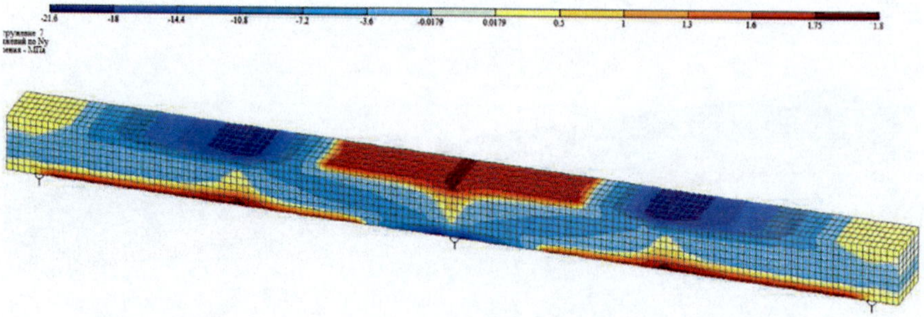

Fig. 6. Intensity isofields of σ_{cx} – stress in the concrete of the beam B-3.

5 Conclusions

Analyzing the results of the research, it can be argued that the presence of dispersed reinforcement in the compressed zone of combined-reinforced continuous beams increases their strength by 15% compared to combined-reinforced beams with dispersed reinforcement throughout the all volume.

It is established that the dispersed reinforcement by steel fibers of the stretched zones of the beam reduces the stresses in the rod reinforcement, and allows to slowing down the development of plastic deformations in the reinforcement during the redistribution of forces in the inseparable beam.

The article presents an algorithm for calculating the combined-reinforced continuous beams by the finite element method using a LIRA SAPR, taking into account the physical nonlinearity of the material. The convergence of the results obtained with LIRA SAPR in comparison with experimental data is in the range of 90%.

References

1. Hameed, A., Mohannad, H.: Influence of steel fiber on the shear strength of a concrete beam. Civil Eng. J. **4**(7), 1501–1509 (2018)
2. Soetens, T., Matthys, S.: Different methods to model the post-cracking behaviour of hooked-end steel fibre reinforced concrete. Constr. Build. Mater. **73**, 458–471 (2014)
3. Kinash, R., Bilozir, V.: Deformational calculation method of bearing capability of fiber-concrete steel bending elements. Czasopismo Techniczne, 8A (2015)
4. Babych, E., Andriichuk, O.: Strength of elements with annular cross sections made of steel-fiber-reinforced concrete under one-time loads. Mater. Sci. **52**, 509–513 (2017)
5. Andriichuk, O., Babich, V., Yasyuk, I., Uzhehov, S.: The impact of the reinforcement percentage on the stress-strain state of the bending steel fiber reinforced concrete elements. In: MATEC Web of Conferences, vol. 230, pp. 02001 (2018)
6. Dvorkin, L., Dvorkin, O., Zhitkovsky, V., Ribakov, Y.: A method for optimal design of steel fiber reinforced concrete composition. Mater. Des. **32**(6), 3254–3262 (2011)
7. Bilozir, V.: Vplyv nyzhidnoi vitky diagramy deformuvannia stalefibrobetonu za roztiagu na nesuchu sdatnist balok. Visnuk Lvivskogo natsionalnogo agrarnogo universytetu. Seria: Arhitectura i silskohospodarske budivnutstvo, **16**, 60–64 (2015)
8. Babych, E., Andriichuk, O., Kysliuk, D., Savitskiy, V., Ninichuk, M.: Results of experimental research of deformability and crack-resistance of two span continuous reinforced concrete beams with combined reinforcement. IOP Conf Ser. Mater. Sci. Eng. **708**(1), 012043 (2019)
9. Oliveira, M., Ramos, E., Oliveira, D., Neto, B.: Analysis of influence of concrete element format and properties steel fibers on flexural toughness. Matéria (Rio J.), **23**, 12192 (2018)

Check for updates

Simulation of the Seismic Resistance of Buildings with Account of Unlimited Soil Space

Maria Barabash[1(✉)], Viacheslav Iegupov[2], and Bogdan Pysarevskyi[1]

[1] National Aviation University, Kiev, Ukraine
bmari@ukr.net

[2] S. Subbotin Institute of Geophysics of NAS of the Ukraine, Odessa, Ukraine

Abstract. The paper focuses on methods for dynamic analysis used in modern construction. The real object located in the Comintern district of Odessa region is taken as an example. When the earthquake loads are simulated, the authors use real accelerograms that were taken during the earthquake at the construction site of the object. For this construction site, several accelerograms were generated; they simulate 7 units of magnitude earthquakes from Vrancha zone and local focal zones. Analysis is carried out with obtained three-component accelerograms by the method of spectral analysis, and direct dynamic analysis was carried out. Described technique enables the user to simulate behaviour of the structural system of a building based on the finite elements developed and implemented in LIRA-SAPR program; these finite elements simulate vibration damping, to remedy the shortcomings of spectral and direct dynamic analyses. It also enables the user to solve physically nonlinear problems with soil and do not significantly increase the size, and respectively, time for analysis of the problem.

Keywords: Seismic sustainability · Finite element method · Substructure method · Computer modeling · LIRA-SAPR

1 Introduction

To provide the safety of buildings and structures in earthquake loads, new methods of earthquake-resistant design [1, 2] are developed based on computer simulation of structures' behaviour in future earthquake loads [3]. Such methods take into account building codes of different countries, specific aspects of earthquake-resistant construction, and soil properties.

When behaviour and reaction of building structures to earthquake loads are simulated, it is necessary to take into account many factors, such as distance from the earthquake focus (hypocenter), 3D behaviour of the structure, interaction with the soil, account of physical nonlinearity [4], etc.

Seismic areas with predicted intensity of earthquake as 7, 8 and 9 units of magnitude occupy up to 20% territory of Ukraine. There are many industrial and cultural centres there; the active construction is often underway. It is highly important to design

Z. Blikharskyy (Ed.): EcoComfort 2020, LNCE 100, pp. 26–33, 2021.
https://doi.org/10.1007/978-3-030-57340-9_4

the earthquake-resistant structures in the rational manner, to improve their reliability and structural safety. Seismic sustainability of buildings [5] erected in the seismic regions of Ukraine shows that the actual earthquake loads on the building significantly exceed design loads that were stipulated in regulatory documents.

2 Purpose

Due to analysis of the system 'overground structure – foundation – soil', it is possible to consider the pressure of the soil space on the stress-strain state of the whole building. As a rule, numerical solution of problems based on the FEM involves a limited finite soil region. The question is how to correctly model the infinite half-space of the soil and define boundary conditions to consider the real damping properties of soil [6]. When the earthquake loads are simulated, boundary conditions should provide damping and infinite wave propagation through the soil.

There are two general approaches to model the soil-structure interaction in dynamic analysis. The essence of the direct method is that the soil area is cut off quite far from the building and rigid boundary conditions are imposed.

3 Method

The method of substructures (SBFEM) was proposed by John Wolf [7]. This method is based on the conversion from the Cartesian coordinate system X, Y (in 2D problem) to the isoparametric coordinates ξ, η. Geometry of the region is described by scaling the boundary with the dimensionless radial coordinate ξ that starts from the scaling centre (point O) to the point on the boundary. $\xi = 0$ at the point O and $\xi = 1$ at the boundary (Fig. 1). The SBFEM method is more accurate than the direct method, so the simulated area may be smaller than for the direct method. As a rule, in this method, the exact boundary conditions are expressed in a dynamic stiffness matrix $[S(t)^{\infty}]$.

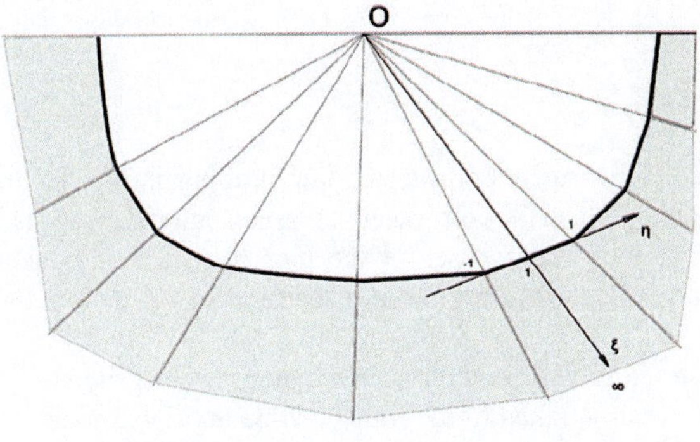

Fig. 1. Isoparametric coordinates ξ, η.

Along the radial line drawn from point O to the node at the boundary, a nodal displacement function is introduced and the equation of displacements is composed.

$$[E^0]\xi^2\{u(\xi)\},_{\xi\xi} + ([E^0] - [E^1] + [E^1]^T)\xi\{u(\xi)\},_{\xi} - [E^2]\{u(\xi)\} + \omega^2[M^0]\xi^2\{u(\xi)\} = 0, \quad (1)$$

where $[E^0]$, $[E^1]$, $[E^2]$ and $[M^0]$ matrices with coefficients.

Connection between the two parts of the soil is provided by the interaction vector

$$\{R_b(t)\} = \int_0^t [M^\infty(\tau)]\{\ddot{u}(t-\tau)\}d\tau, \quad (2)$$

Where $[M^\infty(\tau)]$ - acceleration unit-impulse response matrix, determined as

$$\begin{aligned} &\int_0^t [m^\infty(\tau - t)][m^\infty(\tau)]d\tau + t\int_0^t [m^\infty(\tau)]d\tau + [e^1]\int_0^t\int_0^\tau [m^\infty(\tau)]dtd\tau \\ &+ [e^1]^T\int_0^t\int_0^\tau [m^\infty(\tau)]dtd\tau - \frac{t^3}{6}[e^2]H(t) - t[m^0]H(t) = 0, \end{aligned} \quad (3)$$

where $[m^\infty(\tau)]$ acceleration unit-impulse response matrix in transformed coordinates, $[e^1]$, $[e^2]$ and $[m^0]$ matrices of coefficients, $H(t)$ - Heaviside step function.

Based on Eq. (1) and Eq. (2), the finite elements [8] were developed and implemented in LIRA-SAPR software. These elements simulate interaction between the limited part of soil and the rest part of the half-space: 2-node FE for solving 2D problem, 3-node and 4-node FE - to solve 3D problems.

Spectral analysis of earthquake load is usually carried out. The serious shortcoming of this approach is that it is not possible to take into account physical nonlinearity and the soil.

The paper describes technique for simulation and dynamic analysis; this technique could remedy the shortcomings of the spectral method.

4 Numeric Test

The model of real building is considered; the building is located in the Comintern district of Odessa region, it is considered as seismically dangerous and seismically active zone of Ukraine. The focuses of the most severe earthquakes are located in Vrancha zone and in Dobrudji region at a distance of up to 300 km from the construction site.

For this construction site, several accelerograms were generated; they simulate 7 units of magnitude earthquakes from Vrancha zone and local focal zones (Fig. 2).

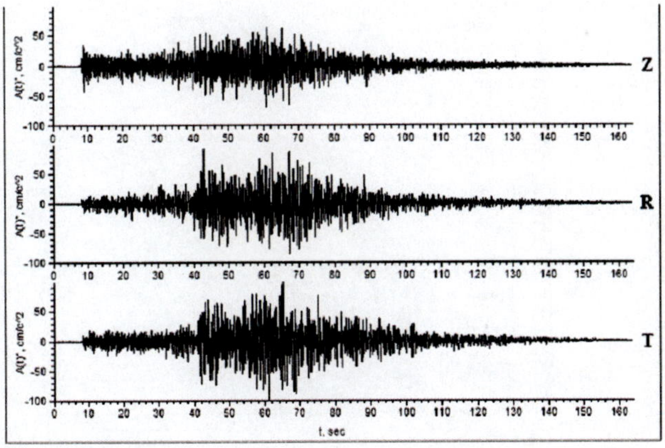

Fig. 2. Graph of three-component accelerogram that simulates the earthquake with 7 units of magnitude.

Based on the data [9], 3D computer model of 13-storey building was generated in the LIRA-SAPR program (Fig. 3). Analysis is carried out with obtained three-component accelerograms by the method of spectral analysis, and direct dynamic analysis was carried out [10].

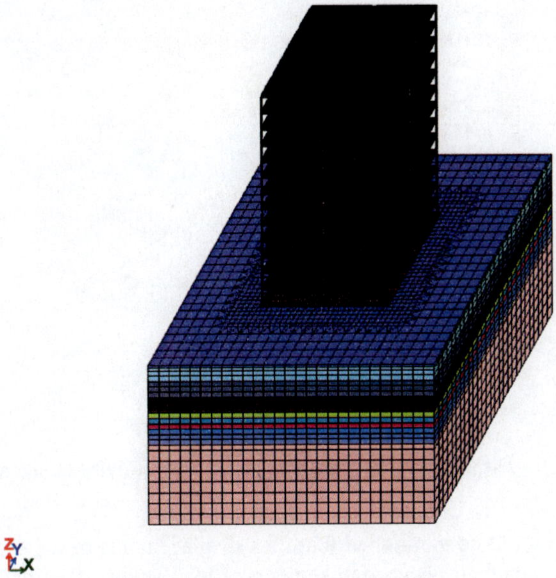

Fig. 3. General view of design model.

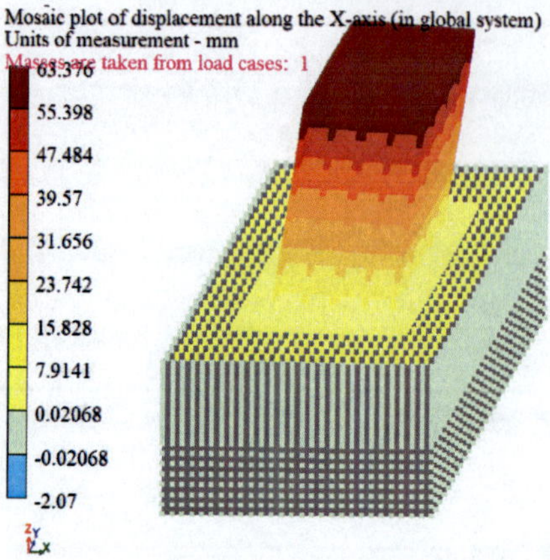

Fig. 4. Displacements along the X-axis in spectral analysis.

Let us compare results of displacements along the X-axis obtained in spectral method (Fig. 4) and max displacements obtained in the direct dynamic method (Fig. 5). There is a significant difference in the stress-strain state of the structure.

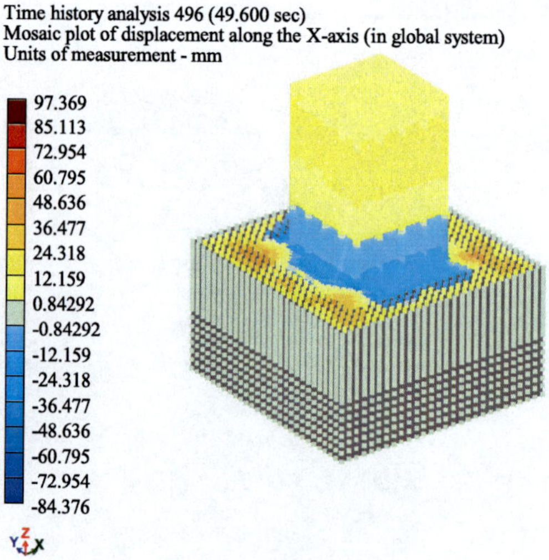

Fig. 5. Displacements along the X-axis in direct dynamic method.

However, it should be mentioned that max displacements of the system in analysis by direct dynamic method occur after the end of load action. The load lasts 41 s, and max displacements are achieved at 49.6 s. This is because when seismic wave reaches the boundary of FE model, it is reflected from the boundary and returns to the structure. This is clearly visible on the graph of displacements for the node (126511) (Fig. 6) of the last storey (Fig. 7) and on the graph of kinetic energy of the system (Fig. 8). After the end of the load action, the energy does not decrease and even tends to increase.

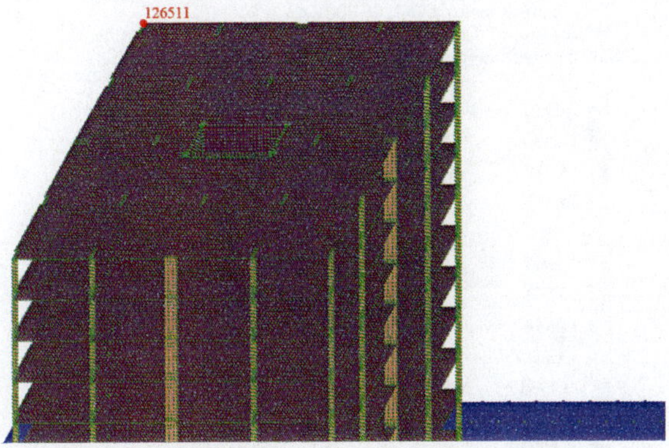

Fig. 6. Location of node 126511 on the model.

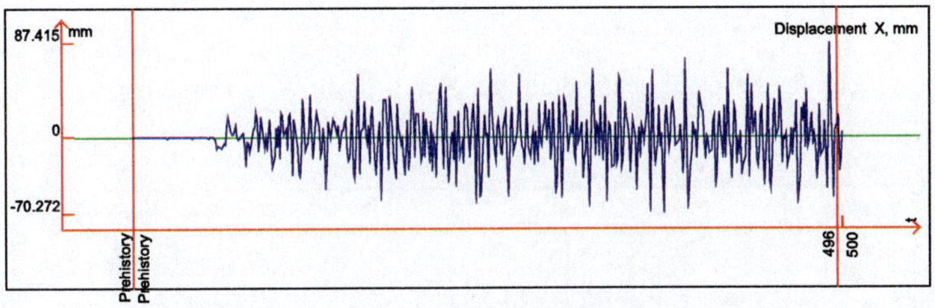

Fig. 7. Displacements along the X-axis at node 126511.

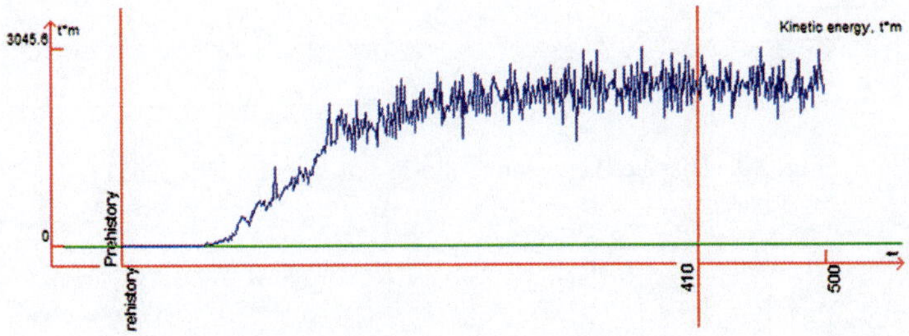

Fig. 8. Kinetic energy for the system.

The results indicate that the system does not work properly. At the end of the load action, instead of kinetic energy attenuation, the energy increases. To eliminate such an unacceptable effect, new FEs were developed in LIRA-SAPR program; they simulate unbounded substructure. When such FEs are applied at the boundary of soil model, displacements correspond to the amplitude values of applied accelerogram (Fig. 9 and 10). Moreover, in Fig. 11 we see that kinetic energy of the system decays when dynamic load is ended, which is natural.

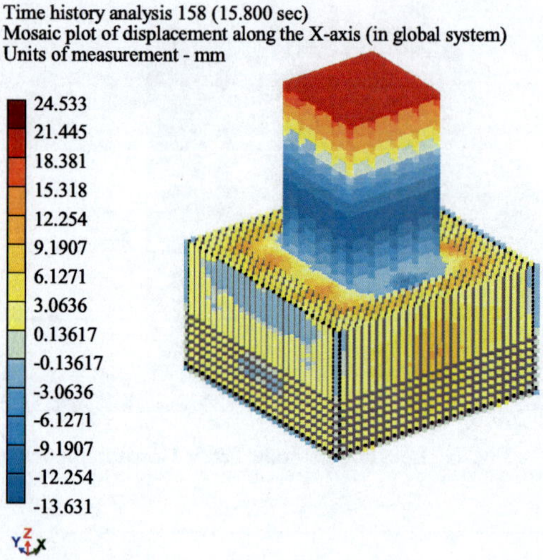

Fig. 9. Displacements along the X-axis in direct dynamic method.

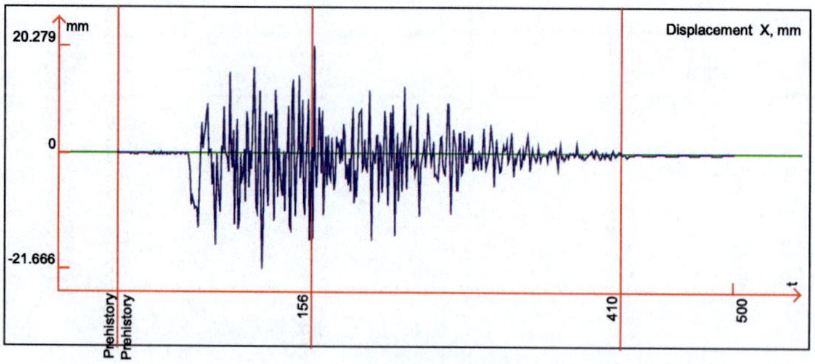

Fig. 10. Displacements along the X-axis at node No. 126511.

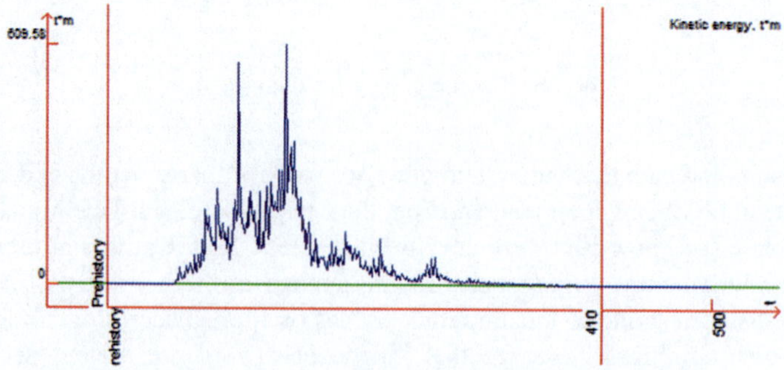

Fig. 11. Kinetic energy of the system.

5 Conclusions

Described technique enables the user to simulate behaviour of the structural system of a building more accurately, to simulate vibration damping, to remedy the shortcomings of spectral and direct dynamic analyses. It also enables the user to solve physically nonlinear problems with soil and do not significantly increase the size, and respectively, time for analysis of the problem.

Together with the technique of seismic microzoning, described technique enables the user to simulate adequate behaviour of buildings and structures in earthquake loads in order to evaluate the seismic resistance.

References

1. Birbraer, A.N.: Raschet konstruktsiy na seysmostoykost. Nauka, St. Petersburg, 255 p (1998)
2. Clough, R., Penzien, J.: Dinamika sooruzheniy. Stroyizadt, Moscow, 320 p (1979)
3. Barabash, M.S., Kostyra, N.O., Pysarevskiy, B.Y.: Strength-strain state of the structures with consideration of the technical condition and changes in intensity of seismic loads. IOP Conf. Ser. Mater. Sci. Eng. **708**, 11 (2019)
4. Fomin, V., Bekirova, M., Surianinov, M., Fomina, I.: Nonlinear dynamic analysis of a reinforced concrete frame by the boundary element method. In: Materials Science Forum 6th International Conference "Actual Problems of Engineering Mechanics" (APEM 2019), vol. 968, pp. 383–395 (2019). ISSN:1662-9752
5. Dorofeev, V.S., Egupov, K.V., Egupov, V.K.: Methods to evaluate the seismic resistance of buildings and structures. Publishing ONMU (Odessa National Maritime University), Odessa, 164 p (2019)
6. Barabash, M., Pisarevskyi, B., Bashynskyi, Y.: Taking into account material damping in seismic analysis of structures. Tech. J. **14**(1), 55–59 (2020)
7. Wolf, J.P.: The Scaled Boundary Finite Element Method. Wiley, Chichester, 364 p (2003)
8. Gorodetsky, A., Pikul, A., Pysarevskiy, B.: Modelirovanie raboti gruntovih masivov na dinamicheskie vozdeystviya. Int. J. Comput. Civil Struct. Eng. **13**(3), 34–41 (2017)
9. Egupov, K.V.: Opredelenie sejsmicheskoj opasnosti ploshhadki stroitel'stva mnogokvartirnogo chetyrehsekcionnogo 14-ti jetazhnogo zhilogo doma so vstroenno-pristroennymi pomeshhenijami social'nogo naznachenija i so vstroenno-pristroennym podzemnym parkingom po adresu: s. Kryzhanovka, ul. Marsel'skaja, Odesskaja oblast' – po dannym sejsmicheskogo rajonirovanija i generirovanija raschjotnyh akselerogramm. Odessa, 79 p (2017)
10. Pikul, A.V., Barabash, M.S.: Problemy modelyuvannya dynamichnykh vplyviv. In: Realizatsiya v PK LIRA-SAPR. Collection of paper abstracts from the International Scientific and Technical Conference devoted to the 90th anniversary of the birth of prof. Yehupov V.K. 'Earthquake engineering in theory and practice', 124 p, ODABA, Odesa, (2016)

Between Tradition and Innovation: The Search for Modern Architectural Forms and Structures in the Design of Wooden Churches

Mykola Bevz(✉)

Lviv Polytechnic National University, Lviv 79013, Ukraine
bevzmist@polynet.lviv.ua

Abstract. Western Ukraine is known for its numerous examples of historic sacral wooden architecture, referred to as "tserkvas" in Ukrainian. In the Halychyna region alone, more than 1,100 wooden tserkvas are listed in government registers. Other regions have an approximately equal number. 16 of these are on the UNESCO World Heritage List, some as joint trans-border nominations of Ukraine and Poland. This historic heritage attests to a long tradition of wooden architecture in the region. Although architectural historians distinguish five primary stylistic groupings that have developed over time, all the groupings share a similar structural approach - horizontal wooden log construction. It would seem that with such a stable historical building tradition in place, and a bountiful supply of timber, the construction of new wooden tserkvas would continue indefinitely. However, that has not been the case. Very few new tserkvas have been built, and all of them have adhered to traditional construction methods. Lately, a perception has formed that the traditional horizontal log construction method has contributed to a lack of innovation and stylistic development in this notable architectural heritage. In light of that, a new initiative was formed with faculty from Vienna Technical University, to research new wooden materials and structural techniques that would permit a design evolution of contemporary tserkvas. An additional goal was to ensure a sustainable, environmentally sympathetic approach, much like the tserkvas of the past.

Keywords: Tserkvas · Wooden churches · Architecture · Construction · Modern and traditional forms

1 Introduction

The inclusion of 16 Ukrainian wooden churches (or tserkvas in Ukrainian) in the UNESCO World Heritage List in 2013 was a recognition of the vibrant historic tradition of wooden architecture in this part of Europe [1]. Wooden places of worship, some from the 13th century, have survived in Norway [2], but their number is relatively small. In contrast, various government registers in Western Ukraine list over 1,100 remaining wooden tserkvas, a wooden architectural heritage of the 16th through the

Z. Blikharskyy (Ed.): EcoComfort 2020, LNCE 100, pp. 34–41, 2021.
https://doi.org/10.1007/978-3-030-57340-9_5

19th centuries. Historically, their numbers were much higher, as determined by registration documents listing 2194 wooden tserkvas in the Lviv diocese at the beginning of the 20th century [3, p. 140].

It would seem that such a stable historical building tradition of wooden places of worship would continue into the present day. Unfortunately, that has not been the case. Of more than 50 new places of worship built in the city of Lviv (the largest city in Western Ukraine) since 1991, the vast majority were built of brick and concrete. One such example was designed by the accomplished architect Mario Botta. These houses of worship were designed for new areas of the city that developed in the second half of the 20th century. All of these brick and concrete structures permit a capacity of over 500 people. Only four of the fifty structures were built of wood.

Interestingly, three were built on the grounds of advanced educational and scientific institutions, such as the Tserkva of St. Oleksiy, built on the academic campus of Lviv Polytechnic National University (Fig. 3). Others, such as the Tserkva of St. Volodymyr was built on the grounds of Lviv University of Forestry (Fig. 4), and the Tserkva of the Blessed Martyrs of Ukrainian Greek Catholic Tserkva was built on the grounds of Ukrainian Catholic University. Although their designs were arrived at with university specialists' engagement, their architectural solutions did not advance the field with advanced technologies, materials, or approaches. These relatively small structures are harmonious in form and continue a traditional approach of sacral Ukrainian architecture, but hold no more than 200 individuals. A similar design approach is common in other new parish tserkvas in the region (Fig. 1 and 2). This lack of technological innovation has prompted a search for new construction solutions utilizing wood so that Ukraine's sacral architecture could evolve to serve a contemporary, large congregation. For that reason, a new initiative was formed with faculty from Vienna Technical University to research new wooden materials and structural approaches that would permit a design evolution of a still relevant historical heritage.

2 A Historical Perspective

The Carpathian Mountains' extensive forests have provided the raw material for constructing Ukraine's wooden architectural cultural heritage for centuries. Timber of high quality was an affordable material not only for the construction of tserkvas but other houses of worship as well, such as synagogues and churches of other denominations. It was also exported far beyond the Carpathian zone, and its use as a building material had several technological advantages. For example, a house of worship could be built quickly, in the span of one or two seasons. Typically, an initial winter phase would encompass the design of the structure and the procurement of the timber, which was then processed to a semi-finished state. Later, in the second summer phase, the tserkva would be built by an experienced group of carpenters. The interior decoration took place gradually over the following years and could stretch over a more extended period. The construction of tserkvas was associated with many different specialists and trades. They included those who harvested wood and prepared it to a condition suitable for construction, to those that cut the wood into elements according to the model and later "assembled" the structure.

The historical tradition of building tserkvas in Ukraine evolved over the centuries, and an extensive body of literature exists to document its various construction and stylistic variations. Writings by various authors from the 20th and 21st centuries, such as M. Dragan, T. Hevryk, Y. Taras, V. Slobodyan, G. Shevtsova, S. Taranushenko, and others [6] are the primary sources. However, there are hardly any publications that research the historical development of such structures from a construction point of view, focusing on the innovative techniques incorporated over time. All of the ecclesiastical wooden buildings erected in Ukraine were built almost exclusively using a horizontal log construction technique, common to the region and beyond. Logs were mostly of coniferous species, and the buildings were raised on stone foundations, with shingles being used to cover the roof. Traditional carpentry and construction techniques were adapted to meet the requirements dictated by a Confession's liturgy and traditions.

Wooden tserkvas in Ukraine display many unique features in terms of building design, structural solutions, decorative schemes, and interior furnishings. Examples include a tripartite ground plan composed of a combination of simple quadrilaterals and octagons (Fig. 5). Original and unique solutions are evident regarding the domes of a quadrilateral or octagonal form, surmounted by specific cupolas (Fig. 5). Other aspects include interior contours of the dome which would which follow its exterior outline (Fig. 6), and a division of the building's interior space with an iconostasis screen.

The Ukrainian tserkva profile has always had a domed shape, with one-, three- or five-domed tserkvas being most common. In addition, all historic tserkvas had one common feature – a horizonal log wall construction. (Fig. 5 and 6). The logs would typically be of varying sizes, and the use of uniform timbers is alien to the Ukrainian tradition. Since there were no attics in tserkvas (Fig. 6), this permitted an open space in the interior. The development of the structural scheme of the building would be carried out in accordance with a three-chamber plan or cruciform plan.

One interesting aspect of tserkva construction was the "lantern" on the top of a dome, which had both a functional and decorative purpose. Its functional purpose was to introduce light and ventilation of the interior. Its decorative role was to act as a soft, gradual transition into the sky, beckoning a spiritual journey. We can notice this difference of forms in the comparison of newly built tserkvas (Fig. 1 and 3) and a traditional solution (Fig. 5).

The architectural form of the tserkva has always been very organic in the unity of exterior and interior. The iconostasis - a wooden screen hung with icons and located between a sanctuary and a nave, was the main functional element of a tserkva's interior. Made of wood and decorated with artistic carvings, it encompassed rows of icons, depicting the symbolic history of salvation. The icons were painted on wide wooden boards or several boards adhered together and connected with a dovetailed spline. Icon painting derived from the Byzantine culture, and their immense popularity in Ukraine led to the development of local schools of iconography [1]. Due to all these interrelated aspects a tserkva in Ukraine has always been an integrated work of art, a unique unity of art and architecture in wood.

3 Technological Possibilities for the Future

The desire to research new and innovative architectural wood designs for contemporary tserkvas was the subject of a joint workshop formed with scientists from Vienna Technical University [4, 5]. It was entitled "A project of a new wooden tserkva for the city of Lviv". One engaging topic of discussion was how to create a modern design solution that would also be in keeping with traditional architectural forms. This aspect necessitated a thorough discussion of current structural issues and their limitations and recent technological advances that could be adapted to future construction.

One crucial point of discussion was the desire that any new architectural and construction designs incorporate a dialog with Ukraine's long cultural heritage of building tserkvas. Including this cultural history in the design process and in a building's structure could generate a genuine source of cultural strength for the congregation. However, being able to differentiate between authentic traditions and imposed influences would be crucial. An unfortunate recent example occurred in the Volyn region, which has a very long history of tserkva construction. During the Russian occupation in the 19th century, foreign architectural features were introduced by edict in tserkva construction. In this instance, a new tserkva was built incorporating these alien architectural forms.

Although a dialog with the past is integral, a dialog with the present is a necessity. Designing a modern tserkva can be a challenging undertaking. The architectural forms of the tserkva require balance and a sound philosophical foundation. For example, Ukrainian tserkvas have a specific elegant shape to their domes and tops. Their flowing lines are always pleasing to the eye and form the silhouette of the tserkva. Such details are often neglected in new construction, using "chopped" or simplified shapes and silhouettes, such as in the following examples (Fig. 1, 2, 3, 4 and 5). Being able to incorporate a domed shape, or many shapes, would be a fundamental, as would be the ground plan of the tserkva.

Present-day technical issues were also discussed. For instance, issues regarding the load-bearing capacity of a tserkva's log walls, such as their bulging, skew, and deformation, could also point to solutions for the future. At present, these issues are addressed by the addition of vertical posts, the so-called "foxes." This method transforms the log structure into a log-frame structure, which permits the stabilization of the tserkvas skeleton, while effectively enabling the repair of individual elements, at times even to change the foundations or individual beams of the walls. This combined method of frame-log bearing structure of the tserkva could be successfully applied in new construction. It would thereby make it possible to increase the height of the log walls, and expand the space of the nave or two other parts of the tserkva, the narthex, and the sanctuary.

Another aspect of wooden sacral architecture that should be taken into account are a structure's acoustics. The Ukrainian religious experience incorporates a substantial body of choral music. Hearing this music in a wooden structure can be especially moving due to a wooden structure's superb acoustics. This feature sets it apart from brick and concrete ones [7]. For example, both the Patriarchal Cathedral in Kyiv or the Tserkva of St. Sophia on the Ukrainian Catholic University in Lviv have poor acoustics and require extensive sound systems. Since construction and finishing materials have a profound influence on an interior's acoustics, it is necessary to incorporate this aspect in the design process. Large tserkvas built of wood, taking acoustics into account, would significantly add to the religious experience.

Fig. 1. New wooden tserkva in Mizhgirya-Potochyna; de-sign of 2002. Photo by M. Syrochman [14].

Fig. 2. New wooden tserkva in Velykyi Bychkiv, architect M. Kravchuk, 2005. Photo by M. Syrochman [14].

Fig. 3. New wooden tserkva in Lviv. Campus of Lviv Polytechnic University. Photo by M. Bevz

Fig. 4. New wooden tserkva on the territory of Lviv University of Forestry. Photo by M. Bevz.

However, the workshop's most valuable insight was the understanding that one does not need to be limited to traditional wood log construction. It is possible to achieve a vast increase in structural scale at almost the same financial cost by using a new frame-space wooden construction. The development of wooden structures today has grown considerably and presents an expanded field of design opportunities.

Fig. 5. The old wooden tserkva in Komarno, 1754. Drawing by M. Bevz and V. Bevz.

New methods of fire protection of buildings and new ways to ensure fire resistance of materials increase the possibilities of using wooden structures safely in architecture. Other technological advances include: frame structures made of large glued or connected multilayer beams; frame structures made from combinations of large and small elements; spatial structures made of modular wooden elements (planar mesh, spatial trusses); combined wooden structures with cable elements and steel belts; structures from the field of "parametric architecture" and many others [4, 5]. These and other types of wooden structures are well known in the construction industry, however their use in tserkva construction has been very limited in Ukraine. Incorporating such technological advances could address many of the shortcomings of traditional building practices and create new design opportunities.

One core goal would be to achieve a larger physical tserkva due to advanced technological design. Due to its nature, such a design would also be able to incorporate a traditional feature of Ukrainian tserkvas, the dome, with an upward-facing open interior, creating a "spiritual sky" (Fig. 6).

There are many examples throughout the world where new wooden structures have been built whose materials and methods could be adapted to Ukrainian tserkvas. For example, many experiments are taking place in Norway, a country with a long tradition of wooden houses of worship. The Bergen Academy of Art and Design is one such center of new possibilities for innovative wooden construction. Professor Petter Bergerud's monograph "Experimental Wooden Structures: Nothing, Or, the Greatness of the Small," published in 2015 [10], highlights many such insights. The name of the book is very symbolic, revealing the vast opportunities provided by the use of this traditional material in current conditions. The specific form of an arc (Fig. 7) highlights an example of the experimental modeling of P. Bergerud's group [11]. Similar experiments of modeling of modern constructive solutions from wooden elements can also be found in Aalto University in Finland [9].

Fig. 6. "Metropol Parasol" – Wooden Structure on the square in Seville. Architect Mayer. Photo by Fernando Alda [12].

Fig. 7. One of the experimental structure of created under promotion of Petter Bergerud [11].

Another example is the unique architectural and urban revitalization project in 2011 in the historic square in the center of Seville, Spain, a country bereft of forests. It was noteworthy that the project competition was won by the German architect Jürgen Mayer, who proposed the construction of a specific wooden open dome covering the area [13] (Fig. 6). It is challenging to categorize this wooden structure. It is both a sculpture and an architectural building. It houses a museum and cafes, and creates a roof over a public space with observation paths and terraces [12]. This creation, however, was matched by Petter Bergerud's sizeable wooden structure (Fig. 7).

Another example of an innovative wooden structure is the multi-curved lattice, the load-bearing structure of the roof of the large office headquarters of the Swatch Group in Biel, Switzerland. This large roof consists of 4.600 wooden beams. It is one of the most significant wooden buildings in the world [8], and the design and construction were very demanding, with construction lasting almost five years. The building has a winding shape and extends along the river through the landscape for more than 240 m in length and 35 m in width. The height is 27 m at the highest point. As an environmentally aware company, the Swatch Group wanted to introduce environmental sustainability into the company's building. Japanese architect Shigeru Ban was the author of the design, and he worked with his favorite material, wood - "the only renewable building material in the world" [8].

4 Conclusions

Modern technology permits the creation of large wooden structures, including houses of worship. Currently, in Ukraine, no new large wooden tserkvas have been built. All of the places of worship, irrespective of denomination, that could accommodate 500 congregants have been built utilizing reinforced concrete technology. Newly built wooden tserkvas are relatively small structures with a capacity of no more than 200 faithful. There are a number of reasons for this, including the underdeveloped wooden construction industry in Ukraine. Additionally, despite its long cultural tradition, the level of wooden architecture education at Ukraine's universities is minimal, including the lack of a separate Department of Wooden Architecture at any university. This results in a lack of experienced architects designing buildings of wood. Ukraine is the largest nation in Europe with a long history of wooden sacral architecture. In light of the above, isn't it time to introduce technological innovations in new wooden architectural construction, create wooden architecture university departments, and educate a new generation of architects – specialists in wood.

References

1. Bevz, M., Czuba, M., Dubyk, Y., Gerych, V., Hajda, M., Marcinek, R., Fortuna-Marek, A., Rulewicz, J., Siwek, A., Slobodian, V., Szałygin, J.: Wooden Tserkvas of Carpathian Regin of Ukraine and Poland. Lviv regional state administration, pp. 3–10. Gerdan Company Ltd, Publishing Company, Kraj (2013)
2. Urnes Stave Church. https://whc.unesco.org/en/list/58. Accessed 03 May 2020
3. Taras, Y.: Sakralna dereviana architektura ukraintsiv Karpat, Lviv. In: NANU, p. 640 (2007)
4. Winter, W.: Holzbauatlas In: "Holzbauatlas", Institut für internationale Architektur-Dokumentation, München (2003)
5. Winter, W., Fadai, A.: Tragwerk und Architektur: Über Logische Strukturen und ressourceneffiziente Materialisierung. Zu den Komplexen Beziehungen zwischen dem, was trägt, und dem, was getragen wird. In: "Die Fakultät für Architektur und Raumplanung/The Faculty of Architecture and Planning", 7; R. Scheuvens (Hrg.); herausgegeben von: Technische Universität Wien (TU Wien); Böhlau Verlag Wien-Köln-Weimar, Wien, S. 69–74 (2016)
6. Vecherskyi, V.: Ukrainski dereviani khramy. Kyiv, Nash chas, p. 270 (2007)
7. Prysiazhnyi, K.: Tserkva doby barokko – muzychnyi instrument? In: Urbanistychno-architekturni problemy mist Halychyny. Red. B. Cherkes, M. Bevz. Derzhavnyi universytet "Lvivska politechnika". Zbirnyk naukowych prats. Lviv – 1996, pp. 126–127 (1996)
8. Swatch: Millimeterwerk in Holz. Holzistgenial.at. Holzbauten, News, 11 October 2019. https://www.holzistgenial.at/blog/swatch-millimeterwerk-in-holz/. Accessed 20 Mar 2020
9. Säie Project Information: Wood programme in architecture and construction. Aalto University. https://www.aalto.fi/en/wood-program. Accessed 02 May 2020
10. Bergerud, P.: Experimental Wooden Structures: Nothing, or, the Greatness of the Small. Kunstoch designhøgskolen. p. 347 (2015)
11. Kjolberg, T.: Experimental wooden structures in Norway. https://www.dailyscandinavian.com/experimental-wooden-structures-in-norway. Accessed 03 May 2020. 28 February 2019/0/1167
12. Metropol Parasol: Largest Wood Structure in the World. In: https://www.iliketowastemytime.com/2012/01/13/metropol-parasol-largest-wood-structure-in-world. Accessed 23 Mar 2020
13. Metropol Parasol - Redevelopment of Plaza de la Encarnación. https://www.lafargeholcim-foundation.org/projects/metropol-parasol-spain. Accessed 02 May 2020
14. Syrokhman, M.: 55 derevianych khramiv Zakarpattia. Hrani-T, Kyiv, p. 88 (2008)

Bearing Capacity of Stone Beam Reinforced by GFRP

Zinoviy Blikharskyy(✉), Taras Bobalo, Andrij Kramarchuk, Borys Ilnytskyy, and Rostyslav Vashkevych

Lviv Polytechnic National University, Lviv 79013, Ukraine
zinovii.y.blikharskyi@lpnu.ua

Abstract. One of the most common issues faced by builders during the reconstruction is the issue of durability and corrosion resistance of structures reinforced with steel reinforcement, especially in the presence of an aggressive environment in special purpose buildings. Annually new materials appear on the market of building materials that can change the idea of standard construction. Glass fiber reinforcement polymer is such the material highly resistant to aggressive environments. More than 40% of buildings are made of fine artificial stone materials (bricks, stones, blocks); it is an environmentally friendly material, which is also resistant to aggressive environments if the manufacturing technology is ensured. The rational combination of a brick and GFRP reinforcement together with maintenance of their reliable joint work will allow to develop more widely the field of this materials' usage. Prefabricated load-bearing ceramic beams are relevant in terms of their economic use in construction, because they can be manufactured directly on construction sites, where it is impossible to provide access for machinery due to lack of space. Therefore it is good solution in terms of technical difficulties in installation of usual reinforced concrete lintels with the use of crane. The use of GFRP reinforcement has a number of advantages comparing with the classic steel reinforcement; coefficient of thermal expansion of reinforcement and concrete are rather close, which prevents formation of cracks when the temperature changes.

Keywords: Bricks · Frameless ceramic block beams · Bearing capacity · Stiffness · Fracture toughness · Experimental samples

1 Introduction

Stone structures are widely spread in residential and industrial construction and are used as load-bearing and fencing structures; perceiving loading from their self-weight, floors and in addition they perform the enclosing functions – namely, thermal and sound insulation [1].

Exploitation requirements for load-bearing structures include the following: strength, rigidity, durability, reliability, maintainability and other properties that determine the quality and exploitation suitability of the structure [2–5]. Reinforced stone structures are widely used in thin structures, floors and light walls in order to increase their bearing capacity; the reinforcement function is mostly ensured by lattices

Z. Blikharskyy (Ed.): EcoComfort 2020, LNCE 100, pp. 42–52, 2021.
https://doi.org/10.1007/978-3-030-57340-9_6

rebar's, rods and rolled steel elements, which do not have sufficient corrosion resistance [6–10]. In some cases during the reconstruction of existing structures could be seen examples of usage of load bearing floors, made of ceramic blocks strengthened with reinforced concrete [11–14].

GFRP reinforcement are fabricated on the basis of glass fibers with the use of polymer binder as the modifier, which ensures its appropriate color [15]. Therefore the high quality of rebar's is ensured. The first advantage of the GFRP rebar's is its corrosion resistance. GFRP rebar's is used in construction for reinforcement of concrete structures and mixed reinforcement of reinforced concrete structures, as well as in the structures, subjected to corrosion actions [16, 17]. Such structures are used for concrete constructions' repair, damaged by aggressive environment (primarily chloride one) [18–23], sewer supports; power lines' supports; in f pool type constructions with wall thickness more than 200 mm; in agricultural objects; in the structures which are exposed to a high thermal conditions; in the construction of thermal houses; as lattices and rods in structures; in the houses' foundations.

Reinforced stone structures are used during the foundations' construction [24], internal and external walls of houses, vaulted ceilings, arches, chimneys, bridges, underground collectors, water towers, elevators and other structures operated in aggressive environments.

The rational combination of a brick and rebar's together with maintenance of their reliable joint work allows to increase parameters of structures' durability and to reduce the cross-section sizes Reinforced stone structures are the structures, reinforced by the rebar's in order to increase their strength.

The reinforcement perceives the internal tensile forces that occur in structures subjected to bending - in floors, lintels, beams (longitudinal reinforcement) [25], as well as increases the strength of compressed structures, such as partitions, columns (transverse reinforcement) [26]. Reinforcement in the form of lattices and rods is located in special grooves and channels of masonry. Reinforced stone structures include also ceramic-blocks beams, reinforcement of which is similar to reinforced concrete beams.

Reinforced stone structures are widely used in Europe, as clay brick structures are easy to install, environmentally friendly and have good strength and durability.

Longitudinal rebar's and in reinforcement (in complexed structures) is used in bearing capacity increase of beam structure, made of ceramic stones, when tensile stresses appear in cross-sections, which exceed masonry tensile strength, in elements with a flexibility of more than 15 [27]. It is also used in thin walls and partitions to increase their stability and strength under the action of transverse forces; in walls and columns, in order to increase the masonry rigidity, its crack and seismic resistance. In order to ensure joint work of masonry with longitudinal rebar's or reinforced concrete it is recommended to use mesh reinforcement.

Reinforced stone structures are wide spread in Hungary, Poland, Ukraine as flooring of reinforced masonry, armor belts, lattices and different beams with the span up to 8 m, despite the technological complexity in such structures' manufacturing, comparing with reinforced concrete.

The purpose of the research was to indicate the expediency of using ceramic beams reinforced with GFRP in construction, the nature of the ceramic block structure with different rebar's types, durability, environmental resistance, destruction nature.

2 Methodology of Experiments

To achieve the research goal, 4 stone beams reinforced with GFRP were designed and tested, with 120 × 250 mm cross section and 2200 mm span.

Research samples were made of ceramic brick of M150 class on the mortar of M150 class. Mortar for research samples was made on the basis of M500 Portland-cement with dense structure. Samples were made by laying of a brick column in a vertical position with through reinforcement. Leveling rails made of metal profiles were fixed at the ends of the structure; verticality of the structure was periodically checked with the help of a water level.

In the stretched zone GFRP rebar of periodic profile, namely: Ø 8 mm ASP in the first case for beams B-1(1) and B-1(2) and Ø 10 mm ASP – in the second case for beams B-2(1) and B-2(2).

Reinforcement in compressed zone of the beams had the periodic profile of Ø 8 mm ASP. Transverse reinforcement was designed of rebar of Ø 4 mm Bp-1, placed in seam at the distance of 1/3 from beams' supports and in each second seam – in the middle of span. Manufactured beams varied by different reinforcement coefficient in stretched zone.

The joint work of GFRP reinforcement and brick was ensured by injecting the space in the cavities between the reinforcement and the brick with a cement-sand mortar that ensured the adhesion of the reinforcement to the brick (Fig. 1).

Fig. 1. A figure caption is always placed below the illustration. Short captions are centered, while long ones are justified. The macro button chooses the correct format automatically.

Simultaneously with manufacturing of samples' series, prism samples were produced with cross-section of 120 × 145 mm and length of 250 mm. Test samples and prisms were endured at 20 °C temperature and 50–70% humidity. This ensured the design strength ensuring during 28 days.

In 28 days beams were located on the test stand, vertically calibrated with the use of level and plastered and micro-indicator chips with 150–200 mm base were fixed, with which further deformation in the area of pure bending was measured.

Tests were performed by stretching of samples on the hydraulic breaking machine with simultaneous recording of the tensile diagram of steel. On the sample near the groove strain gauges were glued, with the use of which strain during the sample loading were measured.

Material characteristics of the test samples are given in Table 1.

Table 1. Characteristics of test samples materials.

Beam parameters	B-1 (1)/(2)	B-2 (1)/(2)
Geometric parameters of cut		
b, mm	120	120
h, mm	250	250
A, sm^2	300	300
Ceramic blocks		
f_k, MPa	7.29	7.29
f_{vko}, MPa	0.9	0.9
$E_{cm} \times 10^3$, MPa	4.3	4.3
Rod reinforcement of stretched zone		
d_s, mm	8	10
f_{yk}, MPa	947	980
A_s, sm^2	0.503	0.785
$E_s \times 10^5$, MPa	0.57	0.57
Class	ASP	ASP
Rod reinforcement of compressed zone		
d_s', mm	8	8
f_{yk} ', MPa	947	947
A_s', sm^2	0.503	0.503
E_s' × 10^5, MPa	0.57	0.57
Class	ASP	ASP
Transverse reinforcement		
d_s, mm	4	4
f_{yd} ', MPa	370	370
A_s', sm^2	0.126	0.126
E_s' × 10^5, MPa	1.7	1.7
Class	A240C	A240C

Testing of the beams' samples with 2000 mm sample was performed with the use of experimental stand. Loading as performed with the use of hydraulic jack of 1000 kN power, located on the distribution traverse, the forces were applied in the upper face of the beam as two concentrated forces applied symmetrically relative to the middle of the beam. The distances between the forces and supports were 650 mm.

Loading on beams was applied symmetrically in stages of 10% from the limit strength with endurance of 30 min on each stage. After that measures were taken by micro indicators and deflectors; additionally formation and opening of cracks were fixed. Load value was controlled by exemplary manometer, calibrated together with a hydraulic pumping station and a jack, as well as the magnitude of the supporting reactions, which were fixed by two annular dynamometers, served as both movable and immovable supports.

Places of measuring devices' location on testing samples' surface and external load application scheme during the experiment are shown on the Fig. 2.

Fig. 2. A figure caption is always placed below the illustration. Short captions are centered, while long ones are justified. The macro button chooses the correct format automatically.

Strain was fixed with the use of clock-type micro-indicators, located with the basis of 200 mm and division price of 0.001 mm. Deformation was measured in the zone of pure bending.

3 Results of the Experimental Study

Experiment was conducted until the yield strength was reached by high-strength rebar or until its destruction. Therefore tested beams are used in the fullest extent.

On the first stage of beams work until the moment of crack formation, elastic deformation take place in both working reinforcement and masonry of compressed and stretched zone. Beam cross-section is the solid elastic-plastic body in which mortar and brick work in tension together with reinforcement.

Results of experimental studies, namely: the largest values of tensile and compressive deformation along the cross section height of the beam in the area of maximum moment, longitudinal and transverse reinforcement, and the value of deflections until cracks in the samples appear are given in Table 2.

In all experimental samples of ceramic bricks at this stage of work, the deformation of the reinforcement and masonry did not differ, because no characteristic cracks were observed due to the loss of adhesion of the reinforcement, which indicates that the reinforcement is deformed evenly along its entire length. It should be noted that the compressive and tensile deformations of the masonry in the area of normal sections are distributed along the height of the element in accordance with the flat sections' hypothesis, and along the cut length- according to the moments' diagram.

The increase in load accompanies the appearance in the stretched zone of inelastic deformations; stresses in the stretched zone of the beam reach the tensile strength of the masonry f_{xk1}. Further increase in the load on the beam is accompanied by the formation of cracks, which is the beginning of the second stage of stress-strain state.

The results of experimental studies, namely: the largest values of deformations of reinforced stone beams with combined reinforcement in stretched and compressed zones, deformations of stretched rebar, the value of deflections until the cracks' formation and load are given in Table 2.

Table 2. Characteristics of test samples materials.

Beam marking	M, kN·m	ε_{30}, ‰	ε_{120}, ‰	ε_{220}, ‰	Deflection f_{max}, mm
B-1	1.55	1.424	0.6	−0.14	1.76
	1.67	1.87	0.68	−0.17	1.95
B-2	2.20	2.34	0.396	−0.06	3.77
	2.35	2.84	0.427	−0.077	3.83

Characteristic is that beams with a higher GFRP percentage cracks are formed later. Corrugated rebar is deformed evenly according to the bending moments diagram.

The largest deflections before the cracks formation were indicated in the second sample with a higher GFRP reinforcement percentage, but it should be noted that the load was also slightly higher. Thus, at the stage from the load application until the normal cracks appearance, stone beams reinforced GFRP work together as monolithic sections. It is important to fix the reinforcement at the ends of the beam, which increases the rigidity of the beam.

With the appearance of normal cracks, the homogeneity of the masonry is disturbed and the stress-strain state along the span is changed. In normal cross-sections with cracks and between cracks, the stress-strain state of bricks and seams is different.

According to the obtained results it could be stated that cracks quickly appear in ceramic- blocks structures at $0.25 \cdot M_{max}$ and take place mostly along the vertical seams of the beams. In the sections between the cracks, the tensile and compressive deformations of the masonry are distributed according to a linear law along the entire section height with the maximum values near the stretched and compressed faces. In cross-section with cracks, the tensile and compressive deformations of the masonry are also distributed according to the linear law, but not along the entire height, but only to the upper point of the crack in the stretched zone.

After the formation of normal cracks and with their development in the span of cut section the longitudinal reinforcement begins to deform intensively. The distribution of deformations along the GFRP reinforcement length, in general, corresponds to the bending moments diagram.

The results of experimental studies, namely: the largest values of deformations of reinforced stone beams with combined reinforcement in stretched and compressed zones, deflections, deformations of stretched rebar until the limit load are given in Table 3.

Table 3. General experimental results for stone beams, reinforced with GFRP reinforcement at the loading of 0,65Mmax, in the zone of maximum bending moment action.

Beam marking	$0{,}65M_{max}$, kN·m	ε_{30}, ‰	ε_{120}, ‰	ε_{220}, ‰	Deflection f_{max}, mm
B-1	3.3	3.982	1.475	−0.297	7.11
	3.5	4.172	1.692	−0.324	7.24
B-2	4.8	5.888	2.076	−0.276	9.57
	4.9	5.970	2.087	−0.292	9.70

It should be noted that the yield strength is reached by rebar of ASP class when strain in it is equal to $\varepsilon_{s,ASP}= 1.781 \cdot 10^{-2}$, which are rather high deformations for beam structures without pre-stressing, thus the destruction of samples is expected before the limit strain is reached in GFRP reinforcement.

Formation and development of cracks cause sharp increase in deformations of structures cross-sections and deflections. Therefore the crack formation moment was taken at the point, which corresponds to break on the deflection graph.

The issue of ceramic block structures crack resistance is a topical issue nowadays. Small number of research has been conducted in this area. In addition, the normative regulations do not specify the method of calculating cracks for such composite materials as ceramic blocks, including bricks with cement mortar. Therefore, when performing experimental studies, special attention was paid to cracks. The test results allowed to conduct a comparative analysis of the crack resistance of ceramic beams reinforced with GFRP.

If the reinforcement percentage is increased, the beams strength according to normal cracks formation increases. Also, when the load is applied, the number of normal cracks and their density increases. It should be noted that part of the cracks was formed in the area of maximum transverse force action.

A further increase in load causes an increase in stresses in the compressed zone, as well as the development of cracks in the stone beams. Finally the moment is reached, when the sample does not resist the forces' action and the destruction takes place.

The ceramic block beams have been designed in such a way that a possible loss of serviceability should occur when the limit of excessive deflections is reached.

The experimental values of the moments of crack formation, the maximum allowable width of crack opening and the value of bearing capacity are given in Table 4.

Table 4. Comparison of moments of crack formation and maximum crack opening.

Beam marking	Beginning of crack formation		Maximum crack opening		Bearing capacity
	M^{d}_{crc}, kN·m	a^{d}_{T}, mm	M, kN·m	a^{d}_{T}, mm	Physical destruction, kN·m
B-1	2.85	0.05	3.5	0.40	5.45
	2.75		3.7		5.24
B-2	3.4	0.05	5.37	0.40	5.82
	3.5		5.45		5.34

Destruction of beams B-1 occurred at load level of 0.78–0.81 M_{max}, for beams B-2 this loading was equal to 0.72–0.79 M_{max}. This bending moment (M_{max}) was indicated a premature breaking and pulling out of the GFRP reinforcement near the support, which is due to insufficient anchoring of the rebar, as well as the fragility of the GFRP reinforcement at bending.

Destruction of beams B-1 occurred at load level of 0.78–0.81 M_{max}, for beams B-2 this loading was equal to 0.72–0.79 M_{max}. This bending moment (M_{max}) was a premature breaking and pulling out of the GFRP reinforcement near the support, which is due to insufficient anchoring of the rebar, as well as the fragility of the GFRP reinforcement at bending.

Taking into consideration the same nature of the sudden rupture, even before reaching the limit strength of the reinforcement, it could be concluded that the determining factor that caused the rupture were not only tensile stress but also shear. Also, one of the failure criteria could be considered the reaching of the maximum allowable

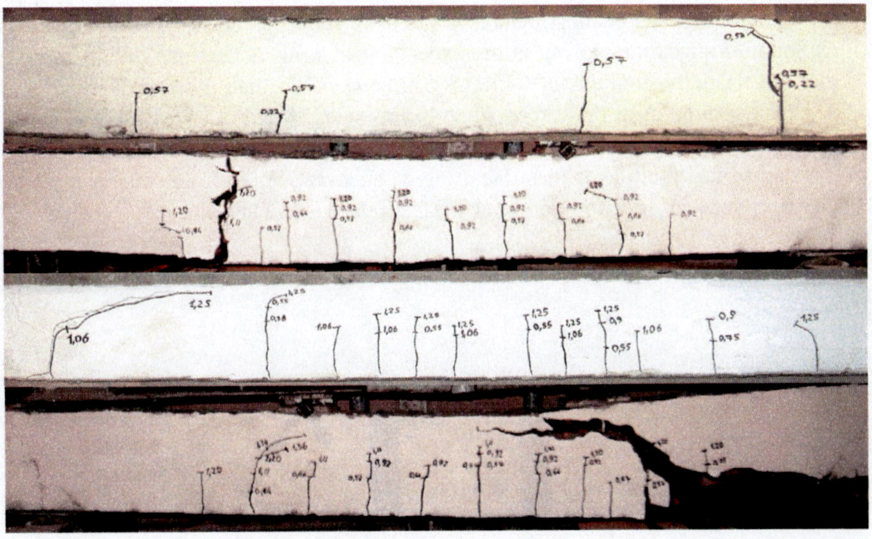

Fig. 3. Character of beams destruction.

deformations of the experimental beams, normal cracks with a width of more than 0.4 mm between the masonry. The failure of all beams took place in the place of the highest bending moment under concentrated forces, where the maximum transverse force also acts, additionally in these the rebar anchoring is minimum (see Fig. 3).

In order to ensure the anchoring strength of the rebar following conditions should be ensured the minimum protective layer thickness of the mortar from the rebar to upper face of the masonry should be 15 mm and when filling cavities, the minimum protective layer of the composite reinforcement coating made of mortar must be 20 mm or equal to the diameter of the reinforcement.

In the areas of Ø 8 mm rebar passaging through the cavities of bricks with 12 mm weight, which were not filled with cement-sand mortar, these requirements were not ensured. Therefore, the length of the anchoring cannot be taken as a section of reinforcement that passes through the brick. The length of anchoring includes only vertical seams t = 10 mm between ceramic bricks, the sum of such sections is not a sufficient length of anchoring for reinforcement Ø 8 mm ASP.

Since the nature of the test beams' failure is similar, and taking into consideration that they were destroyed in the area of maximum transverse forces, could be indicated the inexpediency of reinforcement with collars of such structures. In order to ensure the strength of inclined sections it is better to reinforce the support areas with welded mesh.

4 Conclusions

1. According to the results of experimental ad theoretical research, it was indicated that stone beams, reinforced with GFRP rebars can perceive loading.
2. When designing such structures, special attention should be paid to the zones of shear force action in order to prevent shear deformation.
3. When reinforcing beam structures with GFRP reinforcement, the main criteria of exploitation properties losing is the exceeding of limit deflections due to the fact that the elasticity modulus of GFRP is 5 times smaller than those for steel bars.
4. Crack formation in reinforced stone beams with GFRP mostly occurs in seams between bricks. Therefore, in order to ensure higher crack resistance it is recommended to use multilayer manufacturing of such structures from several masonry layers in vertical direction and bricks offset relative to each other.

References

1. Klymenko, Y., Grynyova I., Kos Z.: The method of calculating the bearing capacity of compressed stone pillars. In: Lecture Notes in Civil Engineering, vol. 47, pp. 161–167 (2019). https://doi.org/10.1007/978-3-030-27011-7_20
2. Bobalo, T., Blikharskyy, Y., Kopiika, N., Volynets, M.: Serviceability of RC beams reinforced with high strength rebar's and steel plate. In: Lecture Notes in Civil Engineering, vol. 47, pp. 25–33 (2020). https://doi.org/10.1007/978-3-030-27011-7_4

3. Karpiuk V., Somina Y., Maistrenko O.: Engineering method of calculation of beam structures inclined sections based on the fatigue fracture model. In: Lecture Notes in Civil Engineering, vol. 47, pp. 135–144 (2019). https://doi.org/10.1007/978-3-030-27011-7_17
4. Khmil, R., Tytarenko, R., Blikharskyy, Y., Vashkevych, R.: Influence of load level during strengthening of reinforced concrete beams on their reliability. IOP Conf. Ser. Mater. Sci. Eng. **708**(1), 012054 (2019)
5. Krainskyi, P., Blikharskyy, Y., Khmil, R., Vegera, P.: Crack resistance of rc columns strengthened by jacketing. In: Lecture Notes in Civil Engineering, vol. 47, pp. 195–201 (2020). https://doi.org/10.1007/978-3-030-27011-7_25
6. Blikhars'kyi, Y.Z.: Anisotropy of the mechanical properties of thermally hardened A500s reinforcement. Mater. Sci. **55**, 175–180 (2019). https://doi.org/10.1007/s11003-019-00285-0
7. Kharchenko, Y.V., Blikharskyy, Z.Y., Vira, V.V., Vasyliv, B.D.: Study of structural changes in a nickel oxide containing anode material during reduction and oxidation at 600 °C. In: Springer Proceedings in Physics, vol. 221 (2019). https://doi.org/10.1007/978-3-030-17759-1_42
8. Kharchenko, Y., Blikharskyy, Z., Vira, V., Vasyliv, B.: Vasyliv, B: Study of nanostructural changes in a Ni-containing cermet material during reduction and oxidation at 600 °C. Appl. Nanosci. (2020). https://doi.org/10.1007/s13204-020-01391-1
9. Lobodanov, M., Vegera, P., Blikharskyy, Z.: Planning experiment for researching reinforced concrete beams with damages. In: Lecture Notes in Civil Engineering, vol. 47, pp. 243–250 (2020). https://doi.org/10.1007/978-3-030-27011-7_31
10. Zhang, Q., Mol'kov, Y.V., Sobko, Y.M. et al.: Determination of the mechanical characteristics and specific fracture energy of thermally hardened reinforcement. Mater. Sci. **50**, 824–829 (2015). https://doi.org/10.1007/s11003-15-9789-9
11. Azizov, T.N., Kochkarev, D.V., Galinska, T.A.: New design concepts for strengthening of continuous reinforced-concrete beams. IOP Conf. Ser. Mater. Sci. Eng. **708**(1), 012040 (2019). https://doi.org/10.1088/1757-899X/708/1/012040. IOP Publishing
12. Bobalo, T., Blikharskyy, Y., Vashkevich R., Volynets M.: Bearing capacity of RC beams reinforced with high strength rebars and steel plate. In: Matec Web of Conferences, vol. 230, p. 02003 (2018). https://doi.org/10.1051/matecconf/201823002003
13. Krainskyi, P., Blikharskyy, Y., Khmil, R., Blikharskyy, Z.: Experimental study of the strengthening effect of reinforced concrete columns jacketed under service load level. In: Matec Web of Conferences, vol. 183, p. 02008 (2018). https://doi.org/10.1051/matecconf/201818302008
14. Krainskyi, P., Vegera, P., Khmil, R., Blikharskyy, Z.: Theoretical calculation method for crack resistance of jacketed RC columns. IOP Conf. Ser. Mater. Sci. Eng. **708**(1), 012059 (2019)
15. Szymon, C., Di Benedetti, M., Maurizio, G., Zappa, E.: Size effect in FRP RC beams with and without shear reinforcement. In: The 13th International Symposium on Fiber-Reinforced Polymer Reinforcement for Concrete Structures (FRPRCS-13), pp. 1–19 (2017)
16. Blikharskyy, Y., Kopiika, N., Selejdak, J.: Non-uniform corrosion of steel rebar and its influence on reinforced concrete elements' reliability. Prod. Eng. Arch. **26**(2), 67–72 (2020). https://doi.org/10.30657/pea.2020.26.14
17. Karpiuk V.M., Syomina Y.A., Antonova D.V.: Bearing capacity of common and damaged cfrp-strengthened RC beams subject to high-level low-cycle loading. In: Materials Science Forum, vol. 968, pp. 185–199. Trans Tech Publications Ltd (2019). https://doi.org/10.4028/www.scientific.net/MSF.968.185
18. Gariboldi, E., Naumenko, K., Ozhoga-Maslovskaja, O., Zappa, E.: Analysis of anisotropic damage in forged Al–Cu–Mg–Si alloy based on creep tests, micrographs of fractured specimen and digital image correlations. Mater. Sci. Eng. A **652**, 175–185 (2016)

19. Kos, Ž., Klimenko Y.: The development of prediction model for failure force of damaged reinforced-concrete slender columns. Tehnički vjesnik, **26**(6), 1635–1641 (2019). https://doi.org/10.17559/TV-20181219093612
20. Kramarchuk, A., Ilnytskyy, B., Lytvyniak, O., Famulyak, Y.: Strengthening prefabricated reinforced concrete roof beams that are damaged by corrosion of concrete and reinforcement. IOP Conf. Ser. Mater. Sci. Eng. **708**(1), 012060 (2019). IOP Publishing
21. Lipiński, T.: Roughness of 1.0721 steel after corrosion tests in 20% NaCl. Prod. Eng. Arch. **15**(15), 27–30 (2017). https://doi.org/10.30657/pea.2017.15.07
22. Lobodanov, M., Vegera, P., Blikharskyy, Z.: Influence analysis of the main types of defects and damages on bearing capacity in reinforced concrete elements and their research methods. Prod. Eng. Arch. **22**(22), 24–29 (2019). https://doi.org/10.30657/pea.2019.22.05
23. Zatkalíková, V., Markovičová, L.: Influence of temperature on corrosion resistance of austenitic stainless steel in cl − containing solutions. Prod. Eng. Arch. **25**(25), 43–46 (2019). https://doi.org/10.30657/pea.2019.25.08
24. Kramarchuk, A., Ilnytskyy, B., Lytvyniak, O.: Arrangement of the foundations under the new hotel in Lviv. In: Matec Web of Conferences, vol. 183, p. 02007. EDP Sciences (2018)
25. Bobalo, T., Blikharskyy, Y., Kopiika, N., Volynets, M.: Theoretical analysis of RC beams reinforced with high strength rebar's and steel plate. IOP Conf. Ser. Mater. Sci. Eng. **708**(1), 012045 (2019)
26. Vatulia, G., Lobiak, A., Chernogil, V., Novikova, M.: Simulation of performance of CFST elements containing differentiated profile tubes filled with reinforced concrete. In: Materials Science Forum, vol. 968, pp. 281–287. Trans Tech Publications Ltd (2019). https://doi.org/10.4028/www.scientific.net/MSF.968.281
27. Pavlikov, A., Kosior-Kazberuk, M., Harkava, O.: Experimental testing results of reinforced concrete beams under biaxial bending. Int. J. Eng. Technol. **7**(3.2), 299–305 (2018)

Influence of the Percentage of Reinforcement on the Compressive Forces Loss in Pre-stressed RC Beams Strengthened with a Package of Steel Bars

Taras Bobalo, Yaroslav Blikharskyy(✉), Nadiia Kopiika, and Myhailo Volynets

Lviv Polytechnic National University, Lviv 79013, Ukraine
yaroslav.z.blikharskyy@lpnu.ua

Abstract. Pre-stressed reinforced concrete structures are widely used nowadays in construction industry. Such structures are rather cost-effective and give an opportunity to implement project ideas which are impossible without the use of pre-stressing technology. The issue of metal consumption reduction is highly topical. One of the possible solutions is the usage of steel-concrete elements with external reinforcement. The effectiveness of the use of reinforced concrete structures with external reinforcement has been repeatedly noted at international construction symposia and conferences. Application of advanced technologies and acceleration of scientific and technological progress in the construction field leads to increase in the reinforced concrete structures' usage and expansion of application fields. Moreover, advanced technologies increase their efficiency and cost-effectiveness. This article describes results of research pre-stressed steel-concrete beams, strengthened with the package of steel bars with different ratio of sheet and rod reinforcement in the zone of pure bending moment action. The purpose of this research was definition of reinforcement percentage influence on loss on compressive forces' loss in pre-stressed reinforced concrete beams strengthened with a package of steel bars. In addition the aim was to evaluate effectiveness of steel bars' usage in combined reinforcement. The practical significance of the experiments is to study the loss of compressive forces in pre-stressed bending elements with tape and rod reinforcement, taking into account the influence of different ratios of reinforcement areas within the combined reinforcement and development of design and calculation proposals for such structures.

Keywords: Combined reinforcement · High-strength rebar · Steel concrete · External reinforcement · Pre-stressed rebar

1 Introduction

Reinforced concrete structures are widely used nowadays in construction industry [1–5]. During their exploitation in such structures could take place problems with their exploitation, in particular their damages [6–8] and reinforcement corrosion [9–13]. For today, great amount of works are dedicated to analysis of influence of reinforcement

Z. Blikharskyy (Ed.): EcoComfort 2020, LNCE 100, pp. 53–62, 2021.
https://doi.org/10.1007/978-3-030-57340-9_7

corrosion on such structures. Therefore the number of articles is dedicated to usage of modern materials in order to ensure reliable RC structures' exploitation [14–17]. Due to the depletion of their resources, many structures need to be restored and strengthened [18]. According to modern technologies, a lot of RC structures are strengthened with the use of various composite materials [19, 20], reinforced concrete jacketing [21], external steel sheets, etc. It is necessary to develop structures, which would be reliably used for a long time [22].

Recent times highly topical was the issue of modern reinforced concrete structures development for capital construction [23]. Capital construction is the important industry, which ensures expansion and continuous improvement of main building funds, creation of their material and technical base. The expansion of application fields and increase in the reinforced concrete structures' volume in the modern construction practice is associated with increased efficiency and cost-effectiveness.

The efficiency of the use of capital investments in building industry can be achieved through the introduction of advanced construction methods and techniques [24], creation of new light and economical materials, further improvement of existing materials, as well as rational spatial and planning solutions.

Based on this, one of the design options that would allow fast and convenient, efficient and economical and therefore successful use of reinforced concrete structures is the use of steel-concrete elements with external reinforcement, strengthened with a package of pre-stressed reinforcement.

2 Research Methodology

The program of experimental research envisages pre-stressed reinforced concrete beams, the stretched zone of which is strengthened with tape and rod reinforcement of periodic profile with the area ratio of 1:1 and 1:0.667.

Investigations of beam elements strengthened with a package of reinforcement (sheet and rod reinforcement of periodic profile) were carried out on experimental models – beams' samples with cross-sectional dimensions of 270 × 135 mm and span of 2700 mm. The total samples' length is 3000 mm. Dimensions are accepted in order to ensure reinforcement percentage within 1.73 … 3.38%, taking into account periodic profile sheet reinforcement with cross-section of 105 × 6 mm.

In all the research beams the sheet reinforcement is made of steel 16G2AF of periodic profile with cross reefs with the main parameter developed by Lviv Polytechnic National University.

Steel bars with periodic profile of A400C class was strengthened by stretching on the power stand. The steel bars were stretched mechanically by means of hydraulic jacks with stress control by means of an exemplary dynamometer of the DOS-50 system and a manometer of the pumping station. Steel bars of the stretched zone was strengthened to the strength limit of the tape reinforcement f_{yd} = 450 MPa, and the rebar of the compressed zone – up to the maximum limit determined by the norms: f_{yd} = 540 MPa.

The study of pre-stress of steel-concrete beams strengthened with a package of reinforcement was carried out in two stages:

- compression by pre-stressing force with holding for 7–90 days under the action of this force;
- short-term bending test.

For each stage methods were developed.

3 Test Specimen Description and Material Properties

Physic-mechanical characteristics of tape steel (yield strength, strength and modulus of elasticity) were determined by testing specially made fish-samples according to norms; steel bars - by testing standard samples on a bursting machine of R-20 mark with simultaneous recording of tensile diagram force - deformation by strain gauges affixed to the test specimen.

The readings of the strain gauges were recorded using an automatic electronic strain gauges.

Physic-mechanical characteristics of concrete (strength, modulus of elasticity, concrete class) were determined by testing cubes with an edge of 150 mm and prisms with a length of 600 mm and 400 mm with a cross section of 150 × 150 mm, respectively, on a hydraulic press of P 250 mark.

The loading speed, experiment conditions, deformation measurements and the determination method of the initial modulus of elasticity of concrete were taken according to the recommendations.

Tensile strength of the concrete was determined by splitting of concrete cylinders on a diametrical plane. Cylinder dimensions: diameter 150 mm, length 150 mm.

In order to fulfil the indicated tasks and purpose of the research 3 beams were developed and designed, material characteristics of the research samples are given in the Table 1.

The presence of pre-stressed reinforcement in the upper zone of the beams is provided in order to ensure the crack resistance of this zone during the transmission of pre-stressing forces of sheet reinforcement from the power stand to hardened concrete (transmission of pre-stressing force was performed 28 days after concreting).

In each of beams in upper compressed zone was designed longitudinal Ø 16 steel bar of A400C class. Stretched zone was reinforced by periodic profile sheet of 105 × 6 mm. Anchoring of sheet reinforcement is provided due to its coupling with concrete and transverse anchor rods. The anchors are Ø 8 A400C transverse bars, placed in pairs along the beam. The transverse bars spacing along the beam was different. In the pure bending zone in all the beams spacing is taken according to design regulations, namely equal to 180 mm. In the zone of transverse forces action depending on the future load level the spacing is following: 135 mm (B-1), 110 mm (beam B-2), 80 mm (beam B-3). Three pairs of transverse rods with the spacing of 50 mm are placed on the supports in all beams to improve the anchoring. The design and reinforcement of experimental specimens are given on the Fig. 1.

The process of frames' manufacturing consisted two stages. On the first stage, transverse rods were welded to the sheet of the periodic profile in the T-section under the flux layer. On the second stage – endings of transverse reinforcement were

Table 1. Characteristics of test samples materials.

Beams' marking		B-1	B-2	B-3
Parameters of beams' cross-section	Width b, mm	135	135	135
	Height h, mm	270	270	270
	Area A, cm^2	364.5	364.5	364.5
	Concrete class (designed)	C40/50	C40/50	C45/55
Heavy concrete				
$f_{ck,\ prism}$, MPa		36.97	39.14	40.65
$E_{cd}{\cdot}10^3$, MPa		38.94	39.25	39.07
Reinforcement of stretched zone - the tape, longitudinal				
$B_s x t_s$, mm		105 × 6	105 × 6	105 × 6
f_{yk}, MPa		450	450	450
$E_s{\cdot}10^5$, MPa		2.1	2.1	2.1
Class		16G2AF	16G2AF	16G2AF
$b_s x t_s$, cm		10.5 × 0.6	10.5 × 0.6	10.6 × 0.6
A_s, cm^2		6.3	6.3	6.3
ρ, tape, %		1.77	1.77	1.77
Reinforcement of stretched zone - the rod, longitudinal				
Ø', mm		2 Ø12	2 Ø16	3 Ø16
f_{yk}', MPa		450	450	450
$E_s'{\cdot}10^5$, MPa		2.0	2.0	2.0
Class		A400C	A400C	A400C
Ø', mm		2 Ø12	2 Ø16	3 Ø16
A_s, cm^2		2.26	4.02	6.03
ρ, rebar, %		0.65	1.16	1.75
$\sum\rho$, %		2.45	2.99	3.57

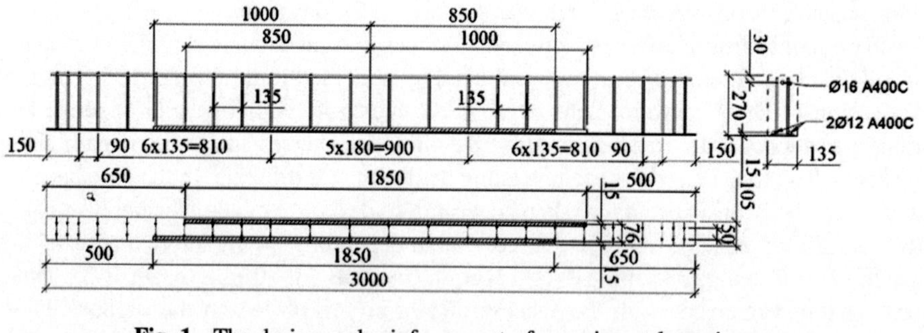

Fig. 1. The design and reinforcement of experimental specimens.

connected with two Ø 8 A240C bars with electric arc welding in the spatial frame. The steel bars were located on top of the sheet and fixed in the design position for the period of concreting with knitting wire.

4 Methods of Beams' Research at Compression by Pre-stressing Forces

Stress losses in accordance with the norms of the mechanical method of tensioning the reinforcement on the stops and the natural hardening of the test specimens consist of:

- first: from relaxation of stresses in reinforcement; deformation of anchors and stand elements; from transient creep;
- second: from settling and creep of concrete under load in time.

Cross section of the investigated beams, strengthened with a package consisting of pre-stressed and unstressed reinforcement.

The presence of unstressed reinforcement in pre-stressed structures influences on losses from transient creep during compression, as well as the course of the processes of creep and shrinkage of concrete.

The research methodology is designed to separate certain types of stress losses.

Among the first losses are studied only those due to transient creep. Other above indicated losses were not considered. In order to avoid losses due to stresses' relaxation sheet reinforcement was strained three times to a level 10% higher than the design and kept in a tense state for 30 … 60 min. After that the reinforcement was strained to the same level and kept for 3 days.

After 3 days of holding the rebar in a tense state for 10–12 h before concreting the tension forces were removed. Immediately before concreting, the reinforcement was stressed to the design level in steps equal to 1/10 of the design stress.

Compression of beams by preliminary compression was carried out on 30–31 days after concreting. The compressive force caused by the pre-stress of the reinforcement on the concrete was transmitted smoothly by means of hydraulic jacks. Simultaneously, the stress-strain state of beams was indicated. Stress losses in the reinforcement during compression were determined by the deformations of the tape reinforcement.

The reinforcement deformations were measured immediately after the full transfer of the compressive force from the reinforcement to the concrete consisted of losses from the transient creep.

Bending of beams at compression was defined according to deformations measured on the compressed and stretched planes of beams according to the Eq. (1).

$$f_0 = \frac{\varepsilon - \varepsilon_1}{h}. \tag{1}$$

Control over stress losses in the tape reinforcement during the time up to the day of the test was carried out by electronic sensors glued at the time of transmission of the pre-tensioning force in the reinforcement from the power stand to the concrete. According to these data, the increase in bending of the beams was determined.

5 Determination of Pre-stressing Losses of Rebar Due to Relaxation

In the case of determining of relaxation losses for different periods of time (stages), when the stress in the reinforcement is not constant, for example, due to the elastic contraction of concrete, it is necessary to apply the method of equivalent time.

The concept of equivalent time method is given on the Fig. 2, when in the moment of t_i the instantaneous deformation of pre-stressed reinforcement takes place, where:

$\sigma^{-}_{p,i}$ - tensile stresses in rebar exactly before ti;
$\sigma^{+}_{p,i}$ - tensile stresses in rebar exactly after ti;
$\sigma^{+}_{p,i-1}$ - tensile stresses in rebar on the previous stage ti;
$\Delta\sigma^{+}_{pr,i-1}$ - the absolute value of relaxation losses at the previous stage;
$\Delta\sigma_{pr,i}$ - the absolute value of relaxation losses at the current stage.

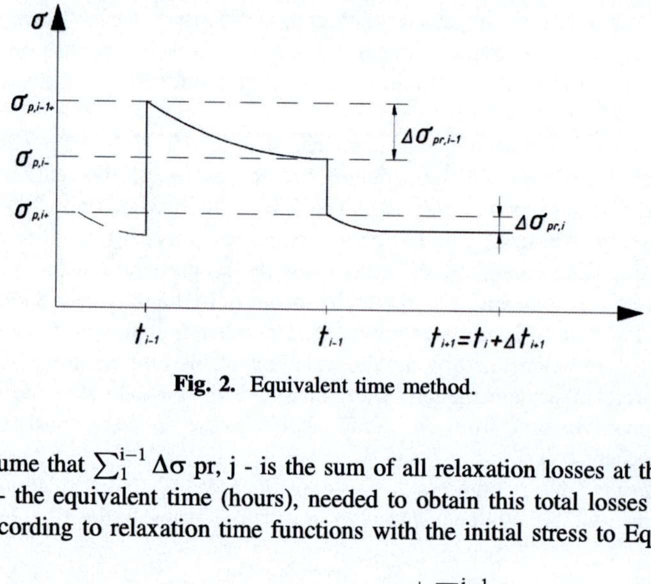

Fig. 2. Equivalent time method.

Let assume that $\sum_{1}^{i-1}\Delta\sigma$ pr, j - is the sum of all relaxation losses at the previous stage, a t_e - the equivalent time (hours), needed to obtain this total losses and which changes according to relaxation time functions with the initial stress to Eq. (2):

$$\sigma^{+}_{p,i}\sum\nolimits_{1}^{i-1}\Delta\sigma_{pr,i}\text{ and with }\mu = \frac{\sigma^{+}_{p,i}\sum_{1}^{i-1}\Delta\sigma_{pr,i}}{f_{pk}}. \tag{2}$$

For example for 2d class of pre-stressed rods t_e, given in Eq. (2), is equal to Eq. (3):

$$\sum\nolimits_{1}^{i-1}\Delta\sigma_{pr,i} = 0{,}66\rho_{1000}e^{9{,}09\mu}\left(\frac{t_e}{1000}\right)0{,}75(1-\mu)\left\{\sigma^{+}_{p,i} + \sum\nolimits_{1}^{i-1}\Delta\sigma_{pr,i}\right\}10^{-5}. \tag{3}$$

After solving the above indicated equation for t_e this formula could be used for calculating of losses due to relaxation at the current stage $\Delta\sigma_{pr,i}$ (if equivalent time is added to interval t_e, which is considered) to Eq. (4):

$$\Delta\sigma_{pr,i} = 0,66\rho_{1000}e^{9,09\mu}\left(\frac{t_e + \Delta t_i}{1000}\right)0,75(1-\mu)\left\{\sigma^{+}_{p,i} + \sum\nolimits_{1}^{i-1}\Delta\sigma_{pr,i}\right\}10^{-5} - \sum\nolimits_{1}^{i-1}\Delta\sigma_{pr,i}. \quad (4)$$

The same principle is used for all three classes of pre-stressed reinforcement.

6 Compression Stress Losses Depending on Reinforcement Percentage

After the experiment following compression stress losses were obtained (see Tables 2 and 3).

Table 2. Experimental values if compression stress losses.

Samples	$\sigma^{+}_{p,i-1}$ 1st day, MPa	$\sigma^{-}_{p,i}$ 28-th day, MPa	$\sigma^{+}_{p,i}$ 28-th day, MPa	$\Delta\sigma_{p,i}$, MPa	$\sigma^{+}_{p,i}$ 56-th day, MPa	$\Delta\sigma_{pr,i}$, MPa
B-1	194.33	181.66	130.34	51.32	112.14	18.2
B-2	206.5	209.66	160	49.66	138.38	21.62
B-3	208	200.71	161.86	38.85	142.43	19.43

Table 3. Values of stress losses in percentages.

Samples	$\rho^{+}_{p,i-1}$ 1-st day, %	$\rho^{-}_{p,i}$ 28-th day, %	$\rho^{+}_{p,i}$ 28-th day, %	$\Delta\rho_{p,i}$, %	$\rho^{+}_{p,i}$ 56-th day, %	$\Delta\rho_{pr,i}$, %
B-1	100	93.5	69.6	23.8	58.3	11.3
B-2	100	97	76	21	66	10
B-3	100	96.5	82	14.5	74.8	7.8

$\rho^{+}_{p,i-1}$ - initial percentage of tension stress in tape reinforcement after concreting;
$\rho^{-}_{p,i}$ - resulting stress percentage in the reinforcement after upper rebar annealing;
$\rho^{+}_{p,i}$ - resulting stress percentage in the tape reinforcement after its full annealing;
$\Delta\rho_{p,i}$ - compression loss percentage;
$\Delta\rho_{pr,i}$ - relaxation losses' percentage.

In the beam B-1 with the reinforcement percentage equal to 26,9% from the total reinforcement of tension zone, compression stress losses were 23,8%.

In the beam B-2 with the rebar reinforcement percentage equal to 39,6%, from the total reinforcement of tension zone compression stress losses were 21,0%.

In the beam B-3 with the rebar reinforcement percentage equal to 49,7%, from the total reinforcement of tension zone, compression stress losses were 14,5%.

According to experimental results the graphic dependency (Fig. 3) was built for compression stress losses depending on non-stressed rebar percentage.

Compression stress losses with increasing of non-stressed rebar percentage were non-linear. Reinforcement of steel-concrete beams with more than 39,6% of non-stressed rebar has given an increase in pre-stress losses.

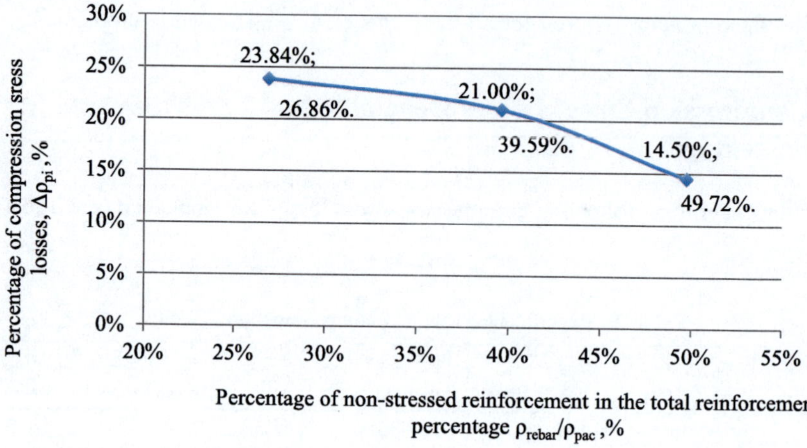

Fig. 3. Compression stress losses depending on non-stressed rebar percentage.

For simplification of pre-stressed beams' design with combined reinforcement by the rebar package was developed the function of compression stress losses depending on non-stressed rebar percentage. Function values were obtained with the use of graphical dependencies approximation (Fig. 3), according to Lagrange method (Eqs. (5), (6)):

$$L_n = \sum\nolimits_{i=0}^{n} y_i \frac{\omega_n(X)}{(X = X_i)\omega_n'(X_i)}, \tag{5}$$

$$\omega_n(X) = (X - X_0)(X - X_1)\ldots(X - X_i)\ldots(X - X_n), \tag{6}$$

where $\{x_i\}_{i=\overline{0,n}}$ - interpolation points, $y_i = \Delta f(x_i),\ i = \overline{0,n}$ - function $\Delta f(x)$ values.

Therefore, in our particular case as interpolation points were taken non-stressed rebar percentage in the total reinforcement percentage ρ_{rebar}/ρ_{pac}, as function values - the percentage of pre-stress losses $\Delta\rho_{pi}$ (Eq. (7)):

$$\Delta\rho_{pi} = -183\left(\frac{\rho_{rebar}}{\rho_{pac}}\right)^2 + 99.48(\rho_{rebar}/\rho_{pac}) + 10.34. \tag{7}$$

The initial formula for convenience was brought to integers. Formula for engineering computation will be the following Eq. (8):

$$\Delta\rho_{pi} = -180\left(\rho_{rebar}/\rho_{pac}\right)^2 + 100\left(\rho_{rebar}/\rho_{pac}\right) + 10. \quad (8)$$

Deviations of the results from the initial formula were less than 0,7%.

Taking into account experimental results it necessary to further conduct study on the beams with higher non-stressed reinforcement percentage. It is expected that according to graphs character they will show lower stress losses.

In addition it is necessary to study beams with lower non-stressed reinforcement percentage in order to identify the minimum effective percentage by non-stressed rebar.

7 Conclusions

Compression stress losses with increasing of non-stressed rebar percentage have non-linear character. At reinforcement of steel-concrete beams with non-stressed rebar above 39.6% could be seen the increase in pre-stress losses; at reinforcement percentage less than 26.86% according to graph character non-stressed rebar do not have sustainable influence on pre-stress losses.

For simplification of pre-stressed beams' design with combined reinforcement by the rebar package was developed the function of compression stress losses depending on non-stressed rebar percentage.

References

1. Azizov, T.N., Kochkarev, D.V., Galinska, T.A.: New design concepts for strengthening of continuous reinforced-concrete beams. IOP Conf. Ser. Mater. Sci. Eng. **708**(1), 012040 (2019). https://doi.org/10.1088/1757-899X/708/1/012040
2. Blikharskyy, Y., Kopiika, N., Selejdak, J.: Non-uniform corrosion of steel rebar and its influence on reinforced concrete elements' reliability. Prod. Eng. Arch. **26**(2), 67–72 (2020). https://doi.org/10.30657/pea.2020.26.14
3. Bobalo, T., Blikharskyy, Y., Kopiika, N., Volynets, M.: Theoretical analysis of RC beams reinforced with high strength rebar's and steel plate. IOP Conf. Ser. Mater. Sci. Eng. **708**(1), 012045 (2019)
4. Krainskyi, P., Blikharskyy, Y., Khmil, R., Vegera, P.: Crack resistance of RC columns strengthened by jacketing. Lecture Notes in Civil Engineering, vol. 47, pp. 195–201 (2020). https://doi.org/10.1007/978-3-030-27011-7_25
5. Pavlikov, A., Kosior-Kazberuk, M., Harkava, O.: Experimental testing results of reinforced concrete beams under biaxial bending. Int. J. Eng. Technol. **7**(3.2), 299–305 (2018)
6. Blikhars'kyi, Z.Y., Obukh, Y.V.: Influence of the mechanical and corrosion defects on the strength of thermally hardened reinforcement of 35GS steel. Mater. Sci. **54**, 273–278 (2018). https://doi.org/10.1007/s11003-018-0183-2
7. Kos, Ž., Klimenko, Y.: The development of prediction model for failure force of damaged reinforced-concrete slender columns. Tehnički vjesnik **26**(6), 1635–1641 (2019). https://doi.org/10.17559/TV-20181219093612
8. Zhang, Q., Mol'kov, Y.V., Sobko, Y.M., et al.: Determination of the mechanical characteristics and specific fracture energy of thermally hardened reinforcement. Mater. Sci. **50**, 824–829 (2015). https://doi.org/10.1007/s11003-015-9789-9

9. Blikhars'kyi, Y.Z.: Anisotropy of the mechanical properties of thermally hardened A500s reinforcement. Mater. Sci. **55**, 175–180 (2019). https://doi.org/10.1007/s11003-019-00285-0
10. Kharchenko, Y.V., Blikharskyy, Z.Y., Vira, V.V., Vasyliv, B.D.: Study of structural changes in a nickel oxide containing anode material during reduction and oxidation at 600 °C. In: Fesenko, O., Yatsenko, L. (eds.) Nanocomposites, Nanostructures, and Their Applications. NANO 2018. Springer Proceedings in Physics, vol. 221. Springer, Cham (2019). https://doi.org/10.1007/978-3-030-17759-1_42
11. Kharchenko, Y., Blikharskyy, Z., Vira, V., Vasyliv, B., Vasyliv, B.: Study of nanostructural changes in a Ni-containing cermet material during reduction and oxidation at 600 °C. Appl. Nanosci. (2020). https://doi.org/10.1007/s13204-020-01391-1
12. Lipiński, T.: Roughness of 1.0721 steel after corrosion tests in 20% NaCl. Prod. Eng. Arch. **15**(15), 27–30 (2017). https://doi.org/10.30657/pea.2017.15.07
13. Zatkalíková, V., Markovičová, L.: Influence of temperature on corrosion resistance of austenitic stainless steel in cl − containing solutions. Prod. Eng. Arch. **25**(25), 43–46 (2019). https://doi.org/10.30657/pea.2019.25.08
14. Karpiuk, V., Somina, Y., Maistrenko, O.: Engineering method of calculation of beam structures inclined sections based on the fatigue fracture model. Lecture Notes in Civil Engineering, vol. 47, pp. 135–144 (2019). https://doi.org/10.1007/978-3-030-27011-7_17
15. Karpiuk, V.M., Syomina, Y.A., Antonova, D.V.: Bearing capacity of common and damaged CFRP-strengthened RC beams subject to high-level low-cycle loading. In: Materials Science Forum, vol. 968, pp. 185–199 (2019). Trans Tech Publications Ltd. https://doi.org/10.4028/www.scientific.net/MSF.968.185
16. Khmil, R., Tytarenko, R., Blikharskyy, Y., Vashkevych, R.: Influence of load level during strengthening of reinforced concrete beams on their reliability. IOP Conf. Ser. Mater. Sci. Eng. **708**(1), 012054 (2019)
17. Vatulia, G., Lobiak, A., Chernogil, V., Novikova, M.: Simulation of performance of CFST elements containing differentiated profile tubes filled with reinforced concrete. In: Materials Science Forum, vol. 968, pp. 281–287 (2019). Trans Tech Publications Ltd. https://doi.org/10.4028/www.scientific.net/MSF.968.281
18. Lobodanov, M., Vegera, P., Blikharskyy Z.: Influence analysis of the main types of defects and damages on bearing capacity in reinforced concrete elements and their research methods. Prod. Eng. Arch. **22**(22), 24–29 (2019). https://doi.org/10.30657/pea.2019.22.05
19. Selejdak, J., Blikharskyy, Y., Khmil, R., Blikharskyy, Z.: Calculation of reinforced concrete columns strengthened by CFRP. Lecture Notes in Civil Engineering, vol. 47, pp. 400–410 (2020). https://doi.org/10.1007/978-3-030-27011-7_51
20. Blikharskyy, Y., Khmil, R., Blikharskyy, Z.: Research of RC columns strengthened by carbon FRP under loading. In: Matec Web of Conferences, vol. 174, p. 04017 (2018). https://doi.org/10.1051/matecconf/201817404017
21. Krainskyi, P., Vegera, P., Khmil, R., Blikharskyy, Z.: Theoretical calculation method for crack resistance of jacketed RC columns. IOP Conf. Ser. Mater. Sci. Eng. **708**(1), 012059 (2019)
22. Sobol, K., Blikharskyy, Z., Petrovska, N., Terlyha, V.: Analysis of structure formation peculiarities during hydration of oil-well cement with zeolitic tuff and metakaolin additives. Chem. Chem. Technol. **8**, 461–465 (2014). https://doi.org/10.23939/chcht08.04.461
23. Tomski, P., Menderak, R.: Contract brewing – production-oriented cooperation in craft brewing industry. Prod. Eng. Arch. **22**(22), 16–23 (2019). https://doi.org/10.30657/pea.2019.22.04
24. Wolniak, R.: Problems of use of FMEA method in industrial enterprise. Prod. Eng. Arch. **23**(23), 12–17 (2019). https://doi.org/10.30657/pea.2019.23.02

Experimental Research Results of the Bearing Capacity of the Reinforced Concrete Beams Strengthened in the Compressed and Tensile Zones

Oleksandr Borysiuk and Yuriy Ziatiuk(✉)

National University of Water and Environmental Engineering, Rivne 33000, Ukraine
y.y.ziatiuk@nuwm.edu.ua

Abstract. Construction is one of the main factors determining the economic development of the state. Introduction of new technologies and materials at the strengthened and restored reinforced concrete structures allows to solve important questions of the building branch. The article describes the experimental research of the work of layers of reinforced concrete beams strengthened by the glued composites in the form of carbon fibers in the tensile zone, and steel fiber concrete in the compressed zone. Namely, tearing the band off in the area between the point of application of the force and the support occurred, when the maximum permissible load had been achieved. The work of the strengthening concrete and the concrete of the beams took place simultaneously without sudden change of deformation. The significant effect of the strengthening system was found in the strengthened beams. Under single load there was an increase of bearing capacity in beams strengthened with fine-grained concrete by 57%. In beams strengthened with steel-fiber concrete by 46%. At low-cycling load for beams strengthened by the fine-grained concrete the bearing capacity increased by 32%, and strengthened with steel-fiber concrete by 69%. For beams strengthened under the loading of fine-grained concrete by 66%, in the beams strengthened with steel-fiber concrete by 65%.

Keywords: Steel fiber concrete · Composites · Load · Beam · Compressed · Tensile zone

1 Introduction

Development of the industrial production, modernization of public and housing stock are connected with reconstruction, expansion, technical re-equipment at the operating enterprises, in residential, administrative and public buildings [1–4]. Reconstruction of buildings or structures is usually accompanied by a change in the load on building structures and their primary structural schemes [5]. All this leads to the need to determine the technical condition of building structures, to determine the residual life of their performance, to decide on their strengthening, restoration or replacement [6].

Z. Blikharskyy (Ed.): EcoComfort 2020, LNCE 100, pp. 63–70, 2021.
https://doi.org/10.1007/978-3-030-57340-9_8

The need to strengthen or restore building structures arises not only during the reconstruction or technical re-equipment, but also due to premature corrosion or mechanical wear [7–11]. All this arouses an increased interest in the problems of strengthening and reconstruction of existing building structures.

For strengthening bending reinforced concrete elements strengthening by topping up is most often applied [12–17]. This method of strengthening uses traditional materials along with modern building ones, such as steel fiber concrete. These materials have a number of advantages over standard fine-grained concrete - greater compressive strength, higher crack resistance and deformability. Accordingly, these properties are valuable in strengthening the compressed zone. Strengthening of the tensile zone of structures with composite bands based on carbon fibers is a universal method.

2 Review of Studies on Strengthening Reinforced Concrete Structures

In European practice, carbon fiber materials of the Swiss company Sika have gained widespread use. Sika Carbodur band and Sika Wrap liner, in particular, can be used to strengthen both sloping and normal sections of bending reinforced concrete structures. In addition, Sika Wrap liner can also be used to anchor Sika Carbodur bands [18].

One of the first to study the work of carbon plastic reinforced concrete structures was U. Meier [19, 20] who, from 1985 on the basis of twenty-six reinforced concrete beams, studied the operation of CFRP bands. The samples had dimensions of 152 × 254 × 2007 mm and were minimally reinforced in the normal Sect. (2Ø8 mm - upper and lower reinforcement) and to the maximum in the cross section (Ø6.35 mm with a step of 216 mm). As a result of testing the control and strengthened samples, the effectiveness of such strengthening was revealed.

P. Ritchie [5] investigated the effectiveness of strengthening based on glass, carbon and aramid fibers, testing sixteen reinforced concrete beams with minimal reinforcement of normal section. Studies showed that depending on the type, quantity and orientation of the strengthening material, the rigidity of such structures increased by 17… 99% and the strength by 40… 97%. As a result of the tests, the effect of anchoring FRP bands was studied. The failure began at the ends of the band, not in the zone of pure bending, where the maximum moment was in the samples tested without anchoring. Three types of anchoring: the first - anchoring the band at the ends with fiberglass angle bars, the second - wrapping FRP plates around the beam at the ends, the last - extending the plate to the support of the beam, were proposed. The third method gave an increase in load-bearing capacity and destruction in the area of pure bending.

Today, the world continues to study the more complex states of strengthened structures performance. For example, R. Al-Mahaidi and A. Hii [12] studied the torsional strength of reinforced concrete beams of the box section, strengthened by glued composite materials in the form of CFRP bands. The peculiarity of their experiment was that in the course of the research with the help of photogrammers the segregation of the external reinforcement before the moment of destruction of the structures was recorded, which made it possible to clearly describe this process. The authors indicate that at present it is necessary to quickly and cheaply repair infrastructure objects, especially bridges. And the fiber-reinforced polymer is a promising material as an external reinforcement to increase the bending and shear strength of reinforced concrete elements. However, for today little attention has been paid to increasing torsional strength and other complex states of operation.

The research of French and Tunisian researchers O. Ben Mekki, D. Siegert [21] and others is quite interesting. The authors have developed a rational design of the span structure of small bridges using modern high-performance materials. During the development of the bridge design with a span of 10.5 m, wooden beams with fiber-concrete paving slabs in the tensile zone and strengthening by carbon fibers in the tensile zone were used. The article pays considerable attention to the connection of fiber-reinforced concrete slabs to wooden beams and gluing carbon materials to the tensile zone. The authors use bolts to fix the slabs, and the fibers are fixed with epoxy resin. According to the results of experimental studies, the design confirmed sufficient resistance to static and repeated loads. The test has resulted in defining the limit vibration parameters for performance in the SLS state. The authors have created light span structures of small bridges that can be erected without using heavy construction equipment.

At the work [22] investigated the efficiency of strengthening the compressed zone of bending reinforced concrete elements with different materials (traditional heavy concrete, steel fiber concrete, concrete using metallurgical waste) under the action of low-cycle loads. The test of the strengthened samples was performed over ten cycles with loading on the eleventh half-cycle to failure. The lower load levels on the cycles were taken 0.3 and 0.4, and the upper – 0.7; 0.8; 0.9. According to the experimental studies, the most effective was the use of traditional heavy concrete for structures that were not exposed to low-cycle impact to strengthening. It was also found a decrease in the efficiency of strengthening by heavy concrete by 9% in the mode of low-cycle loading with a jump in the load beyond planned to the level of 0.9 from the destructive load. As steel fiber concrete is more stably deformed under conditions of repeated loads of different levels, the author considers its use more priority than heavy concrete to strengthen the compressed zone of bending structures that operate under low-cycle loads.

Strengthening reinforced concrete beam structures is quite deeply studied. The authors performed many experimental and theoretical studies and practically examined different methods of strengthening. However, the study of the peculiarities of the work of reinforced concrete structures, strengthened by loads, under the action of low-cycle loads, which is mostly conform to the actual work of structures, has not been studied enough. The urgency of the research is due to the need to study the actual operation of reinforced concrete bending elements, strengthened simultaneously in compressed and tensile zones under the action of low-cycle loads.

3 Experimental Research Results

Reinforced concrete beams with dimensions of 100 × 200 × 2000 mm, concrete class C16/20. Longitudinal working rod reinforcement 2Ø10 A 500C and cross reinforcement Ø6 A240C with a step of 50 mm, except for the zone of pure bending. The top erection reinforcement from a wire of Ø4 Vr-I. Strengthening the tensile zone - Sika® CarboDur® S-512 carbon fiber bands and anchoring them by SikaWrap®-230C/45 liner [18]. Strengthening the compressed zone – half of the beams were strengthened by fine-grained concrete C16/20, the other by steel fiber concrete (SFB 3%) with a concrete thickness of 50 mm. Tests of structures were performed according to the method [18] (Fig. 1).

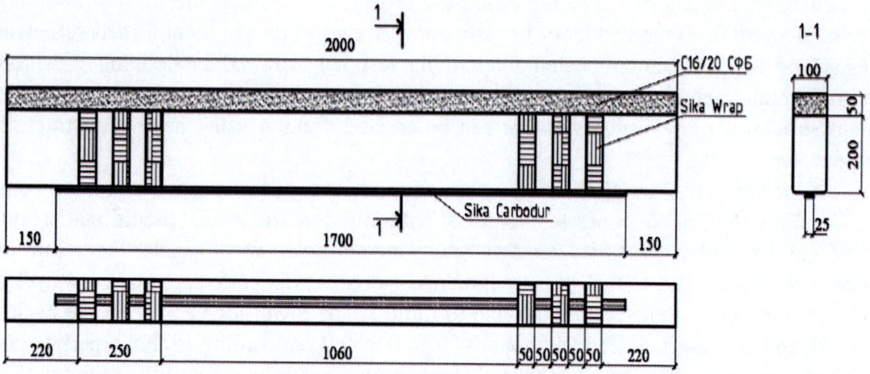

Fig. 1. Scheme of strengthening experimental beams by bands and by concrete

At the first stage, beams were tested for a single static load (B-1 and B-2). At the second stage, beams (BC-5, BC-6,) were tested for low-cycle loads. After low-cycle load testing, the BC-5, BC-6 beams were strengthened under the load of $0.6F_u$ (from the destructive load of B-1, B-2 beams) in the test rig and retested for low-cycle loads. Loads during strengthening were produced by screw jack. After 28 days, the screw jack was replaced by the ordinary one, devices and strain gauges were installed on the body of the beam [18].

During the destruction of the strengthened experimental beams, the band was torn off in the area between the point of application of the force and the support. At the same time, the anchoring system continued to hold the band. At further loading there was a rupture and crumpling of fibers of a liner of anchoring on a band edge and tearing the band off the concrete body practically over the full length between anchoring by the liner. Displacement of the edges of the composite band in the anchoring system took place, followed by disintegration of the compressed zone of concrete [18].

In this case, there is a factor of destruction of the samples, due to the destruction of the contact zone at a distance between the applied vertical forces. The compressed zone of some beams remained practically indestructible (Figs. 2 and 3).

Fig. 2. General view of testing non-strengthned beams

Fig. 3. The location of the devices on the beam strengthened by composite band and fiber concrete

This indicates that the factor of destruction of the samples was the strength of the contact zone from the beginning of the tearing the band off (Table 1).

Table 1. The research results of the bearing capacity of the beams

Beam code	The cross-sectional area of the strengthening element			The area of reinforcement of internal steel bar	Experimental bending moment	Effect of strengthening
	A_f, sm^2	A_b, sm^2	A_{sfb}, sm^2	A_s, sm^2	M^{exp} kN × m	δ, %
B-1	–	–	–	1.57	13.4	–
B-2	–	–	–	1.57	13.0	–
BC-3	–	–	–	1.57	13.5	–
BC-4	–	–	–	1.57	13.8	–
BS-1	0.3	–	50	1.57	20.91	46.7
BS-2	0.3	50	–	1.57	22.46	57.6
BSC-1	0.3	–	50	1.57	21.96	60.9
BSC-2	0.3	50	–	1.57	18.13	32.8
$BSC_{0,6}$	0.3	50	–	1.57	22.72	66.5
$BSC_{0,6}$	0.3	–	50	1.57	22.59	65.5

B – beams without strengthening, tested at single static load;
BC – beams without strengthening, tested at cycle load;
BS - beam strengthened by concrete or steel fiber concrete in the compressed zone of the section and by carbon plastic band in the tensile zone of the section and tested at single static load;
BSC - beam, strengthened by concrete or steel fiber concrete in the compressed zone of the section and by carbon plastic band in the tensile zone of the section and tested at cycle load;
BSC - beam, tested without strengthening by cycle load, strengthened under a load of 0.6*F* of concrete or steel fiber concrete in the compressed zone of the section and carbon plastic band in the tensile zone of the section. and tested again by low-cycle load.

In the second case, another type of destruction of strengthened samples has been revealed. At loads of 0.70 *F* and 0.85 *F*, inclined cracks of the second order appear. When the load increases and goes up to destructive, there is a rapid increase in deformation, and a significant increase in the deflections of the beam. The opening of the inclined crack of the second type is greater than the boundary opening and the crumpling of the support zone of the beam on the support, when the band is torn off simultaneously in the area between the point of application of the force and the support. At that, the strengthening system continues to hold the strengthening band and the entire structure of the beam takes up the load. This is due to the strength of the concrete of the beam on the supports because the band anchoring system did not go beyond the support. It was not possible to fix the band under the support because the beam was under the load in the test rig.

At the operational stage (0.5–0.6 from the destructive force), the main structures of the beams and the strengthening system of the compressed and tensile zones work together. At the last stages of the stress-strain state (0.85 from the destructive force and up), segregation of the strengthening layer in the compressed zone from the concrete of the beam occurs.

During the test, a slight curvature of the fibers of the liner in the direction perpendicular to the longitudinal axis of the beam was observed. After testing the beams, cracking and inspection, the displacement of the band relative to the center of the test beam was descovered. The laminate fibers of strengthening were torn off on both sides of the band edges.

4 Conclusions

Experimental studies of the strength of normal sections, rigidity of bending reinforced concrete beams before and after strengthening by composite materials based on carbon plastics in the tensile zone, fine-grained concrete and steel fiber concrete in the compressed zone, have shown improved functional performance compared to non-strengthed beams.

A significant effect of the strengthening system has been revealed in the strengthened beams. At a single load there was an increase in load-bearing capacity in beams strengthened by fine-grained concrete by 57%, in beams strengthened with steel-fiber concrete by 46%. At low-cycle loading for strengthening by fine-grained concrete by 32%, for strengthening by steel-fiber concrete by 69%. For beams strengthened under the load with fine-grained concrete by 66%, in beams strengthened by steel-fiber concrete by 65%.

References

1. Blikharskyy, Y., Kopiika, N., Selejdak, J.: Non-uniform corrosion of steel rebar and its influence on reinforced concrete elements' reliability. Prod. Eng. Arch. **26**(2), 67–72 (2020). https://doi.org/10.30657/pea.2020.26.14
2. Bobalo, T., Blikharskyy, Y., Kopiika, N., Volynets, M.: Serviceability of RC beams reinforced with high strength rebar's and steel plate. Lecture Notes in Civil Engineering, vol. 47, pp. 25–33 (2020). https://doi.org/10.1007/978-3-030-27011-7_4
3. Khmil, R., Tytarenko, R., Blikharskyy, Y., Vashkevych, R.: Influence of load level during strengthening of reinforced concrete beams on their reliability. IOP Conf. Ser. Mater. Sci. Eng. **708**(1), 012054 (2019)
4. Krainskyi, P., Blikharskyy, Y., Khmil, R., Vegera, P.: Crack resistance of RC columns strengthened by jacketing. Lecture Notes in Civil Engineering, vol. 47, pp. 195–201 (2020). https://doi.org/10.1007/978-3-030-27011-7_25
5. Bobalo, T., Blikharskyy, Y., Kopiika, N., Volynets, M.: Theoretical analysis of RC beams reinforced with high strength rebar's and steel plate. IOP Conf. Ser. Mater. Sci. Eng. **708**(1), 012045 (2019)
6. Blikhars'kyi, Z.Y., Obukh, Y.V.: Influence of the mechanical and corrosion defects on the strength of thermally hardened reinforcement of 35GS steel. Mater. Sci. **54**(2), 273–278 (2018). https://doi.org/10.1007/s11003-018-0183-2

7. Blikhars'kyi, Y.Z.: Anisotropy of the mechanical properties of thermally hardened A500s reinforcement. Mater. Sci. **55**(2), 175–180 (2019). https://doi.org/10.1007/s11003-019-00285-0
8. Kharchenko, Y.V., Blikharskyy, Z.Y., Vira, V.V., Vasyliv, B.D.: Study of structural changes in a nickel oxide containing anode material during reduction and oxidation at 600 °C. In: Fesenko, O., Yatsenko, L. (eds.) Nanocomposites, Nanostructures, and Their Applications. NANO 2018. Springer Proceedings in Physics, vol. 221. Springer, Cham (2019)
9. Kharchenko, Y., Blikharskyy, Z., Vira, V., Vasyliv, B., Podhurska, V.: Study of nanostructural changes in a Ni-containing cermet material during reduction and oxidation at 600 °C. Appl. Nanosci. (2020). https://doi.org/10.1007/s13204-020-01391-1
10. Lobodanov, M., Vegera, P., Blikharskyy, Z.: Planning experiment for researching reinforced concrete beams with damages. Lecture Notes in Civil Engineering, vol. 47, pp. 243–250 (2020). https://doi.org/10.1007/978-3-030-27011-7_31
11. Zhang, Q., Mol'kov, Y.V., Sobko, Y.M., Blikhars'kyi, Y.Z., Khmil', R.E.: Determination of the mechanical characteristics and specific fracture energy of thermally hardened reinforcement. Mater. Sci. **50**(6), 824–829 (2015). https://doi.org/10.1007/s11003-015-9789-9
12. Al-Mahaidi, R., Hii, A.: Bond behaviour of CFRP reinforcement for tarsional strengthering of solid and box. Composites B **38**(5), 720–731 (2007)
13. Blikharskyy, Z., Brózda, K., Selejdak, J.: Effectivenes of strengthening loaded RC beams with FRCM system. Arch. Civil Eng. **64**(3), 3–13 (2018)
14. Bobalo, T., Blikharskyy, Y., Vashkevich, R., Volynets, M.: Bearing capacity of RC beams reinforced with high strength rebars and steel plate. In: Matec Web of Conferences, vol. 230, p. 02003 (2018). https://doi.org/10.1051/matecconf/201823002003
15. Krainskyi, P., Vegera, P., Khmil, R., Blikharskyy, Z.: Theoretical calculation method for crack resistance of jacketed RC columns. IOP Conf. Ser. Mater. Sci. Eng. **708**(1), 012059 (2019)
16. Krainskyi, P., Blikharskyy, Y., Khmil, R., Vegera, P.: Influence of loading level on the bearing capacity of RC columns strengthened by jacketing. In: MATEC Web of Conferences, vol. 230, p. 02013 (2018). https://doi.org/10.1051/matecconf/201823002013
17. Selejdak, J., Blikharskyy, Y., Khmil, R., Blikharskyy, Z.: Calculation of reinforced concrete columns strengthened by CFRP. Lecture Notes in Civil Engineering, vol. 47, pp. 400–410 (2020). https://doi.org/10.1007/978-3-030-27011-7_51
18. Borysiuk, O., Karavan, V., Sobczak-Piąstka, J.: Calculation of the normal section strength, rigidity and crack resistance of beams, strengthened by carbon-fiber materials. In: AIP Conference Proceedings, vol. 2077, no. 1, p. 020008. AIP Publishing LLC (2019)
19. Meier, U.: Composite for structural repair and retrofitting. In: International Conference on Fiber Composites in Infrastructure 1CC1, 1216, 1202 (1996)
20. Meier, U.: Strengthening of structures using carbon fibre/epoxy composites. Constr. Build. Mater. **9**(6), 341–351 (1995)
21. Ben Mekki, O., Siegert, D.F., Mevel Toutlemonde, L., Goursat, M.: A new composite bridge: feasibility validation and vibration monitoring. Mech. Adv. Mater. Struct. **22**(10), 850–863 (2015)
22. Ritchie, P., Thomas, D., Connelly, G.M.: External reinforcement of concrete beams. Iszng Fiber-Reinforced Plastics. ACI Struct. J. **8**(4), 490–500 (1991)

Experimental Study of Compressed Ceramic Hollow Brick Masonry Structures Strengthened with GFRP Meshes

Serhiy Bula(✉) and Mariana Kholod

Lviv National University, Lviv 79013, Ukraine
serhii.s.bula@lpnu.ua

Abstract. The article presents the results of the experimental study of masonry columns which have been strengthened after high-level axial compression loading with glass fiber reinforced polymer composites. Preliminary damaged ceramic hollow-brick middle-scale models were continuously wrapped with Glass FRP meshes in order to the testing program. The target point of this early exploration is to define the efficiency of the described method of strengthening for considerably affected models after their decompression. The "stress-strain" curves have been compared for confined and unconfined specimens. Reported results demonstrated raised load-bearing and ductility values for confined columns. In addition, the failure mode for wrapping has been noted and described. Control samples have shown brittle failure in contrast to strengthened samples' plain collapse. Any atypical failure mode (due to sharp ceramic broken pieces) has not been observed. The paper also provides conclusions in terms of obtained experimental results and addresses numerical investigations to further research.

Keywords: Masonry columns · GFRP · Mesh · Strengthening · Confinement

1 Introduction

The external wrapping of compressed masonry structures with GFRP (Glass Fiber-Reinforced Polymer) composites becomes more popular owing to applicability in historical buildings [1] and that strengthening method's advantages (handling, growing cost-effectiveness, installation speed, low weight, high strength, etc.). A range of studies has been published in recent years aiming to give more information in this area. Thus, in the frame of recent research [2] two different series of masonry columns were confined with GFRP and CFRP strips bonded to the column with an epoxy resin. Different schemes of FRP wrapping were investigated by means of non-axial compression tests. Other scientists [3] presented experimental research for masonry columns strengthened under the vertical load. All strengthened samples were confined with GFRP straps applied in a different manner (horizontally and spirally) with various overlapping options. The authors also provided load-bearing efficiency analysis of confined columns and the impact of the existing compressive stress in a column during confinement. Some investigations [4] were dedicated to the strengthening of masonry columns with a circular cross-section, confined with glass and basalt FRP systems.

Z. Blikharskyy (Ed.): EcoComfort 2020, LNCE 100, pp. 71–78, 2021.
https://doi.org/10.1007/978-3-030-57340-9_9

Glass and basalt FRP composites (sheets and grids) were used with different strengthening schemes (including complete jacketing and discontinuous FRP strips) basing on different bonding type (including epoxy resin and polymer/cement-based mortar). Studies reported in paper [5] showed the outcomes of experimental tests for columns subjected to non-axial compression load. Three confinement types have been experimentally analyzed in order to evaluate and compare the effectiveness of glass FRP, carbon FRP, and basalt FRP laminates wrapping. Further investigations were performed by explorers [6] for full-scale columns using: GFRP sheets, discontinuous and continuous GFRP wrapping and internal carbon FRP bars. The experimental results were presented and compared with the results obtained from the author's experimental tests on medium scale specimens (using the same materials). Canadian scientists [7] studied steel-reinforced and plain masonry columns strengthened by spraying them with GFRP. Minor increases in strength and large increases in strain capacity were achieved with both types of columns. In addition to the mentioned above researches where it was dealt particularly with GFRP, a range of authors [8–11] reported noteworthy experimental results of masonry columns investigations with other FRP materials.

Considering the lack of studies when masonry columns are damaged significantly before strengthening, this paper presents early experimental results as a part of a more complex program meaning to research strengthened by GFRP jacketing masonry structures exposed to mechanical and temperature [12] actions under variable loading levels and materials.

2 Materials and Methods

2.1 Materials

Test Program is described briefly in Table 1. Specimens are labeled respectively to the stage of testing (without strengthening ("n"), before ("d") or after ("s") strengthening.

Table 1. Testing program

Specimen's label	Strengthening	Description
S 1n and S 2n	None	Control sample
S 1d and S 2d	None	Damaged
S 1 s and S 2 s	GFRP mesh	Strengthened after damaging

As specimen's material the hollow (volume of holes is 28%) ceramic bricks [13] were used with unit dimensions 250 mm × 120 mm × 88 mm. Bonding mortar was manufactured in-site by mixing Portland cement and sand with a mass ratio of (1:6). Lime was not used in the mortar therefore plasticizer was applied instead to improve the mortar workability. Mechanical properties were defined according to the national standards. Mortar strength was recorded by results of compression test for 28-day standard cubes 70 mm × 70 mm × 70 mm. The compression strength for bricks was received by testing bricks from series used in the specimens. Average (10 tests) compressive strength reported as f_m = 5.70 MPa for mortar [14] and f_b = 6.31 MPa for brick [15]. Reinforcing system included the glass fiber mesh [16] and two-component ready-mixed fiber-reinforced repair mortar [17]. Basing on producer's data mesh properties were taken as follows: mesh size - 12.7 mm × 12.7 mm; tensile strength – 1300 MPa; elastic modulus 72 GPa; ultimate strain – 1.8%.

For this early exploration as specimens were used four masonry square columns made by assembling bricks with mortar joints. Each brick's layer had an orthogonal direction to the top and bottom layers. The specimen's parameters was as follow: height ~800 mm, cross-section - 250 mm × 250 mm, mortar thickness - ~10 mm.

2.2 Methods

Masonry columns have been tested under axial compression provided by means of the hydraulic press (max. capacity −1250 kN). Specimens were axially loaded with a 10-min pause on every loading step to achieve full crack development and stabilization. Longitudinal and transversal deformations were measured by mechanical strain gages during compression test with tolerance 0.01 mm and 0.001 mm respectively. Gages location and test set–up scheme are represented in Fig. 1a. Two columns ("s" series) were loaded up to failure as control samples. The other two columns ("d" series) were subjected to ~80% of ultimate loading and staged for 20 min.

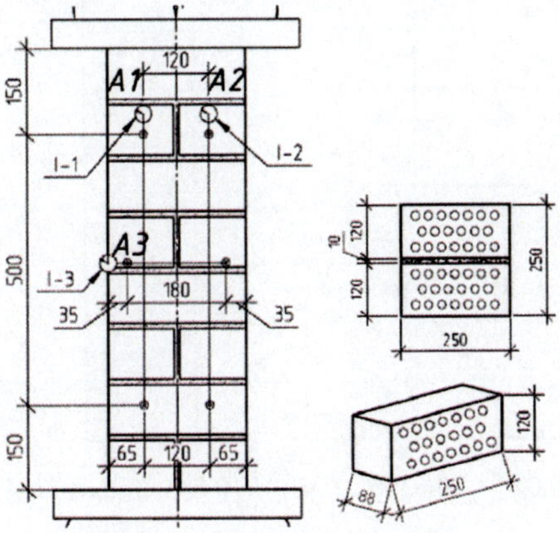

(a)

(b)

Fig. 1. Test set-up scheme: (a) gages location; (b) reinforcing after damaging.

After that the specimens were unloaded and confined with continuous GFRP-mesh wrapping. Preparation and strengthening application procedure was realized in accordance with producer recommendations. All surfaces were cleaned carefully to ensure good adhesion. External column's corners were rounded (r ~ 20 mm) to avoid mesh cutting and stresses concentrating. When the first layer of repair mortar (~5 mm) was applied the fiber-glass mesh (2 layers) was placed and pressed down into mortar while it was still wet. The mesh was lightly smoothed with respect to mortar adhere considering overlapping in strap joints due to product technical sheets. Reinforcing mesh has been completely covered with the second layer of repair mortar with the same thickness as the first one (Fig. 1b).

3 Results and Discussions

3.1 Control Samples

Specimens within "n" series (see Table 1) have been assumed as control samples and were tested without any reinforcing in aim to define columns' ultimate strength experimentally. Since ceramic bricks (which tend to rapid deconstruction) were used in this masonry, hereinafter the first macro crack occurrence was accepted as the failure start point. Owing to material peculiarities this moment could be easily defined from "stress-strain" diagram (Fig. 2) at the point where curve slope changes sharply. Longitudinal strains were measured for each column's side (labeled "A", "B", "C", "D"), consequently macro crack developing was controlled not only visually but analytically as well.

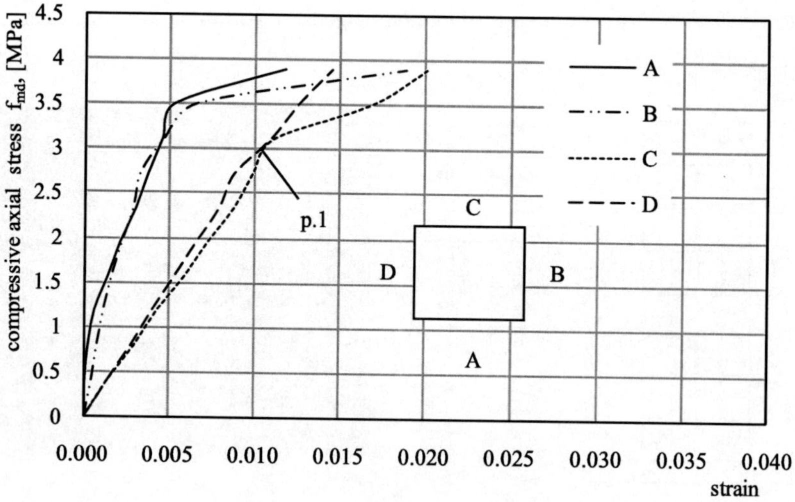

Fig. 2. Experimental "stress-strain" diagram for unconfined specimens ("n" series)

As shown above the maximum stress value reached f_{max} = 3.8 MPa. The ultimate stress was noted as f_{md} = 3.02 MPa ($\sim$0.79 f_{max}) which was confirmed by macro crack rising on side "C" (see p. 1 on Fig. 2). At 0.9 f_{max} loading level cracks widely spread through other faces ("A", "B") thus model's brittle failure followed up. The strain value for the ultimate stress level was reported as ε_{md} = 0.010.

Using f_{md} value as control compressive strength target loading level for the next series ($f_{0.8md}$ = 2.4 MPa) has been obtained. Specimens included to “d” series were subjected to this compressive loading due to the testing program. To justify such an approach and to avoid using samples with significantly different mechanical properties “stress-strain” curves for these two series (“n” and “d”) were compared (Fig. 3). Related curves were similar thus it could be stated that further reference between those two series is reasonable. After specimens’ unloading crack patterns for each column’s side were investigated Fig. 4.

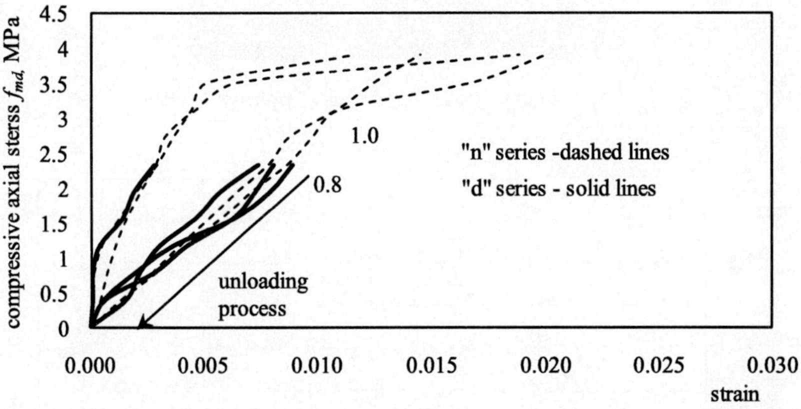

Fig. 3. Experimental “stress-strain” diagram for unconfined specimens (“d” series)

After columns strengthening by applying methods described in Sect. 2.2 specimens were considered as “s series” and the new tests for these confined models were performed (Fig. 5). No significant cracking in repair mortar was not observed up to level f_{mcd} = 2.5 MPa. After passing this stage the first initial cracks were detected (~0.1 mm) with plain developing till the stress level f_{mcd} = 3.52 MPa. This level overstepping was marked by the cracking process with openings up to 0.3 mm and confinement mortar crumpling as well. At the maximum stress level f_{mcd} = 4.2 MPa samples’ collapse (caused by GFRP mesh tensile rupture) has occurred.

Curve’s slope changing relates to the macro crack occurrence in the repair mortar surface (see p. A in Fig. 5). Following the previous assumption it was treated as ultimate strength level f_{cmd} = 3.52 MPa. Ultimate strain for confined series was adjusted as ε_{cmd} = 0.013. For stress analysis given above (see Fig. 2, Fig. 5) the mean value for twins-samples in each series was used. Variation coefficient for “n” and “s” series was 8.13% and 7.14% respectively.

The model’s appearance after testing is presented in Fig. 6. Strengthened samples had been showing slow and plain damage process until the moment when confinement lost integrity. Mesh rupture started in the most concentrated zones near to the column’s corners following jacket mortar separating initiated from the sample’s mid-height (Fig. 6). Subsequently, the failure process progressed more sharply due to no

Fig. 4. Crack patterns after damaging by axial loading

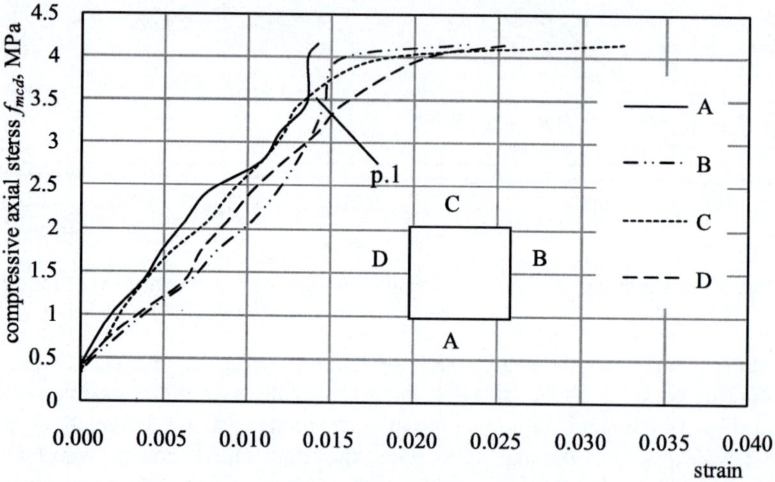

Fig. 5. Experimental "stress-strain" diagram for confined specimens ("s" series)

Fig. 6. Failure mode and appearance of strengthened specimens

restrictions for abandoned masonry core expansion. Such kind of failure mode was typical for all specimens.

4 Conclusions

This paper presents the results of investigation for confined middle-scale masonry structures produced with hollow ceramic bricks after subjecting to a high level (up to 80% of ultimate strength) compression. Strengthening with GFRP meshes was provided after models' decompression. Provided early research draws to some conclusions listed below. Initially, a significant strengthening effect due to the level of preliminary damages and applied type of GFRP mesh was not expected. And obtained results confirmed that in a numerical manner. Thus, according to test data, GFRP jacketing demonstrates values of normalized ultimate stress f_{cmd}/f_{md} = 1.15 and ultimate axial strain $\varepsilon_{cmd}/\varepsilon_{md}$ = 1.3. Generally, jacketed models presented more uniform models' deformation. This study shows that despite full samples' unloading GFRP-mesh applying could be not suitable enough for such a "damages-materials" combination and leads to more efficient strengthening techniques or materials (e.g. CFRP or BFRP) in terms of reasonable repairing approach. However, high-level preliminary loading was not dramatic and, finally, the strengthening effect was still reported.

Further research will be addressed to numerical analysis of obtained results and studies with different pre-loading levels as well.

Acknowledgments. All materials used for strengthening of specimens in this study have been granted by Mapei, UA.

References

1. Valluzzi, M.: Strengthening of masonry structures with fibre reinforced plastics: from modern conception to historical building preservation. In: D'Ayala, D., Fodde, E. (eds.) Structural Analysis of Historic Construction: Preserving Safety and Significance, pp. 33–45. Taylor & Francis Group, London (2008)
2. Cascardi, A., Lerna, M., Micelli, F., Aiello, M.A.: Discontinuous FRP-confinement of masonry columns. Front. Built Environ. **5**(147), 1–7 (2020)
3. Galić, J., Vukić, H., Kalafatić, I.: Masonry columns behaviour analyses due to a different mode of confinement with GFRP straps. Eng. Power: Bull. Croatian Acad. Eng. **13**(4), 24–32 (2018)
4. Aiello, M.A., Micelli, F., Angiulli, R., Corvaglia, P.: Masonry circular columns confined with glass and basalt fibers. In: Proceedings of CICE 2012 6th International Conference on FRP Composites in Civil Engineering, CD ROM Rome (2012)
5. D'Ambra, C., Di Ludovico, M., Balsamo, A., Prota, A., Manfredi, G.: Confinement of tuff and brick masonry columns with FRP laminates. In: Proceedings of the 3rd Conference on Mechanics of Masonry Structures Strengthened with Composite Materials - MURICO 3, Venice, pp. 232–240 (2008)
6. Micelli, F., Di Ludovico, M., Balsamo, A., Manfredi, G.: Mechanical behaviour of FRP-confined masonry by testing of full-scale columns. Mater. Struct. **47**(12), 2081–2100 (2014)

7. Shaheen, E., Shrive, N.G.: Sprayed glass fibre reinforced polymer (SGFRP) masonry columns under concentric and eccentric loading. Can. J. Civ. Eng. **34**(11), 1495–1505 (2005)
8. Corradi, M., Grazini, A., Borri, A.: Confinement of brick masonry columns with CFRP materials. Compos. Sci. Technol. **67**(9), 1772–1783 (2007)
9. Fossetti, M., Minafò, G.: Strengthening of masonry columns with BFRCM or with steel wires: an experimental study. Fibers **4**(2), 15 (2016)
10. Valdés, M., Concu, G., de Nicolo, B.: FRP strengthening of masonry columns: experimental tests and theoretical analysis. Key Eng. Mater. **624**, 603–610 (2014)
11. Murgo, F., Mazzotti, C.: Masonry columns strengthened with FRCM system: numerical and experimental evaluation. Constr. Build. Mater. **202**, 208–222 (2019)
12. Bula, S., Kholod, P., Bogdan, S., Sadlovska, M.: Strengthening of subjected to fire masonry structures with GFRP meshes (TM «Mapei»). Bulletin of Lviv Polytechnic National University. Series "Theory and Building practice", no. 888, pp. 18–28 (2018)
13. Vistovytska Keramika. http://www.vistovytska-keramika.com.ua/index.php/uk/. Accessed 18 May 2020
14. DSTU B.2.7-239:2010 (EN 1015-11:1999, NEQ). Building mortars. Methods of test. Minregionbud, Kyiv (2010)
15. DSTU B.2.7-248:2011. Masonry bricks. Compressive and bending strength testing. Minregionbud, Kyiv (2012)
16. Mapei. Products & Solutions. Mapegrid 120. http://www.mapei.com/it/en/products-and-solutions/products/detail/mapegrid-g-120. Accessed 18 May 2020
17. Mapei. Products & Solutions. Planitop HDM Maxi. http://www.mapei.com/it/en/products-and-solutions/products/detail/planitop-hdm-maxi. Accessed 18 May 2020

Simplified Method for Determining the Energy Efficiency of Window Blinds in the Field

Vsevolod Buravchenko(✉), Oleg Sergeychuk, and Serhii Kozhedub

Kyiv National University of Construction and Architecture, Kiev 03037, Ukraine
vburav4@gmail.com

Abstract. Calculation of heat gain from solar radiation is a necessary part of simulating the energy balance of a building, which is an important aspect of the architectural design of modern energy-efficient buildings. Window blinds are an effective measure of the control of light and heat influence of insolation in buildings. Efficiency of blinds depends on many factors, such as position, material, distance from the sun and parameters of the microclimate and exploitation mode of the rooms where they are used. In the summer of 2019 a group of employees of KNUCA explored the sunscreens of the new office building in Kyiv with a high percentage of glazing. Efficiency of protection from glare and overheating was estimated from measured microclimate parameters in two similar rooms of the building. This article is devoted to describing methods of measurement and calculation of efficiency of controlled present window blinds in the existing buildings.

Keywords: Solar control · Solar protection devices · Energy efficiency · Solar gain · Window blinds

1 Introduction

Calculation of consumption of energy and other resources is an important aspect of modern architectural design. Actual national standards of Ukraine [3–5] and other countries demand incorporating the effect of solar gain onto the energy balance of buildings. Calculation of influence of solar radiation on needs for lighting, cooling and heating requires considering the position and size of translucent structures, physical properties of glass and solar-protection devices. While most researchers [7, 8, 10, 11, 13–18, 20–22] agree that external awnings or blinds are more effective than internal ones for the task of protecting the rooms from overheating, it would be wrong to ignore the properties of the latter in the calculation.

In the summer of 2019 a group of employees of KNUCA were invited to estimate the efficiency of the window blinds in the new office building in Kyiv. The building has a high ratio of glazing and its microclimate and energy balance are heavily influenced by insolation. To prevent glare effect and to reduce overheating and overloading the system of air conditioning, windows of the building are equipped with internal remotely controlled blinds of synthetic cloth with metallic coating. The external layer

Z. Blikharskyy (Ed.): EcoComfort 2020, LNCE 100, pp. 79–86, 2021.
https://doi.org/10.1007/978-3-030-57340-9_10

of the glass in the insulated units is a tinted glass which also adds to the factor of solar protection but was installed mostly for architectural reasons.

The energy of solar radiation enters rooms through several ways: visible and infrared rays, convection from the materials that absorb heat etc. [1, 2, 9], so the factor of effectiveness of solar protection devices needs to consider not only one component, but the entire part of energy flow coming into rooms.

2 Target

The purpose of the work was to consider the effectiveness of window blinds utilised in the building for protection from overheating without disassembling the equipment and disturbing the working processes in the building as little as possible and to propose a simple method of measuring and calculating the efficiency of the blinds and similar solar protection devices in the field.

3 Methods

The chosen method of measurement required using 2 identical rooms with similar volumes, functions, orientations, ratios of glazing, blinds, furniture and decoration materials. The choice of available rooms was limited due to security and the exploitation regime in the building, so 2 nearly identical gyms on the 8th floor with windows facing South-south-west (rooms 914 and 915) (Fig. 1) were chosen for measurements.

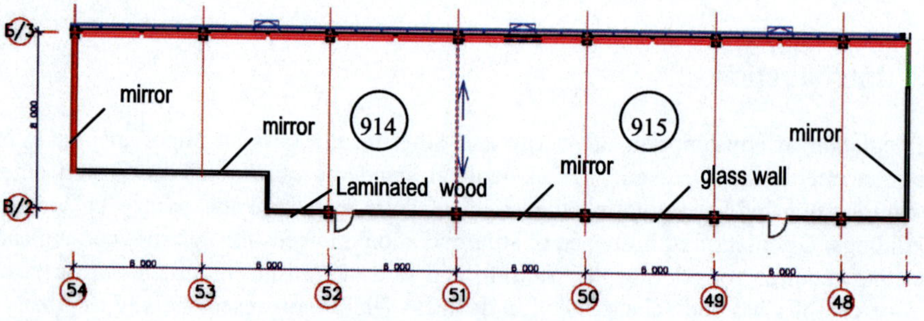

Fig. 1. Plans of rooms 914 and 915.

Measurements were conducted during 2 non-consecutive days with mostly sunny weather: the 10th and 31st of August. At the end of the day preceding the measurements ventilation systems in both rooms were turned off. In the morning, before the insolation of the rooms started to take place, thermal sensors were set: 2 on the floor and 2 on the windowsill of each room. In one of the rooms blinds were open, in the other they were shut throughout the day. On the 10th of August shades in room 915 were open and in

914 were shut; on the 31st of August room 915 was shaded to minimize the influence of the differences of the rooms in length, shape and materials on the results.

The measurements were conducted every 30 min from 10:00 local time (the approximate time when the sun in August starts insolating the façade) until 17:00 when the shadow cast from an additional storey on the South-western part of the building started to cover the windows of room 915 which set the rooms in unequal conditions and made further measurements useless. The results of measurement of temperature were recorded in a mobile hardware-and-software complex for examination of enclosing structures in the field conditions. At the same time illuminance in 3 points on the floor of both rooms was measured using a light meter CEM DT-8809A.

To calculate the factor of energy efficiency of window blinds, the following formula was proposed:

$$\eta = \left(1 - \frac{Q_{\Sigma 2}}{Q_{\Sigma 1}}\right) \cdot 100\%; \qquad (1)$$

where Q_1, Q_2 – quantities of heat exceeding the temperature of cooling according to [5] in the rooms where blinds are used and where they are not used respectively, calculated according to (2).

$$Q_{\sum} = \sum\nolimits_{t1}^{t2} \Delta\theta \cdot c \cdot \rho \cdot V; \qquad (2)$$

where

c – specific heat of the air, which equals 1005 J/(kg•K);
ρ – density of air, which equals 1,2 kg/m^3;
V – volume of air in the room, m^3;
t_1, t_2 – start and end of the period of measurements, which are 10:00 and 17:00 local time (GMT +3) respectively;
Δθ – excess of the temperature of air in the room over the temperature of cooling calculated according to (3):

$$\Delta\theta = \theta_i - \theta_c; \qquad (3)$$

where θ_i – measured value of the temperature of the air (average of the readings of the sensors), °C;

θ_C – temperature of cooling (temperature to which air needs to be cooled in the period of overheating), taken by Tab. 16 of [5]. For office and sport rooms is +24 °C;

Since formulas for numerator and denominator include the same values of factors of c, ρ and V, they are redundant, and the formula (1) looks like (4):

$$\eta = \left(1 - \sum\nolimits_{t1}^{t2} \Delta\theta_2 / \sum\nolimits_{t1}^{t2} \Delta\theta_1\right) \cdot 100\% \qquad (4)$$

4 Results

Values of the measured temperature of the air in the rooms on both days of measurement are displayed in (Fig. 2).

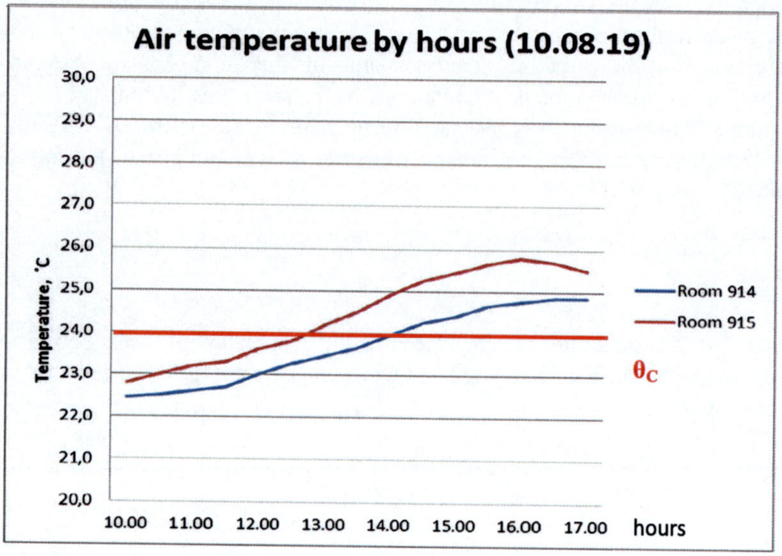

a)

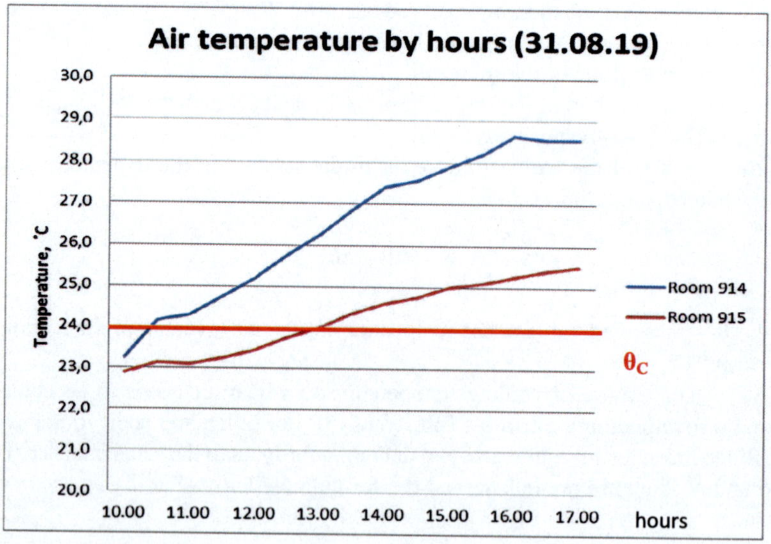

b)

Fig. 2. Temperature of the air in the rooms 914 and 915 by hours: *a*) on 10.08.2019, *b*) on 31.08.2019.

Values of the measured illuminance in both rooms on both days of measurement are displayed in (Fig. 3).

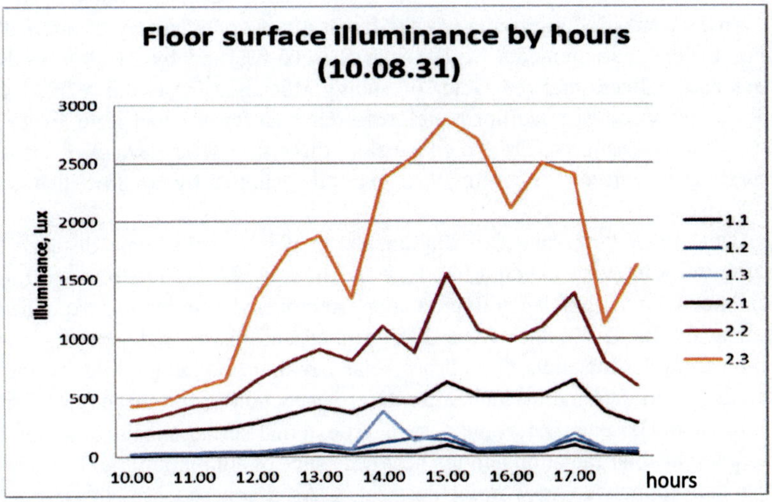

a)

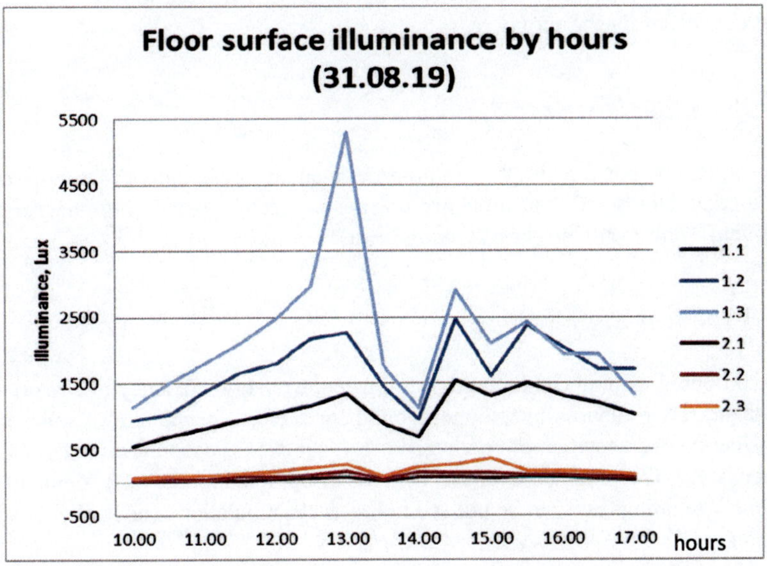

b)

Fig. 3. Illuminance from natural light on the floor in rooms 914 and 915 by hours: *a*) on 10.08.2019, *b*) on 31.08.2019.

One can see that in both cases the air temperature grows steadily through accumulation of heat and reaches its highest values about 16:00 in unshaded rooms and slightly later in shaded rooms. This slight delay occurs possibly because blinds reduce not only incoming solar radiation but also losses of heat from the room by radiation. At the same time illuminance reaches its highest values at 13:00 and for the rest of the day changes in an irregular way due to the appearance of *cumulus* clouds in the sky.

When calculated by formula (4), the factor of energy efficiency of internal window blinds by the measurements of 10.08.2019 is 65,8% and by 31.08.2019 is 79,1%. These data indicate that the factor of energy efficiency (η) is not a fixed value but depends on weather conditions and requirements for microclimate of the rooms, however it is possible to draw a conclusion, that using solar devices of this type will decrease energy needs for cooling throughout the summer by 60–80%, that is decrease 2,5–5 times.

At the same time values of illuminance in points at the same distance from the glazing in rooms with lowered blinds were lower by 97–98% than in the rooms with open blinds. Lower efficiency in protection from overheating may be explained by heat reaching the room by other ways than infrared and visible radiation, including convection from the materials that absorb solar radiation: blinds themselves and the aluminium window frames which after brief inspection seem inadequate for climate conditions in Ukraine. Dark blue coating on external surfaces leads to a high factor of absorption of solar radiation and the heat resistance of aluminium frames is presumably lower than demanded by current norms [3]. As a consequence, the temperature of internal surfaces of window frames reached +50 °C which led to increased heat output and convection in the rooms.

5 Scientific Novelty

This article proposes a simple and practical method of estimation of energy efficiency of window blinds and their influence on energy needs for cooling by measurements in the field using available sensory equipment.

6 Practical Significance

The proposed method can be used to estimate the energy efficiency of already installed solar protection devices like shutters, blinds or screens for the aim of solar protection considering the specifics of room geometry, functions, decoration materials, glazing systems etc. The obtained data can be used to optimize modes of systems of climatisation in buildings and can be utilised in life-cycle simulations and further architectural designs.

7 Conclusions

The conducted research confirmed that window blinds with metallic coating are an effective means of protection of rooms from overheating, a fact that cannot be neglected in simulations of energy balance of the buildings, particularly those with a high ratio of glazing. On the other hand, blinds do not solve the problem completely and additional air cooling may still be required for effective use of the rooms.

Blinds are also effective in protecting the rooms from glare which is important for rooms with long-term presence of people. Illuminance in the shaded room remained throughout the day below 300 lux a requirement for administrative and office rooms which make up a majority of the rooms of the building. As a result, using the blinds in buildings leads to the need for artificial lighting. Whether reduction in energy consumption for cooling exceeds the energy need for lighting and the total influence of solar protection devices on energy balance of the building needs further measurements and a proper life-cycle simulation.

References

1. Anuranjan, S., Sudhir, K.: Heat transfer through glazing systems with inter-pane shading devices. Energy Technol. Policy **1**, 23–34 (2014)
2. Cuevas, C., Fissore, A., Fonseca, N.: Natural convection at an indoor glazing surface with different window blinds. Energy Build. **42**, 1685–1691 (2010)
3. DBN V.2.6-31:2016. Heat insulation of buildings. Minregion, Kyiv, Ukraine (2017)
4. DSTU-N B V.2.2-27:2010. Instruction for calculation of insolation in civil objects. Ministry of Construction and Regional Development of Ukraine, Ukrarkhbudinform, Kyiv, Ukraine (2010)
5. DSTU B.A.2.2-12:2015. Energy efficiency of buildings. Method for calculation of energy use for heating, cooling, ventilation, lighting and hot water supply of the buildings. Ukrarhbudinform, Kyiv, Ukraine (2015)
6. Galasiu, A.D., Atif, M.R.: Field-performance of daylight-linked lighting controls and window blinds. In: 5th European Conference on Energy-Efficient Lighting - Right Light 5 Proceedings, Nice, pp. 1–14 (2002)
7. Grimm, F.: Energieeffizientes Bauen mit Glas. Verlag Georg D.W. Callwey Gmbh & Co, Muenchen (2004)
8. Harkness, E., Mehta, M.: Solar radiation control in buildings. Stroyizdat, Moscow, USSR (1984)
9. Kotey, N.A., Barnaby, C.S., Wright, J.L.: Solar gain through windows with shading devices. ASHRAE Trans. **115**(2), 18–30 (2009)
10. Lechner, N.: Heating, Cooling, Lighting: Sustainable Design Methods for Architects, 4th edn. Wiley, Hoboken (2015)
11. Marcus, T., Maurice, E.: Buildings, Climate, Energy. Gidrometeoizdat, Leningrad, USSR (1985)
12. Naylor, D., Shahid, H., Harrison, S.J.: A simplified method for modelling the effect of blinds on window thermal performance. Int. J. Energy Res. **30**, 471–488 (2006)
13. Pidhirny, O.L., Sergeichuk, O.V., Shchepetova, I.M.: Translucent encasements of buildings. Vitryna, Kyiv, Ukraine (2005)
14. Renckens, J.: Façades Architecture. Fascination in Aluminium and Glass. FAECF, Bockenheimer (1998)
15. Schittich, C., Stalb, G., Balkow, D.: Glass Construction Manual. Institut fur Internationale Dokumentation GmbH, Munchen (1999)

16. Sergeichuk, O.V., Buravchenko, V.S.: Specificities of introduction of solar protection devices into Energy Passport of Building. Prykladna Heometriya ta Inzhenerna Grafika **47**, 164–170 (2010)
17. Sergeichuk, O.V., Buravchenko, V.S.: Specificities of method of calculation of solar gains in national supplement to DSTU B EN ISO 13790. Energoefektyvnist v budivnytstvi ta arkhitekturi **6**, 267–272 (2014)
18. Sergeichuk, O.V.: Methods and means of solar protection of rooms. Vitryna **1**(11), 24–31, **2**(12), 34–38 (2001)
19. Shapoval, S., Zhelykh, V., Venhryn, I., Kozak, K.: Simulation of thermal processes in the solar collector which is combined with external fence of an energy efficient house (2020)
20. Shteynberg, A.Ya.: Solar Protection of Buildings. Budivel'nyk, Kyiv, USSR (1986)
21. Solar shading for low energy buildings. ES-SO, Meise, Belgium (2012)
22. Szokolay, S.V.: Introduction to architectural science. The Basis of Sustainable Design. Stroyizdat, Moscow, USSR (1984)

Methods of Reinforcing for Engineering Restoration of Architectural Monuments

Olena Chernieva[1](✉), Gennadiy Plahotny[1], and Matej Babič[2]

[1] The Odessa State Academy of Civil Engineering and Architecture, Odessa, Ukraine
spring.cherneva@gmail.com
[2] Faculty of Information Studies, Novo Mesto, Slovenia

Abstract. The article presents the survey results of the architectural monument-St. Alexis Church in Bayramaly (Turkmenistan). The main causes of structural damage have been analyzed. The purpose of the meticulous task was to study the technical condition of the structural elements, establish the causes of damage, evaluate the possibility of eliminating the root cause of the defects and to renovate the historical architectural monument to its original grandeur, with bare minimum changes to the original design and/or structural replacements. The original structural analysis was carried out on the said structures and measures for their recovery were proposed. The article details the path to restoration of monuments and proposes a new approach (method) to reinforcement of walls. This method addresses the problem of development of cracks that appear due to the shearing of walls as a result of uneven foundation soil deformations. This method of reinforcement of walls, allows anchors to work as a dowel joint to withstand high shear force.

Keywords: Arches · Architectural monument · Cracks · Damage · Damaged wall areas · Foundations

1 Introduction

The destinies of most of the architectural monuments in the world have been impacted by historical events, with the passage of time. The forces of nature have also seriously influenced the conservation and maintenance of such buildings. Created by the genius of master architects as a symbol of faith and reverence, many of these edifices have often become victims of neglect, public apathy, vandalism, ignorance and poor cultural appreciation of the masses. Non-destructive testing techniques are widely applied for diagnostic investigation of the architectural monuments which was noticed in [1, 2].

2 Historical Reference

In 90-ies of XX century the group of Specialists from the Odessa State Academy of Civil Engineering and Architecture carried out an exhaustive survey to estimate the robustness and the architectural status of the following three major monuments, namely

Z. Blikharskyy (Ed.): EcoComfort 2020, LNCE 100, pp. 87–94, 2021.
https://doi.org/10.1007/978-3-030-57340-9_11

- St. Alexis Church in Bayramaly city, the Old Nisa and the Iranian mosque in Chardzhou city (Turkmenistan). The purpose of the meticulous task was to study the technical condition of the structural elements, establish the causes of damage, evaluate the possibility of eliminating the root cause of the defects and to renovate the historical architectural monuments to their original grandeur, with bare minimum changes to the original design and/or structural replacements. The original structural analysis was carried out on the said structures and measures for their recovery were proposed. This article is devoted to the first object of this expedition – the St. Alexis church in Bayramaly city, Turkmenistan.

The historical reference and archival search were carried out in the archives of Mary, Ashkhabad, and in the marine archives of St. Petersburg. According to archival documents, the temple project was ordered by His Royal Highness, Tsar Nicholas II in honor of his heir and son, Alexy. To attract talent and with the idea of getting the best design, the Emperor called for a nationwide competition across all the Republics of Russia. The competition was well received by the public and several hundred projects were submitted, in deference to the orders of His Royal Highness, the Tsar.

In the competition, the project submitted by Chokhorov K., who was the fourth-year student of the St. Petersburg Institute of Civil Engineering, was proclaimed the winner (see Fig. 1). The project was implemented during 1910–1914 in the Murgab oasis of the Kara-Kum desert in the city of Bayramaly.

Fig. 1. The Chokhorov's temple project - archive search results.

3 Consequences of Negative Human Activity

Since the 1920's the gradual destruction of the Temple began. First, the crosses, domes and wooden cover structures were demolished. The premises of the temple was used as a canteen and warehouse and during the war, as a camp of the Prisoners of War. In the late 90s the building was transferred for the use by the diocese.

As illustrated in the plan view in Fig. 1 above, the open hall of the church was designed with four arch projections, one in each direction, giving it the resemblance of the cross. On the east side of the temple there is an apses and on the west side there is a staircase with an exit to the second floor, where the choir transept is located.

3.1 Civil Engineering Excavation Findings

According to the results based on digging of pits and drilled wells the substructure of the foundation was observed to be made up of dense tertiary loam and clay of considerable thickness. Average values of the physical-mechanical properties of the foundation soil have been detailed in Table 1. There were strip foundations under the walls, made of clay bricks on a cement-sand mortar, having a foundation depth of up to 1.2 m, and strip width of 3.0 m. The bearing walls of the building were made of clay bricks on a cement-sand mortar of 770 cm thickness.

Table 1. Average values of physic-mechanical properties of the foundation soil

Name of properties	Units	Name and layer number			
		1	2	3	4
Thickness	m	Clay loam	Clay	Clay loam	Clay
Natural moisture	%	1.5	2.0	2.0	4.5
Soil density	g/sm^3	20	25	20	18
Dry soil density	g/sm^3	1.86	1.83	1.91	1.82
Deformation modulus	MPa	1.54	1.47	1.59	1.62
Internal friction angle	degrees	8.0	8.0	14.0	16.0
Specific cohesion	MPa	18	17	22	17

3.2 Exterior Inspection Findings

Cross arches were made of cast reinforced concrete, reinforced with space frames. The inner hall's floor construction of the temple is in-situ concrete slab. Initially, the rafter system consisted of timber, roofing battens and sheathing. The roof covering was made of metal sheets.

An instrumental-visual survey revealed that part of the building was lost, including:

- dome with a cross over the central part of the building;
- spire with a cross in the belfry;
- roof covering and rafter system timber, roofing battens and sheathing;

- exterior cement-sand plaster walls, window and door fillings, frescoes of icons from majolica;
- bells from the belfry;
- corbel arch above the entrances was partially lost.

A visual structural survey revealed a number of deformed vertical and inclined cracks, 1 to 3 mm thick in the bearing elements. On the main cracks in the bearing walls fixed beacons (60 pcs) were observed.

3.3 Second Survey Findings

As a part of the study, the extensions to the premises constructed at a later period were opened and cleaned. Upon conducting a second survey of the wall beacons six months later, no further dynamics of cracks and development of deformations were noticed. The exception to this was the formation of additional hair line cracks in the right corner of the building near the staircase. After a subsequent careful survey to identify the causes of progressive damage it was noticed that in that specific area of the building there existed a partially preserved underground passage. Upon executing a well-planned cementing activity in the underground passage, further deformation and development of cracks were successfully arrested.

4 Recommendations for Building Renovation

After studying the results of visual and instrumental surveys and archival materials, the group of Specialists from the Odessa State Academy of Civil Engineering developed a scientific and restoration report [3] consisting of:

a) preliminary work section and
b) repair and restoration project

This scientific-restoration report of the Church's renovation was handed over to the Turkmenistan's Ministry of Culture and after examination the recommendations were successfully implemented at the beginning of the 21st century (see Fig. 2).

Following specific restoration related civil works were carried out in conjunction with the recommendations made in the report:

1. For strengthening the deformed bearing walls, individual cracks were cemented with cement-sand mortar to prevent further expansion after a thorough cleaning and washing operation.
2. Wherever the damaged wall areas were found (Figs. 3 and 4) to have a network of fine/hair-line cracks or single cracks of considerable depth, the restoration of the bearing capacity of the walls were restored by re-laying part of the wall with a more durable mortar than what was originally used in the old masonry.
3. To increase the stability of the walls, horizontal strands with screw thread endings were passed into the pre-drilled holes in the bearing walls.
4. Successful joint packing was achieved by systematically carrying out the following activity:

Fig. 2. The church's condition before restoration.

Fig. 3. Damaged part of the apse.

a. Installation of channels on the sides of the building, fixed by flanges to the wall with tension of the horizontal strands maintained by means of nuts located at the ends.
b. Finally, the required tension was set with double threaded couplings in the middle section of the horizontal strands, which were technically segmented into 3 parts, by employing a torque wrench.
c. Consequent to the setting of the mortar in place, the strands were dismantled.

Fig. 4. The southern porch of the temple - condition before restoration.

5 Patent and Research Work

5.1 Introduction to Research

Damaged structures are continuously being investigated in Ukraine. As part of the process, damaged reinforced concrete constructions [4–6] damaged reinforced concrete columns [7] have already been tested by various researchers.

The problem of restoration of monuments in Ukraine using different techniques [8] is usually common, especially considering the fact that bearing walls are made of "saw" stones of limestone-shell rock in limy mortar. The issue of shear strengthening of masonry walls is being considered all over the world [9–11].

Based on the test results of 15 prototypes of stone walls, a patent for utility model No. 28855 "Method of wall joints' reinforcing" was developed and submitted. This patent was finally approved and granted by the Odessa State Academy of Civil Engineering. This method addresses the problem of development of cracks that appear due to the shearing of walls as a result of uneven foundation soil deformations.

5.2 Implementation Methodology

Initially, the boreholes (ø32–36 mm) are required to be drilled in the damaged wall at an angle of 35 to 60° every 0.8 to 1.2 m. The boreholes have to be filled with polymer-cement mortar. Subsequently, they should be clogged with internal anchors (ø16 to 18 mm) made of ribbed bars. Finally, the reinforcement areas need to be closed with a cement-sand mortar (see Fig. 5).

This method of reinforcement of walls, allows anchors to work as a dowel joint to withstand high shear force. In this arrangement, the crack does not open further due to the laying of the brickwork on the dowels. Anchor passes not only in the mortar joint, but also through the stone. Driving anchors at an angle of 35–50° makes it possible to

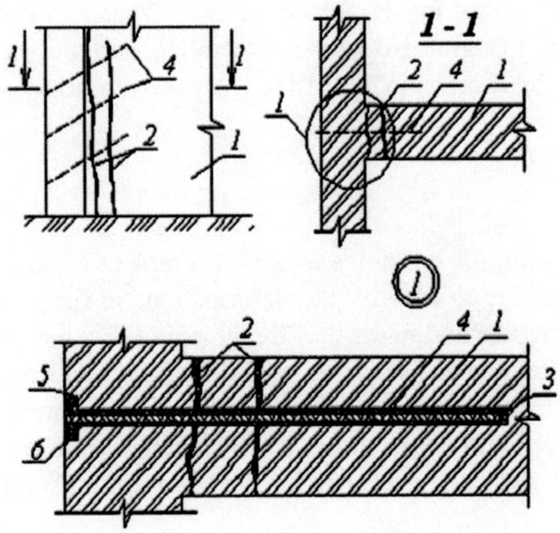

Fig. 5. Method of wall joints' reinforcing: 1 - damaged wall; 2 - crack in the corner part of the wall; 3 - borehole; 4 - internal anchors; 5 - recess in the wall; 6 - finishing with a cement-sand mortar.

withstand the maximum shear force. If anchors are driven at an angle of less than 35 or more than 50°, the efficiency of the anchors to withstand shear are greatly reduced, as they no longer work as dowels (see Fig. 6).

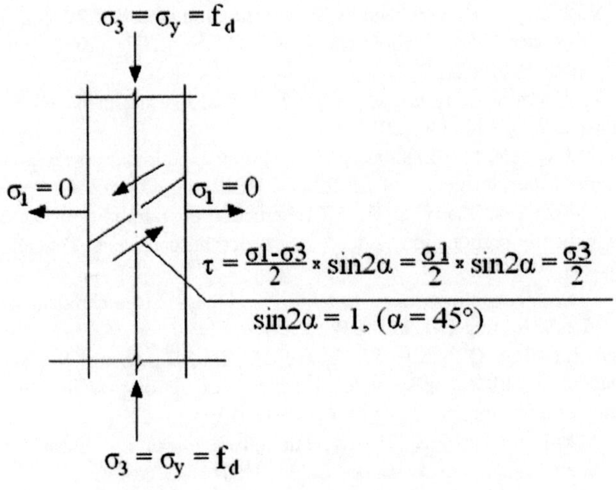

Fig. 6. Graphic justification of the decision.

This method is simple and much less expensive to execute, and does not require the use of complex or extraordinary heavy equipment; it is a relatively uncomplicated and successful formula that eliminates localized deformations.

The Department of Reinforced Structures and Transport Facilities at the Odesa State Academy of Civil Engineering & Architecture is currently carrying out extensive research devoted to the determination of the bearing capacity of damaged structures.

6 Conclusions

The condition monitoring of buildings is one of the most important tasks for modern engineers. It is imperative to identify the deformations in time and eliminate the root cause. For a further understanding of the strengthening time-tested buildings, it would be necessary to employ modern scientific methods and if required, conduct specialized research. Great architectural wonders constructed over the centuries stand testimony to our civilization, culture and engineering achievements. It is our moral obligation and duty to preserve and patronize such master pieces for the future generations.

References

1. Binda, L., Lualdi, M., Saisi, A.: Non-destructive testing techniques applied for diagnostic investigation: Syracuse cathedral in Sicily, Italy. Int. J. Arch. Herit. **1**, 380–402 (2007)
2. Lacidogna, G., Manuello, A., Niccolini, G., Carpinteri, A.: Acoustic emission monitoring of Italian historical buildings and the case study of the Athena temple in Syracuse. Arch. Sci. Rev. **58**(4), 290–299 (2015)
3. Lisenko, V., Plachotny, H., Matvienko, G., Chlianc, E.: The scientific-restoration report of the St. Alexis Church's renovation in Bayramaly. OSACE, Odesa (1990–1991)
4. Vegera, P., Vashkevych, R., Blikharskyy, Z.: Fracture toughness of RC beams with different shear span. In: Matec Web of Conferences, vol. 174, p. 02021 (2018). https://doi.org/10.1051/matecconf/201817402021
5. Klimenko, Y., Chernieva, O., Ismael, A.M.: Test results of the damaged T-section beams. Tehnički glasnik **7**(4), 344–346 (2013)
6. Lobodanov, M., Vegera, P., Blikharskyy, Z.: Planning experiment for researching reinforced concrete beams with damages. In: CEE 2019. LNCE, vol. 47, pp. 243–250 (2020)
7. Klimenko, E., Korol, N., Korol, I., Kos, Z.: Prediction the durability of the columns on the criterion of concrete carbonation and corrosion of reinforcement. Tehnički glasnik **9**(3), 317–320 (2015)
8. Diachenko, Ye., Voskobiinyk, O., Lomiga, I.: Floors lift technique in old building retrofitting. Metall. Mining Ind. **11**, 18–22 (2016)
9. Li, Z., Chen, L., Fang, Q., Chen, W.S., Hao, H., Zhu, R., Zheng, K.: Experimental and numerical study on CFRP strip strengthened clay brick masonry walls subjected to vented gas explosions. Int. J. Impact Eng. **129**, 66–79 (2019)
10. Vasconcelos, G., Lourenco, P.B.: Experimental characterization of stone masonry in shear and compression. Constr. Build. Mater. **23**(11), 3337–3345 (2009)
11. Raouf, A.M., Saeed, J.A.: In-plane shear strengthening of masonry walls after damage. Sulaimani J. Eng. Sci. **7**(1), 65–85 (2020)

Comparison of Bitumen Modified by Phenol Formaldehyde Resins Synthesized from Different Raw Materials

Yuriy Demchuk(✉), Volodymyr Gunka, Iurii Sidun, and Sergii Solodkyy

Lviv Polytecnic National University, Lviv 79013, Ukraine
yuriy_demchuk@ukr.net

Abstract. Phenol formaldehyde resins were obtained by polycondensation of a mixture of phenols (technical or "raw" phenols) obtained from the phenolic fraction of coal tar. The resulting resin approximately consists of 1/3 of phenol and 2/3 of cresols. The obtained resins were compared with industrial phenol formaldehyde resins, the raw material of which is pure (synthetic) phenol. The possibility of using the obtained resins as modifiers of road petroleum bitumen has been established, especially as adhesive promoters to bitumen. It is shown that the addition of these resins significantly increases the adhesion to the surface of granite rubble and glass. Adhesion to the glass surface increases for PhCR-F from 33 to 87%; for Iditol from 33 to 90% and adhesion to the crushed stone surface for PhCR-F from mark 3 to mark 5; for Iditol from mark 3 to mark 5. This indicates that PhCR-F and Iditol can used as adhesive additives for oil road bitumen

Keywords: Bitumen · Phenol · Cresol · Phenol formaldehyde resin · Modifier · Adhesion promoters

1 Introduction

Oil bitumen is the main organic binder for the construction of asphalt roads, airfield, sidewalks, etc., and that is about 85% of its total consumption. According to [1], more than 90% of the 5.2 million km European highways are surfaced with asphalt concrete.

But there are some problems in bitumen using, the most acute of them are the heat resistance and adhesion properties of commercial bitumen are not high enough (even if they meet the requirements of regulatory documents) [2].

One of the most promising ways to improve the binder's quality for obtain pavements with improved properties is their modification by polymeric materials [2, 3]. However, the modifiers use is limited due to their significant cost. Therefore, it is important to find inexpensive substances that would improve the bitumen properties, primarily adhesive.

It is known that phenol-formaldehyde resins increase the oil bitumen adhesion properties and are effective modifying additives [4–6].

Z. Blikharskyy (Ed.): EcoComfort 2020, LNCE 100, pp. 95–102, 2021.
https://doi.org/10.1007/978-3-030-57340-9_12

The paper [4] describes a number of studies on the production and use of phenol-formaldehyde resins as oil road bitumen modifiers. It is established that the addition 2 wt% of phenol-formaldehyde resin into bitumen leads to an increase bitumen softening temperature by the R&B method from 49.1 to 51 °C, a slight decrease in bitumen penetration (plasticity) at 25 °C from 62 m $\cdot$ 10^{-4} to 60 m $\cdot$ 10^{-4}, and an increase in kinematic viscosity at 135 °C from 0.35 to 0.44 PA $\cdot$ s.

The studies [5] have shown that Bakelite optimal content (resol type phenol-formaldehyde resin) in PMB is 1.75 wt%. At this amount of Bakelite addition, there is an increase in stability by Marshall from 20 to 24.2 kN, an increase in the softening temperature by the R&B method from 52 to 61 °C, and also a decrease in bitumen penetration at 25 °C from 49 m $\cdot$ 10^{-4} to 39 m $\cdot$ 10^{-4}.

The authors [6] were found that 2 wt% of Bakelite addition (resol type phenol-formaldehyde resin) into bitumen leads to increase in the softening temperature of bitumen by the R&B method from 58 to 66 °C, decreases the penetration (plasticity) of bitumen at temperature 25 °C from 62 m $\cdot$ 10^{-4} to 20 m $\cdot$ 10^{-4}, as well as an increase in kinematic viscosity at 135 °C from 0.4 to 0.98 PA $\cdot$ s.

Therefore, it is observed that addition phenol-formaldehyde resins in some cases can significantly increase the bitumen heat resistance (softening temperature) in others, this effect is not high.

However, phenol-formaldehyde resins obtained from synthetic phenol are not widely used as polymer modifiers, also due primarily to their high cost [7].

In recent years, the Department of Chemical Technology of Oil and Gas Processing, Lviv Polytechnic National University is conducted researches to obtain relatively cheap and effective oil bitumen modifiers from coal cooking liquid products [8, 9].

Previous studies [10–13] showed the phenol-cresol-formaldehyde resin, obtained from phenolic fraction of coal tar and synthesized by polycondensation of "raw" phenols with formaldehyde, as an effective modifier of road oil bitumen. In addition, these works show that the addition of PhCR-F to oil road bitumen, primarily, significantly increases the adhesion to the glass and gravel surfaces and provides a complete, irreversible bond between bitumen binder and stone material.

Figure 1 presents the block diagram for obtaining the phenol fraction and its application for resin production.

To date, no comparison of novolac phenol-formaldehyde resins obtained from technical (phenol and cresols mixture) and "pure" phenols as a road oil bitumen modifier has been performed. This paper presents a compare between bitumen modified with phenol-cresol-formaldehyde resin (PhCR-F) obtained from raw coal coking products and bitumen modified with phenol-formaldehyde resin obtained from pure (synthetic) phenol (Iditol).

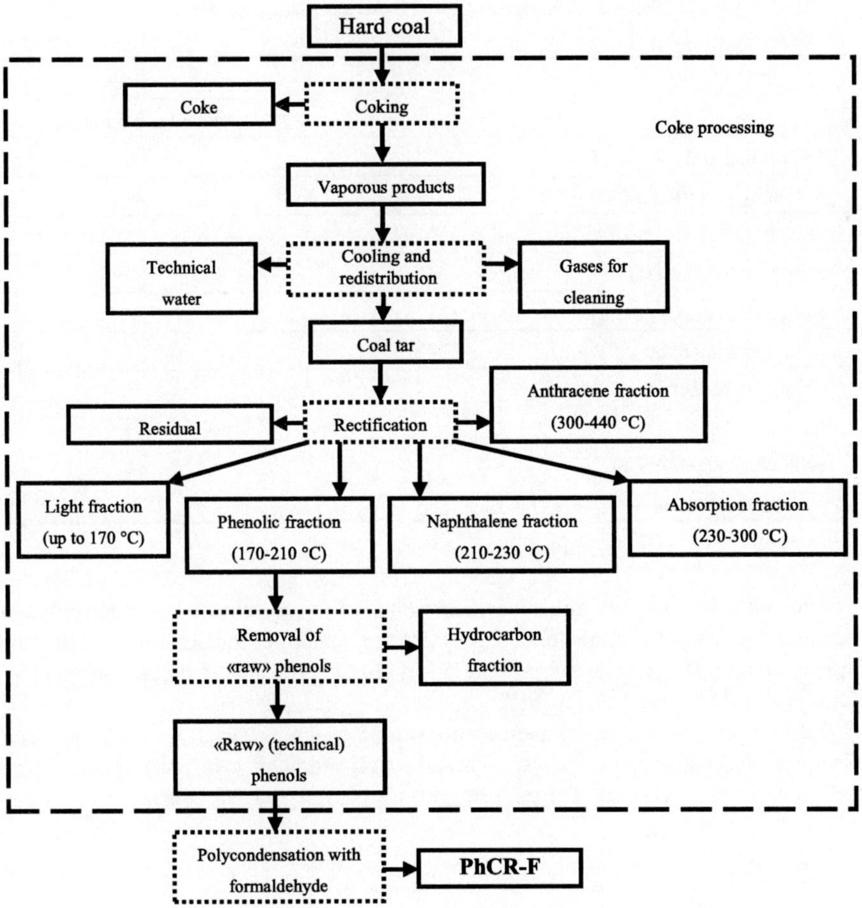

Fig. 1. Scheme for extracting the phenol fraction and PhCR-F.

2 Experimental

2.1 Initial Materials

To obtain the modified bitumen, we used the oxidized bitumen 70/100 produced by PJSC «Transnational financial and industrial oil company Ukrtatnafta» (Kremenchuk, Ukraine). Characteristics of bitumen are given in Table 1.

Table 1. Characteristics of oxidized road bitumens.

Index	Values
	Oxidized bitumen
Penetration at 25 °C ($m \cdot 10^{-4}$)	70
Softening point R&B (°C)	46
Ductility at 25 °C ($m \cdot 10^{-2}$)	63
Adhesion to glass (%)	33
Adhesion to gravel (mark)	3
Fraas breaking point (°C)	-18
Change in properties after heating:	
mass loss (%)	0.1
residual penetration (%)	93
change in softening temperature (°C)	2

The phenol-cresol-formaldehyde resin (PhCR-F) was obtained *via* the polycondensation of "raw" phenols obtained from the phenolic fraction of coal tar and formaldehyde under conditions presented in [11–14]. The PhCR-F softening temperature by the R&B method is 105 °C.

Table 2 presents the component compositions of the extracted raw phenols determined by chromatographic analysis. The adhesion promoter (PhCR-F) obtained from such raw materials consist approximately of 1/3 phenol and 2/3 cresol.

Table 2. Composition of «raw» phenols.

Component	«Raw» (technical) phenols (% wt)
phenol	33.551
o-cresol	14.913
m- and *p*-cresol	37.914
indan	0.120
indene	0.353
benzene	6.719
toluene	0.701
naphthalene	2.323
Non-identified	3.407
Total	100.000

For comparison, there was used phenol-formaldehyde resin of novolac type (Iditol) with a softening point by the R&B method −92 °C and a free phenol content 0.1 wt%.

2.2 Mixing Bitumen with PhCR-F and Iditol

The polymer modified bitumen (PMB) was prepared as follows: the definite amount of bitumen was heated to a fixed temperature, and then a modifier (1,0 wt%) was added. The mixture was obtained at 190 °C for 1 h. The process conditions were chosen on the basis of results, described in [10–14].

2.3 Raw Material and Products Analyses

The breaking point was determined according to the standard described in [15], the ductility was set at 25 °C [16], penetration at 25 °C [17], and softening point [18]. Adhesion to the glass surface and rubble was determined by requirements of Ukrainian regulatory documents [19] and [20], respectively, as described in [21, 22].

3 Results and Discussion

The main characteristics of the primary oxidized bitumen 70/100 and bitumen containing 1 wt% PhCR-F and Iditol were compared to confirm the positive effect, as presented in Table 3.

Table 3. Comparison of the main characteristics of pure and modified bitumen.

Index	70/100	70/100 + 1% modifier	
		PhCR-F	Iditol
Penetration at 25 °C ($m \times 10^{-4}$)	70	68	50
Softening point (ball & ring method) (°C)	46,0	48,0	54,8
Ductility at 25 °C ($m \times 10^{-2}$)	63,0	46,0	52,5
Fraas breaking point (°C)	−18	−18	−17
Adhesion to glass (%)	33	87	90
Adhesion to gravel (mark)	3	5	5
Homogeneity	Homogeneous	Homogeneous	Homogeneous

Table 3 shows that addition 1 wt% of phenol-cresol-formaldehyde resin (PhCR-F), obtained from the coal tar phenolic fraction, into the original bitumen the softening temperature increases by 2 °C (from 46 to 48 °C), penetration decreases by 2 points (from 70 to 68 $m \times 10^{-4}$). The adding to the original bitumen 1% of industrial phenol-formaldehyde resin (Iditol), obtained from pure (synthetic) phenol, the softening temperature increases by 8.8 °C (from 46 to 54.8 °C), penetration decreases by 20 points (with 70 to 50 m × 10-4). Therefore, at Iditol using there is more intense increase in heat resistance and decrease in plasticity of the obtained bituminous composition. In our opinion, chemical modification of oxidized bitumen takes place during Iditol (reactoplast) modification, and physical modification during PhCR-F. The cresols presence in PhCR-F (Table 2), especially ortho- and meta-cresols, greatly reduces resin reactivity. This is due to the fact that potentially reactive ortho- and meta-positions in PhCR-F are blocked by methyl substituents.

It is also worth to note, that PhCR-F and Iditol addition significantly improves bitumen adhesion properties. Adhesion to the glass surface increases (for PhCR-F from 33 to 87%; for Iditol from 33 to 90%; Table 2 and Fig. 2) and adhesion to the crushed stone surface (for PhCR-F from mark 3 to mark 5; for Iditol from mark 3 to mark 5; Table 2). This indicates that PhCR-F and Iditol can also be used as adhesive additives for oil road bitumen.

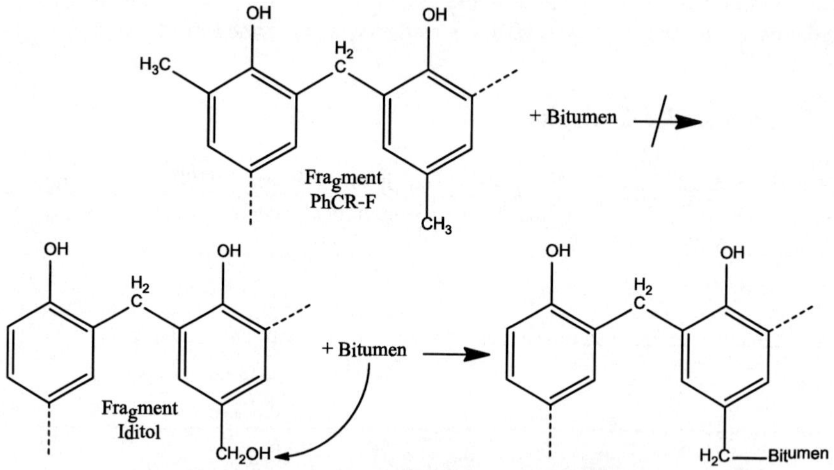

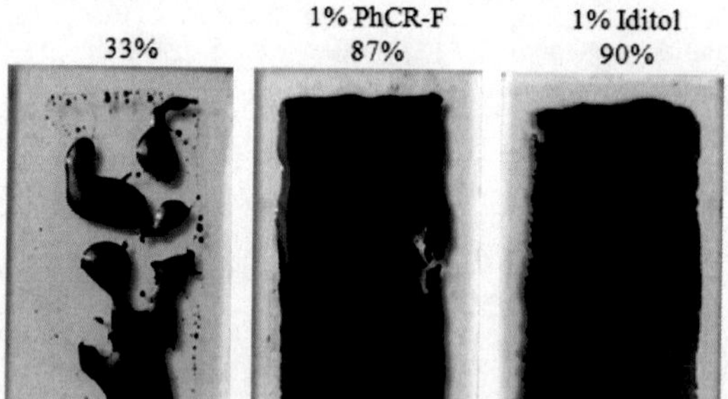

Fig. 2. Results of determining adhesion to glass surface.

Therefore, it has to be noted that for bitumen heat resistance improving (to increase the softening temperature) should be used novolac phenol-formaldehyde resins synthesized only from pure phenol, such as Iditol. If phenol-formaldehyde resins are used as adhesive additives to the bitumen, the cresol presence in the resins structure does not adversely affect the bitumen adhesion to mineral materials. That is, PhCR-F and Iditol should be used as adhesive additives to bitumen.

4 Conclusions

There was performed a compare between bitumen modified with novolac phenol-cresol-formaldehyde resin, the raw material of which are "raw" phenols isolated from the coal tar phenolic fraction (PhCR-F) and bitumen modified with novolac phenol-formaldehyde resin, which raw material is pure (synthetic) phenol (Iditol).

It was found that adding 1 wt% of PhCR-F to the original bitumen, the softening temperature increases by 2 °C (from 46 to 48 °C), penetration decreases by 2 points (from 70 to 68 m $\times$ 10^{-4}), at Iditol using, the softening temperature increases by 8.8 °C (from 46 to 54.8 °C), penetration decreases by 20 points (from 70 to 50 m $\times$ 10^{-4}). Therefore, when Iditol reactoplast is added, chemical modification of oxidized bitumen takes place, and with PhCR-F - physical (acts as thermoplastics). The passive PhCR-F effect on bitumen is due to the cresol's presence (approximately 2/3), especially ortho- and meta-cresols, in the structure. This is because potentially reactive ortho- and meta-position in PhCR-F are blocked by methyl substituents.

Bitumen modification by PhCR-F and Iditol significantly improves their adhesive properties towards mineral materials. Adhesion to the glass surface increases (for PhCR-F from 33 to 87%; for Iditol from 33 to 90%) and adhesion to the crushed stone surface (for PhCR-F and Iditol from mark 3 to mark 5). This indicates that PhCR-F and Iditol can be also used as adhesive additives for oil road bitumen.

Therefore, to improve the bitumen heat resistance should be used novolac phenol-formaldehyde resins synthesized only from pure phenol, such as Iditol. If these resins are used as adhesives additives for bitumen, this condition is not mandatory.

References

1. Kowalski, K.: Eco-friendly materials for a new concept of asphalt pavement. Transp. Res. Procedia **14**, 3582–3591 (2016)
2. Pyshyev, S.: Polymer modified Bitumen: Review. Chem. Chem. Technol. **10**(4s), 631–636 (2016)
3. Hrynchuk, Y.: Epoxide of rapeseed oil-modifier for bitumen and asphalt concrete. Pet. Coal **61**(4), 836–842 (2019)
4. Çubuk, M.: Rheological properties and performance evaluation of phenol-formaldehyde modified bitumen. J. Mater. Civ. Eng. **26**(6) (2014)
5. Gupta, A.: Comparative study of conventional and bakelite modified bituminious mix. Int. J. Civ. Eng. Tech. **10**(3), 1386–1392 (2019)
6. Saha, S.: Characterization of bakelite-modified bitumen. Innovative Infrastruct. Solutions **2**(3) (2017)

7. Yanru, X.: Research status, industrial application demand and prospects of phenolic resin. Royal Soc. Chem. **9**, 28924–28935 (2019)
8. Pyshyev, S.: Production of indene-coumarone resins as bitumen modifiers. Petrol. Coal **57**(4), 303–314 (2015)
9. Pyshyev, S.: Oil and gas processing products to obtain polymers modified bitumen. Int. J. Pavement Res. Tech. **10**(4), 289–296 (2017)
10. Demchuk, Y.: Obtaining the modifiers of road bitumen from phenol fraction of coal tar. Coal Chem. J. **5**, 23–28 (2017)
11. Demchuk, Y.: Effect of phenol-cresol-formaldehyde resin on adhesive and physicomechanical properties of road bitumen. Chem. Chem. Technol. **12**(4), 456–461 (2018)
12. Gunka, V.: The selection of raw materials for the production of road bitumen modified by phenol-cresol-formaldehyde resins. Petrol. Coal **60**(6), 1199–1206 (2018)
13. Pyshyev, S.: Influence of the amount of catalyst on the process of receiving modifiers of road bitumens from phenol fraction of coal tar. Coal Chem. J. **4**, 27–33 (2019)
14. Pyshyev, S.: Development of mathematical model and Identification of optimal conditions to obtain phenol-cresol-formaldehyde resin. Chem. Chem. Technol. **2**, 212–217 (2019)
15. CEN (European Committee for Standardization): Bitumen and bituminous binders. Determination of the softening point. Ring and Ball method. DSTU EN 1427:2018 (2018)
16. CEN (European Committee for Standardization): Bitumen and bituminous binders. Determination of needle penetration. DSTU EN 1426:2018 (2018)
17. CEN (European Committee for Standardization): Bitumen and bituminous binders. Determination of the tensile properties of bituminous binders by the tensile test method. DSTU EN 13587:2018 (2018)
18. CEN (European Committee for Standardization): Bitumen and bituminous binders. Determination of the Fraas breaking point. DSTU EN 12593:2018 (2018d)
19. DSTU (National Standard of Ukraine): Viscous road oil bitumens. The method to determine the index of engagement with the surface of glass and rock materials. DSTU B V.2.7-81-98 (1998)
20. DSTU (National Standard of Ukraine): Bitumen and bituminous binders. Method for determining adhesion to crushed stone. DSTU 8787:2018 (2018)
21. Gunka, V, Sidun, I, Solodkyy, S, Vytrykush, N.: Hot asphalt concrete with application of formaldehyde modified bitumen. In: Blikharskyy, Z., Koszelnik, P., Mesaros, P. (eds.) Proceedings of CEE 2019, CEE 2019, LNCE, vol. 47, pp. 111–118. Springer, Cham (2020)
22. Sobol, K., Blikharskyy, Z., Petrovska, N., Terlyha, V.: Analysis of structure formation peculiarities during hydration of oil-well cement with zeolitic tuff and metakaolin additives. Chem. Chem. Technol. **8**, 461–465 (2014)

The Influence of Concrete Structure on the Destruction of Reinforced Concrete Bended Elements

Dorofeyev Vitaliy[1], Pushkar Natalia[2](✉), and Zinchenko Hanna[2]

[1] Odessa National Maritime University, Odessa, Ukraine
[2] Odessa State Academy of Building and Architecture, Odessa, Ukraine
PushkarN@ukr.net

Abstract. When ecological effect of the environment on the structural constructions the changing structure of concrete influences definitely on the cracking, bearing capacity and durability. The mechanism of the formation of the composite building materials structure – the concretes with the formation of interfaces on the contact plane, inclusions, and matrix is described. Fine and coarse aggregate are considered as inclusions, and mortar and cement mortar as a matrix. Such a selection of structural levels makes it possible to establish the occurrence of dangerous defects, which size is larger than the size of the constituents of the element's structure. Therefore, the size of the defect, safe at one structural level, becomes dangerous at a lower level. So a crack that is safe for concrete on large aggregates is destructive for cement stone. The questions of structure formation in concrete and reinforced concrete products on micro - and macro levels are considered.

The method of accounting for the initial (technological) damage to the iron reinforced concrete bending elements is presented. The tests of reinforced concrete beams under the action of low-cycle load are made, and their results are presented. The nature of crack formation and the development of cracks under the action of an external load in flexible concrete elements, depending on the technological damage have been established. The crack depth was determined depending on the load, using ultrasonic equipment. It is established that cracks from an external load develop from technological ones, and along their paths.

Keywords: Concrete · Structure · Damage · Beams · Load · Cracking

1 Introduction

Composite building materials are heterogeneous materials, which properties are formed as a result of rather complex processes of interaction of the initial components with the formation of structures that are transformed over time [1–6]. The logical chain "composition - technology - structure - level of properties" allows you to highlight the structure as the main factor that determines the quality indicators of the finished product. It is known that the destruction of any material and structure occurs by its separation into parts by the edges of cracks. This poses the problem of identification of

Z. Blikharskyy (Ed.): EcoComfort 2020, LNCE 100, pp. 103–111, 2021.
https://doi.org/10.1007/978-3-030-57340-9_13

the crack nucleation causes in coarse heterogeneous materials, studying the conditions for their development in a structured medium and their influence on the physical-mechanical characteristics of the materials and the operational characteristics of structures.

In analyzing the causes of the manifestation of the desired properties, the material is considered as heterogeneous, consisting of individual structural elements, interacting through the interface [7–13]. On the interface, the redistribution of strains and stresses between individual components and structures occurs under the action of technological and internal influences on the material, as well as external loads and influences. The composite building structures are considered as specially organized composite materials, the interaction of individual components and structural elements of which provides the functional purpose of the structure [14–17]. Therefore, the structure of the construction includes the whole variety of material's structure: the cracks at the micro and macro levels, pores and capillaries. It is proposed to determine the heterogeneity of the structure through surface damage by technological defects.

2 The Analysis of the Literary Sources

The previous studies have established the theoretical principles of structure formation of building materials and structures [18]. It was determined that during the formation of the material structure in micro - and macro levels on the borders of the matrix material and filler inhomogeneities arise, which leads to the formation of cracks process [19] (Fig. 1).

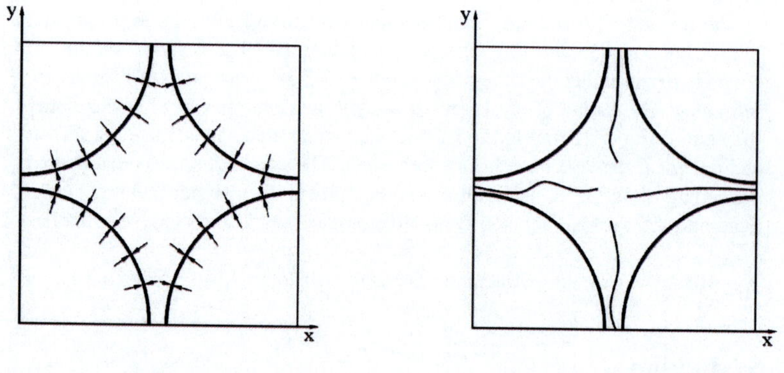

a b

Fig. 1. The mechanism of distribution of deformations (a) and the nature of crack formation at the macro level (b).

The similar conclusions were also made in the papers [20, 21].

In the general case, in composite materials and structures, several characteristic types of damage can be distinguished, which differ in the formation mechanisms (Fig. 2).

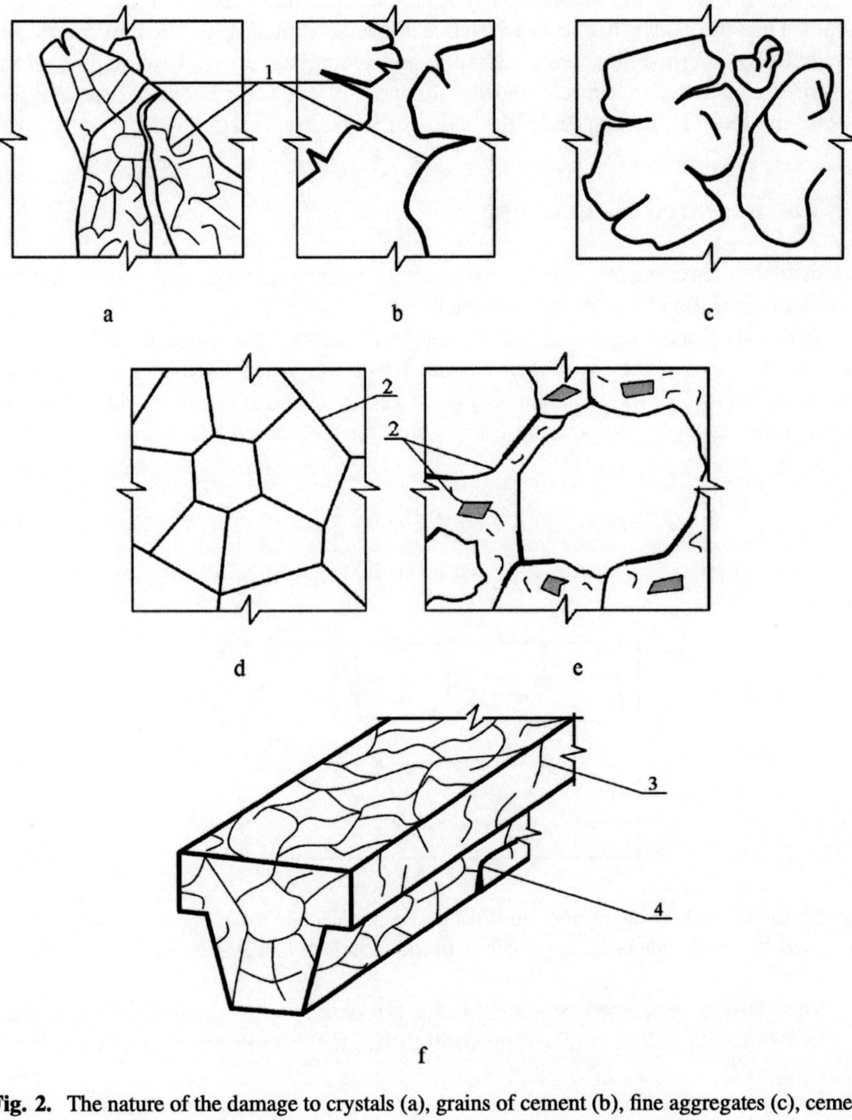

Fig. 2. The nature of the damage to crystals (a), grains of cement (b), fine aggregates (c), cement stone (d), concrete (e) and structure (f): 1 – cracks in the initial components; 2 – cracks formed during the structure formation of the materials; 3 – cracks formed due to geometric design features; 4 – operational cracks.

The defects of the individual components of the material can be considered as occasional, the quantitative and qualitative composition of which can be predicted only by special methods; therefore, they are excluded from further analysis. The object of our analysis is the defects that occur during the technological processing of building materials and structures. Such defects relate to technological, initial or hereditary defects, and they are present in the material before the application of operational loads and influences. It is assumed that the cracks that occur in the material are automatically structural cracks, and determine its deformability, crack resistance and bearing capacity. Therefore, we set the task of studying the appearance and development of cracks in bending structural elements, and their destruction under the action of low-cycle loads.

3 The Research Methodology

A quantitative assessment of concrete damage by technological defects was carried out by determining the extent of surface cracks.

Cracks appeared when using water solution of tannins that allows to detect and fix the cracks. The concrete, damaged by the defect was determined by measuring the length of surface cracks odometer along the geodesic line (Fig. 3) and by the calculation of the damage coefficient by Eq. (1), that for the studied beams constituted 1,05 and 1.03.

$$K_d = l_{cr}/l \quad (1)$$

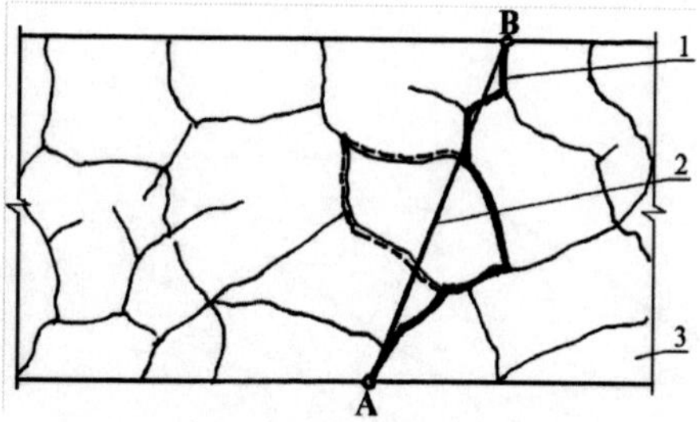

Fig. 3. The method for the coefficient of damage determination: 1 – technological crack along the boundaries of the blocks l_{cr}; 2 – geodesic line l; 3 – selected area on the surface of concrete.

When testing reinforced concrete beams, we used the ultrasonic method of non-destructive testing that also allows to control the crack development under the loads action [22, 23].

To determine the depth of the technological cracks we used an ultrasonic device UK-14P. The range of time measuring of ultrasound spreading was UT –20…8800 mcs. The range of the front duration of the first arrival of the received signal is 3…30 mcs, and the absolute sensitivity of the device is not less than 110 dB (decibels). The scheme for measuring the depth of the crack is shown in Fig. 4.

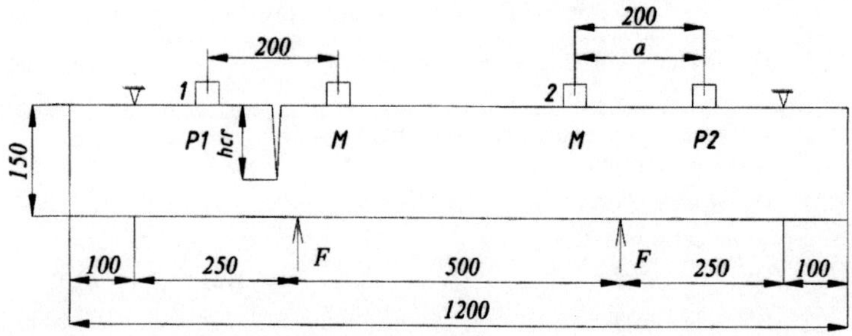

Fig. 4. The scheme for measuring the depth of the crack: 1 – measurement base on concrete through a crack (position of M-P1 sensors); a – measurement base on intact concrete (position of M-P2 sensors).

The device automatically calculates the crack depth using the Eq. (2):

$$h_{cr} = \frac{a}{2} \cdot \sqrt{\left(\frac{t_1}{t_a}\right)^2 - 1} \tag{2}$$

The appearance and development of cracks was fixed using markers.

Beams were loaded in steps of 500 kgf cyclically, with exposure at each stage and subsequent unloading. Two twin-beams were tested, their durability characteristics are given in Table 1.

Table 1. The characteristics of the experienced beams.

A beam's model	b, см	h, см	d, см	f_c, МПа	f_{cd}, МПа	$E_{cm} \times 10^{-5}$, МПа	The return in the days
A1	10,0	15,0	12,9	25,0	19,5	2,70	205
A2	10,0	15,1	13,0	25,1	19,3	2,60	225

On one beam A1 while testing, the devices, according to Fig. 4, were installed; on a twin-beam A2 the visual supervision on the cracks development was conducted.

4 The Research Results

The cracks parameters, obtained in the result of beam's A1 testing, is given in Table 2.

The two beams were destructed under the load of 2750 kgf on the sloped section (Fig. 5). According to the tests' results, given in Table 2, the graph of the dependence of crack depth on load was built (Fig. 6).

The studies have shown that on the first stage of loading the cracks developed at a depth of up to 70 mm, and after the load shedding they closed. Under the load of 1000 kgf the crack grew to 71 mm and after load shedding had a depth of 19 mm.

Table 2. The results of beam testing.

The load F, kgf	The device's UK–14P indicators, mks		The crack's height, mm
0	The measuring on pure concrete (the layout of the sensor MP1)	t_1 = 54	8.6
0	The measuring on pure concrete (the layout of the sensor MP2)	t_a = 53,8	–
500	The load supply	t_1 = 66,1	70.59443
500	The exposure under the load during 5 min	t_1 = 65,8	69.62667
0	The load's discharge	t_1 = 54	8.6
0	In 5 min. after the load's discharge	t_1 = 54	8.6
500	The load's return on the previous stage	t_1 = 66,1	70.59443
1000	The load's increase up to 1000 kgs	t_1 = 66,3	71.2347
1000	In 5 min. after the load's increase (the exposure under the load)	t_1 = 66,3	71.2347
0	The load's discharge	t_1 = 55	19.3339
0	In 5 min. after the load's discharge	t_1 = 55	19.3339
1000	The load's return on the previous stage	t_1 = 66,3	71.2347
1500	The load's increase up to 1500 kgs	t_1 = 67,4	74.69095
1500	In 5 min. after the load's increase	t_1 = 67,4	74.69095
0	The load's discharge	t_1 = 56	27.4674
0	In 5 min. after the load's discharge	t_1 = 55,6	24.52288
1500	The load's return on the previous stage	t_1 = 68,6	78.348
2000	The load's increase up to 2000 kgs	t_1 = 68,8	78.94712
2000	In 5 min. after the load's increase	t_1 = 69,3	80.43286
0	The load's discharge	t_1 = 57	33.79313
0	In 5 min. after the load's discharge	t_1 = 56,7	32.01562
2000	The load's return on the previous stage	t_1 = 70	82.4854
2500	The load's increase up to 2500 kgs	t_1 = 70,8	84.7946
2500	In 5 min. after the load's increase	t_1 = 70,9	85.08065
0	The load's discharge	t_1 = 57,1	34.3672
0	In 5 min. after the load's discharge	t_1 = 57,1	34.3672
2500	The load's return on the previous stage	t_1 = 71,8	87.63076
2750	The beam's destruction	t_1 = 72,8	90.4159

At a load of 2000 kgf, the crack grew to 80 mm, and at load shedding it amounted to 33 mm.

With a load of 2500 kgf per crack, it grew to 85 mm, and with a load shedding – 34.5 mm. Before the destruction, the crack grew to 90 mm, and the beam collapsed due to crushing of the compressed zone of concrete. Also, before the destruction, an inclined crack formed in the beam, which propagated along the trajectory of the main tensile stresses from the support to the force, which can be explained by the influence of the transverse force.

Fig. 5. The characteristic beams' destruction.

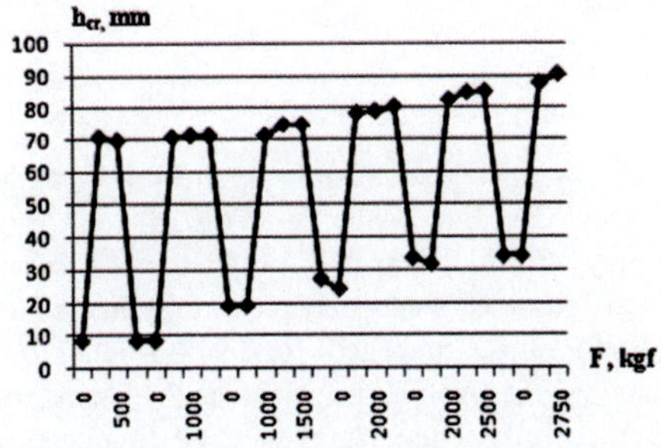

Fig. 6. The dependence of crack depth on load.

The nature of the crack formation and the development of cracks from the action of an external load occurs according to the technological ones in the zone of action of the maximum bending moment. A force crack passes along an energetically advantageous path. The disclosure width of the technological cracks amounted to 0.005…0.3 mm, which was paid attention to in [24], and the crack depth in the zone of the extended working reinforcement was 5.0…8.0 mm. The large depth of the technological cracks in the protective layer of reinforcement can be explained by the influence of shrinkage deformations, especially since the prototypes were tested at the age of 240 days.

Before the destruction of the beams, the development of vertical cracks was stopped, the critical crack began to cross the blocks, and the crushing of the compressed zone of concrete started. At the same time, an inclined crack appeared in both studied samples, passing from the support to the force. This can be explained by the influence of the transverse forces on the nature of fracture.

5 The Conclusions

1. The technological cracks in reinforced concrete beams had a width disclosure of 0.005…0.3 mm with a depth of 5.0…8.0 mm, the maximum depth and width of disclosure were observed in the zone of location of the working reinforcement. The fact of technological fractures presence can be explained by shrinkage deformation.
2. The nature of the spreading of power cracks repeats the "pattern" of technological cracks that pass along an energetically advantageous path and stop the development in a compressed zone due to the influence of transverse forces.
3. Ultrasonic observations showed the presence of invisible (technological) hairline cracks that also occur in the absence of bending moment.
4. The depth of the technological and power cracks, emerging on the surface ranged from 8…87 mm with a concrete protective layer thickness of 20 mm.

References

1. Blikharskyy, Z., Brózda, K., Selejdak, J.: Effectivenes of strengthening loaded RC beams with FRCM system. Arch. Civ. Eng. **64**(3), 3–13 (2018)
2. Bobalo, T., Blikharskyy, Y., Kopiika, N., Volynets, M.: Serviceability of RC beams reinforced with high strength rebar's and steel plate. In: Lecture Notes in Civil Engineering, vol. 47, pp. 25–33 (2020). https://doi.org/10.1007/978-3-030-27011-7_4
3. Krainskyi, P., Blikharskyy, Y., Khmil, R., Vegera, P.: Crack resistance of RC columns strengthened by jacketing. In: Lecture Notes in Civil Engineering, vol. 47, pp. 195–201 (2020). https://doi.org/10.1007/978-3-030-27011-7_25
4. Lobodanov, M., Vegera, P., Blikharskyy, Z.: Planning experiment for researching reinforced concrete beams with damages. In: Lecture Notes in Civil Engineering, vol. 47, pp. 243–250 (2020). https://doi.org/10.1007/978-3-030-27011-7_31
5. Selejdak, J., Blikharskyy Y., Khmil R., Blikharskyy Z.: Calculation of reinforced concrete columns strengthened by CFRP. In: Lecture Notes in Civil Engineering, vol. 47, pp. 400–410 (2020). https://doi.org/10.1007/978-3-030-27011-7_51
6. Vegera, P., Vashkevych, R., Blikharskyy, Z.: Fracture toughness of RC beams with different shear span. In: MATEC Web of Conferences, vol. 174, p. 02021 (2018). https://doi.org/10.1051/matecconf/201817402021
7. Blikharskyy, Z., Vashkevych, R., Vegera, P., Blikharskyy, Y.: Crack resistance of RC beams on the shear. In: Lecture Notes in Civil Engineering, vol. 47, pp. 17–24 (2020). https://doi.org/10.1007/978-3-030-27011-7_3
8. Bobalo, T., Blikharskyy, Y., Kopiika, N., Volynets, M.: Theoretical analysis of RC beams reinforced with high strength rebar's and steel plate. IOP Conf. Ser. Mater. Sci. Eng. **708**(1), 012045 (2019)
9. Khmil, R., Tytarenko, R., Blikharskyy, Y., Vashkevych, R.: Influence of load level during strengthening of reinforced concrete beams on their reliability. IOP Conf. Ser. Mater. Sci. Eng. **708**(1), 012054 (2019)
10. Krainskyi, P., Vegera, P., Khmil, R., Blikharskyy, Z.: Theoretical calculation method for crack resistance of jacketed RC columns. IOP Conf. Ser. Mater. Sci. Eng. **708**(1), 012059 (2019)

11. Vegera, P., Khmil, R., Vegera, P.: Shear strength of reinforced concrete beams strengthened by PBO fiber mesh under loading. In: MATEC Web of Conferences, vol. 116, p. 02006 (2017). https://doi.org/10.1051/matecconf/201711602006
12. Krainskyi, P., Blikharskyy, Y., Khmil, R., Blikharskyy, Z.: Experimental study of the strengthening effect of reinforced concrete columns jacketed under service load level. In: MATEC Web of Conferences, vol. 183, p. 02008 (2018). https://doi.org/10.1051/matecconf/201818302008
13. Blikharskyy, Y., Kopiika, N., Selejdak, J.: Non-uniform corrosion of steel rebar and its influence on reinforced concrete elements, reliability. Prod. Eng. Arch. **26**(2), 67–72 (2020). https://doi.org/10.30657/pea.2020.26.14
14. Blikhars'kyi, Y.Z.: Anisotropy of the mechanical properties of thermally hardened A500s reinforcement. Mater. Sci. **55**, 175–180 (2019). https://doi.org/10.1007/s11003-019-00285-0
15. Kharchenko, Y.V., Blikharskyy, Z.Y., Vira, V.V., Vasyliv B.D.: Study of structural changes in a nickel oxide containing anode material during reduction and oxidation at 600 °C. In: Fesenko, O., Yatsenko, L. (eds.) Nanocomposites, Nanostructures, and Their Applications. NANO 2018. Springer Proceedings in Physics, vol. 221. Springer, Cham (2019). https://doi.org/10.1007/978-3-030-17759-1_42
16. Kharchenko, Y., Blikharskyy, Z., Vira, V., Vasyliv, B., Vasyliv, B.: Study of nanostructural changes in a Ni-containing cermet material during reduction and oxidation at 600 & °C. Appl. Nanosci. (2020). https://doi.org/10.1007/s13204-020-01391-1
17. Zhang, Q., Mol'kov, Y.V., Sobko, Y.M., et al.: Determination of the mechanical characteristics and specific fracture energy of thermally hardened reinforcement. Mater. Sci. **50**, 824–829 (2015)
18. Dorofeyev, V.S., Vyrovoy, V.N.: Technological damage to building materials and structures. The place of masters, Odessa (1998)
19. Szeląg, M.: The Influence of Cement Composite Composition on the Geometry of their Thermal Cracks. Lublin University of Technology, Lublin (2017)
20. Malakhov, V., Vykydanets, S., Pushkar, N.: Influence of quantity and quality of filler on technological damage of reinforced concrete beams. In: Proceedings of the fib Symposium "Concrete - Innovations in Materials, Design and Structures", vol. 47, pp. 328–334 (2019)
21. Szeląg, M., Fic, S.: Analysis of the development of cluster cracks in the cement paste modified by microsilica. Constr. Archit. **14**(4), 117–127 (2015)
22. Jones, R., Fakeoaru, I.: Non-Destructive Testing Methods for Concrete. Stroyizdat, Moscow (1974)
23. Popova, A.I.: The definition of the quality of a concrete layer of a concreted tube. Transp. Storage Petrol. Prod. Hydrocarbons **1**, 28–33 (2013)
24. Znaychenko, P.A.: The determination of crack depth by ultrasonic method in the lining of transport tunnels of large cross section. Min. Inform. Anal. Bull. Sci. Tech. J. **1**, 34–37 (2006)

Ukraine Energy Transition in Light of the EU Experience

Nataliia Fialko(✉) and Mykola Tymchenko

Institute of Engineering, Thermophysics of NAS of Ukraine, Kiev, Ukraine
nmfialko@ukr.net

Abstract. The results of the analysis of the Ukraine's starting positions to implement the energy transition from thermal power with fossil carbon-based and nuclear fuels as the main energy resources to low-carbon energy based on alternative, mainly renewable energy sources, are presented. The initial energy-economic conditions of this transition are considered. It is noted that Ukraine has powerful, but morally and physically obsolete energy sector with outdated indicators which do not meet modern requirements for efficiency, environmental friendliness, safety and investment attractiveness. A certain amount of attention is paid to coverage of the starting legislative conditions for the "energy transition" in Ukraine. At the same time, the particular importance of the set of regulatory documents that make up the so-called Fourth Energy Package ("Clean Energy for all Europeans" package) is indicated. Energy transition strategies features of the number of EU member-states (Germany, France, Denmark) are highlighted. The possibility of the presence in the EU member-states of significantly different models of this transition in terms of the use of nuclear generation, various renewable energy sources, etc. is indicated. The rating data of the EU member-states and Ukraine by some energy indicators are given. The analysis of the concept of the "energy transition" in the context of ensuring long-term energy security in the conditions of fulfilling the requirements of SDGs is carried out. It is noted mutual influences and links between the concepts of "energy transition" and "energy security" (at the last case also at the economic and state levels).

Keywords: SDGs · CoP-21 · Energy transfer · Renewables · Energy efficiency

1 Introduction

At the Paris Climate Conference (CoP-21), a consensus view was adopted by representatives of international and national organizations that the new economy should be low-carbon, therefore it requires an energy transition (ET) from thermal energy, which is based on combusting of fossil carbon-containing fuels and energy sources (FES) to low-carbon power [1]. ET global roadmap (until 2050) is presented in [2]. Energy efficiency as one of the main factors of ET is considered in [3]. ET political economy problems mainly at single country level are discussed in the compendium [4] which was compiled and written by world lead scientists.

Z. Blikharskyy (Ed.): EcoComfort 2020, LNCE 100, pp. 112–117, 2021.
https://doi.org/10.1007/978-3-030-57340-9_14

According to new concepts, traditional power energy has largely exhausted its development potential. Its continued existence with keeping of the intensity of FES combusting is one of the main reasons for the current acyclic global warming [5]. More in-depth justification of the latest stage of current climate warming is presented in [6].

Ukraine, as the associate member of the EU, should be guided in its development by the logic of the Paris Agreement and the ET concept. This determines the priority relevance of the analysis of the starting conditions of the energy transition in Ukraine. Particular interest is the performing of this analysis in the context of relevant experience of the EU countries.

2 The Aim of the Work

The aim of the study is to analyze the starting conditions for the energy transition in Ukraine in the context of the EU experience.

3 Research Methodology

The work uses data obtained on the basis of the methodology of the International Energy Agency, which is the standard for the OECD countries. A comparative analysis of the energy indicators of Ukraine and the countries-member of the European Union is also used.

4 Results

4.1 Energy Transition

The concept of Energy Transition. In the definitions of ET, a certain pluralism is observed, as well as different degrees and levels of elaboration of this issue. Further, under the "energy transition" as applied to a single country, we mean a complex variable concept in the context of ensuring long-term national energy security at sustainable growth. ET expresses the idea of changing technological eras, when alternative, mainly renewable energy sources are used as key FES instead of fossil fuels (carbon-containing and nuclear sources).

The advantage of the above definition is, first of all, in that it links the concept of "energy transition" with the concept of "national energy security". The latter, in turn, correlates with the concept of a higher level of "national economic security". At the next upper system level, it is assumed that there is a relationship between the concept of "energy transition" and the concept of "national security".

The concept of "energy transition" at the level of energy security, has a very specific technological expression, which reveals the physical processes of the energy generation and its flows. This concept is significantly modified depending on the characteristics of national energy, economic and state kinds of security.

Energy Transition in Germany (Energiewende). Back in the 80s of the 20th century Germany proposed the concept of energy transition (Energiewende) and since then the German state has been purposefully implementing it [7]. As the most economically developed EU country, Germany was able in the 21st century fundamentally and comprehensively change both the technological base of energy and its structure. For example, in 2022, it is planned to close the last of the nuclear power units. The termination of coal heat generation is also envisaged. In Bottrop (Ruhr coal field), the last coal mine was closed in 12/22/2018. The four largest German companies EnBW, E. ON, RWE and Vattenfall noticeably reduce energy generation at TPP and are reprofiling for RES generation. Germany's energy industry is approaching a carbon-free state every year. Share of RES-generation in net public electricity generation in Germany in January, February, March, April, May 2020 was, respectively 48.2% (47.94 TWh), 61.8% (45.9 TWh), 56, 8% (44.22 TWh), 60.3% (34.88 TWh), 58.2% (34.64 TWh) [8]. On average, for the indicated 5 months of 2020, the share of RES was 56.7%.

Energy Transition in France. Since 2015, France has been in force special law No. 2015-992 of August 17, 2015, relating to the energy transition. French law does not ban completely nuclear generation as the German one. It only limits the share of the allowed capacity of nuclear energy (to 63.2 GW, article L. 311-5-5). Also, this law not only strengthen existing nuclear safety requirements but also establishes a number of new nuclear safety demands and rules.

By the example of the two leading EU countries, it is clear that the issue of the mandatory abandonment of nuclear generation in the case of the development of ET is not unambiguous. In addition to France (81.8% share of nuclear generation in the energy balance in 2018), nuclear power plants (NPP) dominated in the energy balances of yet seven EU countries in 2018: Slovakia (50%), Sweden (46.9%), Slovenia (40%), Hungary (34.8%), Belgium (31.9%), Finland (26 1%), Spain (20.1%). NPP took second place in the energy balance in three European countries: Bulgaria (42.1%), Czech Republic (41.1%), and Great Britain (18.5%). Share of NPP in Germany in Q1 2020 was 11.83%. In Ukraine in 2018, 54.3% of all electricity purchased by the state-owned enterprise "Energorynok" (as the only buyer of Wholesale Electricity Market) was produced at NPP.

In many other EU countries, the energy policy for ET seems less effective than in Germany and not as "ecologically patriotic" as in France. But at the same time, the EU has set ambitious aims and objectives, for example, to reduce GHG emissions to almost zero by the 50s of this century. In nowadays ETs are organizationally formatting within EU as National Energy and Climate Plans (NECPs) from 2021 to 2030. By other words NECPs are the framework for EU Member States to outline their ETs.

4.2 Starting Conditions for the Development of ET Plans in Ukraine. Features of the Starting Position of Ukraine in Relation to EU Countries

Starting Legislative Conditions for Ukraine. Since 7 August 2015 Ukraine has status of associated member-state of the European Union. But in order to obtain full membership, Ukraine needs to bring the national legislative framework in line with the

European regulatory system. An important source of law for Ukraine as secondary legislation are the Directives, Regulations, EU decisions, which are an integral part of the acquis communau-taire.

In terms of the energy transition for Ukraine, the complex of regulatory documents that form the so-called Fourth Energy Package of the EU is of particular importance.

Starting Energy and Economic Conditions. The preparation and implementation of the energy transition in Ukraine requires the formulation and solution of a number of complex and costly economic and technological problems. Ukraine possesses, although powerful, but morally and physically obsolete energy, whose indicative indicators do not meet modern requirements for efficiency, environmental friendliness, safety and investment attractiveness. This fact is illustrated, in particular, through the integrated index of electric intensity of gross domestic product (GDP), which combines the main macroeconomic GDP indicator W_{GDP}, electricity generation W_e and electric intensity (in the form of power intensity indicator W_e/W_{GDP}) EU member states and Ukraine (Fig. 1). As can be seen from the Figure, the value of power intensity indicator in Ukraine reaches 1.44 kWh/Euro, which is 6–8 times higher than the values characteristic, for example, of Germany (0.18 kWh/Euro) or France (0.24 kWh/Euro).

Models of the energy transition in different European countries have individual features. So, most countries of the northern energy profile focus on wind farms, especially their offshore sector (for example, FRG, Denmark). The UK besides wind power has also a large share of PV-systems and plans to invest heavily in both of these types of RES generation in the coming years.

Danish analysts [10] testify that in Denmark and the FRG – two countries with the largest share of wind power in Europe in national electrical balances – in the conditions of the competitive spot market in the daily cycle, time intervals appeared with negative prices for electric energy. In 2018, the accumulated total annual number of such hours with negative prices amounted to 133 in the FRG and 91 h in Denmark.

In general, the small Danish energy market with two price zones is in the shadow of the large German market. In view of this, most of the Danish peak wind power is "locked up" inside the country. This circumstance leads to the fact that Denmark is looking for new ways not only to sell, but also to accumulate excess peak energy in both daily and seasonal cycles. And this, perhaps, is the main message from small Denmark to the whole world of RES generation.

5 Scientific Novelty and Practical Significance

For the first time, the starting positions of Ukraine for the implementation of the energy transition under the Fourth Energy Package of the EU are considered. A comparative analysis of these positions with the peculiarities of the energy transition for a number of leading EU countries is carried out.

The practical value of the work is determined by the important significance that the issues of the energy transition have for the energy practice of Ukraine.

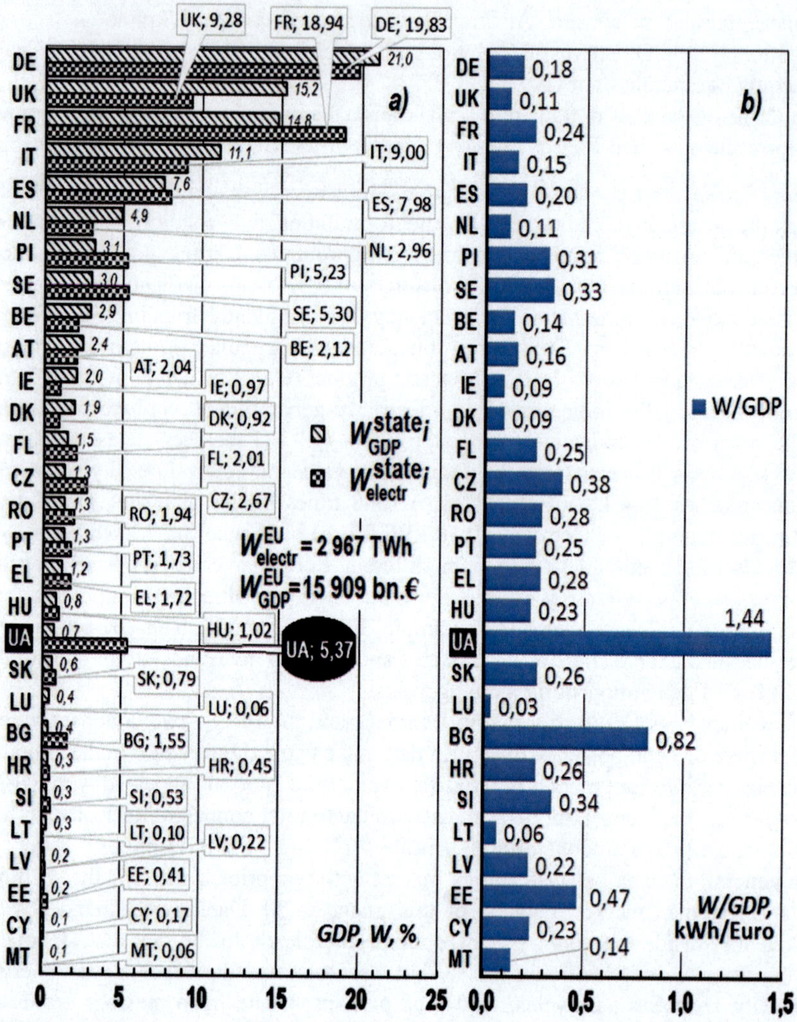

Fig. 1. Rankings of EU member states and Ukraine by some energy indicators in 2018. ***a)*** – breakdown of EU countries and Ukraine according to their share (%) of electricity production W_e and share (%) of GDP in EU total (sorted by GDP); ***b)*** breakdown by GDP power intensity indicator (unsorted). *Source: Eurostat (online data).* Country codes meet the standard ISO 3166-1 alpha- 2: AT – Austria, BE – Belgium, BG – Bulgaria, CY – Cyprus, CZ – Czechia, DE – Germany, FRG, DK – Denmark, EE – Estonia, EL – Greece, ES – Spain, FLPL – Finland, FR – France, HR – Croatia, HU – Hungary, IE – Ireland, IT – Italy, LT – Lithuania, LU – Luxembourg, LV – Latvia, MT – Malta, NL – Netherlands, Pl – Poland, PT – Portugal, RO – Romania, SE – Sweden, SI – Slovenia, SK – Slovakia, UA – Ukraine; UK – United Kingdom.

6 Conclusions

Based on the data above, the following conclusions can be drawn:

1. The analysis of the "energy transition" concept in the context of ensuring long-term energy security in the conditions of fulfilling the requirements of SDGs is carried out. There is a link between the energy transition and the concept of security at higher levels – economic and state.
2. The features of the concepts of energy transition of a number of EU member states are considered. In particular, using Germany and France as the cases of two leading EU countries, the possibilities of radically different strategies of mentioned energy transition in terms of nuclear generation using are shown.
3. The starting conditions in the development of energy transition plans in Ukraine are analyzed. The features of starting legislative and energy-economic conditions are considered. It is shown that the parameter of electric intensity of GDP in Ukraine reaches 1.44, which is 6–8 times higher than the corresponding values for Germany and France. It is noted that it is advisable to carry out the energy transition in Ukraine on the basis of a specially developed document, "National Energy Climate Plan", the analogues of which are currently being formed in the EU with a planning horizon of 10 years.

References

1. Gielen, D., Boshell, F., Saygin, D., Bazilian, M., Wagner, N., Gorinia, R.: The role of RE in the global energy transformation. Energy Strategy Rev. **4**(24), 38–50 (2019)
2. IRENA. Global Energy Transformation. A roadmap to 2050, International Renewable Energy Agency. www.irena.org/publications. Accessed 16 May 2020
3. Gielen, D., Boshell, F., Saygin, D., Bazilian, M., Wagner, N., Gorinia, R.: Perspectives for the Energy Transition. The Role of Energy Efficiency. IEA - IRENA, Abu Dhabi (2017)
4. Arent, D., Arndt, Ch., Miller, M., Tarp, F., Zinaman, O. (eds.): The Political Economy of Clean Energy Transitions. Oxford University Press, Oxford (2017)
5. Global Warming of 1.5 °C. An IPCC special report on the impacts of global warming of 1.5 °C above pre-industrial levels. https://www.ipcc.ch/sr15/. Accessed 16 May 2020
6. Fialko, N.M., Tymchenko, M.P., Sherenkovskiy, Ju.V.: Fourth generation of district heating and centralized heating supply systems of Ukraine. In: Blikharskyy, Z., Koszelnik, P., Mesaros, P. (eds.) Proceedings of CEE 2019: Advances in Resource-saving Technologies and Materials in Civil and Environmental Engineering, pp. 74–86. Springer (2019)
7. Pescia, D., Graichen, R, Jacobs, D.: Agora Energiewende and International Energy Transition: Understanding the Energiewende. GmbH, Agora Energiewende, Berlin. https://www.agora-energiewende.de. Accessed 16 May 2020
8. Fraunhofer ISE. Net public electricity generation in Germany in 2020. https://www.energy-charts.de/energy_pie.htm?year=2020&month=2. Accessed 16 May 2020
9. Iovino, F., Tsitsianis, N.: Changes in European Energy Markets. Emerald Publishing Limited, Bingley (2020)
10. Bach, P.-F.: Electricity in Denmark 2018. http://www.pfbach.dk/firma_pfb/references/pfb_danish_electricity_balance_2018_2019_04_06.pdf. Accessed 16 May 2020

Investigation of Preparation Processes of Liquid Feed Mixtures in Rotary Pulsating Apparatus

Valery Gorobets(✉), Viktor Trokhaniak, Ievgen Antypov, and Andrii Serdiuk

National University of Life and Environmental Sciences of Ukraine, Kiev 03041, Ukraine
gorobetsv@ukr.net

Abstract. A new design of the device for preparation of liquid grain feeds using the technology of discrete-pulse energy input is proposed. The essence of this technology is to use in the processing of the feed mixture in the working body, which consists of a rotor having a high speed and a stator with holes for grinding solid components and homogenization of feed when passing through these holes. In the report the experimental research and numerical modeling of processes of hydrodynamics and heat transfer in the rotary-pulsation device at processing of water-grain mix for preparation of liquid grain forages is carried out. The result is the distribution of velocities and temperatures in the channels of the rotor-stator system. It is found that as a result of processing the liquid feed mixture is its homogenization and heating due to the conversion of the kinetic energy of rotation of the rotor into heat energy. An experimental study of the preparation processes of liquid feed mixture in a rotary pulsation apparatus was carried out. The composition and temperature characteristics of feed mixture were studied. The results of numerical simulation are compared with experimental data.

Keywords: Liquid grain feed · Rotary pulsation apparatus · Speed · Temperature · Energy

1 Introduction

Existing devices for the production of feed for livestock in the vast majority are based on the use of technologies for the preparation of solid feed mixtures. The basis of such technologies is the use of hammer rotors, the rotation of which is the grinding of grain or other solid components of the feed. At contact of the surface of hammer with grain owing to shock loadings there is its cracking and formation of the mix having the different size of fractions. The design of the hammer rotors can be different, for example, in the form of cylindrical rollers or disks. The size and geometry of the surface for the hammers are selected depending on the size of the fractions of feed mixture to be obtained at the outlet of the hammer crusher. Crusher designs with airflow can be used to improve the operation of the crushers. The main purpose in the process of preparation is to obtain a feed mixture in which the size of feed particles are optimal for assimilation during feeding of cattle with minimal energy costs per unit of feed mass.

Z. Blikharskyy (Ed.): EcoComfort 2020, LNCE 100, pp. 118–126, 2021.
https://doi.org/10.1007/978-3-030-57340-9_15

In [1–8] studies were conducted to develop new designs of hammer crushers, which have advantages in terms of optimal grain grinding, reduction of energy consumption per unit of output and reducing the share of dust fractions in the feed mixture. In the monograph [1] various types of crushers, their designs are considered, theoretical and experimental data on preparation of firm forage mixes are resulted. In [2] two-stage crusher was developed, which improves the throughput and particle size distribution of the produced feed. The influence of air flows on improving the efficiency of hammer crushers was studied in [3, 4]. Of great importance in the operation of crushers is the amount of specific energy consumption spent on the preparation of feed. These issues were studied in [5, 6], where the optimal designs and modes of operation of two-stage and three-stage roller crushers are proposed. The amount of energy consumption required to prepare a unit mass of feed products when grinding different types of products in a hammer crusher was studied in [7]. The study of the specific energy consumption and fractional composition of grain feed in the disk crusher is devoted to [8].

Known designs of hammer, roller and disc crushers, which are the most common devices for preparing feed, can have a number of disadvantages: large volume and weight; significant specific energy consumption for the preparation of feed products; unbalanced granular composition of feed products (the presence of small fractions that are poorly digested by cattle); dry feed is inferior to liquid feed in its nutritional properties and digestibility for pigs, young cattle, sheep, goats and other animals. In this regard, it is of interest to develop new designs of devices for the preparation of liquid feed mixtures, the principle of which is based on the use of other technologies with lower energy consumption to produce a unit mass of feed product. From this point of view, technologies in which rotary-pulsation mechanisms of product processing are used are promising. Such technologies are used in the processes of preparation of food, pastes, medicines and other products for various purposes.

The main designs of rotary-pulsation devices and the processes that take place in them are considered in the monograph [9]. In [10] hydrodynamics and heat transfer in rotating-pulsating flows, which take place in rotary-pulsating devices, were studied. A number of works are devoted to the use of rotor-pulsation technologies in specific technological processes. In [11] such technologies were used in the processes of grinding and dissolving polydisperse materials, which were studied experimentally and by numerical simulation methods. In [12] new design of rotor-pulsation apparatus with holes of special shape was proposed and the formation of various emulsions in this apparatus was investigated. The study of mechanisms that occur in the rotor-pulsation apparatus in the processing of heterogeneous materials was carried out in [13]. In [14, 15] rotor-pulsation principles were used for preparation of liquid feed mixtures on a grain basis. In [14] the results of researches of hydraulic characteristics of rotor-pulsation device at processing of a grain mix with water are resulted. The work [15] is devoted to mathematical modeling of hydrodynamic and heat processes that occur during the processing of liquid grain feed in the rotary pulsation apparatus. The processes that take place in rotary-pulsation apparatus for the preparation of liquid grain feed are insufficiently studied, and the design of such devices requires further improvement.

2 Materials, Analysis and Numerical Modeling

A new design of the apparatus for preparation of liquid feed grain mixtures, which is based on the use of rotor-pulsation technologies, is proposed. The essence of such technologies is to use a working device consisting of cylindrical rotor and stator, which have holes for the passage of the liquid component and a small gap between the surfaces of rotor and stator. The rotor speed can be 3000–5000 rpm. When the liquid with solid fractions passes through the holes of the rotor and stator there are processes of grinding solid grain fractions, cavitation and turbulence of the flow, dissipation of the kinetic energy rotation of the rotor into heat energy, which leads to heating the treatment mixture, and other processes.

The general scheme of rotary pulsation apparatus (RPA) for the preparation of liquid feed is presented in Fig. 1. It consists of an electric drive (1), working processing chamber (2), nozzle for supplying water to the processing chamber (3), hopper (4), nozzle for re-feeding the mixture into the hopper (5), an outlet pipe for outputting the finished mixtures of installation (6), cover (7) and support valves (8).

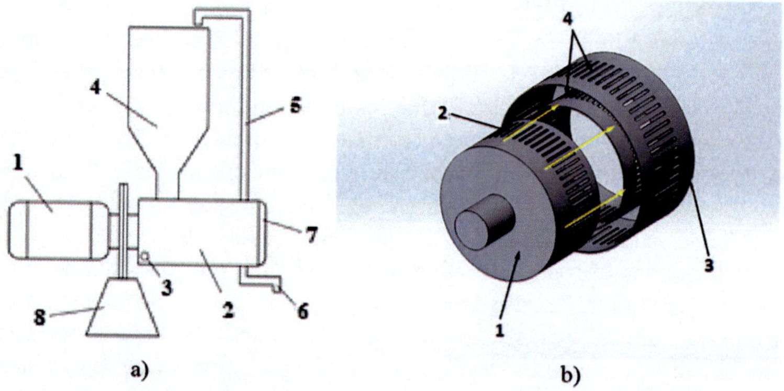

Fig. 1. General scheme of the rotary-pulsation apparatus for the preparation of liquid grain feed (a) and working mechanism of RPA (b)

The design of the working chamber of RPA consists of two coaxial cylinders - fixed stator 2 and movable rotor 1, which have radial holes 2, 4 on the side walls. The rotor speed is in the range of 3000–5000 rpm.

The holes on the rotor and stator have a rectangular shape and are arranged coaxially with each other. The gap between the rotor and stator is chosen to be minimal, which is due to the need to intensify the mixing of components in the mixture, cavitation and turbulence of the flow of liquid feed mixture.

The process of circulation of liquid mixture is due to the centrifugal forces that occur during the rotation of rotor. Subsequently, the liquid mixture through the pipeline 5 enters the hopper 4, from which it enters the working chamber 2, where the grain is crushed and the feed mixture is processed. The liquid feed mixture passes through the working chamber 2 several times until the size of the crushed grain particles is optimal for assimilation by cattle. The apparatus provides a nozzle 5 for water supply and an outlet nozzle 6 for draining the finished feed. Thus, the feed mixture must undergo several processing cycles, which are determined in the process of experimental and numerical studies to obtain the desired consistency of liquid grain feed.

2.1 Numerical Modeling of Transfer Processes in RPA for Preparation of Liquid Grain Feeds

In Fig. 2 schematically shows the working chamber of RPA for preparation of liquid grain feed. All elements of working chamber are made of aluminum. The initial temperature of the liquid feed mixture and the temperature on the surface of metal body was chosen to be +20 °C. The rotor rotates at a speed of 3000 rpm.

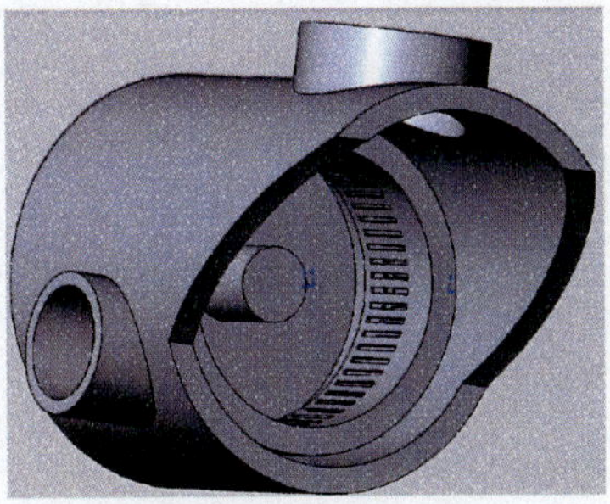

Fig. 2. General view of RPA working chamber system

Numerical modeling studies the feed mixture of water and grain in the ratio 3:1, 6:1 and 10:1. Thermophysical properties of water, which depend on temperature, are taken from known tabular data. Thermophysical properties of grain are considered independent of temperature and equal: density 770 kg/m^3; thermal conductivity coefficient 0.11 W/(m °C); specific heat 1550 kJ/(kg °C).

Thermophysical properties for the mixture of water/grain (Y), taking into account all the necessary thermophysical values, is calculated by the formula $Y = \frac{10 \cdot x_1 + 1 \cdot x_2}{x_1 + x_2}$, where x is a thermophysical quantity, indices 1 and 2 are respectively water and grain.

The consumption of liquid feed mixture depends on the geometric parameters of working chamber, the speed of rotation for rotor, the size of the holes on the rotor and stator, as well as the value of hydraulic resistance when passing the liquid mixture through the pipelines. In numerical studies, feed consumption through the working chamber varied from 0.5 to 1.5 kg/s based on the analysis of the modes of operation of the known rotary-pulsation apparatus and the results of research for the developed experimental setup. The dynamic viscosity of liquid feed ranged from 0.2 to 2.1 Pa · s.

The finite element method (FEM) based on the ANSYS FLUENT application package is used to numerically calculate the problems of hydrodynamics and heat transfer. The maximum size of grid throughout the volume does not exceed $1 \cdot 10^{-3}$ m and size of the thickening of grid near the grooves is $5 \cdot 10^{-4}$ m. The number of elements of grid is about 4.5 million. The quality of grid by orthogonal quality is 0.161.

2.2 The Results of Numerical Modeling of Transfer Processes in RPA

The mathematical model is based on the Navier-Stokes equation [16] and energy transfer equation for convective flows. The k-ε turbulence model was used in the calculations [17]. As a result of numerical calculations in ANSYS FLUENT application package, all dynamic and thermal characteristics of the liquid feed mixture when passing it through the rotor-stator system were obtained. The results of the calculations are presented in Fig. 3–4.

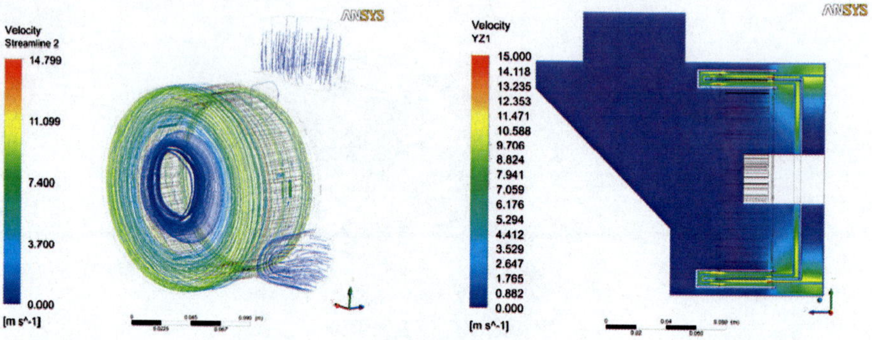

a) b)

Fig. 3. Speed distribution in rotor-stator system: a - 3D format; b - longitudinal section

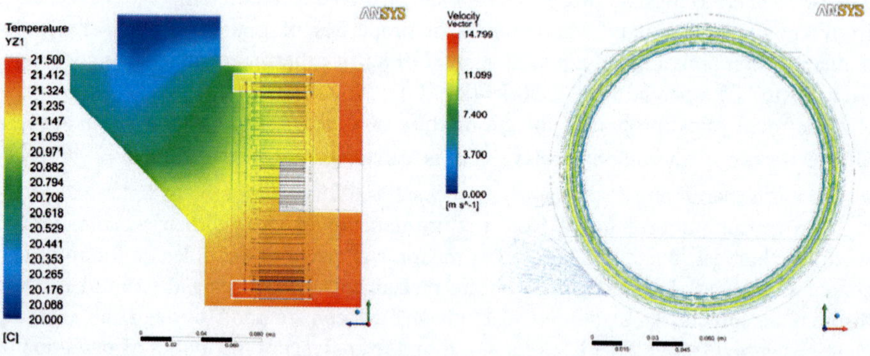

a) b)

Fig. 4. Temperature distribution in feed mixture for longitudinal section (a) and cross section (b) in the rotor-stator system

In Fig. 3 shows the speed distributions in the channels of working area in the rotor-stator system. The maximum values of flow rate of the feed mixture are observed in the channels between the rotor and stator, where these values can exceed 7 m/s.

Temperature distributions in the liquid mixture in cross section of the channel are shown in Fig. 4. The highest values of temperatures in the liquid feed mixture take place in the areas adjacent to the surface of the end rotor plane and may exceed 21°C.

This indicates a dissipation of kinetic energy of rotation of rotor and an increase in the temperature of feed mixture during its processing. As a result, the feed mixture is heated, which is especially important in winter and allows you not to heat the feed mixture during this period.

As a result of numerical simulations, the fields of pressures, velocities and temperatures in the liquid feed mixture were obtained, which made it possible to choose the RPA design used to develop an experimental sample of such an apparatus when varying the geometric dimensions of rotor and stator.

3 Experimental Studies

A general view of the experimental RPA sample for the preparation of liquid grain feed is presented in Fig. 5. In addition to the electric drive, working chamber, hopper, pipe for water supply to the working chamber, pipe for re-supplying the mixture to hopper, outlet pipe for withdrawing the finished mixture from the installation (see Fig. 1) also includes measuring devices. Working volume of RPA - 0,15 m^3. Studies of feed mixture consisting of water and millet in a ratio of 3:1 were performed. The total weight of mixture was 8 kg. In the course of experimental researches measurements of the following parameters were carried out: expenses of the liquid forage mix passing through the working chamber; the temperature of feed mixture at different times; the amount of energy used to prepare a unit of feed; size distribution of grain particles in the feed mixture at different times.

Fig. 5. General view of the experimental setup

Temperature measurement was performed with a thermocouple UT320A, included in the set of UNI-T multimeter with a measuring range from −50 to +1300 °C. The electricity consumed by the engine was determined by the magnitude of operating current and voltage, respectively, by an ammeter and voltmeter. The flow rate of feed mixture in RPA was measured with a flow meter. To determine the particle size of the grains in feed mixture, the feed mixture was taken at intervals of 5 min and the samples of mixture were examined under a microscope XS-5520 with binocular nozzle.

In Fig. 6a shows the change in the average particle size of the grain over time during their grinding when passing through the holes of rotor and stator. Experimental data have shown that to obtain a liquid feed mixture in which the average grain particle size is 150–200 μm, it is necessary to process the mixture in RPA for 25–30 min. In this case, according to the measurements, the feed mixture passes through the working chamber 30–35 times or has 30–35 processing cycles. The study of the particle size distribution of feed mixture under a microscope shows a uniform particle size distribution of crushed grain and grain fractions of very small size are almost absent. This composition of feed mixture, in contrast to dry feed, which contains a large amount of feed dust, improves its digestibility by livestock.

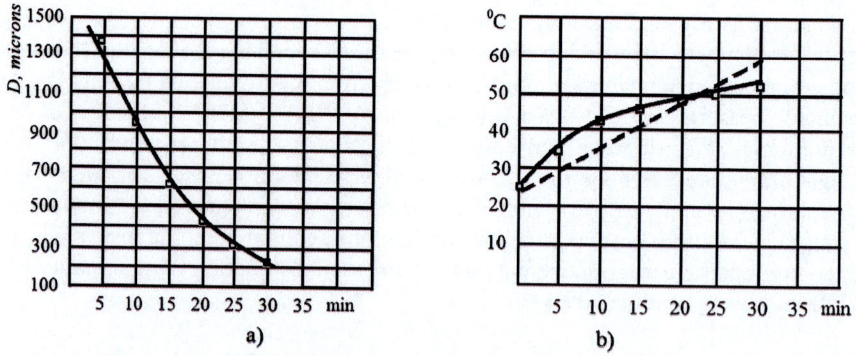

Fig. 6. Depending of average particle grain size (a) and temperature of liquid grain mix (b) on time: ■ - experimental data; - - - - - - numerical calculation

In Fig. 6b shows the dependence of the temperature change of feed mixture on time.

Analysis of obtained dependence shows that the maximum increase in the temperature of feed mixture is observed in the initial period of grinding the grain mixture. This is due to more intensive processes of friction and grinding of grains in the holes of rotor and stator and the release of a significant amount of heat energy. In the future, the average particle size of the grain decreases, which leads to a decrease in heat energy. As follows from Fig. 6b, the temperature of the feed mixture can exceed 50 °C, which allows to reduce energy costs for heating the feed in the winter. In Fig. 6b also shows the temperature dependence, which is obtained as a result of numerical simulation of studied processes. This dependence is monotonous because the mathematical model does not take into account the mechanical interaction of the walls of rotor and stator with the individual particles of grain mass included in the feed mixture. The difference between experimental and numerical data does not exceed 15%.

4 Conclusion

1. The new design of the device for preparation of liquid grain forages based on use of rotor-pulsation technologies is offered.
2. A mathematical model has been developed and numerical modeling of liquid feed preparation processes in the working chamber of rotor-pulsation apparatus has been carried out. The distributions of velocities and temperatures in the channels of working chamber are obtained. It is shown that the processes of preparation of feed mixtures are accompanied by an increase in feed temperature.
3. An experimental study of preparation process for liquid grain mixtures on an experimental sample of rotary pulsation apparatus. The dependence of the change in the average grain particle size depending on the processing time of feed mixture was found. The dependence of the temperature of the feed mixture on the time of its processing is obtained. The experimental data are compared with the results of numerical simulations.

References

1. Sysuev, V.A., Alyoshkin, A.V., Savinyh, P.A.: Kormoprigotovitelnie mashini. Teoriya, razrabotka, eksperiment (Feed Preparation Machines. Theory, Development, Experiment). V.2. NIISH Severo-Vostoka, Kirov (2009)
2. Sysuev, V., Ivanovs, S., Savinyh, P., Kazakov, V.: Movement and transformation of grain in two–stage crusher. Engineering for rural development, Jelgava, pp. 22–25(2015)
3. Xuan, C., Cao, L., Pei, W., Ma, Y., Han, D.: Development on a hammer mill with separate sieving device. Telkomnika, Indonesian J. Electric. Eng. **10**(6), 1151–1156 (2012)
4. Nwadinobi Chibundo Princewill: Development and performance evaluation of improved hammer mill. J. Sci. Eng. Res. **4**(8), 159–164 (2017)
5. Guritno, P., Haque, E.: Relationship between energy and size reduction of grains using a threeroller mill. Trans. Am. Soc. Agric. Eng. **4**, 1243–1249 (1994)
6. Savinyh, P., Aleshkin, A., Nechaev, V., Ivanovs S.: Simulation of particle movement in crushing chamber of rotary grain crusher. In: 16th International Scientific Conference Engineering for Rural Development, Jelgava, pp. 309–316 (2017)
7. Tumuluru, J.S., Tabil, L.G., Song, V., Iroba, K.L., Meda, V.: Grinding energy and physical properties of chopped and hammer–milled barley, wheat, oat, and canola straws. Biomass Bioenerg. **60**, 58–67 (2014)
8. Vaculik, P., Maloun, J., Chladek, L., Poikryl, M.: Disintegration process in disc crushers. Res. Agric. Eng. **59**(3), 98–104 (2013)
9. Promtov, M.A.: Pulsatsionnie apparati rotornogo tipa: teoriya i praktika (Rotary type pulsation devices: theory and practice). Mashinostroyenie, Moscow (2001)

10. Basok, B.I., Avramenko, A.A., Pirozhtnko, I.A.: Hidrodinamika, tploobmen I effekti drobleniya vo vraschatelno–pulsiruyuchsih potokah (Hydrodynamics, heat transfer and effects of crushing in rotating pulsating flows). Expres, Kiev (2012)
11. Kukhlenko, A.A., Orlov, S.E., Ivanova, D.B., Vasilishin, M.S.: Process of dissolution of polydisperse materials in a unit with a rotary pulsation apparatus. J. Eng. Phys. Thermophys. **88**(1), 23–34 (2015)
12. Erenkov, O.Yu., Lopushanskii, I.Ya., Yavorskaya, E.V.: A new design of rotary pulsation apparatus. Chem. Petrol. Eng. **54**, 890–893 (2019)
13. Dolinskiy, A.A., Ivanitskiy, G.K., Obodovich, A.N.: Ispolzovaniye mekhanizmov DIVE pri rotorno–pulsatsionnoy obrabotke geterogennih sred (Use of DPEI mechanisms in rotor-pulsation processing of heterogeneous media). Promishlennaya teploteknika **30**(4), 5–13 (2008)
14. Obodovich, A.N., Limar, A.Yu.: Researches hydraulic characteristics of a rotor-pulsation apparatus in processing of water-grain mix. Eastern-Euro. J. Enterp. Tech. **1**(7(67)), 19–22 (2014)
15. Gorobets, V.G., Trokhaniak, V.I., Serdyuk, A.M.: Processes in rotor-pulsing apparatus for preparation of liquid feed. Energetika i avtomatika **5**, 22–29 (2019)
16. Schlichting, H.: Boundary-Layer Theory. McGraw Hill Book Company, New York (1979)
17. Launder, B.E., Spalding, D.B.: Lectures in Mathematical Models of Turbulence. Academic Press, London (1972)

Investigations of Compact Recuperators Acoustic Properties

Bohdan Gulay, Iryna Sukholova(✉), Oleksandra Dzeryn, and Volodymyr Shepitchak

Lviv Polytechnic National University, Lviv 79013, Ukraine
isukholova@gmail.com

Abstract. The article is devoted to solving the urgent task of increasing the efficiency of air distribution in a room. The aim of this work is the investigation of Prana recuperator acoustic properties at air jets leakage, which flow out from outlets under varying conditions. In this article relationship between such factors as a noise level, sound frequency and the different noise decrease measures at Prana recuperator action in a room has been determined. There are considered some sources of noise: Prana recuperator action, noise in a room and outside and aerodynamic noise of the air jet flowing out from the outlet. There have been designed the charts of sound level difference dependence from the sound's frequency and noise level. Obtained curves were approximated by equations. Comparing charts for the different noise decrease measures at Prana recuperator action in a room, we shall note that noise level difference is the highest at using of the cover device. Installation of the Prana recuperator does not impair the sound-proofing ability of the structure to the air noise of traffic flows, and the installation in front of the Prana of acoustic screens does not give a tangible result.

Keywords: Prana recuperator · Acoustic properties · Air-Jet · Noise level · Air distribution · Air velocity

1 Introduction

The environment of the modern city is often unpleasant to its inhabitants when viewed at higher noise levels. This harms the psychological and emotional state of the person, worsening human health [2, 9]. In this part, with this factor, the completed building structures remain high to noise [20]. However, some factors influence on the hygienic comfort of a room: air temperature in the air, its humidity and air temperature of toxic gases, CO_2 temperature [4]. Maintaining these parameters in rooms with a large number of people (classrooms, meeting rooms, etc.) is impossible in the absence of mechanical supply and exhaust ventilation systems [16, 17]. However, these systems are energy intensive, since a significant energy amount is required to heat the incoming ambient air. Therefore, in recent years for the ventilation of such rooms, there are used inflow-and-exhaust ventilation units with heat recovery of exhaust air, for instance, Prana heat exchangers. These installations have a wide range of performance and are distinguished by some design solutions.

Z. Blikharskyy (Ed.): EcoComfort 2020, LNCE 100, pp. 127–133, 2021.
https://doi.org/10.1007/978-3-030-57340-9_16

2 Analysis of Recent Research and Problem Statement

In ventilation systems of the premises, the normalized air velocity and temperature in the working area of the room should be provided [18], as well CO_2 concentration [5] and a noise level [21]. Today, there are a large number of different designs of air distribution with a high intensity of attenuation of velocity and temperature of the incoming air [7, 15]. Mathematical [8] and physical [10] modeling of air distribution by these specified air jets has been carried out. One way to increase turbulence is using of swirled [3, 12] and compact jets [19]. These characteristics are quite consistent with Prana devices. They form swirled-compact air jets of low enough attenuation coefficients of velocity *m* and temperature *n*.

There is an unresolved issue of ensuring a proper noise level in a room equipped with a Prana recuperator. It is necessary to find out the influence of the Prana heat exchanger on the sound insulation characteristics of the exterior structure and to determine the noise level when the heat exchanger is operating in different modes.

For installation of the Prana heat exchanger in the outer wall, a through-hole with a diameter of more than 162 mm is provided, into which the working module is installed. This results in deterioration of the sound insulation characteristics of the structure. As it is known that for high sound insulation the room from airborne noise should not be allowed in the walls of cracks, openings and leakage of the elements. In particular, the design of a ventilation valve in a modern multi-chamber window significantly degrades its sound insulation characteristics.

This necessitates the study of the Prana recuperators operation effect on acoustic state of the premises in which they are used, the functioning of them at night in residential premises and the compliance of acoustic characteristics with the regulatory requirements.

The calculated analytical and graphical dependencies for the determination of the acoustic parameters of ventilation systems are considered and analyzed [11, 13, 14]. In them, the determination of the resulting noise level in a room can be carried out according to the known equations with the substitution in them of the corresponding values [1, 6].

On the basis of the analysis of the literature data on the patterns of acoustic properties of ventilation systems we summarize: it is necessary to solve the complex problem of comfortable conditions providing in the room in aerodynamic and acoustic aspects.

3 The Purpose and Objectives of the Study

The purpose of the work is to solve the complex problem of providing comfortable conditions in the room in aerodynamic and acoustic aspects.

To achieve this goal, it should be completed the following research objectives:

- to evaluate the acoustic condition of residential and office premises for various purposes during the operation of Prana installations with heat recovery;
- to evaluate the performance of Prana recuperator accessories with acoustic screens to reduce ambient noise.
- to determine the sound power levels of the Prana device at different operating modes;

– to evaluate the acoustic condition of residential and office premises for various purposes, depending on the mode of operation, taking into account the combined effect of external noise.

4 Methods, Materials and Research Results

One of the Prana recuperator modes is Prana-150 (Fig. 1a). This model delivers 115 m^3/h and 105 m^3/h of exhaust air. There is a possibility of supply airflow regulating. The Prana-150 design features 9 fan modes, which can be switched by remote control.

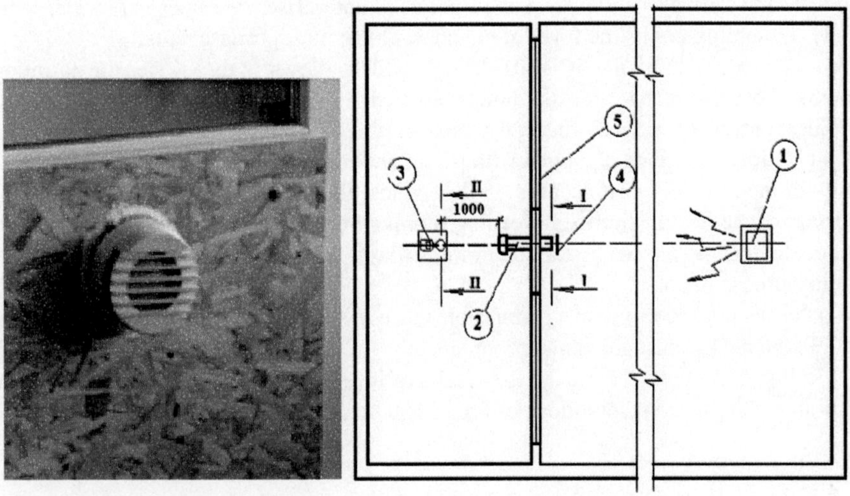

a) b)

Fig. 1. a) - Prana-150; b) - scheme of the experimental setup: 1 - noise source, 2 - Prana -150, 3 - UNI-T UT-352 sound level meter; 4 - acoustic screen; 5 - partition of a plastic window system

Acoustic measurements were carried out on the elements (structures) of building objects using samples of full size (walls, ceilings, floors, ceilings, roofs, windows, etc.), and engineering equipment (ventilation systems) (Fig. 1b).

Experimental studies were conducted under the following conditions.

The sound insulation of the walls in which the windows and a door are located is determined by the sound insulation of these structures, which is lower than the sound insulation of the massive wall. Therefore, the studies were conducted in a room separated by a partition along the length, made of a two-chamber metal-plastic window system with a door. Prana-150 device installing in the door is used in household objects: private homes, apartments, office premises, educational and preschool institutions, other household objects. A noise source is installed in front of the recuperator.

The noise source used was an acoustic system with high frequency response, low background noise and low nonlinear distortion. For further analysis, the audio signal (noise) was stored on electronic media, allowing the time scale to be scaled to the wavelength. The study used an audio signal with geometric mean frequencies of 31.5; 63; 125; 250; 500; 1000; 2000; 4000; 8000 Hz and two typical types of urban noise.

The UNI-T UT-352 sound level meter was used to measure the noise level with a sound range of 30–130 dB, a frequency range of 31.5–8000 Hz, and a measurement error of ±1.5 dB. Noise measurements were carried out before and after the door before and after Prana 150 was installed.

The intensity of noise during the operation of the heat exchanger Prana depends on the mode of its operation and is regulated by the documents of the Ministry of Health of Ukraine:

- in modes V - X it is 55 dB, which allows its use in the halls of cafes, canteens, bars, restaurants; premises for the reception of citizens, trading halls, passenger halls of airports and railway stations, reception points of enterprises of domestic service, sports halls, swimming pools, the foyer of cinemas, clubs, multipurpose halls;
- in modes II, III, IV it is 40–45 dB, which allows it to be used in dormitory rooms (day), residential rooms of hotels less than 3 stars (day), libraries, administrative rooms, conference rooms, offices, workspaces and offices of research and development organizations, premises equipped with personal computers and business equipment;
- in mode I, the noise level is 35 dB, which allows it to be used in the offices of doctors of medical institutions, sleeping rooms of category I housing (day); dormitory rooms (night), 4-star and 5-star hotel rooms (day), less than 3-star hotel rooms (night), multi-purpose halls;
- for sleeping rooms with a permissible noise level of 30 dB at night and below, using of a Prana heat exchanger is inappropriate.

In Fig. 2, 3, 4 and 5 there are the results of investigations of noise level decrease in a room at the different conditions of Prana-150 recuperator action.

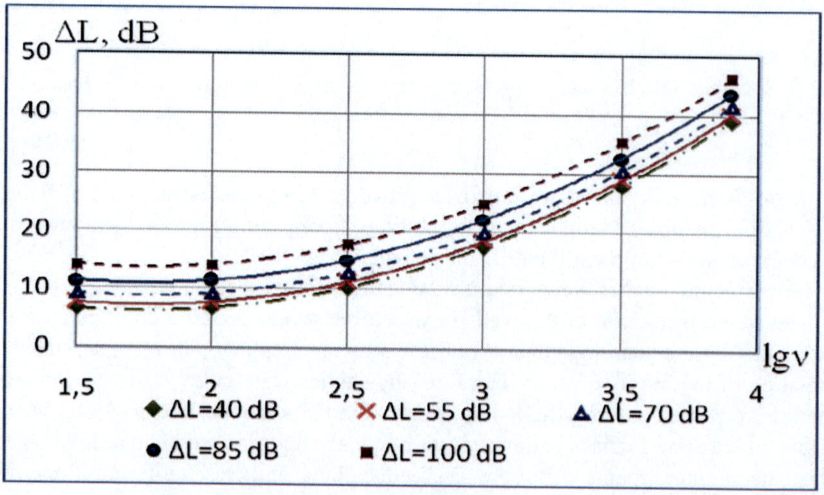

Fig. 2. Decrease of the noise level by the structure before installing the heat exchanger Prana-150.

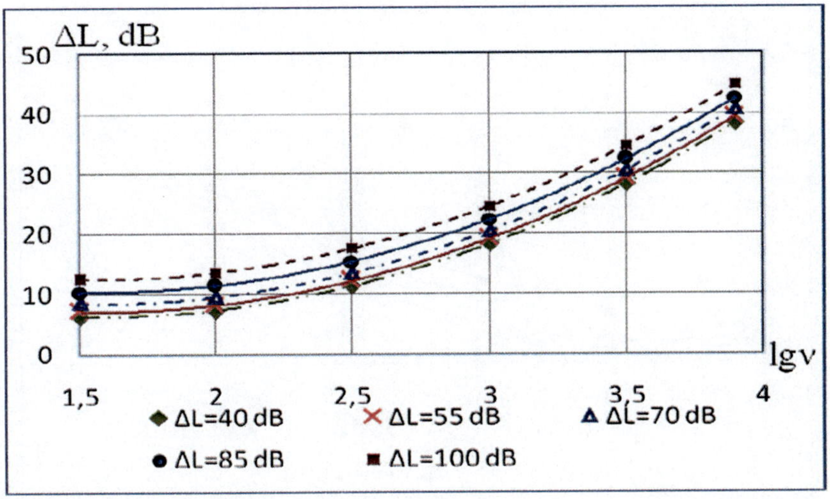

Fig. 3. Decrease of the noise level by the structure when the heat exchanger Prana-150 with a soundproof screen in the form of a cap is installing.

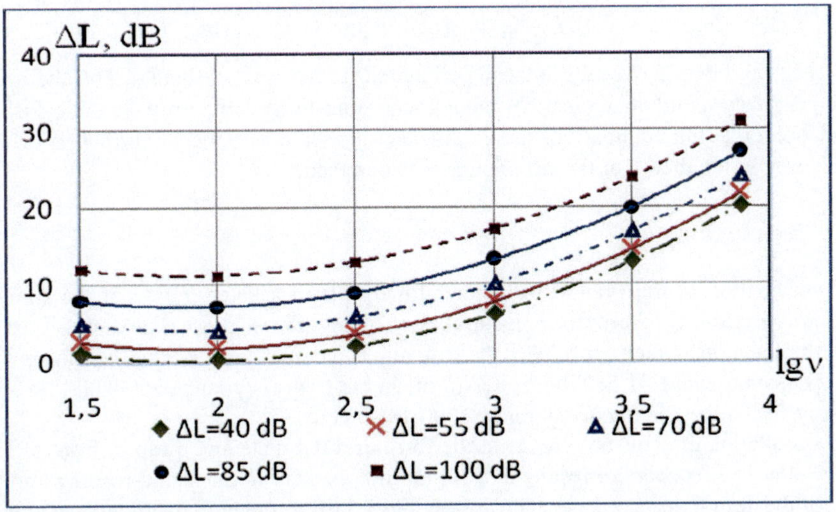

Fig. 4. Decrease of the noise level by the structure when the heat exchanger Prana-150 with a soundproof screen in the form of a flat plate is installing.

The charts (Fig. 2, 3, 4 and 5) were approximated by Eqs. (1), (2), (3) and (4):

$$\Delta L = 7 - 0,06L + 0,0013L^2 - 3,3(\lg \nu - 2) + 8,5(\lg \nu - 2)^2; \quad (1)$$

$$\Delta L = 6 - 0,034L + 0,001L^2 - (\lg \nu - 2) + 7(\lg \nu - 2)^2; \quad (2)$$

$$\Delta L = 1,5 - 0,025L + 0,001L^2 - 4(\lg \nu - 2) + 7(\lg \nu - 2)^2; \quad (3)$$

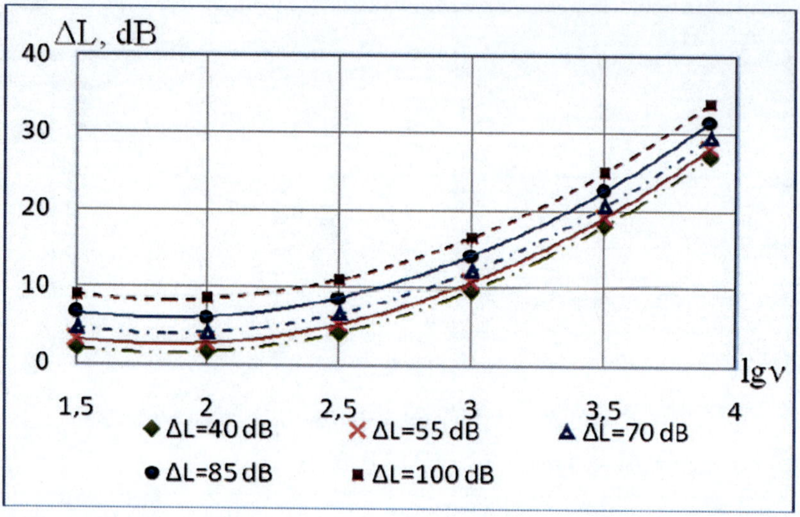

Fig. 5. Decrease of the noise level by the structure when the heat exchanger Prana-150 is installing.

$$\Delta L = 1 - 0,07L + 0,0014L^2 - 4(\lg \nu - 2) + 7(\lg \nu - 2)^2; \quad (4)$$

The validity of the results of the experimental studies was substantiated by checking the adequacy of the mathematical model according to the criteria of Student, Fisher and Kohren at the boundary of the confidence interval $\alpha = 0.95$. Confirmation of the adequacy of the theory of the experiment was obtained.

5 Conclusions

1. Installation of the Prana-150 recuperator results in a deterioration of the sound-proofing ability of airborne noise structures. At an external noise level of 40–100 dB, the sound insulation capacity of the structure is reduced by an average of 5 dB in the frequency range 31.5–250 Hz, by 7.5 dB in the frequency range 250–2000 Hz and by 13.5 dB in the frequency range 2000–8000 Hz.
2. Installation of acoustic screens in the form of a flat plate and a cap in front of the Prana-150 recuperator results in a slight improvement in the sound-reading ability of the design in accordance with 2.5 dB and 5.5 dB throughout the frequency range.
3. Installation of the Prana-150 recuperator practically does not impair the sound-proofing ability of the structure to the air noise of traffic flows, and the installation in front of the Prana of acoustic screens in the form of a flat plate and a cap does not give a tangible result.

References

1. Lau, A.S.H., Huang, X.: The control of aerodynamic sound due to boundary layer pressure gust scattering by trailing edge serrations. J. Sound Vibr. **432**, 133–154 (2018)

2. Lai, D., Qi, Y., Liu, J., Dai, X., Zhao, L., Wei, S.: Ventilation behavior in residential buildings with mechanical ventilation systems across different climate zones in China. Build. Environ. **143**, 679–690 (2018)
3. Holyoake.: Diffuser Performance Data Sheet, Ceiling Fixed Pattern Radial Swirl Diffuser, Model CFP Radial Induction Swirl Diffuser (2006)
4. Kapalo, P., Vilceková, S., Domnita, F., Voznyak, O.: Determine a methodology for calculating the needed fresh air. In: The 9th International Conference "Environmental Engineering", Vilnius, Lithuania Selected Papers, Section: Energy for Buildings (2014). eISSN 2029-7092/eISBN 978-609-457-640-9
5. Kapalo, P., Vilcekova, S., Voznyak, O.: Using experimental measurements the concentrations of carbon dioxide for determining the intensity of ventilation in the rooms. Chem. Eng. Trans. **39**, 1789–1794 (2014). ISBN 978-88-95608-30-3; ISSN 2283-9216
6. Zhao, K., Alimohammadi, S., Okolo, P.N., Kennedy, J., Bennett, G.J.: Aerodynamic noise reduction using dual-jet planar air curtains. J. Sound Vibr. **432**, 192–212 (2018)
7. Lee, K., Jiang, Z., Chen, Q.: Air distribution effectiveness with stratified air distribution systems. ASHRAE Trans. **115**(2), 1–16 (2009)
8. Lukáč, P., Al-Rabeei, S.A.S.: Possibilities of using numerical methods to optimize the flow in the cold aisle in the data centre. Interdiscip. Theory Pract. Journal Present. Interdiscip. Approaches Various Fields **17**, 17–22 (2018). ISSN 2344-2409
9. Zhang, M., Kang, J., Jiao, F.: A social survey on the noise impact in open-plan working environments in China. Sci. Total Environ. **438**, 517–526 (2012)
10. Kobayashi, T., Sugita, K., Umemiya, N., Kishimoto, T., Sandberg, M.: Numerical investigation and accuracy verification of indoor environment for an impinging jet ventilated room using computational fluid dynamics. Build. Environ. **115**, 251–268 (2017)
11. Katinas, V., Marciukaitis, M., Tamasauskiene, M.: Analysis of the wind turbine noise emissions and impact on the environment. Renew. Sustain. Energy Rev. **58**, 825–831 (2016)
12. Voznyak O., Korbut V., Davydenko B., Sukholova I.: Air distribution efficiency in a room by a two-flow device. In: Blikharskyy, Z., Koszelnik, P., Mesaros P. (eds.) Proceedings of CEE 2019. CEE 2019. Lecture Notes in Civil Engineering, vol 47. Springer, Cham (2020)
13. Voznyak O., Myroniuk K., Sukholova I., Kapalo P.: The impact of air flows on the environment. In: Blikharskyy Z., Koszelnik P., Mesaros P. (eds) Proceedings of CEE 2019. CEE 2019. Lecture Notes in Civil Engineering, vol 47. Springer, Cham (2020)
14. Choi, W., Pate, M.B.: An evaluation and comparison of two psychoacoustic loudness models used in low-noise ventilation fan testing. Build. Environ. **120**, 41–52 (2017)
15. Grimitlin M.I.: Air distribution in the room. Issue 3, adapted and supplemented, publication AVOK North-West, 320 p (2004). (in Russian)
16. Gumen, O.M., Dovhaliuk, V.B., Mileikovskyi, V.O.: Determination of the intensity of turbulence of streams with large-scale vortices on the basis of geometric and kinematic analysis of macrostructure. J Lviv Polytech. Nat. Univ. Ser. Theory Build. Pract. **844**, 76–83 (2016). (in Ukrainian)
17. Dovhaliuk, V.B., Mileikovskyi, V.O.: Efficiency of organization of air exchange in heat-stressed premises in compressed conditions. J. Build. Ukraine **3**, 36 (2007). (in Ukrainian)
18. Dovhaliuk V.B., Mileikovskyi V.O.: Estimated model of non-isothermal stream, which is laid out on a convex cylindrical surface. In: Ventilation, Illumination and Heat and Gas Supply: Scientific and Technical Collection, Kyiv, KNUBA, no. 12, pp. 11–32 (2008). (in Ukrainian)
19. Dovhaliuk V.B., Mileikovskyi V.O.: Analytical studies of the macrostructure of jet currents for calculating energy-efficient systems of air distribution. Energy efficiency in construction and architecture, no. 4, pp. 11–32 (2013). (in Ukrainian)
20. Yudin, Y.: Noise Control at Work. Directory, Moscow (1985). (in Russian)
21. Yudin, Y.: Noise protection. Directory, Moscow (1974). (in Russian)

Effect of Nano-TiO_2 and ETS Antifungal Agent Addition on the Mechanical and Biocidal Properties of Cement Mortars

Marko Hohol[1(✉)], Vira Lubenets[2], Olena Komarovska-Porokhnyavets[2], and Myroslav Sanytsky[1]

[1] Department of Building Production, Lviv Polytechnic National University, S. Bandera str. 12, 79013 Lviv, Ukraine
gogol.lviv@gmail.com

[2] Department of Technology of Biologically Active Substances, Pharmacy and Biotechnology, Lviv Polytechnic National University, S. Bandera str. 12, 79013 Lviv, Ukraine

Abstract. The scope of the study is to test the influence of nano-TiO_2 and ETS biocide addition on the mechanical and fungicidal properties of cement mortars. It is shown that the use of doped titanium dioxide and thiosulphanate structure modifier of the antifungal agent ETS is one of the ways to obtain an effective composite antifungal additive to cement mortars. The possibility of developing such complex additives to achieve a synergistic effect in the field of fungicidal coatings has been investigated. The use of nanosized doped titanium dioxide creates a photocatalytic layer on the surface of the material, which is activated by UV or visible light, while the biocide ETS modifier works in the dark, also extending the spectrum of action to other living organisms. Modified with such an additive, cement mortar became 19% stronger than the control sample. The change of the surface microstructure, the appearance of a significant number of mesopores and macropores, which creates a larger surface, thus increasing the photocatalytic activity, was confirmed. A comparison of the growth of mold colonies on control and modified samples is shown.

Keywords: Mechanical and fungicidal properties · Nano-TiO_2 and ETS biocide addition · Photocatalysis · Microstructure · Nanomodification

1 Introduction

At the present stage, the problems of sustainable construction and the environment have become the main factors worldwide. In the process of construction of a new generation of eco-comfortable buildings, as well as for restoration, finishing work and decoration of facades, modern building and finishing materials based on multicomponent cements are successfully used [1–3]. A new generation technology which is used in civil engineering to improve the development and properties of materials is nano technology, involving the use of nano-liquids and nano-additives [4–6].

One of such nano-modified materials is tiocement - a high-tech cement which provide the photocatalytic and biocidal properties of cement mortars. The main

Z. Blikharskyy (Ed.): EcoComfort 2020, LNCE 100, pp. 134–141, 2021.
https://doi.org/10.1007/978-3-030-57340-9_17

modifier in this cement are titanium dioxide nanoparticles, which provide adsorption of harmful components of the environment (smoke, organic matter, oil, carbon monoxide, nitrogen, etc.) and under the action of ultraviolet and visible light neutralize them [7]. This makes it possible to ensure a healthy indoor climate, such as offices, kindergartens, schools and apartments, which is an important factor in human health. At the same time, the quality of people's lives, (especially in big agglomerations) has become seriously impaired by factors such as air pollution and diseases caused by microorganisms [8, 9].

One of the main causes of poor indoor air quality can be molds that grow on the surfaces of building materials or are part of the dust. The presence of microscopic fungi in the room can cause allergies, bronchial asthma and other respiratory diseases. In addition, some fungi secrete harmful secondary metabolites (mycotoxins) that can adversely affect the human body. This problem is especially relevant in rooms with high humidity and limited insolation.

Therefore, the development of building materials with antifungal properties is a topical issue. The problem of fungal infection of the walls in the premises requires comprehensive methods of solution. Despite the regulation of ventilation, temperature and humidity, it is necessary to use materials that will neutralize the colonies of fungi on its surface. Recently, increasingly for water treatment, wastewater and air purification new types of photocatalysts are used. Surfaces coated with photocatalysts exhibit self-cleaning and antibacterial properties. The addition of photocatalyst such as TiO_2/N to gypsum mortar can create coatings with the air cleaning properties and improved technical properties [10].

In a studies [11, 12] it was shown that nanosized TiO_2 is able to delay the onset of colonization or quantitatively reduce the surface coverage of fungal colonies. Inorganic substances and organic pollutants undergo photocatalytic oxidation. Photocatalytic titanium dioxide has antimicrobial properties, which has been confirmed for gram-negative and gram-positive bacteria (including *Escherichia coli*, *Staphylococcus aureus*), yeast, cyanobacteria, some protozoa and viruses [13]. In order to increase the photocatalytic efficiency, titanium dioxide is doped with various elements. Modified titanium dioxide has a number of properties that significantly improve the physico-mechanical, chemical, photocatalytic and biocidal properties of cement mortars [14–16].

In combination with titanium dioxide, the antifungal agent (biocide) ETS (S-ethyl-4-amino-benzenethiosulfonate) was used, which is able to neutralize strains of fungi and bacteria regardless of the presence of a light source [17, 18]. It is known that most mortars contain lime (calcium hydroxide), which has antifungal action. However, over time, the structure of the mortar undergoes a process of carbonization, which leads to the conversion of calcium hydroxide to calcite (change in pH of the solution from 12.7 to 8.5), as a result of which the surface may be exposed to fungal colonies. As a result, there is a need for a component that will protect the surface from the development of fungal crops throughout its life [19, 20].

2 Experimental Program

2.1 Raw Materials

Composite Portland cement CEM II/BM (S-P-L) 32,5R (manufacturer PJSC "Ivano-Frankivskcement") based on Portland cement clinker of normalized mineralogical composition (wt%: C_3S - 61.8; C_2S - 14.25; C_3A - 7.20; C_4AF - 11,85) was used for this study. Composite Portland cement CEM II/B-M (S-P-L) 32,5R contains 35 wt% of the main components (granular blast furnace slag (S), natural zeolite (P) and limestone (L)), with the true density of 3,0 g/cm^3, specific surface (according to Blaine) of 380 m^2/kg. Fine sand was used as aggregate. Cement-sand mortar with a nominal composition of 1:3 and a water-cement ratio of 0.50 was used as a control. As a plasticizer additive, a new generation superplasticizer based on polycarboxylates ether MasterGlenium Ace 430 (RSE), (BASF, Germany) was used. As a nanomodifier, titanium dioxide co-doped with sulfur (S) and carbon (C) (TiO_2 S, C) was used. TiO_2 S, C consists of 97% of the anatase form and has a particle size of 10-30 nm (synthesized in Lviv Polytechnic National University, Ukraine). As an antifungal agent in cement mortar was used ETS - S-ethyl-4-amino-benzenethiosulfonate.

To determine the fungicidal properties of the mortars surface was used sterile Czapek-Dox medium of the following composition: K_2HPO_4 - 0,3 g; $KH_2PO_4 \cdot 3H_2O$ - 0,7 g; $MgSO_4 \cdot 7H_2O$ - 0,5 g; $NaNO_3$ - 2,0 g; KCl - 0,5 g; $FeSO_4 \cdot 7H_2O$ - 0,01 g; saccharose - 30,0 g; agar-agar - 20,0 g; distilled water - 1000 ml (pH = 6 ± 0,5). Cultures of *Aspergillus niger* BKMF - 1119 and *Penicillium chrysogenum* BKMF - 245 were used in microbiological tests.

2.2 Experimental Process

Physico-mechanical properties of cement mortars with antifungal properties were determined in accordance with current standards and generally accepted methods. To determine the strength of the solution were prepared samples of cement-sand mortar size 20 × 20 × 20 mm with a ratio of cement/sand 1:3 and water/cement ratio of 0,5. Samples in the molds were kept for 24 h while providing temperature and humidity. After dismantling and labeling, the samples were placed in a desiccator for storage before testing after 7; 28 and 90 days. Dispersion of titanium dioxide and ETS biocide was performed in an ultrasonic bath for efficient particle distribution in water. Two compositions of cement-sand mortar were selected for the experiment: control (non-additive) and modified with 2.0 wt% TiO_2 S, C and 0.05 wt% ETS.

To study the fungicidal properties of the modified cement mortar, cylindrical samples with a diameter of 32 mm and a thickness of 5 mm were prepared. In determining the antifungal properties used a method that establishes the presence of fungicidal properties in the coatings and the assessment of fungal resistance of the coating in the presence of an additional power supply according to the degree of surface destruction according to GOST 9.050-75. The prepared samples were decomposed into Petri dishes with nutrient medium and inoculated with a suspension of fungal spores (microbial load 10^6 spores/ml). Cups with modified and control samples were placed in a desiccator, which was placed in a thermostat with an air temperature of 29 ± 2 °C and a relative humidity of more than 90%. The duration of the experiment since the establishment of the regime in determining the fungal resistance by the degree of destruction of the coating surface is 28 days, with an intermediate assessment of the fungal resistance of the coating every 2 days. SEM images were taken under VEGA3 TESCAN microscope (Czech Republic). Also, the study of fungal resistance of cement mortar was performed on samples with a size of 50 × 50 mm and a thickness of 5 mm, which was applied water-emulsion paint EcoCristal IP-131 for external work with the addition of 1% fungicidal component of biocide ETS ("IP-131 + 1% ETS") and without its contents ("IP-131"), according to the above method.

3 Results and Discussion

Nanomodification of the solution with the addition of titanium dioxide provides an increase in its strength at an early and design age. Experimental studies have shown that for cement mortars with water/cement ratio of 0.5 and flowability of 180 mm, the compressive strength after 28 and 90 days is 20.4 and 23.5 MPa, respectively. After modification of mortar with a complex biocide additive, the strength after 28 days was 21,3 MPa and after 90 days was 27.9 MPa (Fig. 1). An increase in the strength of the sample with a complex addition of titanium dioxide and ETS biocide at the design age by 19% compared to the control sample was achieved. This increase in strength is explained by the nano-core effect, i.e. the ability of titanium dioxide nanoparticles to inhibit the propagation of microcracks, thereby strengthening the cementing matrix.

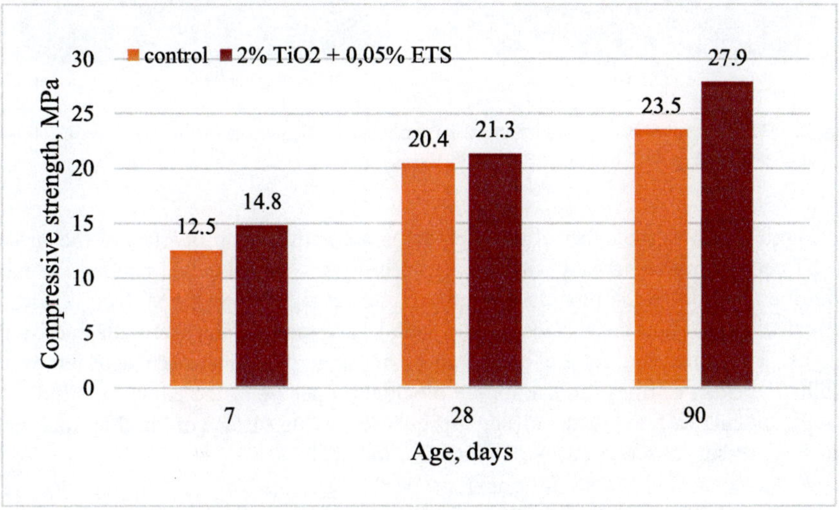

Fig. 1. Compressive strength of cement mortars

Figure 2 shows the SEM image of the modified and control samples of cement mortars. It should be noted that the biocidal modifier fills the pores in the structure of the mortar (Fig. 2a), which is reflected in the SEM image of the samples. The presence of meso- and macropores increases the specific surface area of the mortar, due to which more nanoparticles of titanium dioxide can enter the photocatalysis reaction. This change in the microstructure of the surface of the finishing mortars increases the biocidal action of the complex additive of titanium dioxide and ETS.

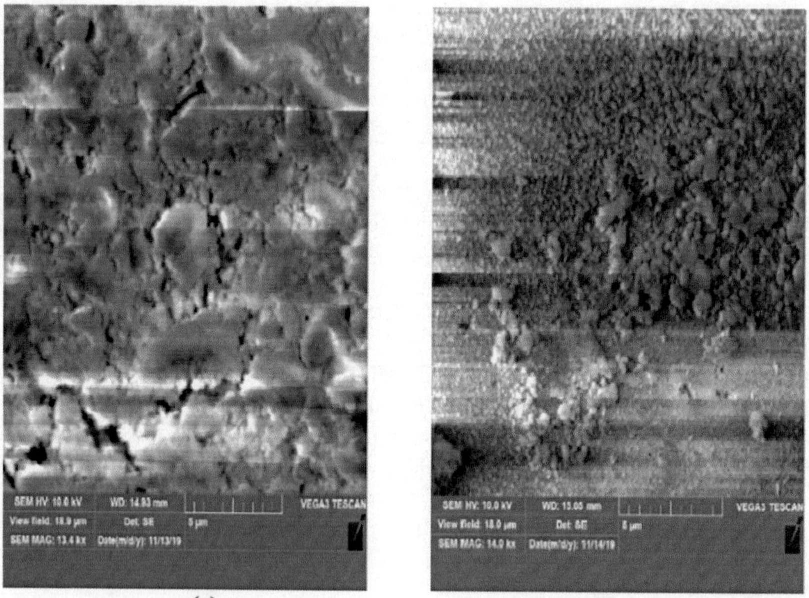

(a) (b)

Fig. 2. SEM-image of the microstructure of samples of cement mortars: a) 2 wt% TiO_2 S, C + 0,05% ETS biocide, b) control non-additive sample.

Figure 3 shows the effect of the biocide on the antifungal properties of the surface of the cement mortar treated with paint. According to the obtained results, the paint from the addition of the biocide created an antifungal layer on the surface, while the untreated surface and the surface treated with conventional paint were affected by the fungus. The difference in the colonization of fungi on the surface determines the qualitative effect of thiosulfonate on the biocidal properties of the paint. This indicates that the biocide ETS exhibits antifungal properties on the surface of building materials, which increases its scope in the construction industry.

When determining the antifungal properties, samples of cement mortars were photographed, which were placed in a nutrient medium and inhabited by colonies of the fungus *Aspergillus niger*. Figure 4 shows the difference in the infection of surfaces of cement mortars with the fungus A.niger. According to the visual assessment, it is seen that the sample with a complex additive (2 wt% TiO_2 S, C and 0.05 wt% biocide

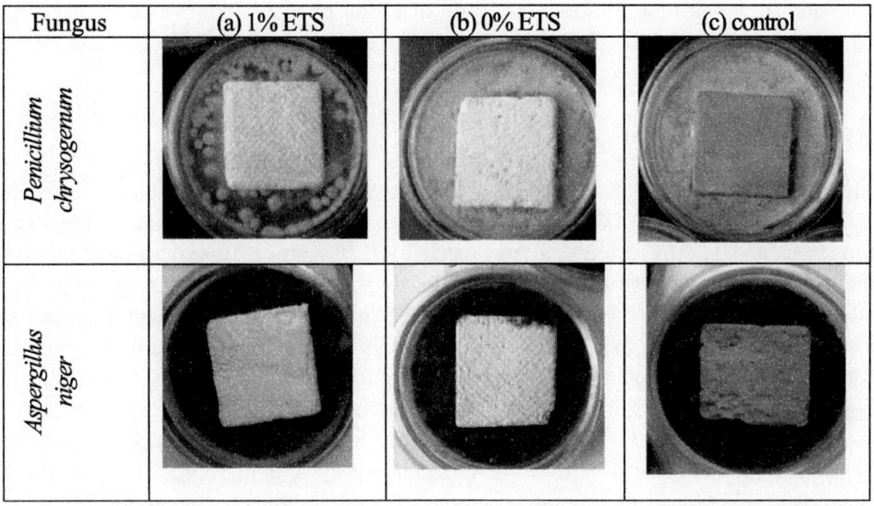

Fig. 3. Visualization of fungal resistance of mortar samples against colonies of *P. chrysogenum* and *A. niger.*

ETS) provides fungicidal properties of the surface. (Fig. 4a), while surface of control sample was covered with fungus (Fig. 4b) This result indicates the absence of anti-fungal properties of unmodified cement mortar.

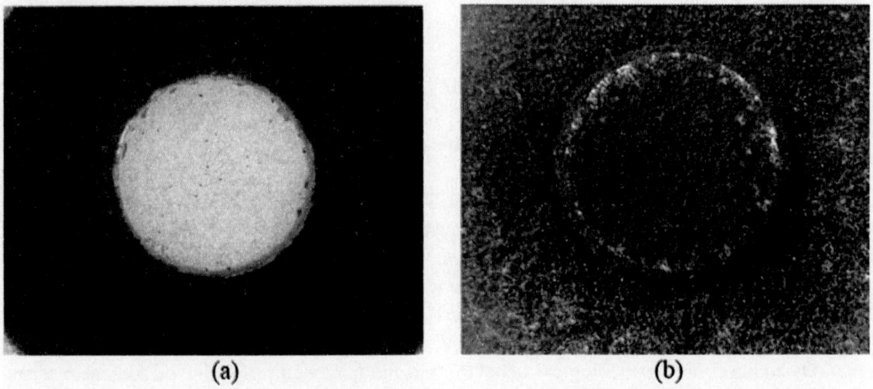

Fig. 4. Image of samples during the study of fungicidal properties: a) sample with 2 wt% TiO_2 S, C + 0.05% biocide ETS b) control non-additive sample.

Thus, the complex addition of titanium dioxide and biocide ETS gives the surfaces of cement mortars biocidal properties, also having a positive effect on the physico-mechanical, chemical and rheological properties. The study was conducted on the most common strains of fungi, but the spectrum of action of such a composite additive can be much wider, which is of additional scientific interest in this study.

4 Conclusions

Modification of cement mortars with a complex additive containing nanosized doped TiO_2 S, C and modifier of biocide ETS allows to obtain finishing mortars with biocidal properties and improved physical and mechanical properties. It was investigated that the use of TiO_2 S, C and ETS increases the strength of cement mortars by 19%. The effect of the additive on the change of the surface microstructure of cement mortars was also investigated. It is established that the complex modifier increases the specific surface area and promotes the formation of macro- and mesopores. Nanoparticles of doped titanium dioxide have been shown to accelerate surface biocidal processes under the influence of UV or visible light, and the ETS biocide modifier works to neutralize biological contaminants regardless of the presence of a light source. As a result, a synergistic effect of combining the properties of both modifiers for integrated control of fungal colonies under any indoor conditions is achieved.

References

1. Venkat, R.N., Rajasekhar, M., Vijayalakshmi, K., et al.: The future of civil engineering with the influence and impact of nanotechnology on properties of materials. Procedia Mater. Sci. **10**, 111–115 (2015)
2. Krivenko, P., Sanytsky, M., Kropyvnytska, T., Kotiv, R.: Decorative multi-component alkali activated cements for restoration and finishing works. Adv. Mater. Res. **897**, 44–49 (2014)
3. Sanytsky, M., Kropyvnytska, T., Kotiv, R.: Modified plasters for restoration and finishing works. Adv. Mater. Res. **923**, 42–47 (2014)
4. Kropyvnytska, T., Semeniv, R., Kotiv, R., Kaminskyy, A., Gots, V.: Studying the effect of nano-liquids on the operational properties of brick building structures. Eastern-Eur. J. Enterp. Technol. **5/6**(95), 27–32 (2018)
5. Krivenko, P., Sanytsky, M., Kropyvnytska, T.: The effect of nanosilica on the early strength of alkali-activated portland composite cements. Solid State Phenom. **296**, 21–26 (2019)
6. Kropyvnytska, T., Sanytsky, M., Rucinska, T., Rykhlitska, O.: Development of nanomodified rapid hardening clinker-efficient concretes based on composite Portland cements. EEJET **6**(6), 38–48 (2019). https://doi.org/10.15587/1729-4061.2019.185111
7. Senff, L., Hotza, D., Lucas, S., Ferreira, V., Labrincha, J.: Effect of nano-SiO_2 and nano-TiO_2 addition on the rheological behavior and the hardened properties of cement mortars. Mater. Sci. Eng. **532**, 354–361 (2012)
8. Sikora, P., Cendrowski, K., Markowska-Szczupak, A., Horszczaruk, E., Mijowska, E.: The effects of silica/titania nanocomposite on the mechanical and bactericidal properties of cement mortars. Constr. Build. Mater. **150**, 738–746 (2017)
9. Augustyniak, A., Sikora, P., Jablonska, J. et al.: The effects of calcium–silicate–hydrate (C–S–H) seeds on reference microorganisms. Appl. Nanosci. (2020). https://doi.org/10.1007/s13204-020-01347-5

10. Zając, K., Rucińska, T., Morawski, A.W., Janus, M.: Photocatalytic gypsum plasters – studies of air cleaning properties and selected technical parameters. Cement-Wapno-Beton **1**, 10–19 (2019)
11. Kądziołka, D., Rokicka, P., Markowska-Szczupak, A., Morawski, A.W.: Influence of titanium dioxide activated under visible light on survival of mold fungi. Med. Pr. **69**, 59–65 (2018). https://doi.org/10.13075/mp.5893.00652
12. Markowska-Szczupak, A., Tomaszewska, M., Morszczyzna, A., Morawski, A.W.: Studies on antifungal properties of photocatalytic paints. Przem. Chem. **93**(5), 766–770 (2014)
13. Aflori, M., Simionescu, B., Bordianu, I., Sacarescu, L., Varganici, C., Doroftei, F., Nicolescu, A., Olaru, M.: Silsesquioxane-based hybrid nanocomposites with methacrylate units containing titania and/or silver nanoparticles as antibacterial/antifungal coatings for monumental stones. Mater. Sci. Eng., B **178**, 1339–1346 (2013)
14. Russa, M.F., Macchia, A., Ruffolo, S.A., Leo, F.D., Barberio, M., Crisci, G.M., Urzì, C.: Testing the antibacterial activity of doped TiO_2 for preventing biodeterioration of cultural heritage building materials. Int. Biodeter. Biodegr. **96**, 87–96 (2014)
15. Ivanov, S., Barylyak, A., Besaha, K., Bund, A., Bobitski, Y., Wojnarowska-Nowak, R., Yaremchuk, I., Kus-Liśkiewicz, M.: Synthesis, characterization, and photocatalytic properties of sulfur- and carbon-codoped TiO_2nanoparticles. Nanoscale Res. Lett. **11**(1), 1–12 (2016). https://doi.org/10.1186/s11671-016-1353-5
16. Gómez-Ortíz, N., De la Rosa García, S., González-Gómez, W., Soria-Castro, M., Quintana, P., Oskam, G., Ortega-Morales, B.: Antifungal coating base on $Ca(OH)_2$ mixed with ZnO/TiO_2 nanomaterials for protection of limestone monuments. Appl. Mater. Interfaces **5**, 1556–1565 (2013)
17. Lubenets, V., Vasylyuk, S., Monka, N., Bolibrukh, K., Komarovska-Porokhnyavets, O., Baranovych, D., Musyanovych, R., Zaczynska, E., Czarny, A., Nawrot, U., Novikov, V.: Synthesis and antimicrobial properties of 4-acylaminobenzene-thiosulfoacid S esters. Saudi Pharm. J. **25**, 266–274 (2017)
18. Lubenets, V., Stadnytska, N., Baranovych, D., Vasylyuk, S., Karpenko, O., Havryliak, V., Novikov, V.: Thiosulfonates: the prospective substances against fungal infections. In: Érico Silva de Loreto, E., Moraes Tondolo, J. (eds.) IntechOpen, London. (2019). https://doi.org/10.5772/intechopen.84436
19. Martyrosian, A., Pakholiuk, O., Semak, B., Komarovska-Porokhniavets, O., Lubenets, V., Pambuk, S.: New technologies of effective protection of textiles against microbiological damage. Nanosistemi Nanomateriali Nanotehnologii **17**(4), 621–636 (2019)
20. Martirosyan, I., Pakholiuk, O., Semak, B., Lubenets, V., Peredriy, O.: Investigation of wear resistance of cotton-polyester fabric with antimicrobial treatment. In: Tonkonogyi, V., Ivanov, V., Trojanowska, J., Oborskyi, G., Edl, M., Kuric, I., Pavlenko, I., Dasic, P. (eds.) InterPartner 2019. LNME, pp. 433–441. Springer, Cham (2020). https://doi.org/10.1007/978-3-030-40724-7_44

Non-uniformity of Water Inflow into Pressure Collector-Pipeline Depending on the Values of Reynolds Criterion and of Inflow Jets Angles

Vasyl Ivaniv[1], Volodymyr Cherniuk[1,2](✉), and Vasyl Kochkodan[1]

[1] Lviv Polytechnic National University, Lviv 79013, Ukraine
v.cherniuk@ukr.net
[2] Catholics University of Lublin Named After John Paul II, Lublin, Poland

Abstract. Results of experimental investigation of the influence of Reynolds criterium Re_D on the non-uniformity in the work of pressure collector-pipeline (CPs) for different values of the angle β of jet inflow into the fluid stream in CP are presented. The angle β was assigned the following values: 0°; 45°; 90°; 135°; 180°. The value of Reynolds criterion Re_D was varied within the range from 5211 to 28321. In has been found that the non-uniformity of distribution of the working heads in the CP and those of water inflow into the CP along the path decreased with the increase in the Reynolds criterion Re_D. By means of variation of the values of the angle β, the reduction of non-uniformity of the working heads distribution along the CP was achieved up to 44.49%, and that of water inflow into the CP along the path up to 9.07% (as compared to the case of orthogonal jet inflow into the main stream of collector-pipeline).

Keywords: Pressure collector-pipeline · Angle of jet inflow · Non-uniformity of stream along the path

1 Relevance of the Investigation

Pressure collector-pipelines (CPs) with inflow of fluid along the path are used in different branches of engineering. The fluid inflow through the holes in the CP wall along the path is variable. It increases with approaching the estuary of the CP. The cause of this is the decrease in piezometric head along the stream in CP caused by hydrodynamic resistances. However, in the overwhelming majority of technological processes there is need in uniform fluid inflow along a CP.

Some polymer substances effectively reduce the hydrodynamic resistance of turbulent stream in pipelines, with this, they also reduce local hydrodynamic resistances [1, 2]. Polyacrylamide additives are used to purify drinking water [3–5]. When applied in small doses, it does not pollute the environment [6–8]. This is an operative method of regulation of work of pressure pipeline with variable fluid flow rate along the path. However, it is effective only under the presence of hydrodynamically active additives in the stream.

Z. Blikharskyy (Ed.): EcoComfort 2020, LNCE 100, pp. 142–149, 2021.
https://doi.org/10.1007/978-3-030-57340-9_18

The fluid inflow into pressure collector-pipeline along the path is influenced by its geometric characteristics, by hydrydynamic parameters of the stream in the CP, and those of the inflowing jets. In the known publications, there are presented results of investigation of work of only such CPs into which the jets are inflowing at right angle [9–11]. However, it is the value of the angle β between the direction of the flowing of the main stream in the CP and that of the inflowing through inlet nozzles to it jets that essentially influences the non-uniformity of inflow along the path. The invented by us way of regulation of fluid inflow into pressure CP along the path is asserted by the patent № 115840, Ukraine [12]. In practice, collector-pipelines work under different values of Reynolds criterion. The influence of this criterion on work of CP has not been investigated before.

Aim of the work is to establish the influence of the value of Reynolds criterion Re_D on non-uniformity of water inflow into a pressure pipeline along the path under different values of the angle β of jet inflow.

2 Planning of Experiment

The schematic diagram of the stand is presented in the article [13]. In this article, there is presented only a simplified diagram for calculation of pressure collector-pipeline (Fig. 1).

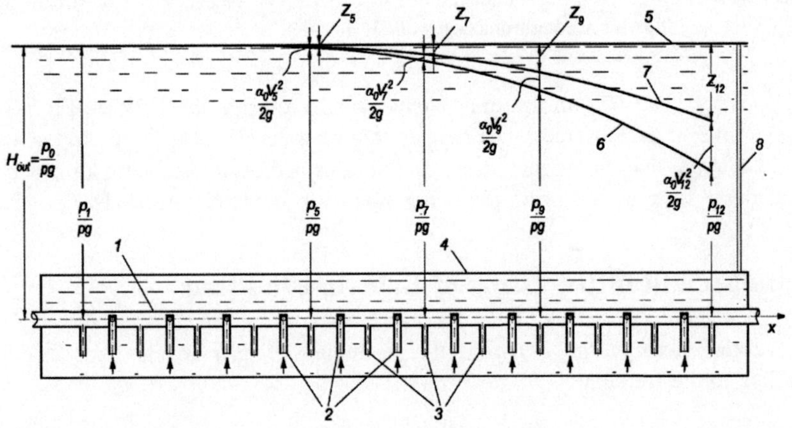

Fig. 1. Schematic diagram for calculation of collector-pipeline: 1 – CP; 2 – inlet nozzles; 3 – connections for pulse holes from piezometers; 4 – transparent hermetically sealed cylinder; 5 – water level corresponding to the head in the transparent cylinder; 6 – piezometric curve plotted for water stream inside CP; 7 – curve of full working head for the stream in CP; 8 – piezometer (In this diagram, there is no conventional presentation of pulse hoses which join connections 3 with 12 piezometers which are in the piezometers board)

Pressure CPs whose inner diameter D = 11,28 mm, and which had eleven inlet nozzles of $d = 4.83$ mm have been investigated. The distances between the nozzles were equal to 160 mm. The nozzles (Fig. 2) had been installed with the possibility of their rotation about their longitudinal axes in order to regulate the values of the angle β between the stream in the CP and the inflowing jets (Fig. 3).

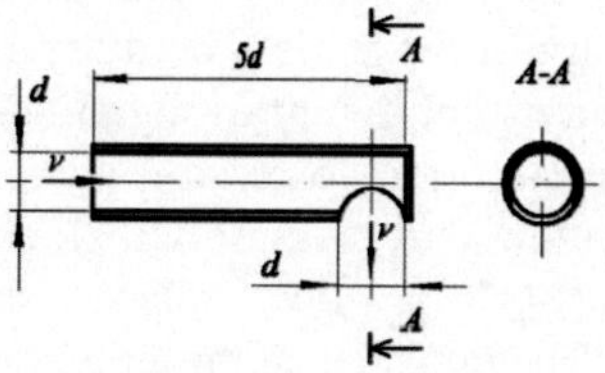

Fig. 2. Inlet cylindrical nozzles with lateral inflow of jet

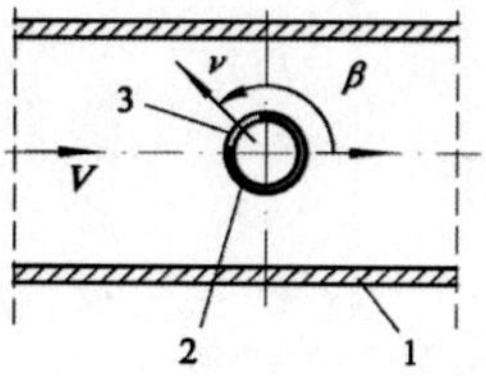

Fig. 3. Diagram for reference of angle β of jet inflow: 1 – wall of CP; 2 – inlet nozzle with connection angle $\beta = 135^o$; 3 – outlet hole of the nozzle; V – mean speed of stream in CP; v – ditto for jet inflowing through nozzles into CP

The water head H_{out} in the transparent cylinder outside the CP (see Fig. 1) was varied from 300 to 2300 mm. The temperature of water $T = 14\text{–}21$ °C. In each individual CP (from five investigated CPs), the nozzles had been installed with $\beta = const.$ The angles β were assigned the following values: 0°; 45°; 90°; 135°; 180°.

3 Mathematical Processing of Experimental Data

The working head at the ith (from the beginning of CP) nozzles was calculated according to the formula:

$$Z_i = H_{out} - \frac{p_i}{\rho g} - \frac{\alpha_0 V_i^2}{2g}, \tag{1}$$

where H_{out} is the actual head outside the collector-pipeline; $p_i/\rho g$ is the piezometric head at the ith nozzles; $\alpha_0 V_i^2/2g$ is the kinetic-energy head of water stream in the CP before the ith nozzles.

The water inflow into the CP through the ith nozzles was calculated theoretically, depending on the working head Z_i at it:

$$q_i = \mu_i \omega \sqrt{2gZ_i}, \tag{2}$$

where μ_i is the coefficient of the flow rate of the nozzles, its value $\mu = f(\mathrm{Re}_d)$ for the investigated nozzles we have determined experimentally [14]; ω is the cross-section area of the nozzles; g is the gravity acceleration; Z_i is the working head at the ith nozzles.

The water flow rate in the CP in the range line of the kth nozzle, being guided by the expression (1) and (2), we calculated according to the formula:

$$q_k = \sum_{i=1}^{k} q_i + \mu_k \omega \cdot \sqrt{\left[2gH_{out} - \frac{2p_k}{\rho} - \alpha_0 \left(\sum_{i=1}^{k-1} q_i/\Omega\right)^2\right]}, \tag{3}$$

where, $\sum_{i=1}^{k-1} q_i$ is the water flow rate in the range line of the CP before the kth nozzles; μ_k is the coefficient of flow rate of the inlet nozzle whose ordinal number is k; ω is the cross-section area of the nozzle; H_{out} is the head of fluid outside the collector-pipeline; $p_k/\rho g$ is the piezometric head at the kth nozzle; a_0 is the Coriolis' acceleration, $a_0 = 1,05$; Ω is the cross-section area of The CP.

The value of the Reynolds criterion was calculated for the range line of the CP after the last (eleventh) nozzle.

The kinematic viscosity of water was determined depending on its temperature.

The non-uniformity of distribution of working head Z_i along the CP:

$$\eta_z = Z_m / Z_{beg}, \tag{4}$$

where Z_m in the maximal head at the nozzle (in our experiments it was the head at the last nozzle), $Z_m = Z_{end} = Z_{11}$; Z_{beg} is the head at the first nozzle, $Z_{beg} = Z_1$ of the CP [15].

The non-uniformity of water inflow q_i into the CP:

$$\eta_q = q_m / q_{beg}, \tag{5}$$

where, q_m is the maximal inflow through a nozzle (in our experiments it was the inflow through on the last nozzle); $q_m = q_{end} = q_{11}$; q_{beg} is the inflow through the first nozzle, $q_{beg} = q_1$ [15].

4 Results of the Experimental Investigation

The distributions of working heads Z_i and that of water inflow q_i, through individual nozzles along pressure pipelines for jet inflow angle β equal to 90° and 135° are presented in Fig. 4 and Fig. 5.

In each of Fig. 4 and Fig. 5, the dependences $Z_i = f(l)$ and $q_i = f(l)$ are presented for a CP whose ratio of inner diameters d/D = 0,428 for three different values of Reynolds criterion. The values of Re_D which are presented in the foresaid figures were calculated for cross-sections which are situated at the end of the collector-pipeline, after the eleventh inlet nozzle. The value varies within the range of 5211 through 28321.

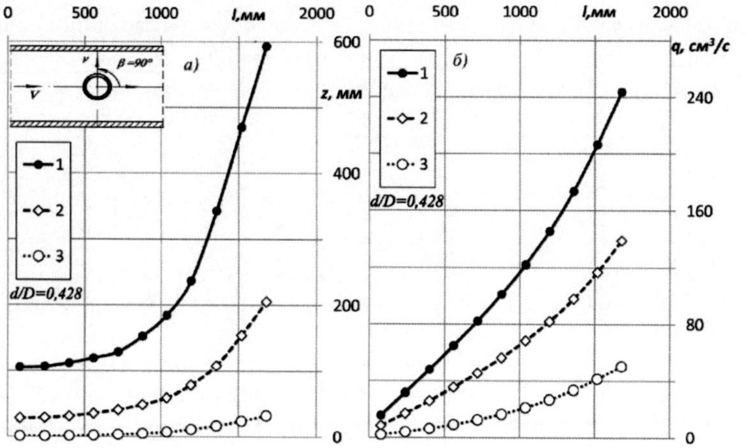

Fig. 4. Distributions of (a) working heads Z_i and (b) water inflow q_i through individual nozzle along CP with the angle $\beta = 90^o$ for different values of Reynolds criterion Re_D: 27764 – (1); 14894 – (2); 6354 – (3)

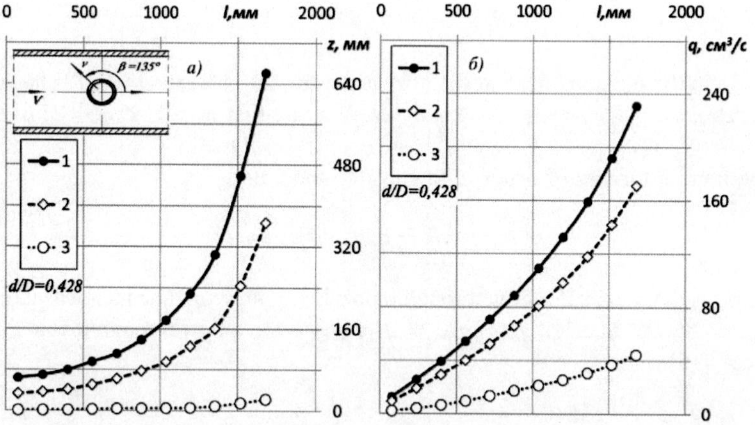

Fig. 5. Distributions of (a) working heads Z_i and (b) water inflow q_i through individual nozzle along CP with the angle $\beta = 135^o$ for different values of Reynolds criterion Re_D: 24569 – (1); 18201 – (2); 5499 – (3)

The graphic dependences $\eta_Z = f(\mathrm{Re}_D)$ and $\eta_q = f(\mathrm{Re}_D)$, (Fig. 6) indicate that the non-uniformities η_z of distribution of working heads of CP and the non-uniformity η_q of water inflow into CP along the path decrease with the increase in Re_D (Fig. 6). The cause of this in the following. Greater values of Reynolds criterion Re_D take place when the water flow rate (and correspondingly water speed V_D) are greater: in this case, the energy consumption is greater as well. Thus, the working head Z_i along the CP, in particular Z_1 at the first nozzle (see Formula (1) and Fig. 1). Therefore, the water inflow through the first nozzle into CP increases: $q_1 = \mu_1 \omega \sqrt{2gZ_1}$. As a consequence, with the increase in $q_1 = q_{beg}$, the non-uniformity η_q decreases (see Formula (5)).

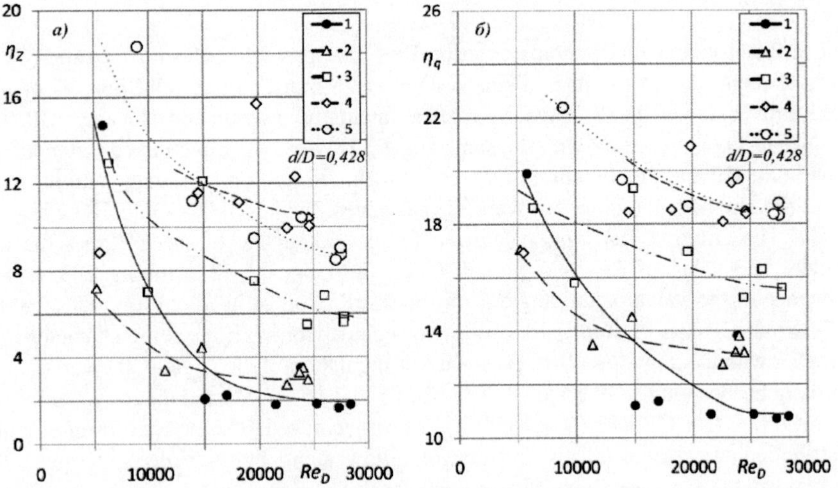

Fig. 6. Non-uniformity (a) of distribution of working heads in CP η_z and (b) of water inflow into CP η_q depending on Reynolds criterion ReD for different values of jet inflow angle β: 0° – (1); 45° – (2); 90° – (3); 135° – (4); 180° – (5)

The non-uniformity η_z of working heads and the non-uniformity η_q of inflow essentially depend on the angle β of jet inflow. To confirm this, we present comparison of value of η_z and of η_q for close values of the criterion Re_D (Re_D = 5211–6354). The effectiveness of regulation of non-uniformity in work of CP by means of variation of the angle β has been calculation in comparison with that of CP in which all inlet nozzle had been installed at the angle $\beta = 90^o$ (Table).

From Table 1, it can be seen that the installation of inlet nozzles at the angle β of jet inflow equal to 0°; 45°; 135° and 180° ensure reduction of non-uniformity of distribution of working heads Z_i and that of water inflow q_i through individual nozzles into CP along the path as compared to a CP whose all nozzles had been installed at the angle $\beta = 90^o$. The smallest non-uniformity q_i of water inflow to the CP was obtained at $\beta = 135^o$. The obtained results are consistent with the previously obtained data for collector-pipelines for which $d/D = 0.398$.

Table 1. Effectiveness of regulation of non-uniformity in work of pressure CP by variation of angle β in comparison with a CP whose all inlet nozzles are at angle $\beta = 90^o$

Non-uniformity of distribution	Angle β			
	Relative change, $\psi = \frac{\eta_{90^o} - \eta_{i^o}}{\eta_{90^o}} \cdot 100\%$			
	0°	45°	135°	180°
Working heads η_z	13.51	44.49	31.81	9.88
Inflow along the path η_q	6.71	8.48	9.07	5.05

5 Conclusion

For different values of Reynolds criterion $\mathrm{Re_D}$, five pressure collector-pipeline (CPs), each of them was of its inner diameter D = 11.28 mm and had inlet nozzles of d = 4.83 mm (d/D) = 0,428, have been experimentally investigated. In each CP, the nozzles had been installed with the same angle β between the direction of water stream flow in the CP and the direction of the jet inflowing through inlet nozzles into it. In the five different CPs, the angles β were of the values: 0°; 45°; 90°; 135°; 180°, respectively. The criterion $\mathrm{Re_D}$ was calculated for the range line of the CP after the last nozzle. The value of Re_D was varied within the range of 5211 throng 28321. The influence of the value of the angle β on the distribution of head and on that of water inflow into CP was established by means of comparison with the same characteristics for a CP with the angle $\beta = 90^o$. By means of regulation of the angle β, the variation of working heads along a CP up to 44.49% (for CP with $\beta = 45^o$) and that of water inflow up to 9.07% (for CP with $\beta = 135^o$) have been achieved. The non-uniformity of the distribution of working heads and that of inflow along the path decreases with the increase in the value of Reynolds criterion Re_D.

References

1. Povkh, I.L., Chernyuk, V.V.: Experimental investigation of the influence of polyacrylamide admixtures on the resistance of diffusers. Eng.-Phys. J. **51**(3), 357–361 (1986)
2. Povkh, I.L., Chernyuk, V.V.: Resistance of confusors in turbulent flow of water with additives of poliakrilamide. Eng.-Phys. J. **57**(5), 709–712 (1989)
3. Abramov, N.N.: Vodosnabzhenie: Uchebnik dlia vuzov. Moskva, Stroiizdat, p. 440 (1982)
4. Pohrebennyk, V., Dzhumelia, E., Korostynska, O., Mason, A., Cygnar, M.: X-Ray fluorescent method of heavy metals detection in soils of mining and chemical enterprises. In: Proceedings of 9th International Conference on Developments in eSystems Engineering, DeSE 2016, no. 7930667, pp. 323–328 (2017)
5. Styskal, O., Ishchenko, V., Petruk, R., Pohrebennyk, V., Kochanek, A.: Assessment of chlorinated water impact on phytoplankton. In: 16th International Multidisciplinary Scientific Geoconference, SGEM2016, Vienna, AUSTRIA, pp. 373–380 (2016)

6. Mitryasova, O., Pohrebennyk, V., Cygnar, M., Sopilnyak, I.: Environmental natural water quality assessment by method of correlation analysis. In: International Multidisciplinary Scientific GeoConference Surveying Geology and Mining Ecology Management, SGEM, vol. 2, pp. 317–324 (2016)
7. Pohrebennyk, V., Petryk, A.: The degree of pollution with heavy metals of fallow soils in rural administrative units of Psary and Płoki in Poland. In: International Multidisciplinary Scientific GeoConference Surveying Geology and Mining Ecology Management, SGEM, vol. 17, no. 52, pp. 967–974 (2017)
8. Karpinski, M., Pohrebennyk, V., Bernatska, N., Ganczarczyk, J., Shevchenko, O.: Simulation of artificial neural networks for assessing the ecological state of surface water. In: International Multidisciplinary Scientific GeoConference Surveying Geology and Mining Ecology Management, SGEM, vol. 18, no. 2.1, pp. 693–700 (2018)
9. Smyslov, V.V.: K raschetu sbornykh truboprovodov. Gidravlika i gidrotehnika, Nauchalno-tehnschnyi sbornik. Vypusk 30. Kiev, Tehnika, pp. 60–65 (1980)
10. Kravchuk, A.M.: Hidravlika zminnoi masy napirnykh truboprovodiv tekhnichnykh system: avtoreferat dysertacii na zdobuttia naukovoho stupenia doktora tekhnichnykh nauk, spetsialnist 05.23.16. Kyiv, 35 p. (2004)
11. Voloshchuk, V.A.: Doslidzhennia hidravlichnykh oporiv i hidravlichni rozrakhunky truboprovodiv z dyskretno zminnymy vytratamy uzdovzh potoku: dysertacia kandydata tekhnichnych nauk 05.23.16. Rivne, Rivnenskyi derzh. tekhn. un-t. 217 p. (2001)
12. Patent na vynakhid № 115840 Ukraina, MPK G05D 7/00, F17D 1/02, F17D 1/08. Sposib rehuliuvannia shliakhovoi vytraty ridyny v truboprovodakh z nasadkamy. Cherniuk, V.V., Ivaniv, V.V. (Ukraina); Natsionalnyi universytet “Lvivska politekhnika”. Biuleten №. 24, Ukrainskyi instytut intelektualnoi vlasnosti, 5 p. (2017)
13. Cherniuk, V., Ivaniv, V.: Influence of values of angle of jet-joining on non-uniformity of water inflow along the path in pressure collector-pipeline. In: 10th International Conference “Environmental Engineering”. Vilnius Gediminas Technical University Lithuania, 27–28 April 2017, 7 p. (2017). eISSN 2029-7092/eISBN 978-609-476-044-0; Article ID: enviro.2017.073. https://doi.org/10.3846/enviro.2017.073
14. Ivaniv, V., Cherniuk, V.: Influence of jet-to-main stream turning angle in fluid flow from cylindrical nozzle of collector-pipeline on flow coefficient. Czasopismo Inżynierii lądowej, środowiska i architektury. J. Civ. Eng. Environ. Archit. **XXXIII**(63), 229–238 (2016). (p-ISSN 2300-5130), (e-ISSN 2300-8903). JCEEA
15. Smyslov, V.V., Ezerskiy, N.O.: Gidravlicheskiy raschet perforirovannyh tsilindricheskih truboprovodov s razdachei rashoda. Gidravlika i gidrotekhnika. Vypusk **30**, 52–59 (1980)

Examining the Interdependence of the Various Parameters of Indoor Air

Peter Kapalo[1], Maria Sulewska[2], and Mariusz Adamski[2(✉)]

[1] Technical University of Kosice, 042 00 Kosice, Slovakia
[2] Bialystok University of Technology, 15-351 Bialystok, Poland
mariusz.adamski@pb.edu.pl

Abstract. In this paper, the partial results of the investigation of the interrelationship of individual parameters of indoor air - temperature, relative humidity and CO_2 concentration, obtained during the experimental measurement, are presented. The experimental measurements have been performed in a selected room, where three separate lessons with three different groups of people have been conducted. During these measurements, the processes of temperature, relative humidity and CO_2 concentration have been recorded. Also, people involved in this investigation have made their own evaluation. They, have been asked for opinion about the internal air quality at the beginning and at the end of the experimental measurement.

Keywords: Experimental measurement · Temperature · Relative humidity · Concentration CO_2 · Questionnaire · Interdependence measurement

1 Introduction

The residents of a big city average spend more than 80% of their life indoors.

In paper [1] were carried out measurements of indoor air temperature, relative humidity and CO_2 concentrations as well as were determined CO_2 production by students and teacher during various physical activities. The highest increase of CO_2 was recorded during harder physical activity (running on the spot, squats, right and left side lunges, and rotating of the hips). Regarding CO_2 production by the respondents, it can be seen [2] that it visibly increased with increasing physical activity. The work [3] contains an overview of the results of indoor air quality tests in Poland and ranges of CO_2 concentration in rooms. The papers [4, 5] present a simplified model of CO_2 concentration based on experimental research in classrooms equipped with stack ventilation systems. The test was conducted in six classrooms in the building of the Faculty of Civil and Environmental Engineering of the Białystok University of Technology in north-east Poland. In all classrooms, a linear increase in the CO_2 concentration during the classes was observed. According to paper [6] related to school buildings located in two different climates: Białystok (Poland) and Belmez/Córdoba (Spain) the CO_2 concentration in first 45 min met requirements of regulations, but with medium occupation of places in classrooms. In paper [7] were first time presented the results of CO_2 concentration measurements in a classroom in Ukraine. The works [7, 8]

Z. Blikharskyy (Ed.): EcoComfort 2020, LNCE 100, pp. 150–157, 2021.
https://doi.org/10.1007/978-3-030-57340-9_19

presented some of the results of a field study about the perception of air quality in the selected classroom in a high school in Ukraine.

In the work [9] the method for intensity ventilation determining of the indoor premises was developed on the basis of the measured values of carbon dioxide which was verified also by another experimental measurements. The resulting values of ventilation intensity rate obtained by calculation from the measured values of carbon dioxide were compared with the results of calculations executed according to the laws and standards, current in Slovakia.

Results of experimental investigations of air supply into the room by swirl jets are presented in paper [10]. It was established that an increase in the angle results in increase of the attenuation coefficients and in decrease of the resistance coefficient of two-flow air distributor (TFAD). The optimum angle of the plates is determined considering aerodynamic and energy aspects.

The effect of temperature, humidity, air velocity and CO_2 concentration [11] were tested in regard to the human behaviour in indoor environment with natural ventilation, i.e. without a ventilation device, where ventilation is forced or an air conditioning.

A methodology for determination the volumetric air flow rate [12] is presented on the basis of the results obtained from experimental measurements of carbon dioxide concentrations, which were carried out inside an apartment house lived by a standard family. The required fresh air flow rate was determined in an occupied room, based on carbon dioxide concentration measurements, in order to maintain a comfortable level of indoor air quality.

When indoor levels of CO_2 are high, people tend to be less satisfied with indoor air quality [2, 10], report more acute health symptoms, work slightly slower, and are more often absent from work or school.

The carbon dioxide production was examined [12] separately for men and women, for persons of different mass, for persons of different ages and also was analysed the carbon dioxide production during a sedentary and physical activities. In parallel with the production of carbon dioxide is presented the monitoring of the human pulse and blood pressure. All these parameters are monitored together with relative humidity and indoor air temperature.

Measurements of indoor air temperature [13], relative humidity and carbon dioxide concentrations were carried out as well as CO_2 production by students and teacher during various physical activities were determined. Two classrooms of Technical University of Kosice were selected for the research. Results of objective measurements confirmed strong correlation between CO_2 concentration and occupancy for all measurements. Recommended value for indoor CO_2 level according to [2, 14] (1,000 ppm) was exceeding in 60.9%. Results of this study showed the insufficient ventilation intensity in classrooms as well as obvious rise of CO_2 concentration during the exams.

Studies performed in educational buildings concluded [13] that indoor environmental quality of classrooms is often inappropriate and may be linked with insufficient level of ventilation [15–17].

From 2014 to 2017, through the cold season, 37 experimental measurements of carbon dioxide concentration were done inside a selected university classroom [18] during the same type of activity, namely written examination. The relationship between the body mass of the students and the intensity of carbon dioxide releases inside the experimental classroom was determined by using statistical analysis (least-square estimation of the regression line). The resulting equation may be used to determine in an easier way the required outdoor air volume flow rate inside a university classroom.

According to the medical experts from human health organizations inadequate temperature in the rooms where people are staying can lead to deterioration of health and well-being [19]. On the basis of statistical analysis of the test results [20], it was found that in the analysed age group 10 and 15 students present in the room, the perceived temperature declared in the room is not affected by the respondent's gender. The research [21] was carried out in a didactic teaching room located in the building of the Faculty of Civil and Environmental Engineering in Poland. Tests on the temperature were carried out simultaneously with questionnaire surveys. The conclusion is that the temperature sensations of young (and probably healthy) people who do not do physical work are similar, regardless of gender.

The thermal comfort assessments of men and women were similar and overlapped. The results of this study confirm that under the same thermal conditions about 85% of respondents assess thermal comfort as good, and about 15% of respondents assess thermal comfort as bad. The test results presented [21] are similar to the results of tests carried out by other authors in other climatic conditions.

Mechanical ventilation systems in Central Europe require the use of ventilation heat recovery systems. Counter current recuperators [22, 23] are more resistant to frost than cross flow recuperators. Spiral recuperators with the longitudinal counter current flow [23, 24] refunds the capital expenditure within one or two years in Polish climatic conditions. Investigation results [25] of condensation phenomena and frost problems in the ventilation heat recuperators are presented.

2 Materials and Methods

In order to investigate the interdependence of indoor air parameters: temperature, relative humidity and CO_2 concentration, an experimental measurement was performed in a selected room. A total of 3 experimental series of measurements were performed. All measurements took place in the same room but at a different time and with different people. Prior to each experimental measurement, the room was ventilated by natural ventilation through windows and doors. During the stay of people in the room, subjective evaluations of indoor air quality were also performed in the form of questionnaires. After the measurement, an analysis of the interaction of individual parameters of indoor air and perceived air quality was performed.

2.1 Characteristics of the Selected Room

The experimental room is located on the fourth floor of a 7-story building. The windows in the room are facing north, so direct sunlight does not affect the course of experimental measurements during the day. The room is 5.9 m long, 6.3 m wide and 3.3 m hight. The room volume is about 123 m^3. When measured in March 2019, the average outdoor air temperature was 9 °C.

2.2 The Characteristic of Occupants

Experimental measurements were divided into periods. The first period lasted from 10:05 to 11:05 and was attended by 5 people (100% men) in average age 21 years, average weight 77 kg and average height 180 cm. The second period lasted from 11:15 to 12:45 and was attended by 16 people (81% men; 19% women) in average age 20 years, average weight 71 kg and average height 178 cm. The third period lasted from 13:15 to 14:55 and was attended by 10 people (30% men; 70% women) in average age 23 years, average weight 64 kg and average height 167 cm.

3 Results and Discussions

3.1 Measured Indoor Air Parameters

In the first period of experimental measurement, when the room temperature was 5 people, the air temperature increased from 21.4 °C to 21.7 °C. The temperature increased by 0.3 °C, which is an increase of 1.4%. Relative humidity increased from 37.2% to 41.6%. The relative humidity increased by 4.4%, which is an increase of 11.83%. The CO_2 concentration increased from 1.175 ppm to 2.038 ppm. The increase in CO_2 concentration was 863 ppm, which is an increase of 73.45%.

In the second period 16 people were in the room. The air temperature increased from 21.0 °C to 22.6 °C. The temperature increased by 1.6 °C, which is an relative increase by 7.62%. Relative humidity increased from 34.0% to 48.2%. The relative humidity increased by 14.2%, which is an increase of 41.76%.

CO_2 concentration increased from 1.284 ppm to 3.649 ppm. The increase in CO_2 concentration was 2,365 ppm, which is an increase of 184.19%.

In the third period, there were 10 people in the room. The air temperature increased from 21.3 °C to 21.8 °C. The temperature increased by 0.5 °C, which is an increase of 2.35%. Relative humidity increased from 32.6% to 44.1%. The relative humidity increased by 11.5%, which is an increase of 35.28%. The CO_2 concentration increased from 1.024 ppm to 2.233 ppm. The increase in CO_2 concentration was 1.219 ppm, which is an increase of 119.04%.

3.2 Results of Subjective Evaluation

All people in a room participated in the experimental measurements are tested in a subjective assessment of indoor air quality. At the beginning of the stay in the room, each person filled out a questionnaire in which he commented on the temperature in the room, the perception of the smell and whether they are generally satisfied with the quality of the air in the room. To assess the impact of the indoor air temperature on good mood in the room are available five options to choose one's option from: Cool, Moderate cool; Acceptable temperature; Moderate hot and Hot. To assess the impact of the state of air in room are available four options: Fresh; Acceptable; Moderate smell and Stuffy air. At the end of their stay people were surveyed again with the same questions in new questionnaires in order to compare changes in comfort parameters in the room.

Surveys show that in the first measurement period 40% of people noticed an increase in temperature and 20% of people noticed a deterioration in air quality. In the second measurement period 37.5% of people stated a rise in temperature and 62.5% of people felt a decrease in air quality. During the third measurement period, 60% of people noticed an increase in temperature and 90% of people noticed a deterioration of air quality. The third group consisting of 70% women responded the most to the increase in temperature and smell in the room.

3.3 Examining the Interdependence of the Parameters of Indoor Air

The interdependence of indoor air parameters were investigated using statistics. The subject of this study are the courses of measured air parameters in the room: temperature, relative humidity and CO_2 concentration. According to physics, increasing temperature lowers our relative humidity and also reduces volumetric CO_2 concentration slightly. This means that the relative humidity and CO_2 concentration recorded by measuring instruments are lower than their actual values.

All recorded measurements in particular periods of the experiment show that the values of temperature, relative humidity and CO_2 concentration change due to the people staying in a room. Figure 1 presents the relationship between the values of relative humidity and the values of air temperature constructed on the basis of measured data during the stay of people in all three periods of experimental measurements.

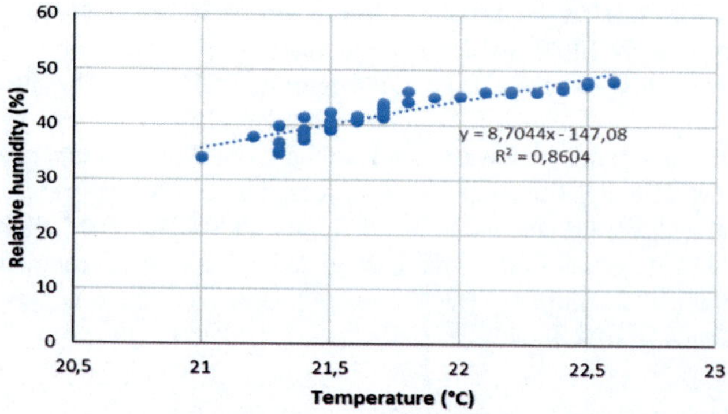

Fig. 1. The relative humidity versus temperature – all three periods.

Figure 2 documents the dependence of the CO_2 concentration values on the air temperature values; Fig. 3 presents the dependence of the CO_2 concentration values on the relative humidity values of the indoor air based on the measured data during the stay of people in all three periods of the experimental measurements.

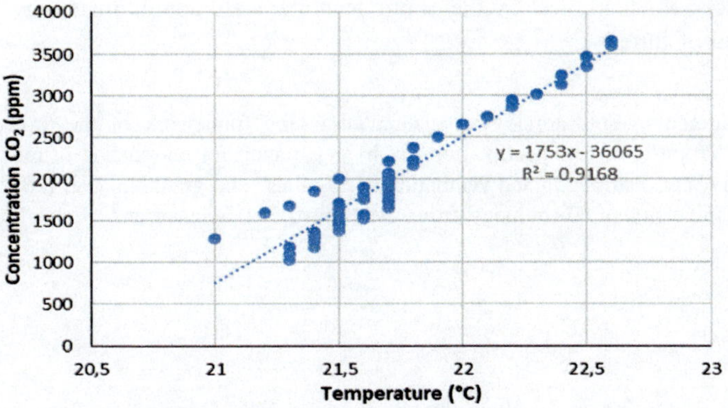

Fig. 2. The concentration CO_2 versus temperature – all periods.

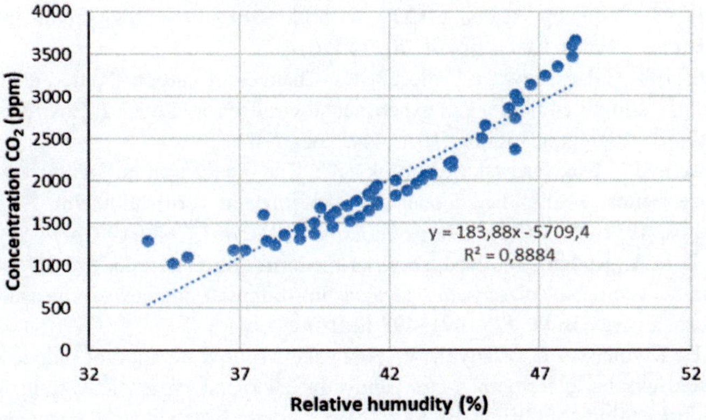

Fig. 3. The concentration CO_2 versus relative humidity – all periods.

4 Conclusion

Nowadays in modern buildings, where air exchange is provided by a demand-controlled ventilation (variable air volume systems), the CO_2 concentration sensor is used as a part of the control. The CO_2 concentration sensors significantly increase the cost of a ventilation system. This expensive sensor will have to be replaced by a more affordable sensor, which will also reliably perform the required function.

1. In individual measurement series, which lasted up to 90 min, no stable CO_2 concentration values were observed. The CO_2 concentration increased.
2. The numerical values of temperature and humidity changed less than the concentration of carbon dioxide. Therefore, the proportionality coefficients in the equations for CO_2 concentration have significant values:
   a. Around 1753 in an approximation CO_2 equation depending on temperature,
   b. Approximately 184 in CO_2 approximation equation depending on air humidity.

3. The CO_2 concentration in the room is quite well approximated by the linear function of time.

Acknowledgments. This article was elaborated in the framework of the projects: VEGA 1/0697/17 "Proposal of technical platform of hygienic audit for elimination of microbiological pollution in water distribution and ventilation in hospitals" and grant No. WZ/WBiIS/1/2019 of Bialystok University of Technology from the Ministry of Science and Higher Education of Poland.

References

1. Kapalo, P., Meciarova, L., Vilcekova, S., Burdova, E.K., Domnita, F., Bacotiu, C., Peterfi, K.E.: Investigation of CO_2 production depending on physical activity of students. Int. J. Environ. Health Res. **29**(1), 31–44 (2019). https://doi.org/10.1080/09603123.2018.1506570
2. Adamski, M.: Pomiary stężenia CO_2 w pomieszczeniu mieszkalnym w zabudowie jednorodzinnej. Rynek Istalacyjny **4**, 70–74 (2014)
3. Teleszewski, T., Gładyszewska-Fiedoruk, K.: Changes of carbon dioxide concentration in class-rooms - simplified model and experimental verification. Polish J. Environ. Stud. **27**(5), 2397–2403 (2018). https://doi.org/10.15244/pjoes/77074
4. Teleszewski, T., Gładyszewska-Fiedoruk, K.: The concentration of carbon dioxide in conference rooms: a simplified model and experimental verification. Int. J. Environ. Sci. Technol. **16**, 8031–8040 (2019). https://doi.org/10.1007/s13762-019-02412-5
5. Krawczyk, D.A., Rodero, A., Gładyszewska-Fiedoruk, K., Gajewski, A.: CO_2 concentration in naturally ventilated classrooms located in different climates – measurements and simulations. Energy Build **129**, 491–498 (2016)
6. Kapalo, P., Klymenko, H., Zhelykh, V., Adamski, M.: Investigation of indoor air quality in the selected ukraine classroom – case study. In: Blikharskyy, Z., Koszelnik, P., Mesaros, P. (eds.) Proceedings of CEE 2019. CEE 2019. Lecture Notes in Civil Engineering, vol 47. Springer (2020). https://doi.org/10.1007/978-3-030-27011-7_21
7. Kapalo, P., Voznyak, O., Klymenko, H., Zhelykh, V., Adamski, M.: Perception of air quality in the selected classroom, construction of optimized energy potential. Budownictwo o zoptymalizowanym potencjale energetycznym **8**(2), 77–84 (2019). https://doi.org/10.17512/bozpe.2019.2.09
8. Kapalo, P., Vilcekova, S., Voznyak, O.: Using experimental measurements the concentrations of carbon dioxide for determining the intensity of ventilation in the rooms. Chem. Eng. Trans. **39**, 1789–1794 (2014)
9. Voznyak, O., Korbut, V., Davydenko, B., Sukholova, I.: Air distribution efficiency in a room by a two-flow device. In: Proceedings of CEE. Advances in Resourse-saving Technologies and Materials in Civil and Environmental Engineering, vol. 47, pp. 526–533. Springer (2020)
10. Kapalo, P., Sedláková, A., Košicanová, D., Voznyak, O., Lojkovics, J., Siroczki, P.: Effect of ventilation on indoor environmental quality in buildings. In: The 9th International Conference "Environmental Engineering", Vilnius, Lithuania. Selected Papers, eISSN 2029-7092/eISBN 978-609-457-640-9 Section: Energy for Buildings (2014)

11. Kapalo, P., Vilceková, S., Domnita, F., Bacotiu, C., Voznyak, O.: Determining the ventilation rate inside an apartment house on the basis of measured carbon dioxide concentrations - case study. In: The 10th International Conference on "Environmental Engineering", Vilnius, Lithuania, Selected Papers, pp. 30–35 (2017)
12. Kapalo, P., Domniţa, F., Bacoţiu, C., Spodyniuk, N.: The impact of carbon dioxide concentration on the human health - case study. J. Appl. Eng. Sci. **8**(21), 61–66 (2018)
13. Kapalo, P., Mečiarová, L., Vilčeková, S., Krídlová Burdová, E., Domnita, F., Bacotiu, C., Péterfi, K.E.: Investigation of CO_2 production depending on physical activity of students. Int. J. Environ. Health Res. **29**(1), 31–44 (2018)
14. Von Pettenkofer, M.: Über den Luftwechsel in Wohngebäuden. Der J.G. Cotta'schen Buchhandlung, München (1858)
15. Daisey, J.M., Angell, W.J., Apte, M.G.: Indoor air quality, ventilation and health symptoms in schools: an analysis of existing information. Indoor Air **13**(1), 53–64 (2003)
16. Dorizas, P.V., Assimakopoulos, M.N., Helmis, C., Santamouris, M.: Analysis of the indoor air quality in Greek primary schools. https://repository.edulll.gr/edulll/retrieve/11398/3584_1.60_%CE%94%CE%97%
17. Goyal, R., Khare, M.: Indoor–outdoor concentrations of RSPM in classroom of a naturally ventilated school building near an urban traffic roadway. Atmos. Environ. **43**(38), 6026–6038 (2009)
18. Kapalo, P., Domnita, F., Bacotiu, C., Podolak, M.: The influence of occupants' body mass on carbon dioxide mass flow rate inside a university classroom – case study. Int. J. Environ. Health Res. **28**(4), 432–447 (2018)
19. Dziubanek, G., Spychała, A., Marchwińska-Wyrwał, E., Rusin, M., Hajok, I., Ćwieląg-Drabek, M., Piekut, A.: Long-term exposure to urban air pollution and the relationship with life expectancy in cohort of 3.5 million people in Silesia. Sci. Total Environ. **580**, 1–8 (2017)
20. Sulewska, M.J., Gładyszewska-Fiedoruk, K., Sztulc, P.: Analysis of the results of empirical research and surveys of perceived indoor temperature depending on gender and seasons. Environ. Sci. Pollut. Res. **25**(31), 31205–31218 (2018)
21. Gładyszewska-Fiedoruk, K., Sulewska, M.J.: Thermal comfort evaluation using linear discriminant analysis (LDA) and artificial neural networks. Energies **13**(3), 538 (2020)
22. Adamski, M.: Optimization of counterflow plate type ventilation heat exchangers. In: Heat transfer and renewable sources of energy, Szczecin, pp. 31–38 (2002)
23. Adamski, M.: Longitudinal spiral recuperators in ventilation systems of healthy buildings. In: Healthy Buildings: Creating a Healthy Indoor Environment for People, HB 2006, Lisboa, Portugal, 4 June 2006 through 8 June 2006, Code 94622, vol. 4, pp. 341–344 (2006)
24. Adamski, M.: Mini longitudinal flow spiral recuperator. In: Healthy Buildings Europe 2017, HB 2017, 2 July 2017 through 5 July 2017, Code 138756. Lublin University of Technology Campus Lublin, Poland (2017)
25. Adamski, M., Kiszkiel, P.: Condensation phenomena and frost problems in the air heat recuperators. In: 101 EUROTHERM Seminar - International Conference on Transport Phenomena in Multiphase Systems, HEAT 2014, Krakow, Poland, 30 June 2014 through 3 July 2014, Code 109487 (2014)

The Analysis of Heat Consumption in the Selected City

Peter Kapalo[1] and Mariusz Adamski[2(✉)]

[1] Technical University of Kosice, 042 00 Kosice, Slovakia
[2] Bialystok University of Technology, 15-351 Bialystok, Poland
mariusz.adamski@pb.edu.pl

Abstract. The higher standard of living, the longer time of stay in buildings, the higher requirements for the creation of the internal environment of buildings using technical equipment of buildings and the climate change on the planet and the climate change on the planet are causing an increase in energy consumption in buildings. One of the reduction options is to thermal insulation buildings with an optimized thickness of thermal insulation. Before measures is need to take consider the building is used, the properties of building materials, the climatic conditions of country, the rate of use of renewable energy sources, the standard of modernization of the system of remote energy supply of buildings, the standard of automatic control of technological equipment points and mainly return on investment, especially in older buildings. So far, the most effective measure in order to reduce the energy intensity of the whole system is thermal insulation of buildings. In city settlements, it is possible to insulate individual buildings in successive steps, which allows the input investment costs to be decompose over several years. In this article is documented the results of energy consumption of buildings in a selected city, where there is a gradually thermal insulation of buildings. The put emphasis is on for heat consumption for heating and hot water.

Keywords: Building · Thermal insulation · Heat · Heating · Domestic hot water · Water consumption

1 Introduction

Energy consumption has dramatically increased in buildings [1] over the past decade due to population growth, more time spent indoors, increased demand for building functions and indoor environmental quality, and global climate change. Building energy use currently accounts for over 40% of total primary energy consumption in the U.S. and E.U. On the other hand, the climate changes significantly impact building energy performance [1], particularly in space heating and cooling. Improvements on building envelope and ventilation play an important role in reducing space heating and cooling consumption levels. The interaction between intermittent occupancy and thermal mass has a significant impact on overall energy use [2] too. In the study [3] has been examined the impact of climate change on thermal comfort conditions and on energy demand for heating and cooling in homes in selected cities in Brazil. The results

Z. Blikharskyy (Ed.): EcoComfort 2020, LNCE 100, pp. 158–165, 2021.
https://doi.org/10.1007/978-3-030-57340-9_20

of the study show that annual energy demand will increase between 56%–112% in 2050. The way buildings are used significantly impacts on energy consumption. The research [4] has been concerned on the processes taking place in the building envelope. The temperature, relative humidity and heat flux have been measured at different points of the renovated facade. Older buildings in cities are responsible for a significant use of energy, which results in a negative impact on the environment. The research work [5] seeks to find the optimal scenario for the renovation of institutional buildings with respect to energy consumption. According to [6] one of the easy and effective way to conserve the energy is through residential building insulation.

In the study [7] have been worked a review of the main commercialized insulation materials (conventional, alternative and advanced) for the building sector through a holistic and multidisciplinary approach, considering thermal properties, acoustic properties, reaction to fire and water vapor resistance. The environmental issues were also taken into account by means of Life Cycle Assessment approach.

A literature review on determining the optimum thickness of the thermal insulation material in a building envelope and its effect on energy consumption has been carried out in study [8]. The study [9] provides a procedure for determining the optimal thickness of thermal insulation to be applied to the sheath. Using the optimal insulation thickness [10] energy consumption could be reduced by 46.6% and CO_2 emissions by 41.53%. Similarly CO_2 emission [11] could be reduced by 50% due to use the optimal insulation thickness. The effects of a thermal insulation layer in the outer walls [12] on annual cooling loads and the annual maximum demand for cooling have been presented. The quality of heat distribution in the city has a significant effect on the heat consumption. According to [13], precise prediction of heat demand is crucial for optimising district heating systems. Energy consumption patterns represent a key parameter in developing a good mathematical model to predict heat demand.

The article [14] and [15] presents a description of contemporary developments and findings related to the different elements needed in future, that involves meeting the challenge of more energy efficient buildings as well as the integration of district heating into a future smart energy system based on renewable energy sources.

The paper [16] provides a perspective on the development of future district heating systems and technologies and their role in future smart energy systems. Focus is on the important role of the next generation of district heating and cooling technologies. The impact of thermal insulation of buildings in a selected city on energy consumption and population reaction has been described.

2 Aim

The aim of the research is to analyze the heat consumption in Košice in the Slovak Republic, where individual buildings are gradually isolated thermally year by year; subsequent reactions of consumers and heat producers, observation the course of heat consumption in residential buildings supplied with heat by the central heat supply system. The price of heat rises year by year.

3 Materials and Methods

The development of the population in the city, development of apartment construction and their thermal insulation, heat and hot water consumption in buildings were analysed in this research. Heat and hot water are supplied by a central supply system.

The analysed data in this article related to heat supply were taken from publicly available information provided by the annual reports from 2000 to 2018 of the company TEHO [17], which provides heat and hot water to households in the city. Data on the number of inhabitants and the number of dwellings in the city are taken from the official website of the city of Košice [18] and from the statistical office [19].

4 Results and Discussions

4.1 The Development of the Population

The subject of the research is the city which has approximately 234,000 inhabitants. Nowadays most of the population lives in apartment buildings supplied by heat from the heating plant through a pipe distribution system. Number of buildings and building construction system are closely related with the history of the city and the development of the population of the city. The largest increase in population in the city occurred between 1961 and 1980 [18, 20]. This increase was due to the development of heavy industry in the city. During this period, ironworks and manufacturing plants were built. The smallest increase in population was in the years 1991 to 2018, when several important production plants ceased or reduced production.

4.2 Increase in the Number of Flats

Most of the buildings were built in 1961–1990 in town. During this period, it has been the largest increase in the population of the city. In 1991–2000 and next years, the increase in the number of buildings constructed has been minimal and during this period the existing residential buildings have been renovated to reduce the energy consumption of buildings [21]. Until 1950, local heating system for solid fuel (coal and wood) have been mainly used for heating of residential buildings. In the years 1950–1961 have been started building residential buildings with boiler rooms in the basements of houses or central boiler rooms supplying heat and hot water to residential nearby located buildings. In 1967 the heat supply system has begun to change. The district boiler rooms in the individual housing estates have been liquidated and heat was supplied to the entire city from one central heat source. Nowadays the heat producer for the city also produces electricity by cogeneration units.

4.3 Average Monthly Exterior Air Temperatures in the Heating Seasons

The maximum and minimum average temperature values are given in weather-atlas [22]. Values of average monthly exterior air temperatures in the heating seasons are presented in Fig. 1. According to the assumptions [23], the values are assumed for the degree-days calculations as follows: The average temperature inside t is: 19.0 °C weighted average of internal calculated temperatures (according to room volume); Reference temperature t_{em}: 13.0 °C according to decree No. 194/2007 (average daily outside air temperature for start and end of heat supply).

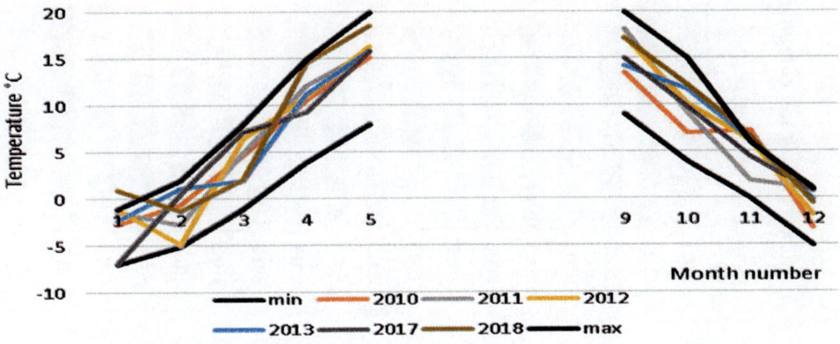

Fig. 1. The average monthly exterior air temperatures in city [22].

4.4 Energy Demand Analysis

The numerical data [19] assumed for the calculations are illustrated in the Figs. 2 and 3. The analyses covered the years 2010–2018 and are carried out based on the Statistica program.

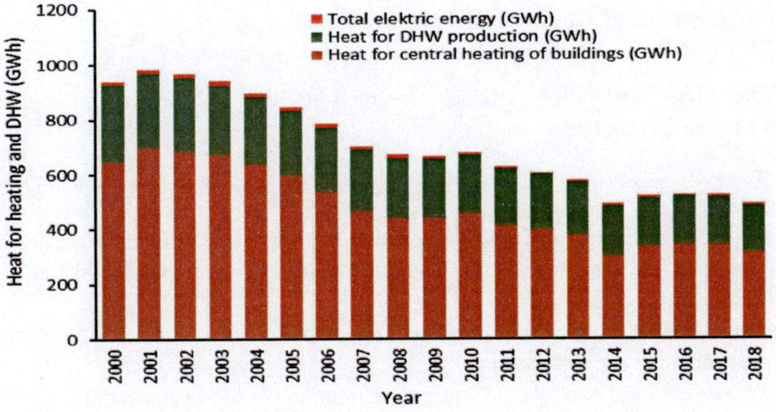

Fig. 2. Heat demand for heating, hot water and electricity demand [19].

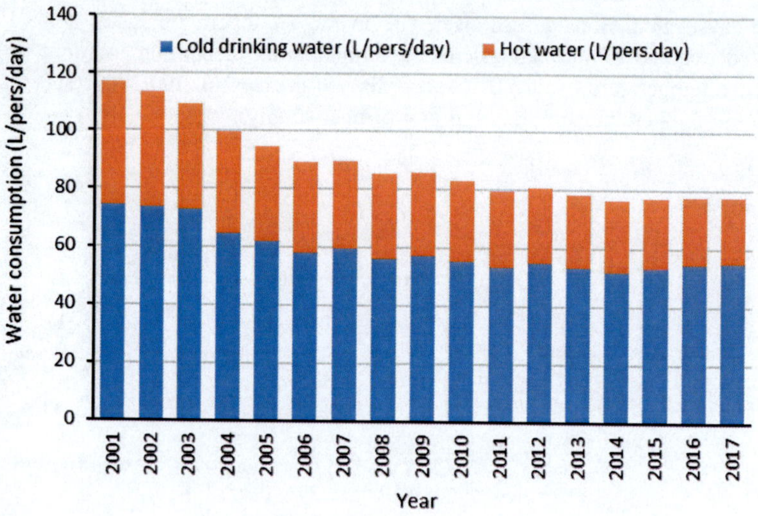

Fig. 3. Daily hot and cold water consumption per capita [19].

List of independent variables taken into account for individual years:

- Year
- DD – number of degree days
- heating [GWh] – heat for central heating of buildings
- DHW [GWh] - heat for domestic hot water
- electricity [GWh] – total electric energy
- DHW [m^3] - domestic hot water consumption
- CDW [L/pers/day] – daily cold water consumption per capita
- HW [L/pers/day] – daily hot water consumption per capita

Correlations between the variables presented in Table 1 show that in the analysed period of time 2010–2018:

- there is no statistically significant relationship between the year and number of degree days,
- there is a significant relationships between year and heat for domestic hot water [GWh] and [m^3] and electricity. These variables do not depend statistically significantly on degree days,
- heating [GWh] – heat for central heating of buildings is statistically significant depend on year and degree days,
- CDW [L/pers/day] – daily cold water consumption per capita statistically significant depend on degree days only and does not depend statistically significant on other variables,
- HW [L/pers/day] – daily hot water consumption per capita does not depend statistically significant on degree days (Fig. 4).

Table 1. Correlations between variables.

	Year	DD	Heating [GWh]	DHW [GWh]	Electricity [GWh]	DHW [m3]	CDW [L/pers/day]	HW [L/pers/day]
Year	1,00	−0,27	**−0,83**	**−0,98**	**−0,85**	**−1,00**	−0,09	**−0,98**
DD	−0,27	1,00	**0,76**	0,43	0,55	0,31	**0,90**	0,35
Heating [GWh]	**−0,83**	**0,76**	1,00	**0,90**	**0,90**	**0,85**	0,58	**0,87**
DHW [GWh]	**−0,98**	0,43	0,90	1,00	**0,88**	**0,99**	0,24	**0,98**
Electricity [GWh]	**−0,85**	0,55	**0,90**	**0,88**	1,00	**0,87**	0,44	**0,92**
DHW [m^3]	**−1,00**	0,31	**0,85**	**0,99**	**0,87**	1,00	0,13	**0,99**
CDW [L/pers/day]	−0,09	**0,90**	0,58	0,24	0,44	0,13	1,00	0,19
HW [L/pers/day]	**−0,98**	0,35	**0,87**	**0,98**	**0,92**	**0,99**	0,19	1,00
CDW + HW [L/pers/day]	**−0,80**	**0,72**	**0,96**	**0,88**	**0,93**	**0,83**	0,65	**0,87**

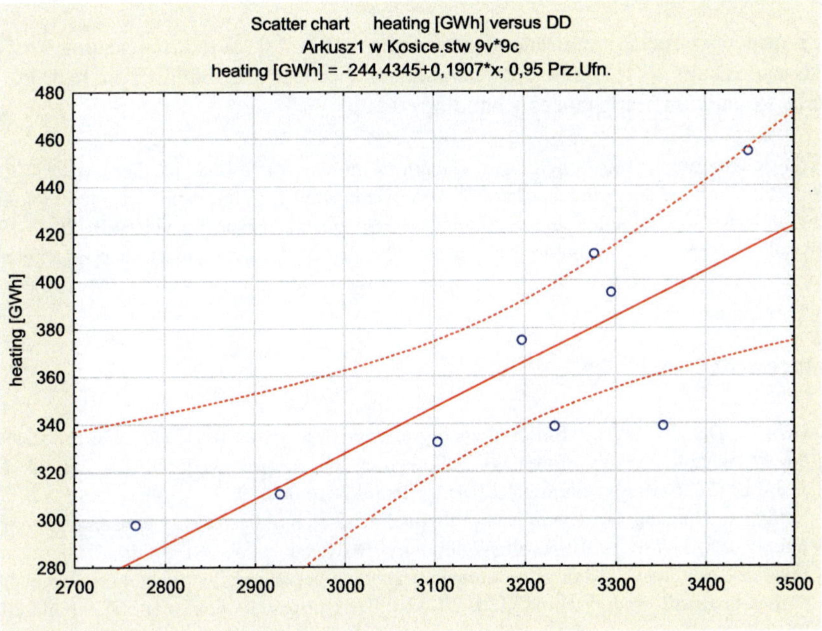

Fig. 4. Heat for central heating of buildings versus degree days.

Figure 5 presents values of heat for central heating and sum of the heat for central heating and heat for domestic hot water for 1-degree day over the years.

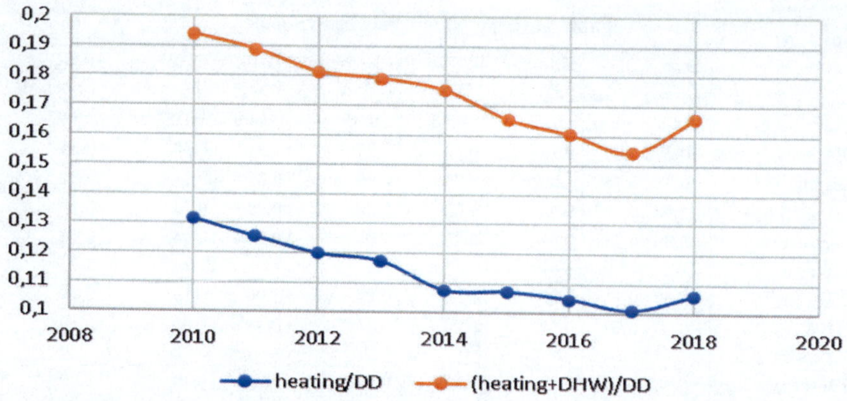

Fig. 5. Heat for central heating and sum of the heat for central heating and heat for domestic hot water for 1-degree day over the years.

5 Conclusion

Such successive reduction of energy requirement of buildings reduces heating network loads and results to less efficient operation of the heat distribution system from the source to individual appliances - buildings.

Acknowledgments. This article was elaborated in the framework of the project VEGA 1/0697/17 "Proposal of technical platform of hygienic audit for elimination of microbiological pollution in water distribution and ventilation in hospitals' and grant No. WZ/WBiIS/1/2019 of Bialystok University of Technology from the Ministry of Science and Higher Education of Poland.

References

1. Cao, X., Dai, X., Liu, J.: Building energy-consumption status worldwide and the state-of-the-art technologies for zero-energy buildings during the past decade. Energy Build. **128**, 198–213 (2016). https://doi.org/10.1016/j.enbuild.2016.06.089
2. Reilly, A., Kinnane, O.: The impact of thermal mass on building energy consumption. Appl. Energy **198**, 108–121 (2017). https://doi.org/10.1016/j.apenergy.2017.04.024
3. Invidiata, A., Ghisi, E.: Impact of climate change on heating and cooling energy demand in houses in Brazil. Energy Build. **130**, 20–32 (2016). https://doi.org/10.1016/j.enbuild.2016.07.067
4. Colinart, T., Bendouma, M., Glouannec, P.: Building renovation with prefabricated ventilated façade element: a case study. Energy Build. **186**, 221–229 (2019). https://doi.org/10.1016/j.enbuild.2019.01.033
5. Sharif, S.A., Hammad, A.: Simulation-based multi-objective optimization of institutional building renovation considering energy consumption, life-cycle cost and life-cycle assessment. J. Build. Eng. **21**, 429–445 (2019). https://doi.org/10.1016/j.jobe.2018.11.006

6. Paraschiv, L.S., Paraschiv, S., Ion, V.I.: Increasing the energy efficiency of buildings by thermal insulation. Energy Proc. **128**, 393–399 (2017). https://doi.org/10.1016/j.egypro.2017.09.044
7. Schiavoni, S., D'Alessandro, F., Bianchi, F., Asdrubali, F.: Insulation materials for the building sector: a review and comparative analysis. Renew. Sustain. Energy Rev. **62**, 988–1011 (2016). https://doi.org/10.1016/j.rser.2016.05.045
8. Kaynakli, O.: A review of the economical and optimum thermal insulation thickness for building applications. Renew. Sustain. Energy Rev. **16**(1), 415–425 (2012). https://doi.org/10.1016/j.rser.2011.08.006
9. Evin, D., Ucar, A.: Energy impact and eco-efficiency of the envelope insulation in residential buildings in Turkey. Appl. Therm. Eng. **154**, 573–584 (2019). https://doi.org/10.1016/j.applthermaleng.2019.03.102
10. Dombaycı, Ö.A.: The environmental impact of optimum insulation thickness for external walls of buildings. Build. Environ. **42**(11), 3855–3859 (2007). https://doi.org/10.1016/j.buildenv.2006.10.054
11. Çomaklı, V., Yüksel, B.: Environmental impact of thermal insulation thickness in builings. Appl. Therm. Eng. **24**(5–6), 933–940 (2004). https://doi.org/10.1016/j.applthermaleng.2003.10.020
12. Bojic, M., Yik, F., Sat, P.: Influence of thermal insulation position in building envelope on the space cooling of high-rise residential buildings in Hong Kong. Energy Build. **33**(6), 569–581 (2001). https://doi.org/10.1016/S0378-7788(00)00125-0
13. Ma, Z., Li, H., Sun, Q., Wang, Ch., Yan, A., Starfelt, F.: Statistical analysis of energy consumption patterns on the heat demand of buildings in district heating systems. Energy Build. **85**, 464–472 (2014). https://doi.org/10.1016/j.enbuild.2014.09.048
14. Lund, H., Østergaard, P.A., et al.: The status of 4th generation district heating: research and results. Energy **164**, 147–159 (2018). https://doi.org/10.1016/j.energy.2018.08.206
15. Shapoval, S., Zhelykh, V., Vehryn, I., Kozak, K.: Simulation of thermal processes in the solar collector which is combined with external fence of an energy efficient house. In: Blikharskyy, Z. et al. (ed.) Procedings of CEE 2019. CEE 2019. Advances in Resource-Saving Technmologies and Materials in Civil and Environmental Engineering. LNCE, vol. 47, pp. 510–517. Springer (2020). http://doi.org/10.1007/978-3-030-27011-7
16. Lund, H., Duic, N., Østergaard, P.A., Mathiesen, B.V.: Future district heating systems and technologies: on the role of smart energy systems and 4th generation district heating. Energy **165**(A), 614–619 (2018). https://doi.org/10.1016/j.energy.2018.09.115
17. TEHO: Annual reports from 2000 to 2018. Thermal management. Tepelne hospodarstvo s.r.o., Kosice www.teho.sk
18. https://sk.wikipedia.org/wiki/Ko%C5%A1ice?veaction=edit&section=16
19. Statistical Office of the Slovak Republic, Bratislava (2020). https://slovak.statistics.sk/wps/portal/
20. http://datacube.statistics.sk/#!/view/sk/VBD_DEM/om7101rr/v_om7101rr_00_00_00_sk
21. https://www.siea.sk/bezplatne-poradenstvo/publikacie-a-prezentacie/zateplovanie-a-vymena-okien-v-bytovych-domoch/
22. https://www.weather-atlas.com/en/slovakia/kosice-climate#temperature
23. https://vytapeni.tzb-info.cz/tabulky-a-vypocty/103-vypocet-denostupnu

Assessment of Thermal Insulation Properties of Envelope Structures of a Burgher House in Kosice

Dusan Katunsky(✉) and Jana Katunska

Technical University of Kosice, 042 00 Kosice, Slovakia
dusan.katunsky@tuke.sk

Abstract. The article presents a basic overview of the most frequently used and proven interventions in historic buildings, focusing on the details of a burgher house in Kosice. In the case of renovation and reconstruction of an old building, it is necessary to take into account specific facts. A normal building is approached differently and a building that has historical value is different. The Energy Efficiency Act exempts historic buildings from certification. Not all old historic buildings are subject to monument protection. Methods of covert isolation that preserves the authenticity of cultural monuments and real estate in monument areas are a current challenge of heritage practices. The sustainability of the operation of historic buildings ultimately means the preservation and appropriate use of the heritage fund. The list of interventions that do not endanger the monumental values or use of the building is gradually expanding. This is mainly due to modern, increasingly sophisticated materials and technologies that are coming into practice.

Keywords: Historical building · Burger house · Temperature measurement · In situ · Heat flux

1 Introduction

The common theme of the works dedicated to historic buildings is their significant restoration. However, their direction is divided into research on materials that are commonly available and used new materials and simulations, which are considered a new means of better understanding of the events taking place in building construction. Their connection with in situ measurements is perceived as a basic tool in the study of the building structure of historic buildings and a more effective design of their restoration. Based on energy requirements, historic buildings undergoing renovation must also show an improvement in the thermal properties of packaging structures. Therefore, it is necessary to address critical details here as well. Satisfactory values can be achieved by various available means, including correct design and evaluation of critical details by simulation in combination with in situ or laboratory measurements (see Fig. 1) [1–3]. However, it is necessary to follow the basic rules that apply to the monumental, historical value of the building, while paying attention to the basic requirements set by a major renovation and proven methods and procedures of

Z. Blikharskyy (Ed.): EcoComfort 2020, LNCE 100, pp. 166–173, 2021.
https://doi.org/10.1007/978-3-030-57340-9_21

construction research. Other main topics of this problem are especially research into structures and the internal environment in historic buildings. Last but not least, it is a necessary issue for the protection of monuments, [4] energy retrofit of historical buildings [5–7]. There are some new methods of planning and modeling the reconstruction of a historical building [8, 9].

The diagram in Fig. 1 shows the basic differences between the restoration and reconstruction of a historic building. It shows what all plays an important role in these adjustments.

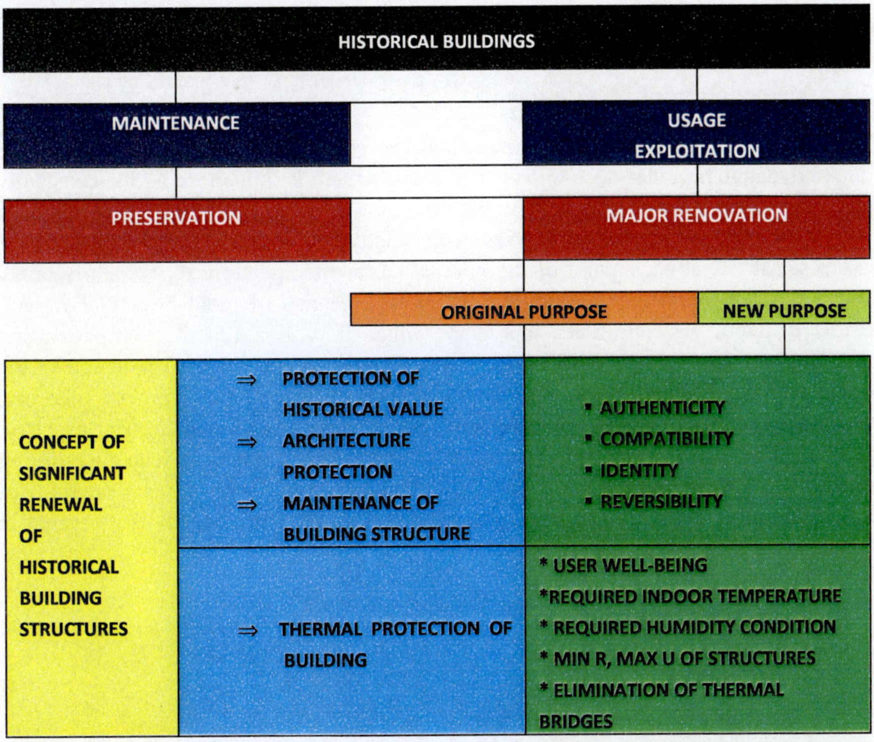

Fig. 1. Requirements for restoration and reconstruction of historic buildings

When renovating a historic building, the condition is the use of materials that are verified and their impact on a specific type of historic building material is known. The verification of materials is focused not only on the impact on the construction of the building, but also on the improvement of its thermal-technical properties and at the same time on the elimination or elimination of possible failures. The alternatives and suitability of the used thermal insulation layer from the point of view of monument protection are different, but in each case the aim of its use is to improve the thermal-humidity behavior of the building structure. It is not necessary to ensure airtightness in a historic building, as is the case with low-energy houses [10].

The question is whether it is necessary to insulate and improve the thermal properties of the historic structure at all. The construction of historic buildings has one basic characteristic. It is a solid wall and has a high heat capacity. This most influences the thermal-technical properties of the building structure. The historic building envelope is characterized by a large accumulation, which has an impact mainly on the change in the internal surface temperature over time. The minimum value of the internal surface temperature is specified in the standard STN 730540. Monitoring of the surface temperature in real terms in situ shows the actual behavior of the building envelope near reality. The recorded data can be used to create a numerical model for the simulation.

2 Thermo-Technical Assessment of Burgher House

2.1 Methodology of Measurement

The aim of the experimental measurement was to record changes in surface temperature due to dynamic boundary conditions. The measurement is influenced by the constantly changing conditions of the indoor and outdoor environment. According to STN 730540, ceilings and floors in areas with relative humidity $\varphi_i \leq 80\%$ must be expressed in °C at each point of the internal surface temperature θ_{si}, which is safely above the dew point temperature and eliminates the risk of mold growth. For standardized indoor air conditions, this is the value of $\theta_{si,80} = 12.6$ °C. Simplified stationary conditions of the outdoor and indoor environment are used to evaluate the hygienic criteria for the building envelope. Such an assessment does not take into account the effect of material parameters such as density and specific heat capacity. Sensors based on the type of tip resistance without PT, Ntc and NiCr with a measuring range from −50 to +125 °C and with a hundred resolutions with a linear accuracy of ±0.05 K are used for temperature measurement. The heat flux density is measured in the range −260 to + 260 W/m^2.

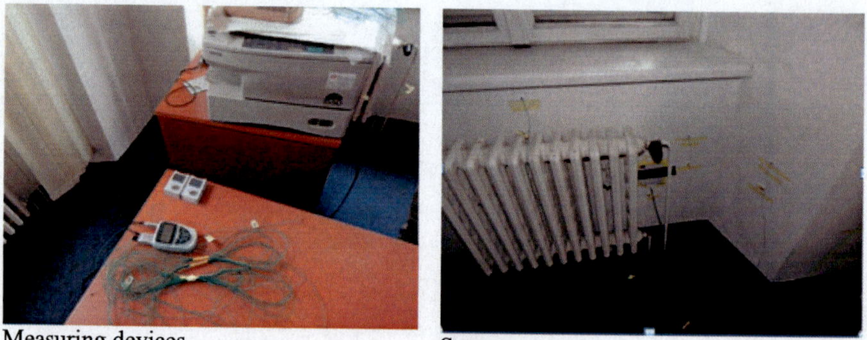

Measuring devices Senzors

Fig. 2. Historical buildings of burgher house in Kosice internal view.

Monitoring of the historic façade of the burgher house was in the winter. The surface temperature was recorded on the building envelope, where significant architectural properties and a variety of building materials could lead to the formation of thermal bridges (see Figs. 2 and 3). From an architectural point of view, details are important as well as critical thermal bridges from the point of view of building physics. Measurements were performed on the building envelopes (Fig. 2). The monitored points were placed to measure critical segments of the building envelope. The aim was to capture the expected temperature field of deformation of the building envelope. The measurements were short-term (7 days) and were performed in January and February. The measurement consisted of recording the temperature of the inner surface using temperature sensors and heat flux density. The data from the measurement of the inner surface temperature can then be used for a numerical model. In this way, it is possible to analyze all types of building envelopes in terms of material solutions and specific geometry (Fig. 3).

2.2 Thermal Technical Measurements

As already mentioned, the basic criterion for evaluating the building envelope from the point of view of thermal technology is the internal surface temperature, the hygienic criterion. Thermal bridges are characterized by a reduction in the internal surface temperature of the building envelope. As the surface temperature of the building envelope decreases, steam condenses and mold forms. Restoring the historic building envelope is a complex process. In many cases, restoration is accompanied by a lack of knowledge of the building envelope. In particular, there is a lack of knowledge that needs to be examined quickly and non-destructively in the building envelope (geometry, composition of the building envelope, thermo-physical properties of building materials) [11].

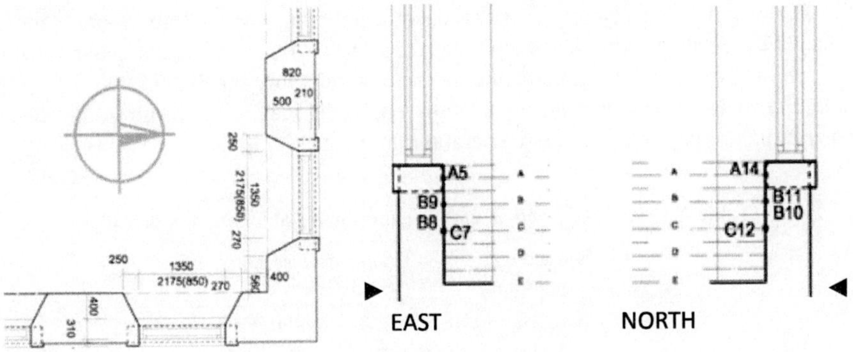

Floor plan Cross sections

Fig. 3. Historical buildings of burgher house in Kosice measured points

A stone ledge (tuff) was identified in the envelope of the building. The width of the bar was approx. 250 mm and a height of approximately 150–200 mm. On the outside, the stone ledge is covered with stucco. The sensors were placed north and east of the building envelope in the same places (see Fig. 3). A comparison of the surface temperature of the sensor with a different orientation to the sides of the world showed that instead of a stone parapet ledge, the surface temperature A5 (north side) is higher than A14 (east side) within 4 days. The temperature difference between the north and east orientation is 0.20 K. From the comparison of surface temperatures measured at the point of receding parapet masonry B8 and B11, the value of measured surface temperatures on the north wall below the east wall is about 0.5 K. The highest internal surface temperature is recorded on sensors B9, B10 (east) increase the thickness of the building envelope and increase the surface in suitable boundary conditions (higher internal heated area). In terms of the requirements of the hygiene criteria, the average values of the temperature on the inner surface are above the level required at the lowest surface temperature. The value of A5 decreased below the hygienic criteria. The measurement results can be seen (see Fig. 4 and 5) (Table 1).

Table 1. Comparison of measured temperatures in selected points in east and north facade

	Boundary conditions		Orientation to the cardinal					
	θai	θae	NORTH			EAST		
	°C	°C	A5	B8	B9	A14	B11	B10
Θaverag	17,6	−2,4	14,6	14,8	15,7	14,8	15,2	15,5
Θmin.	15,6	−8,7	12,1	12,6	13,7	13,1	13,5	13,8
Θmax.	18,8	+1,4	16,2	16,0	16,8	15,9	16,3	16,5

Points can be seen in Fig. 3

The heat flux density on the perimeter of a building with a northern orientation is affected by a reduction in thermal resistance, probably due to the stone parapet, which was defined by a higher value of the thermal conductivity coefficient. On the eastern side, a decrease in heat flux density is observed, which may be the result of the effect of solar radiation on reducing the temperature gradient.

Table 2. Measured internal surface temperatures and heat flux densities

	Boundary conditions			Temperature sensors		
	θai	θae	B8	C7	B11	C12
	°C	°C		q (W/m^2)	°C	q (W/m^2)
Θaverag	17,6	−2,4	14,8	19,5	14,8	17,6
Θmin.	15,6	−8,7	12,6	12,4	13,1	15,6
Θmax.	18,8	+1,4	16,0	23,8	15,9	18,8

Medium hi = 7,1 W/m^2K, hi = 6,5 W/m^2K
Rsi = 0,14 m^2K/W, Rsi = 0,15 m^2K/W

The value of the internal surface resistance of the building envelope (R_{si}) expressed as mean values calculated from the heat transfer coefficient hi and for the observed period is 0.14 m^2K/W (sensor B8, C7 (q)) located on the building envelope with a north-facing facade) and 0.16 m^2K/W (sensors B11v, C12 (q) located on the east-facing facade). The sensors were placed above the window ledge and in the place of the receding parapet masonry. The monitored street is oriented to the east and north side. The results of measurements in the area of receding parapet masonry may be affected, in addition to secondary interventions in the ventilation grille of the original gas and hot water heating equipment tempered in the adjacent room.

From the point of view of the requirement of the hygienic criterion, the average measured values of the temperature on the inner surface are higher than the lowest required surface temperature required by the standard of 12.62 °C. The value of A5 has been reduced to the limit of the hygiene criterion.

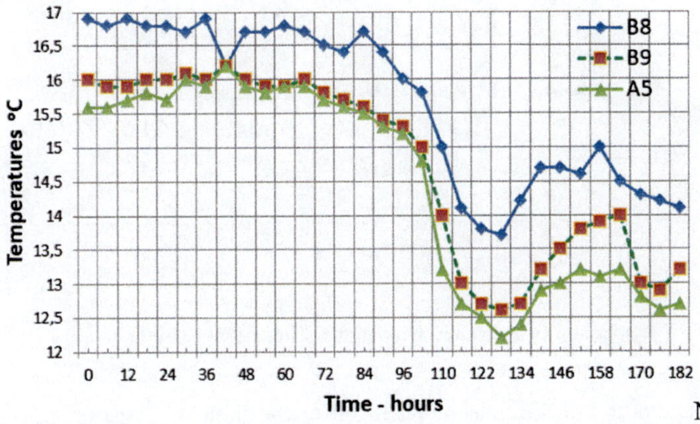

North A5, B8, B9

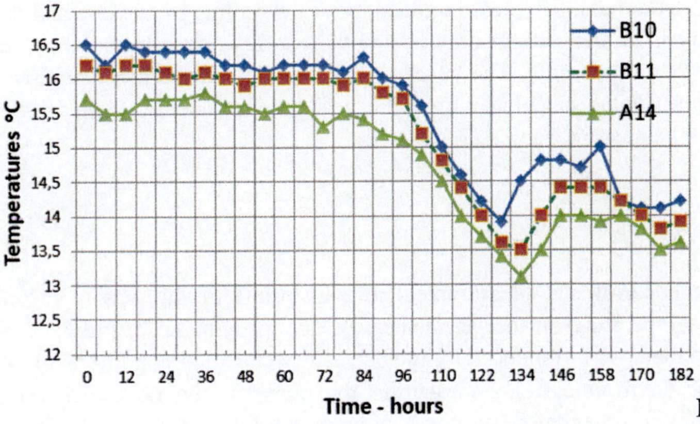

EastA14,B11, B10

Fig. 4. Comparison of measured temperatures in selected points

The internal surface temperatures measured at points 7 and 1 - at the point above the window ledge showed lower values compared to the surface temperature B7 measured at the point below the window ledge. The difference of the measured surface temperatures is 0.6 K. Sensors located at critical points on the window sills 2 and in the contact floors and masonry of the sills recorded lower values than the surface temperature sensors 7 located on the surface of the masonry sills.

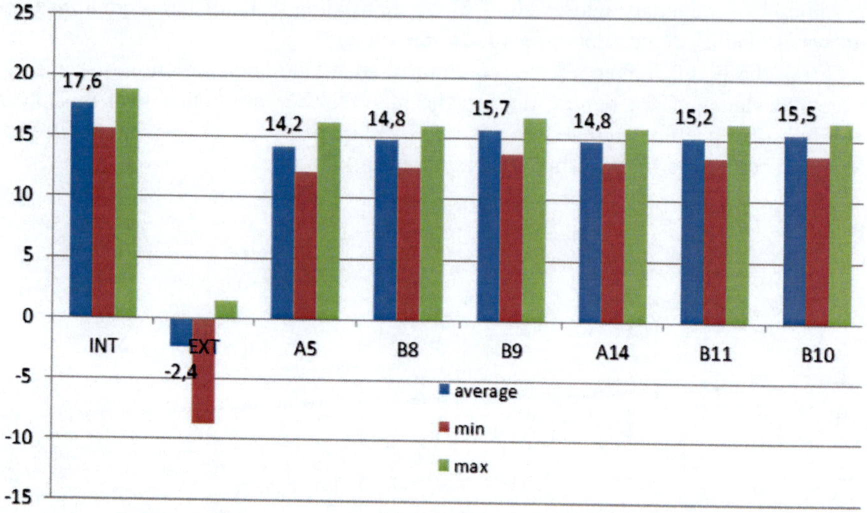

Fig. 5. Comparison of measured temperatures in selected points

The surface temperatures in the masonry sill fall to the limit of hygienic criteria during the entire measurement. The heat flux density of the building envelope and the window sill masonry is influenced not only by the thermal resistance but also by the difference in temperature gradients. The value of the internal surface resistance (R_{si}) expressed as the mean value calculated from the heat transfer coefficient h_i for the observed period can be seen in Table 2.

3 Conclusion

The purpose of the experimental measurement in a historic building was to record changes in surface temperature on the perimeter wall due to dynamic boundary conditions. The measurement was influenced by the constantly changing conditions of the indoor and outdoor environment. By measuring the changes, the internal surface temperatures of the building envelope in non-stationary conditions were recorded.

Based on the results of measurements, it is necessary to expand knowledge about the material composition of historic buildings and material parameters to more accurately define changes in surface temperature. Surface temperature sensors were placed on the north and east facades in the building of the burgher house. The measured temperature of the inner surface fell below the hygienic criteria only at the site of the Roman stone slabs on the northern perimeter of the building. The reduction of the surface temperature at the point above the window ledge (town house) is not very significant in comparison with the surface temperature measured at the point below the window ledge. This can be justified by the profiling of the window bar - cooling by enlarging the facade. The unknown composition of the masonry above the window opening can assume the existence of a window lintel with various thermo-physical parameters, such as the material forming the building envelope.

The low surface temperature in the parapet masonry can be caused by gas heating of the air vents, which was created in one of the phases of the building development of building and other building influences associated with the adaptation of the building. The overall impact of the façade profiling on the building envelope can be described as minimal. Material or significant geometric thermal bridges have a higher effect on surface temperatures. In all cases, the massive construction envelopes show that the drop in the outside air temperature still meets the hygienic criteria. Appropriate selection of the boundary conditions of the indoor environment can keep the indoor surface temperature above the value of the hygienic criteria.

References

1. Pisello, A.L., et al.: Energy refurbishment of historical buildings with public function: pilot case study. Energy Proc. **61**, 660–663 (2014)
2. Ferretti, V., Bottero, M., Mondini, G.: Decision making and cultural heritage: An application of the multi-attribute Value Theory for the reuse of historical buildings. J. Cult. Heritage **15** (6), 644–655 (2014)
3. Labovská, V., Katunský, D.: In situ monitoring of internal surface temperature of the historic building envelope. Sel. Sci. Papers-J. Civ. Eng. **11**(1), 77–84 (2016)
4. Clementi, F., Pierdicca, A., Formisano, A., Catinari, F., Lenci, S.: Numerical model upgrading of a historical masonry building damaged during the 2016 Italian earthquakes: the case study of the Podestà palace in Montelupone (Italy). J. Civ. Struct. Health Monit. **7**(5), 703–717 (2017)
5. Ascione, F., De Rossi, F., Vanoli, G.P.: Energy retrofit of historical buildings: theoretical and experimental investigations for the modelling of reliable performance scenarios. Energy Build. **43**(8), 1925–1936 (2011)
6. Mazzarella, L.: Energy retrofit of historic and existing buildings. The legislative and regulatory point of view. Energy and Build. **95**, 23–31 (2015)
7. Ciulla, G., Galatioto, A., Ricciu, R.: Energy and economic analysis and feasibility of retrofit actions in Italian residential historical buildings. Energy Build. **128**, 649–659 (2016)
8. Radziszewska-Zielina, E., Śladowski, G., Sibielak, M.: Planning the reconstruction of a historical building by using a fuzzy stochastic network. Autom. Constr. **84**, 242–257 (2017)
9. Saba, M., Quiñones-Bolaños, E.E., López, A.L.B.: A review of the mathematical models used for simulation of calcareous stone deterioration in historical buildings. Atmos. Environ. **180**, 156–166 (2018)
10. Katunský, D., Nemec, M., Kamenský, M.: Airtightness of buildings in Slovakia. In: Advanced Materials Research, vol. 649, pp. 3–6. Trans Tech Publications Ltd., Switzerland (2013)
11. Katunská, J., Katunský, D., Labovská, V.: Selected problems of thermal insulation of historical buildings. Sel. Sci. Papers-J. Civ. Eng. **14**(1), 67–74 (2019)

Performance Analysis of the Small-Scale Refrigeration System Using Natural Refrigerants and Their Mixtures

Mykhailo Khmelniuk(✉), Oleksii Ostapenko, and Olga Yakovleva

Odessa National Academy of Food Technologies, Odessa 65039, Ukraine
hmel_m@ukr.net

Abstract. Refrigeration system holds an important role in many commercial and household applications The optimal utilization of energy saving technologies in commercial and household refrigeration systems is a key issue for the rational use of energy resources. The scope of this study is an experimantal analysis of the small-scale refrigeration system efficiency using HFC, hydrocarbons and their mixtures as the most promising working substances in terms of environmental indicators. For analysis purposes were selected refrigerants R32, R152a and R134a in combination with hydrocarbons R600a and R290. Based on analysis results, it was concluded that the targeted modification of ozone-safe refrigerants based on synthetic refrigerants with natural hydrocarbon components is an effective means of creating a new class of alternative azeotropic working fluids in refrigeration technology.

Keywords: Natural refrigerants · Small-Scale refrigeration systems · Energy efficiency

1 Introduction

The proposed alternative working fluids based on mixtures as the possible replacement for existing HFC refrigerants. Refrigerant selection in each individual case involve both a forecast of the consequences of the test and confirmation of the forecast at a real facility. This situation is typical for any case of retrofit replacement, since compressors and units that would be specially designed and manufactured by industry for new working fluids does not exist. Therefore, when directly replacing one refrigerant with another, the compressor and other components of the refrigeration system will inevitably work in obviously non-optimal modes. With good reason, it can be argued that in all real cases of direct replacement of a known refrigerant with a new substance, the latter turns out to be in obviously unfavorable conditions of comparison. Nevertheless, comparative experimental studies provide the most objective information that allows us to draw a conclusion about the prospects of certain working fluids, offered as an alternative for existing HFC refrigerants like R134a.

Tests of refrigeration machines provide the most complete information about volumetric and energy indicators, the magnitude and patterns of changes in the overall

Z. Blikharskyy (Ed.): EcoComfort 2020, LNCE 100, pp. 174–181, 2021.
https://doi.org/10.1007/978-3-030-57340-9_22

temperature level of a sealed compressor and unit. In this section, experimental studies of the operational characteristics of small refrigeration machines and hermetic compressors are carried out in order to identify the preference of the proposed refrigerants in the actual operating conditions of refrigeration systems.

2 Experimental Stand and Equipment

To conduct a series of tests of hermetic compressors, an experimental unit that meets was the requirements of international standards was designed. The experimental setup is shown in Fig. 1. In order to increase the measurement accuracy, the flow rate of the refrigerant circulating in the system was determined in two independent ways: by the balance of the calorimeter and by the balance of the condenser.

To smoothly control the input power of the heating pad of the electrocalorimeter, it is envisaged to include a LIPS-35 DC source in the stand. The power of the heating pad and the power consumption of the compressor were determined using the K-506 measuring complex, which is included either in the calorimeter circuit or in the compressor power circuit. The air temperature was measured with a mercury thermometer with a division value of 0.1 K. The installation of copper-constantan thermocouples along the path of the agent through heat exchangers and control devices of the stand with the subsequent output of the temperature values of the agent to the digital printing device made it possible to visually control the output of the test bench and compressor to the specified stationary mode.

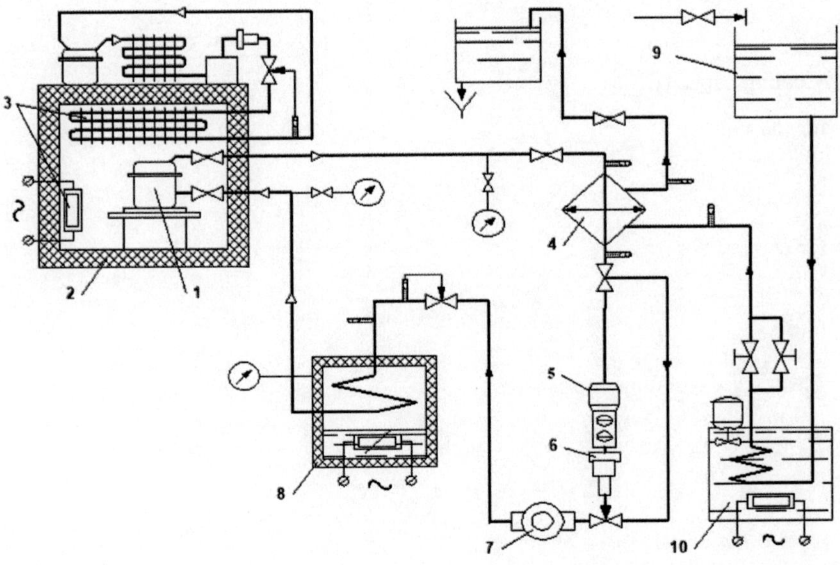

Fig. 1. Experimental unit. 1 – compressor, 2 – chamber, 3 – chamber thermostat, 4 – condenser, 5 – flow meter, 6 – filter, 7 – sight glass, 8 – calorimeter, 9 – pressure tank, 10 – water thermostat.

Thermocouples were also installed inside the compressor at the most characteristic points to study its thermal stress (temperature of the agent, oil, motor windings). As a regulating device, a valve was used that kept the pressure "after itself". The pressure of the refrigerant in the test bench was measured using standard pressure gauges (accuracy class 0.4).

The stand was placed in a thermostatic chamber with a volume of 12 m^3. The temperature in the T_{os} chamber was kept constant and amounted to 305 ± 0.3 K. The energy and other parameters were determined in steady state when the secondary refrigerant temperatures in calorimeters were reached at the levels T_{01} = 255 K and T_{02} = 278 K, which corresponds to the working conditions, respectively, of freezing and refrigerating chambers of a household refrigerator.

3 The Experiment Methodology

The mass flow rate of the refrigerant circulating in the system was determined by two independent methods:

- according to the thermal balance of the electrocalorimeter

$$Ga_{cal} = \frac{N_{el.} + Q_{cal}}{\Delta H_{cal}}, \tag{1}$$

- on the thermal balance of the condenser (in the study of the compressor):

$$Ga_{cond} = \frac{GwCw\Delta Tw}{\Delta H_{cond}} \tag{2}$$

Cooling capacity:

- for the unit

$$Q_o^u = Ga\left(h_{u1}^u - h_{u2}^u\right); \tag{3}$$

- for compressor

$$Q_o^{comp} = Ga\left(h_{comp1} - h_{ev.}\right); \tag{4}$$

Electric refrigeration coefficient:

- for the unit

$$\varepsilon_{el}^u = Q_o^u / N_{el}^u; \tag{5}$$

– for compressor

$$\varepsilon_{el}^{comp} = Q_o^{comp} / N_{el}^{comp}. \tag{6}$$

Compressor feed rate:

$$\lambda = \frac{Gav_1}{V_h}, \tag{7}$$

Electrical efficiency:

$$\eta_{el} = N_t / N_{el}, \tag{8}$$

To take into account all energy losses in the compressor, a technique was used that allows one to evaluate the energy performance of a hermetic compressor during experimental studies. The given technique contains indicators that evaluate certain types of energy losses of a hermetic compressor using the electric efficiency of the compressor:

$$\eta_{el} = \frac{\varepsilon_{el}}{\varepsilon_T} = \frac{N_T}{N_{el}} = \eta_i \eta_{mec} \eta_{mot} \tag{9}$$

The above coefficients from formula (Eq. 9) are described by the following expressions:

$$\eta_i = \frac{N_T}{N_i} = \frac{G_a l_s}{N_i} \tag{10}$$

$$N_{mec} = \frac{N_i}{N_i + N_{mec}} \tag{11}$$

$$N_{mot} = \frac{N_i + N_{mec}}{N_{el}} \tag{12}$$

Electric power N_{el} can be represented as the sum of capacities:

$$N_{el} = N_i + N_{mec} + N_{mot} \tag{13}$$

$$N_{mot} = \Sigma N_{mot} = N_{st} + N_c + N_R + N_F \tag{14}$$

Losses in stator steel N_{st} and mechanical losses N_{mec} can be found by testing the compressor in idle mode. When the compressor is idling, all electric power consumed by the compressor is consumed only for losses, since $N_i = 0$:

$$N_{el\,x.x} = N_{mec\,x.x} + N_{st\,x.x} + N_{c\,x.x} \tag{15}$$

In order to separate these power losses, which are included in expression (Eq. 15), we studied the operation of the compressor in idle mode at various values of the electric voltage U, i.e. to find the experimental dependence:

$$N_{el\,x.x} = N_{mec\,x.x} + N_{st\,x.x} + N_{c\,x.x} = f(U^2) \tag{16}$$

If from the values $N_{el\ x.x}$ at various U, subtract losses in the stator copper wires $N_{c\ x.x}$, then you can get the dependence:

$$N_{mec\,x.x} + N_{st\,x.x} = f(U^2) \tag{17}$$

Power losses in copper stator windings are determined from the dependence:

$$N_{c\,x.x} = I^2_{x.x} R_{x.x} \tag{18}$$

$$N_{mec} = N_{mec\,x.x} \tag{19}$$

Loss value $N_{st\ x.x}$ determined in idle mode of the compressor at voltage $U = 220$ V. From formula (Eq. 16) we find:

$$N_{st\,x.x} = N_{el\,x.x} - N_{mec\,x.} - N_{c\,x.x} \tag{20}$$

Found value $N_{st\ x.x}$ at $U = 220$ V taken as value N_{st} for compressor operating modes, i.e.

$$N_{st} = N_{st} x.x \tag{21}$$

Dependence (Eq. 21) assumes constancy N_{st} at various operating modes of a real compressor. Losses in the rotor as well as indicator losses depend on the parameters of the compressor operating mode and are determined by the ratio:

$$N_R = \frac{N_{el} - N_{st} - N_c}{100} s_j \tag{22}$$

In equation (Eq. 22), the value N_{st}, N_{el} and N_c found when examining the compressor in operating mode. Thus, equation (Eq. 22) contains two unknown parameters N_R and s. To determine them, we study the equation:

$$\frac{s_{x.x}}{N_{elx.x}} = \frac{s_j}{N_{el}} \tag{23}$$

Accepted value $s_{x.x}$ at the end of the calculations, it is specified, since in the end the condition characterized by relation (Eq. 9) must be satisfied.

After calculating N_R from the equation (Eq. 14), the values of indicator losses N_i are found in operating modes. With known losses, using the relations (Eq. 10–12) are calculated η_i, η_{mec}, η_{mot}. As indicated, their work should equal η_{el}. In case of deviations, the value is adjusted $s_{x.x}$, and s_j. Refined Values s_j allow you to calculate the true number of revolutions of the electric motor in a particular operating mode by the ratio:

$$n_j = \frac{(100 - s)n_s}{100} \tag{24}$$

Knowing n_j for each mode, you can finally calculate V_h and λ.

4 Results and Discussion

Changing the concentration of isobutane (R600a) in the mixture was carried out in two directions: the first – when using mineral oil, increasing the concentration of R600a improves the solubility of the mixture with oil (the problem of replacing the refrigerant after repairing the compressor or unit), the second - when using synthetic oil. An increase of the R134a content in the mixture reduces the fire hazard.

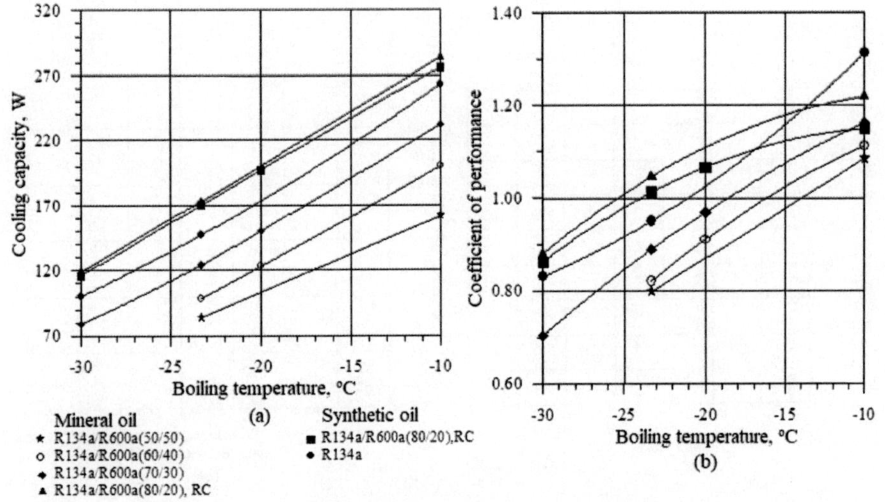

Fig. 2. Compressor cooling capacity (a) and coefficient of performance (b) on a mixture of R134a/R600a of various concentrations and R134a Tc = 55 °C, To = 32 °C

Studies have shown (Fig. 2, a, b) that the addition of R600a to R134a leads to a decrease in cooling capacity and cooling coefficient. However, the compressor power consumption is higher when using synthetic oil in the entire range of boiling points by 3… 5 W. A slight increase in power consumption can obviously be explained by higher viscosity values of the refrigerant-synthetic oil system. Thus, we can conclude that the use of mixtures R134a/R600a and R152a/R600a is most preferable in systems with a hermetic compressor filled with mineral oil. Mixtures based on R32 also have high degree of potential. An analysis of the phase diagrams shows that two features that

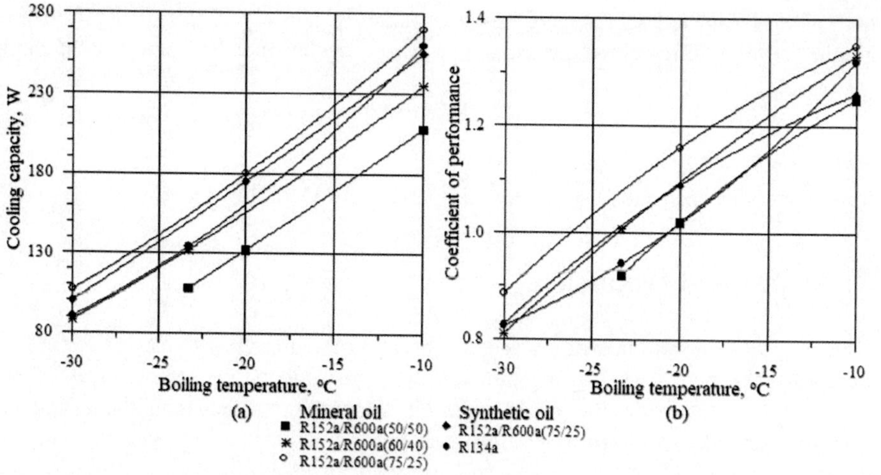

Fig. 3. Compressor cooling capacity (a) and coefficient of performance (b) on a mixture of R152a/R600a and R134a at Tc = 55 °C, To = 32 °C

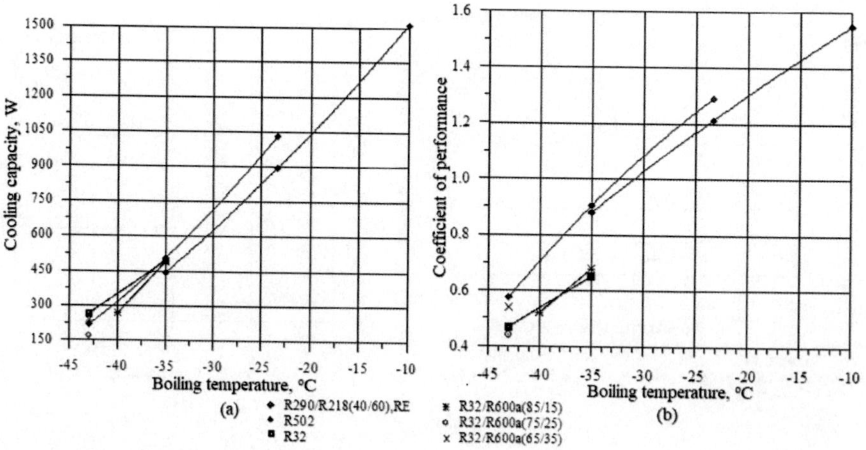

Fig. 4. Compressor cooling capacity (a) and coefficient of performance (b) on a mixture of R32/R600a, R32/R290, R32/R218 and R32 at Tc = 45 °C, To = 32 °C

determine the behavior of the working fluid in the cycle of the refrigeration machine characterize mixtures of R32/R600a, R32/R290 and R32/R218 (Figs. 3 and 4).

Firstly, the mixtures are azeotropic. Secondly, the mixtures separate with the formation of three phases in equilibrium (liquid-liquid-vapor). The greatest interest is the mixture R32/R600a, which has a pressure at the level of R32 and a zone of limited solubility liquid-liquid-vapor at possible (moderately low) boiling points. Obviously, a comparison of the characteristics of such mixtures must be carried out as part of a real refrigeration machine at certain boiling points that realize the maximum energy characteristics.

5 Conclusions

This paper presents experimental investigation of the small-scale refrigeration system. The experiment was conducted using various refrigerants and their mixtures such as: R32, R134a, R152a, and hydrocarbons R600a and R290. The combination of the refrigerants R134a and R600a improves the solubility of the mixture with oil and decreases flammability comparing to pure R600a. The R152a/R600a mixture provides increase in cooling capacity with increasing of R152a concentration. Higher values of COP were achieved using R152a/R600a at 75/25 vol.% concentration. R32 based mixtures allows to achieve highest COP value (5–7% increase) while maintaining same pressure level.

References

1. Stoecker, W.: Industrial Refrigeration Handbook. McGraw-Hill, New York (1998)
2. Zha, S., Ma, Y., Wang, J., Li, M.: The thermodynamic analysis and comparison on natural refrigerants cascade refrigeration cycle. In: 5-th Gustav Lorentzen Conference on Natural Working Fluids, Guangzhou, China, pp. 157–163 (2002)
3. Sun, Z., Youcai, L., Shengchun, L., Weichuan, J., Runqing, Z., Rongzhen, L., Zhikai, G.: Comparative analysis of thermodynamic performance of a cascade refrigeration system for refrigerant couples R41/R404A and R23/R404A. Appl. Energy **184**, 19–25 (2016)
4. Rackett, H.: Equation of state for saturated liquids. J. Chem. Eng. Data **15**, 514–517 (1970)
5. Spencer, C., Danner, R.: Improved equation for prediction of saturated liquid density. J. Chem. Eng. Data **17**, 234–241 (1972)
6. Han, X., Wang, Q., Zhu, Z., Chen, G.: Cycle performance study on R32/R125/R161 as an alternative refrigerant to R407C. Appl. Ther. Eng. **27**(14–15), 2559–2565 (2007)
7. Di Nicola, G., Giuliani, G., Passerini, G., Polonara, F., Stryjek, R.: Vapor–liquid equilibria (VLE) properties of R-32CR-134a system derived from isochoric measurements. Fluid Phase Equilib. **153**, 143–165 (1998)
8. Huber, M., Gallagher, J., McLinden, M., Morrison, G.: NIST standard reference database 23, reference fluid thermodynamic and transport properties (REFPROP), version 9.0. National Institute of Standards and Technology, Thermophysics Division. https://www.nist.gov/programs-projects/reference-fluid-thermodynamic-and-transport-properties-database-refprop. Accessed 15 May 2020
9. EPA Homepage, http://www.epa.gov/ozone/science/ods/classone.html. Accessed 15 May 2020
10. Cavallini, A.: Heat transfer and energy efficiency of working fluids in mechanical refrigeration. Bull. Int. Inst. Refrig. **6**, 4–21 (2002)

The Probabilistic Calculation Model of RC Beams, Strengthened by RC Jacket

Roman Khmil, Roman Tytarenko, Yaroslav Blikharskyy(✉), and Pavlo Vegera

Lviv Polytechnic National University, Lviv 79013, Ukraine
yaroslav.z.blikharskyy@lpnu.ua

Abstract. According to the engineering method of calculation of normal cross section of reinforce concrete beam with use of the basic concepts of probability theory and random function theory, is proposed the probabilistic calculation model of rectangular reinforce concrete beams, strengthened by reinforce concrete jacket under loading. Real-operating conditions of the considered structures are modeled. Therefore, in calculation theory was included number of controlled random parameters, namely: strength of materials (concrete and steel rebar), geometric parameters of a cross section (before and after strengthening), and also the level of load at the moment of strengthening. Developed model was tested, based on the data of previous experimental studies of the stress-strain state of reinforce concrete beams' series, strengthened by reinforce concrete jacket under loading – the values their corresponding qualitative and quantitative reliability indixes (reliability indexes β and probabilities of failure $Q(\beta)$, respectively) are obtained. Analysis of the obtained results was also made.

Keywords: RC beam · RC jacket · Strengthening · Load level · Random parameter · Reliability · Probability of failure

1 Introduction

Nowadays reinforced concrete strutures are one of the most popular construction materials all around the world [1–5]. For today there is the great amount of works, devoted to modern materials, which are used in reinforced concrete structures in order to improve their strength and durability characteristics [6–8]. Reinforcement in RC structures due to various effects is subjected to corrosion damages [9–12]. As reinforced concrete structures with corrosion impairments have reduced bearing capacity, reliable calculation of such structures is highly topical issue [13–15]. Probabilistic methods of RC structures calculation obtain great prevalence, because they allow to assign guaranteed reliability level of the particular structural element at the design stage in the form of probability of its reliability/failure (quantitive reliability estimation). Therefore, taking into consideration increasing of the reconstruction works' volumes all around the world and efficiency and economy of the strengthening method, it could be stated that the issue of reliability estimation of RC structures, strengthened by RC jacket, nowadays is highly topical [16, 17].

Z. Blikharskyy (Ed.): EcoComfort 2020, LNCE 100, pp. 182–191, 2021.
https://doi.org/10.1007/978-3-030-57340-9_23

In turn, taking into account subjectivity parameters for criteriafor reliability ensuring, complexity of mathematical calculation apparatus, as well as control of random paraameters of structures' bearing capacity reserve, durability of bended RC elements, strengthened under loading has recently become the research issue in the world- since the end of XXth century. Herewith only in some works [18–24] the loading factor during the element strengthening of elements with insufficient bending and shear strength was calculated. Moreover, the problem of reliability estimation of beams, strengthened by RC jacket for today was not developed in principle.

It is important to note that the most of considered studies were related to reliability estimation of structures, strengthened by FRP-materials [25–28]. Research data, despite its relevance, contained, in addition, the number of obvious disadvantages. Among them are the following: availability only for one loading level under strengthening, implementation of sizable mathematical calculations, usage of laws of random parameters distribution, which do not fully represent building construction performance and, which is even more important, great deviations of obtained reliability indexes. This could be the problem for implementation in design practice.

Therefore, actual absence of the study of reliability estimation issue for bended reinforced concrete elements, strengthened by reinforced concrete jacket under loading, shows the necessity of further research in this direction, including formulation of probabilistic model for such structures' calculation.

2 The Purpose of Theoretical Research

The purpose of theoretical research, according to engineering calculation method for normal cross-section of bended RC element [29], is development of probabilistic calculation model (reliability estimation) of RC beams of rectangular cross-section, strengthened by RC jacket under loading, which would represent actual conditions of their work.

Achievement of indicated purpose includes performing the following tasks:

- adaptations of the existing method of new design structures' reliability assessment to strengthened RC structures [2, 16];
- implementation in probabilistic calculation model of controled random strength and geometrical parameters of beam bearing capacity reserve before and after strengthening, as well as loading level during strengthening;
- approbation of the developed model (on the basis of experimental data [16], under different load levels during strengthening for different rebar diameters of RC jacket) in order to check its suitability;
- analysis of obtained reliability indexes β and failure probability $Q(\beta)$, as well as developed probabilistic model in general.

In addition is assumed that accepted random parameters are interdependent, their distribution is subjected to normal low.

3 Development of the Probabilistic Calculation Model

Characteristic feature of the proposed calculation model is that as the random parameter is taken the construction bearing capacity by bending moment (residual resource). Herewith, the bending moment due to external load is assumed as predetermined because its change influence on construction is relatively low, compared to changes of strength and geometrical characteristics, which determine residual resource of the construction. Load level during strengthening is taken as the additional variable parameter in residual resource estimation of strengthened construction.

The random value of the boundary bending moment $\tilde{M}_{ult}$ perceived by a beam, strengthened by jacket under an action of load, taking into consideration the presence of reinforcement in the compressed zone of the main section could be written as follows:

*provided $x > h'_{add}$ (see Fig. 1b) and while preservation of the condition $\xi \leq \xi_R$ (after strengthening).

$$\begin{aligned}\tilde{M}_{ult} &= f\left(\tilde{\sigma}_c, \tilde{\sigma}_{c,add}, \tilde{\gamma}^{add}_{c,dis}, \tilde{\sigma}_s, \tilde{\sigma}_{s,add}, \tilde{\gamma}^{add}_{s,dis}, \tilde{\sigma}_{sc}, \tilde{b}, \tilde{b}_{add}, \tilde{d}'_{red}, \tilde{h}'_{add}\right) \\ &= \left(\tilde{\sigma}_c\tilde{b} + 2\tilde{\sigma}_{c,add}\tilde{b}_{add}\tilde{\gamma}^{add}_{c,dis}\right)\left(\tilde{x} - \tilde{h}'_{add}\right)\left(\tilde{d}'_{red} + 0,5\tilde{h}'_{add} - 0,5\tilde{x}\right) \\ &+ \tilde{\sigma}_{c,add}\left(\tilde{b} + 2\tilde{b}_{add}\right)\tilde{h}'_{add}\tilde{\gamma}^{add}_{c,dis}\left(\tilde{d}'_{red} + 0,5\tilde{h}'_{add}\right) + \tilde{\sigma}_{sc}A'_s\left(\tilde{d}'_{red} - a'\right),\end{aligned} \tag{1}$$

where $\tilde{\sigma}_c$, $\tilde{\sigma}_{c,add}$ are the random values of concrete strength for compression of the main and additional sections, respectively (for the ULS); $\tilde{b}$, $2\tilde{b}_{add}$ are the random values of a width of the main and the additional sections, respectively (see Fig. 1); $\tilde{\gamma}^{add}_{c,dis}$ is the random value of usage coefficient of the section of the concrete jacket, which depends on load level on beam before strengthening; $\tilde{h}'_{add}$ is the random value of the additional section height above the upper compressed border of the main section (see Fig. 1); $\tilde{d}'_{red}$ is the random value of distance from the upper compressed border of the main section to the center of weight of the whole stretched steel rebar of beam after strengthening (see Fig. 1); $\tilde{\sigma}_{sc}$ is the random value of steel rebar compressive strength of the main section; A'_s is the area of the compressed steel rebar section (see Fig. 1); a' is the distance from the center of weight of the compressed steel rebar to the upper compressed border of the main section (see Fig. 1); $\tilde{x}$ is the random value of the compressed zone of concrete height (see Fig. 1b), which, in this case, could be found from Eq. (2):

$$\begin{aligned}\tilde{x} &= \frac{\tilde{\sigma}_s A_s + \tilde{\sigma}_{s,add}A_{s,add}\tilde{\gamma}^{add}_{s,dis} - \tilde{\sigma}_{sc}A'_s - \tilde{\sigma}_{c,add}\left(\tilde{b} + 2\tilde{b}_{add}\right)\tilde{h}'_{add}\tilde{\gamma}^{add}_{c,dis}}{\tilde{\sigma}_c\tilde{b} + 2\tilde{\sigma}_{c,add}\tilde{b}_{add}\tilde{\gamma}^{add}_{c,dis}} \\ &+ \tilde{h}'_{add},\end{aligned} \tag{2}$$

where $\tilde{\sigma}_s$, $\tilde{\sigma}_{s,add}$ are the random values of steel rebar tensile strength of the main and the additional sections, respectively; A_s, $A_{s,add}$ are the areas of the section of stretched steel rebar, located in the existing beam and the jacket, respectively (see Fig. 1); $\tilde{\gamma}^{add}_{s,dis}$ is the random value of the usage coefficient of the section of a stretched steel rebar of the jacket, which depends on the load level on a beam before strengthening.

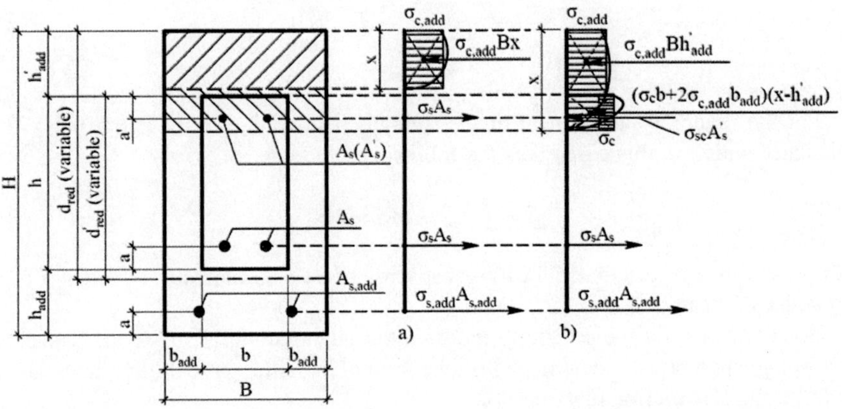

Fig. 1. Scheme of active forces and combined diagram of stress (actual curvilinear and calculation, respectively) in the normal cross section of RC beam, strengthened by RC jacket, at $\xi \leq \xi_R$: a) $x \leq h'_{add}$; b) $x > h'_{add}$.

Substituting Eq. (2) for $\tilde{x}$ in Eq. (1) for $\tilde{M}_{ult}$, with subsequent staged simplification, following is obtained:

$$\begin{aligned}\tilde{M}_{ult} = {} & \left(\tilde{\sigma}_s A_s + \tilde{\sigma}_{s,add} A_{s,add} \tilde{\gamma}^{add}_{s,dis}\right) \tilde{d}'_{red} - \tilde{\sigma}_{sc} A'_s a' \\ & + \tilde{\sigma}_{c,add} \left(0,5\tilde{b} + \tilde{b}_{add}\right) \tilde{h}'^2_{add} \tilde{\gamma}^{add}_{c,dis} \\ & - \frac{0{,}5}{\tilde{\sigma}_c \tilde{b} + 2\tilde{\sigma}_{c,add} \tilde{b}_{add} \tilde{\gamma}^{add}_{c,dis}} \times \left[\begin{array}{c} \tilde{\sigma}^2_s A^2_s + \tilde{\sigma}^2_{s,add} A^2_{s,add} \tilde{\gamma}^{add\,2}_{s,dis} + \tilde{\sigma}^2_{sc} A'^2_s \\ + \tilde{\sigma}^2_{c,add} \tilde{h}'^2_{add} \left(\tilde{b}^2 + 4\tilde{b}\tilde{b}_{add} + 4\tilde{b}^2_{add}\right) \tilde{\gamma}^{add\,2}_{c,dis} \end{array}\right] \\ & - \frac{\tilde{\sigma}_{s,add} A_{s,add} \tilde{\gamma}^{add}_{s,dis} \left(\tilde{\sigma}_s A_s - \tilde{\sigma}_{sc} A'_s\right) - \tilde{\sigma}_s A_s \tilde{\sigma}_{sc} A'_s}{\tilde{\sigma}_c \tilde{b} + 2\tilde{\sigma}_{c,add} \tilde{b}_{add} \tilde{\gamma}^{add}_{c,dis}} \\ & + \frac{\tilde{\sigma}_{c,add} \left(\tilde{b} + 2\tilde{b}_{add}\right) \tilde{h}'_{add} \tilde{\gamma}^{add}_{c,dis} \left(\tilde{\sigma}_s A_s + \tilde{\sigma}_{s,add} A_{s,add} \tilde{\gamma}^{add}_{s,dis} - \tilde{\sigma}_{sc} A'_s\right)}{\tilde{\sigma}_c \tilde{b} + 2\tilde{\sigma}_{c,add} \tilde{b}_{add} \tilde{\gamma}^{add}_{c,dis}}.\end{aligned} \quad (3)$$

Mathematical expectation of boundary bending moment $\bar{M}_{ult}$ could be obtained by substitution of the mathematical expectations of random arguments in the above simplified Eq. (3).

Next, based on the method of statistical linearization [31, 32], and also the norms [30], coefficients D_{x_i} are defined (for identification of a standard of critical bending moment $\hat{M}_{ult}$) in the form of partial derivatives of function $\bar{M}_{ult} = f(x_1, \ldots, x_n)$ by $x_1, \ldots, x_n$ variables:

$$D_{x_i} = \partial \bar{M}_{ult} / \partial \bar{x}_i. \quad (4)$$

Below, the Eq. (5) is written down in order to find the standard of the boundary bending moment $\hat{M}_{ult}$:

$$\hat{M}_{ult} = \sqrt{\sum_{i=1}^{n} \left[(D_{x_i} \hat{x}_i)^2 \right]}, \tag{5}$$

where $\hat{x}_i$ is the standart of x_i variable.

For the reliability assessment of the strengthened beam, the reliability index β is calculated,which in this case takes the following form:

$$\beta = \left(\hat{M}_{ult} - M_{ult} \right) / \hat{M}_{ult}, \tag{6}$$

where M_{ult} is the calculated bearing capacity of the normal cross section of the strengthened beam.

Finally, based on the reliability index β, the quantitative reliability assessment of the strengthened beam is determined (in the form of its failure probability), with the use of the well-known error function $f(\beta)$:

$$Q(\beta) = 0,5 - f(\beta). \tag{7}$$

4 Design of Experimental Samples

The developed model was tested on the experimental samples and materials [16]. RC beams were reinforced in stretched zone of the concrete with 2Ø14 A500C, in compressed – 2Ø8 of A400C class; the transverse reinforcement was Ø8 A500C with spacing of 50 mm. Two metal holder of indicators were welded in the middle of span in the lower zone of a beam on the distance of 200 mm between them for measurement of steel rebar deformations. The design dimensions of RC beams were 100 × 200 2100 mm; the concrete class – C30/35.

RC jacket was reinforced in stretched zone of the concrete with 2Ø8, 2Ø10 or 2Ø12 of A400C class; an additional reinforcement frame of the bottom of jacket was fastened with steel wire. The jacket thickness was 20 mm from the upper border and sides' thickness were 50 mm from the lower border of a beam – for the installation of an additional reinforcement frame. The design length of jacket was 1600 mm; the jacket concrete class – C40/50.

5 Results of Theoretical Study

Next, with use of norms [30] and recommendations [27], for testing of the developed model based on the data of experimental studies [16] are determined statistical characteristics (mathematical expectations and standards) of materials' strength, load levels and geometry of beams' cross sections after strengthening based on the corresponding calculation characteristics (see Table 1).

Table 1. Statistic characteristics $\bar{x}_i$, $\hat{x}_i$ of random parameters $\tilde{x}$.

Characteristics $\bar{x}_i$		Equation for	Value	Unit	Variation coef. V_i	Value	Characteristics $\hat{x}_i$	Value	Unit
$\bar{\sigma}_c$		$f_{cd}/(1-1,64V_c)$	2,5	kN/cm^2	V_c	0,135	$\hat{\sigma}_c$	0,34	kN/cm^2
$\bar{\sigma}_{c,add}$			3,53			0,135	$\hat{\sigma}_{c,add}$	0,48	
$\bar{\sigma}_s$		$f_y/(1-1,64V_s)$	53,86		V_s	0,0437	$\hat{\sigma}_s$	2,35	
$\bar{\sigma}_{sc}$			42,01			0,0594	$\hat{\sigma}_{sc}$	2,5	
$\bar{\sigma}_{s,add}$			42,01			0,0437	$\hat{\sigma}_{s,add}$	1,84	
$\bar{\gamma}^{add}_{c,dis}$	0.3M	$\sum \gamma^{add}_{c,dis(i)}/n$	0,445	–	$V_{\gamma^{add}_{c,dis}}$	0,0107	$\hat{\gamma}^{add}_{c,dis}$	0,0048	–
	0.5M		0,354			0,0143		0,0051	
	0.7M		0,234			0,0385		0,009	
	0.9M		0,074			0,0540		0,004	
$\bar{\gamma}^{add}_{s,dis}$	0.3M		1,158		$V_{\gamma^{add}_{s,dis}}$	0,000	$\hat{\gamma}^{add}_{s,dis}$	0,000	
	0.5M		0,838			0,0024		0,002	
	0.7M		0,526			0,0361		0,019	
	0.9M		0,165			0,0303		0,005	
$\bar{b}$		$\sum x_i/n$	9,92	cm	V_b	0,008	$\hat{b}$	0,08	cm
$\bar{d}'_{red}$	(2Ø8)		18,61		$V_{d'_{red}}$	0,01	$\hat{d}'_{red}$	0,19	
	(2Ø10)		19,04						
	(2Ø12)		19,47						
$\bar{b}_{add}$			2,29		$V_{b_{add}}$	0,145	$\hat{b}_{add}$	0,33	
$\bar{h}'_{add}$			2,29		$V_{h'_{add}}$	0,145	$\hat{h}'_{add}$	0,33	

where in Table 1 V_i – the coefficient of random parameter $\tilde{x}_i$ variation; n – number of elements in the series; equation to finding: $\hat{x}_i = V_i\bar{x}_i$; $M = M_{ult,O}$ – design bearing capacity of normal cross-section of the beam before strengthening.

Thus, in accordance with the model developed above we obtained the values of qualitative and quantitative reliability indixes of strengthened beams – reliability index β_i and probability of failure $Q(\beta)_i$, respectively (Tables 2, 3).

Table 2. Reliability index β_i.

№	Steel rebar of RC jacket	Level of load at the moment of strengthening				
		$0.0M_{ult,O}$	$0.3M_{ult,O}$	$0.5M_{ult,O}$	$0.7M_{ult,O}$	$0.9M_{ult,O}$
1	2Ø8 mm	5.81	–	–	5.64	–
2	2Ø10 mm	5.83	5.76	5.71	5.67	5.59
3	2Ø12 mm	5.86	–	–	5.71	5.62

Table 3. Probability of failure $Q(\beta)_i$.

№	Steel rebar of RC jacket	Level of load at the moment of strengthening				
		$0.0M_{ult,O}$	$0.3M_{ult,O}$	$0.5M_{ult,O}$	$0.7M_{ult,O}$	$0.9M_{ult,O}$
1	2Ø8 mm	3.12×10^{-9}	–	–	8.50×10^{-9}	–
2	2Ø10 mm	2.77×10^{-9}	4.21×10^{-9}	5.65×10^{-9}	7.14×10^{-9}	1.14×10^{-8}
3	2Ø12 mm	2.31×10^{-9}	–	–	5.65×10^{-9}	9.55×10^{-9}

Dependence graph of reliability index β_i depending on the loading level during strengthening ($\times M_{ult,O}$) for different stretched rebar diameters of the jacket is presented below (see Fig. 2).

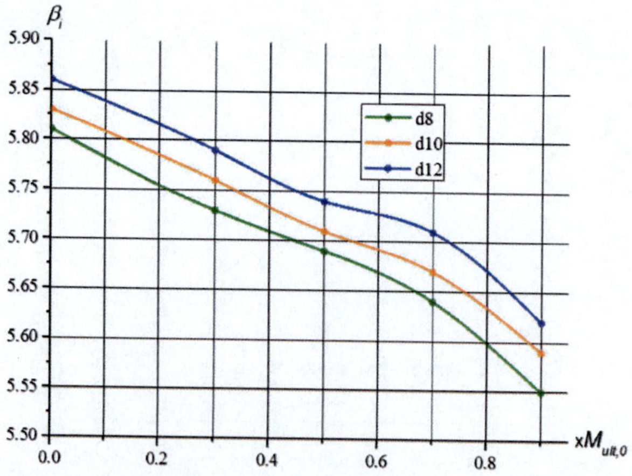

Fig. 2. Dependence of the reliability index β_i on load level during strengthening ($\times M_{ult,O}$) for different diameters of RC jacket stretched reinforcement.

On the basis of obtained results' analysis it could be stated that the proposed model is adequate, because its usage ensures higher values of reliability indeces β = 5,59... 5,86 for strengthened beams (due to incrementation of compressed concrete zone), than corresponding values, proposed in [29] for newly designed constructions. This values could be recommended for further usage as increment of unloading level on beam before strengthening corresponds to higher reliability indeces β_i. In addition, it is obvious that for reduction of deviation range for variation of random parameters coefficients of bearing capacity reserve for RC beams before and after strengthening and load level during strengthening, and therefore obtaining the most objective reliability assessment it is necessary to conduct as many tests of twin elements as possible.

6 Conclusions

Probabilistic calculation model was developed for reliability estimation of RC beams of rectangular cross-section and insufficient bending strength, strengthened with RC jacket under loading. Model provides the possibility to operate with random strength (concrete, steel rebars of existing beam and jacket), geometric (dimensions of main and the additional sections) parametrs of bearing capacity reserve, as well as with load level parameters during strengthening (usage coefficient of stretched rebar and concrete of the jacket). Developed model was tested on the real strengthened beams and the number of reliability parameters were obtained – reliability indexes β and failure probability $Q(\beta)$.

References

1. Blikharskyy, Z., Vashkevych, R., Vegera, P., Blikharskyy, Y.: Crack resistance of RC beams on the shear. In: Lecture Notes in Civil Engineering, vol. 47, pp. 17–24 (2020). https://doi.org/10.1007/978-3-030-27011-7_3
2. Krainskyi, P., Blikharskyy, Y., Khmil, R., Vegera, P.: Crack resistance of RC columns strengthened by jacketing. In: Lecture Notes in Civil Engineering, vol. 47, pp. 195–201 (2020). https://doi.org/10.1007/978-3-030-27011-7_25
3. Lobodanov, M., Vegera, P., Blikharskyy, Z.: Planning experiment for researching reinforced concrete beams with damages. In: Lecture Notes in Civil Engineering, vol. 47, pp. 243–250 (2020). https://doi.org/10.1007/978-3-030-27011-7_31
4. Pavlikov, A., Kosior-Kazberuk, M., Harkava, O.: Experimental testing results of reinforced concrete beams under biaxial bending. Int. J. Eng. Technol. **7**(3.2), 299–305 (2018)
5. Vegera, P., Vashkevych, R., Blikharskyy, Z.: Fracture toughness of RC beams with different shear span. In: MATEC Web of Conferences, vol. 174, p. 02021 (2018). https://doi.org/10.1051/matecconf/201817402021
6. Azizov, T.N., Kochkarev, D.V., Galinska, T.A.: New design concepts for strengthening of continuous reinforced-concrete beams. In: IOP Conference Series: Materials Science and Engineering, vol. 708, no. 1, p. 012040. IOP Publishing, December 2019. https://doi.org/10.1088/1757-899X/708/1/012040

7. Bobalo, T., Blikharskyy, Y., Kopiika, N., Volynets, M.: Theoretical analysis of RC beams reinforced with high strength rebar's and steel plate. In: IOP Conference Series: Materials Science and Engineering, vol. 708, no. 1, p. 012045 (2019)
8. Sobol, K., Blikharskyy, Z., Petrovska, N., Terlyha, V.: Analysis of structure formation peculiarities during hydration of oil-well cement with zeolitic tuff and metakaolin additives. Chem. Chem. Technol. **8**, 461–465 (2014). https://doi.org/10.23939/chcht08.04.461
9. Blikhars'kyi, Y.Z.: Anisotropy of the mechanical properties of thermally hardened A500s reinforcement. Mater. Sci. **55**(2), 175–180 (2019). https://doi.org/10.1007/s11003-019-00285-0
10. Blikharskyy, Y., Kopiika, N., Selejdak, J.: Non-uniform corrosion of steel rebar and its influence on reinforced concrete elements' reliability. Prod. Eng. Arch. **26**(2), 67–72 (2020). https://doi.org/10.30657/pea.2020.26.14
11. Kharchenko, Y., Blikharskyy, Z., Vira, V. Vasyliv, B., Vasyliv, B: Study of nanostructural changes in a Ni-containing cermet material during reduction and oxidation at 600 °C. Appl. Nanosci. (2020). https://doi.org/10.1007/s13204-020-01391-1
12. Lipiński, T.: Roughness of 10721 steel after corrosion tests in 20% NaCl. Prod. Eng. Arch. **15**(15), 27–30 (2017). https://doi.org/10.30657/pea.2017.15.07
13. Kos, Ž., Klimenko, Y.: The development of prediction model for failure force of damaged reinforced-concrete slender columns. Tehn. Vjesnik **26**(6), 1635–1641 (2019). https://doi.org/10.17559/TV-20181219093612
14. Karpiuk, V.M., Syomina, Y.A., Antonova, D.V.: Bearing capacity of common and damaged CFRP-strengthened RC beams subject to high-level low-cycle loading. In: Materials Science Forum, vol. 968, pp. 185–199. Trans Tech Publications Ltd. (2019). https://doi.org/10.4028/www.scientific.net/MSF.968.185
15. Lobodanov, M., Vegera, P., Blikharskyy, Z.: Influence analysis of the main types of defects and damages on bearing capacity in reinforced concrete elements and their research methods. Prod. Eng. Arch. **22**(22), 24–29 (2019). https://doi.org/10.30657/pea.2019.22.05
16. Khmil, R., Tytarenko, R., Blikharskyy, Y., Vashkevych, R.: Influence of load level during strengthening of reinforced concrete beams on their reliability. In: IOP Conference Series: Materials Science and Engineering. TRANSBUD 2019, vol. 708, p. 012054. IOP Publishing (2019)
17. Krainskyi, P., Vegera, P., Khmil, R., Blikharskyy, Z.: Theoretical calculation method for crack resistance of jacketed RC columns. In IOP Conference Series: Materials Science and Engineering, vol. 708, no. 1, p. 012059 (2019)
18. Alsayed, S., Siddiqui, N.: Reliability of shear-deficient RC beams strengthened with CFRP-strips. Constr. Build. Mater. **42**, 238–247 (2013)
19. EN 1992-1-1: Eurocode 2: Design of concrete structures – Part 1-1: General rules and rules for buildings. European Committee for Standardization (CEN), Brussels, 225 (2004)
20. Lima, J., Barros, J.: Reliability analysis of shear strengthening externally bonded FRP models. Proc. Inst. Civ. Eng.: Struct. Build. **164**(1), 43–56 (2011)
21. Pichugin, S., Severin, V.: Reliability of structures under snow load in Ukraine. Snow Eng. **V**, 67–72 (2004)
22. Shahriari, S., Naderpour, H.: Reliability assessment of shear-deficient reinforced concrete beams externally bonded by FRP sheets having different configurations. Structures **25**, 730–742 (2020)
23. Trentin, C., Casas, J.: Safety factors for CFRP strengthening in bending of reinforced concrete bridges. Compos. Struct. **128**, 188–198 (2015)
24. Wang, N., Ellingwood, B., Zureick, A.: Reliability-based evaluation of flexural members strengthened with externally bonded fiber-reinforced polymer composites. J. Struct. Eng.-ASCE **136**(9), 1151–1160 (2010)

25. Blikharskyy, Z., Brózda, K., Selejdak, J.: Effectivenes of strengthening loaded RC beams with FRCM system. Arch. Civ. Eng. **64**(3), 3–13 (2018)
26. Khmil, R., Tytarenko, R., Blikharskyy, Y., Vashkevych, R.: Influence of load level during strengthening of reinforced concrete beams on their reliability. In IOP Conference Series: Materials Science and Engineering, vol. 708, no. 1, p. 012054 (2019)
27. Khmil, R., Tytarenko, R., Blikharskyy, Y., Vegera, P.: Development of the procedure for the estimation of reliability of reinforced concrete beams, strengthened by building up the stretched reinforcing bars under load. Eastern-Eur. J. Enterp. Technol. **5/7**(95), 32–42 (2018)
28. Selejdak, J., Blikharskyy, Y., Khmil, R., Blikharskyy, Z.: Calculation of reinforced concrete columns strengthened by CFRP. In: Lecture Notes in Civil Engineering,vol. 47, pp. 400–410 (2020). https://doi.org/10.1007/978-3-030-27011-7_51
29. DBN V.1.2-14-2018: General principles of reliability assurance and constructive safety of buildings and structures, p. 30. Ministry of Regional Development of Ukraine, Kyiv (2018)
30. DBN V.2.6-98:2009: Constructions of buildings and structures. Concrete and reinforced concrete structures. Basic principles, p. 71. Ministry of Regional Development of Ukraine, Kyiv (2011)
31. Jaynes, E.T.: Probability Theory: The Logic of Science. Cambridge University Press, Cambridge (2003)
32. Pichugin, S.F.: Reliability estimation of industrial building structures. Mag. Civ. Eng. **83**(7), 24–37 (2018)

Operation of Damaged H-Shaped Columns

Yevhenii Klymenko[1(✉)], Zeljko Kos[2], Iryna Grynyova[1], and Olena Maksiuta[1]

[1] Odessa State Academy of Civil Engineering and Architecture, Odessa, Ukraine
klimenkoew57@gmail.com
[2] University North, Varazdin, Croatia

Abstract. During the research, the main causes and consequences of damage to reinforced concrete eccentrically compressed structures of the H-shaped cross-section were established Three main factors that affect the value of the residual bearing capacity of the elements are identified. A three-level three-factor Boks-Benkin plan has been created, which will give statistically substantiated results (within the variation of variable factors) in the research of only 15 experimental columns. Samples of compressed elements were made and tested (according to the experimental plan). As a result of the experiments, data were obtained that allowed to describe the stress-strain state of the damaged H-shaped reinforced concrete columns. The dependences of stresses in reinforcement and relative deformations in concrete on the parameters of damage and load level are obtained. Based on the analysis of scientific proposals and the stress-strain state obtained experimentally, the basic preconditions for calculating the residual bearing capacity of elements damaged during operation of structures are adopted. A system of equations is developed, as a result of the solution of which the value of the residual capacity of the damaged columns is obtained. A system of equations is created and the ways of its solution are planned. Proposals for determining the residual bearing capacity of damaged reinforced concrete compressed columns are based on the main provisions of current regulations and extends their effect to damaged structures of the most common cross section – H-shaped.

Keywords: Reinforced concrete · Columns · H-shaped · Damage · Work · Strength

1 Introduction

One of the most common for the construction of industrial and civil, transport and other buildings and structures are reinforced concrete structures [1, 2]. On the one hand, this is due to the strength and durability of the material, and on the other – due to the relatively low cost. The use of stronger materials leads to more cheap cross-sections of structures. Modern studies of concrete as a material make it possible to obtain high-strength material (class C90/100 and above, high-strength reinforcing steels) [3]. Optimization of the most cross sections of compressed reinforced concrete structures often leads to the use of H-shaped profiles.

Z. Blikharskyy (Ed.): EcoComfort 2020, LNCE 100, pp. 192–201, 2021.
https://doi.org/10.1007/978-3-030-57340-9_24

During operation, all building structures wear out and lose their performance [4, 5]. Due to the active influence of the environment (alternating freezing-thawing, aggressive influence of air gases, etc.), mechanical damage results in partial destruction of the concrete section and corrosion of the reinforcement [6–8]. Defects and damage reduce (sometimes significantly) the serviceability [9] of structures by deteriorating their technical condition and often lead to collapse of buildings and structures.

The most common damage to reinforced concrete structures are: corrosion of the working reinforcement (reduction of its cross-sectional area), breakage of the transverse reinforcement in the compressed element, which reduces the stability of compressed rods, destruction (mechanical or chemical) of the concrete cross section [10–12].

The work of damaged reinforced concrete structures, in which the rods of the working reinforcement received corrosion damage and their cross-sectional area decreased, has been studied in detail [13]. In the calculations of the residual bearing of the eccentrically compressed reinforced concrete structures, the effect of such damage is taken into account directly, i.e. by reducing the cross-sectional area of the rods. However, it should be noted that in the area of damage extracentric compression turns into a more complex form of deformation – oblique extracentric compression. This is due to the fact that the compression plane in the damaged cross-sections does not coincide with the main axes.

The same problems arise when damaging part of the concrete section, when the front of the damage (Fig. 1) is not parallel to the sides of the section.

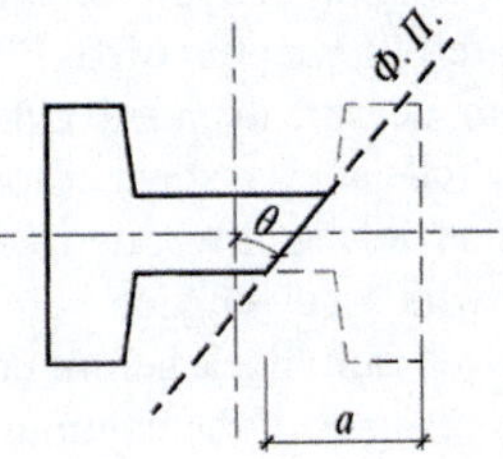

Fig. 1. Damage to part of the concrete section.

The oblique eccentrically compression that occurs at the damaged section (in height) of the column requires other (more general) approaches to determining the bearing capacity. The work of obliquely compressed and bent reinforced concrete elements was studied in detail by Toryanyk M.S. and his students [14, 15], but in these studies oblique compression was realized by creating eccentricities of external force in two planes, and oblique bending - by mismatch of force and one of the main planes.

The stress-strain state and bearing capacity of reinforced concrete elements, which received eccentricities in mutually perpendicular planes, based on asymmetric (relative to the main axes of undamaged section) damage were studied by Klymenko Y.V. and his colleagues [16, 17], however H-shaped profile (as the most complex and general) of compressed elements in the perspective of determining the residual bearing capacity of reinforced concrete elements damaged during operation, was not considered.

Presently, the current regulations do not provide any recommendations for the establishment of one of the most important indicators of serviceability, namely: the residual load-bearing capacity of reinforced concrete compressed elements of the H-profile, damaged during operation.

2 The Purpose of the Study

The purpose of this study is to determine the most significant factors influencing the residual bearing capacity of damaged reinforced concrete columns of H-shaped profile, to develop an experimental plan, to establish a stress-strain state of damaged structures and to develop preconditions for residual bearing capacity calculation.

3 Research Methodology

In order to implement this task, a three-level three-factor plan of the Boks-Benkin experiment was developed. The main variable factors were: the depth of damage a, angle of inclination of the damage front Θ (Fig. 1) and the eccentricity of the application of external force e_0. Limits of factors variation: a = 2–10 cm; $\Theta = 0°–60°$, $e_0 =$ 0–1/4 h.

For experimental researches 15 experimental beams from concrete of class C 25/30 reinforced 4ø12 A 400 (Fig. 2 (a)), which had damages on 1/3 of height of section with parameters according to the plan of experiment were made. Resistance strain gauges (20 mm base) were glued to the longitudinal reinforcing rods in the middle of the height (in the damaged section) and carefully waterproofed (Fig. 2 (b)). During the experiment, these sensors made it possible to measure the relative deformations of steel, followed by determination of stresses in the rods at each stage of loading of the test sample.

The samples were concreted in metal-wooden formwork, in which the wooden elements were hydrophobized with a polyethylene film (Fig. 2 (c)). Damage was simulated by inserting a polystyrene foam block before concreting (Fig. 2 (d)). The main parameters of the experimental samples of the columns are shown in Fig. 3. Testing of

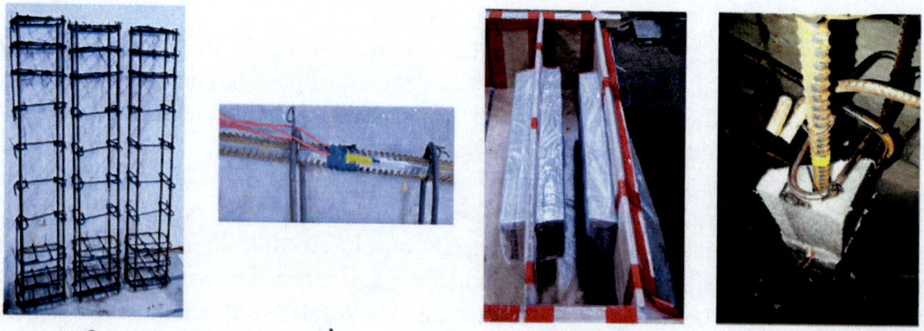

Fig. 2. Samples: (a) frames of test columns; (b) strain gauges; (c) formwork; (d) damage simulation.

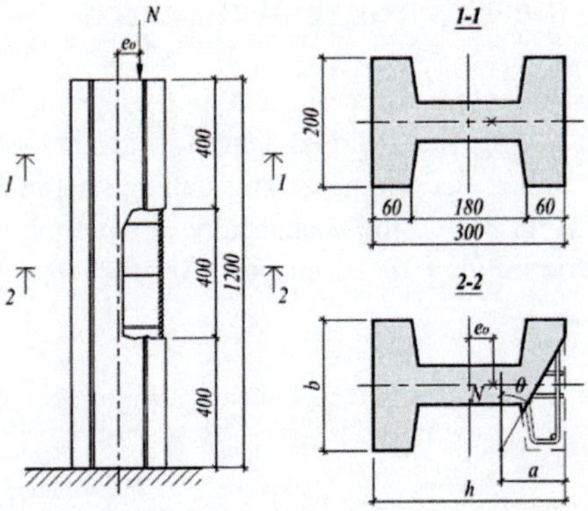

Fig. 3. Parameters of the experimental samples.

prototypes was carried out in the laboratory of the Department of Reinforced Concrete Constructions and Transport Facilities on a hydraulic press with a bearing capacity of 250 tons (Fig. 4).

Before the test on the side faces of the columns (in the middle of the height, i.e. in the damaged section) was glued (evenly around the perimeter) strain gages with a base of 50 mm in the amount of 15–18 pcs. for each column. The readings obtained from these sensors allowed us to describe the stress-strain state of the structures during all stages of loading, including immediately before the depletion of bearing capacity.

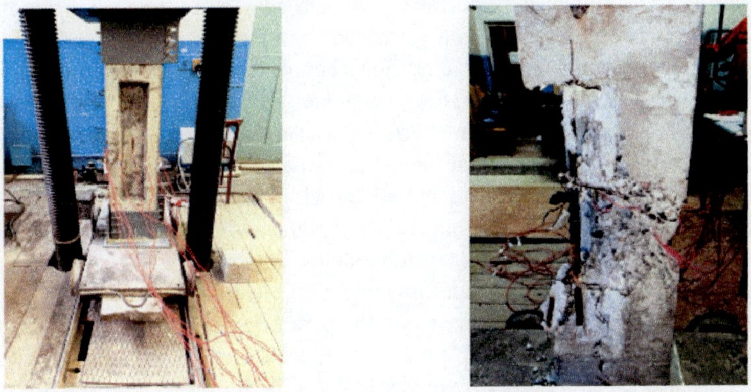

Fig. 4. Test columns during and after tests.

4 The Results of the Experimental Research

Based on the analysis of literature sources, the hypothesis was developed that the most significant factors influencing the residual bearing capacity of damaged reinforced concrete eccentrically compressed elements are: damage depth a, angle of inclination of the damage front Θ (Fig. 3) and the eccentricity of the application of external force e_O. Limits of factors variation: a = 2–10 cm; Θ = 0°–60°, e_O = 0–1/4 h (Table 1).

Table 1. Factors and their limits of using in the testing

Input factors			The levels of variation			Range of variation	The interval of variation
Code	Values	Meas. unit	« −1»	«0»	«1»		
X_1	Angle of inclination of the damage front, Θ	deg.	0	30	60	60	30
X_2	Damage depth, α	cm	2	6	10	8	4
X_3	Relative eccentricity, e_O/h	–	0	1/8	1/4	1/4	1/8

To confirm this hypothesis, a numerical experiment was performed in the environment of the PC LIRA-CAD. The model of experimental samples was made of concrete class C25/30. For reinforcing steel of class A400C, a bilinear relationship between stresses and strains was used.

The calculation is based on the finite element method using as the main unknown displacements and rotations of the nodes of the calculation scheme. In this regard, the idealization of the structure is performed in a form adapted to the use of this method, namely: the system is presented as a set of bodies of standard type (rods and three-dimensional elements), which are called finite elements and attached to them nodes.

The column was divided into finite elements in the form of rectangular paralelepipeds with a face size of 1 to 2 cm, as well as octagonal and hexagonal finite elements in the form of triangular and quadrangular prisms in places where the geometry of the sample required to model the slope of shelves and damage fronts. Concrete was set by physically nonlinear spatial eight-node and six-node isoparametric SE types 236 and 234, respectively, reinforcement - universal spatial core SE type 410, taking into account the geometric and physical nonlinearity.

As a result of the numerical experiment, the parameters of the stress-strain state and the residual bearing capacity of the damaged elements were obtained.

Figure 5 shows the values of normal stresses in the cross sections of the test elements at a load level of 0.95 from the destructive.

The designation of the experimental columns indicates the level of the factor, respectively X1, X2 and X3.

In all considered samples, the neutral line in the damaged section is outside it and has a form close to a straight line. It is parallel to the front of the damage in the case of its parallelism to one of the major axes of the section. In the case of "oblique" damage, there is a rotation of the neutral line relative to the main axes of the section and the front of the damage.

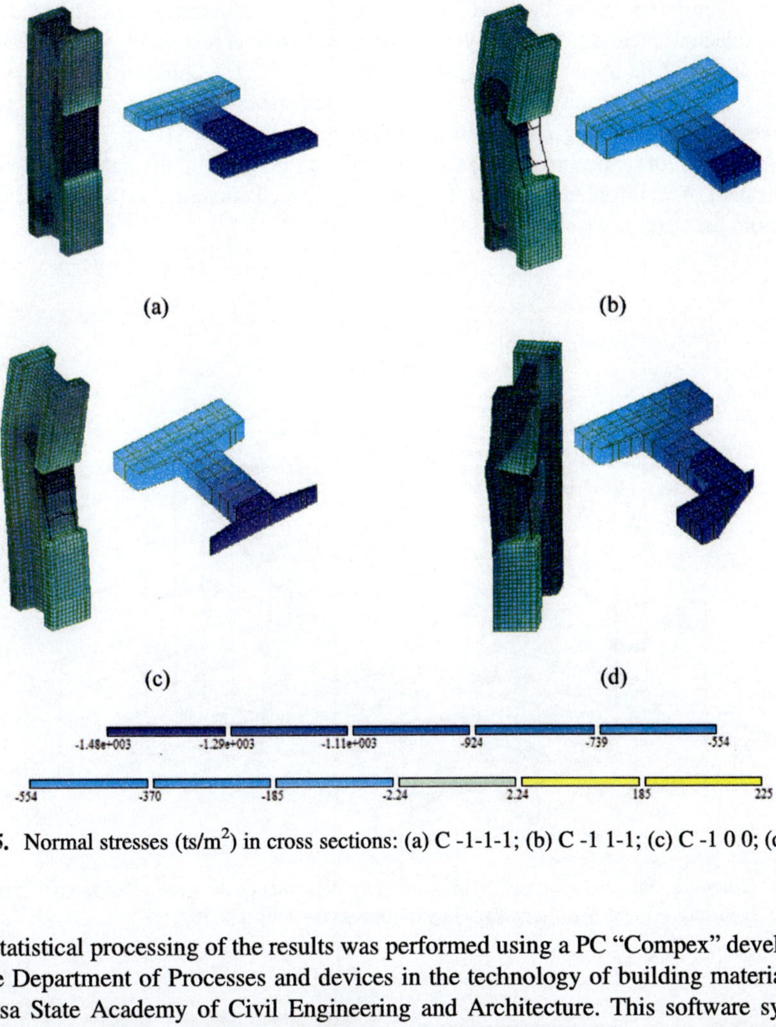

Fig. 5. Normal stresses (ts/m^2) in cross sections: (a) C -1-1-1; (b) C -1 1-1; (c) C -1 0 0; (d) C 0 1 0.

Statistical processing of the results was performed using a PC "Compex" developed at the Department of Processes and devices in the technology of building materials of Odessa State Academy of Civil Engineering and Architecture. This software system allows you to assess the degree of influence of each factor on the test samples. The assessment is based on the calculation of a three-factor experimental-statistical model of the studied factors of variation by the method of least squares.

Statistical processing of the results was performed using a PC "Compex" developed at the Department of Processes and devices in the technology of building materials of Odessa State Academy of Civil Engineering and Architecture. This software system allows you to assess the degree of influence of each factor on the test samples. The assessment is based on the calculation of a three-factor experimental-statistical model of the studied factors of variation by the method of least squares.

Figure 6 shows a graph of the dependence of the residual bearing capacity on the factors under study.

The three-factor three-dimensional diagram (Fig. 6) illustrates the combined effect of all selected factors. The analysis shows that the largest load-bearing capacity N = 73.2 tf will be observed at a chipping angle $\theta = 26°$, a chipping depth a = 2 cm and a relative eccentricity $e_0/h = 0$. The smallest bearing capacity is 25.8 ts and observed in the following combination of factors: $\theta = 0°$, a = 10 cm, $e_0/h = 0{,}2$.

Field tests of experimental samples were carried out in the laboratory of the Department of Reinforced Concrete Constructions and Transport Facilies on a 200-ton hydraulic press (Figs. 7 and 8).

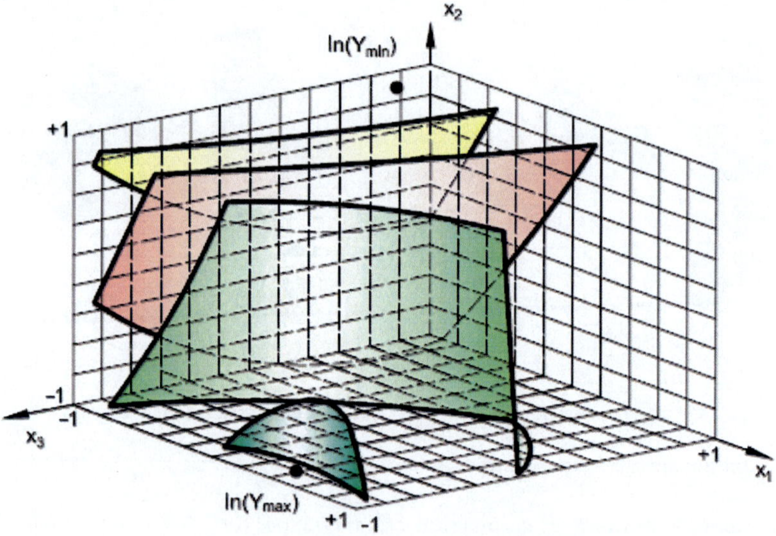

Fig. 6. Three-dimensional graph of bearing capacity dependence on the studied factors. – level 3,5; – level 3,75; – level 4,0; – level 4,25.

(a) (b)

Fig. 7. Tests of experimental columns: (a) C 0-1 0; (b) C 1 1 1.

(a) (b)

Fig. 8. Experimental columns after destruction: (a) C 0 1 0; (b) C 1-1 1.

During the processing of the data obtained during the experiment, the parameters of the stress-strain state were determined (relative deformations of compression and tension of concrete and longitudinal reinforcement and stresses in the reinforcement were determined) (Fig. 9).

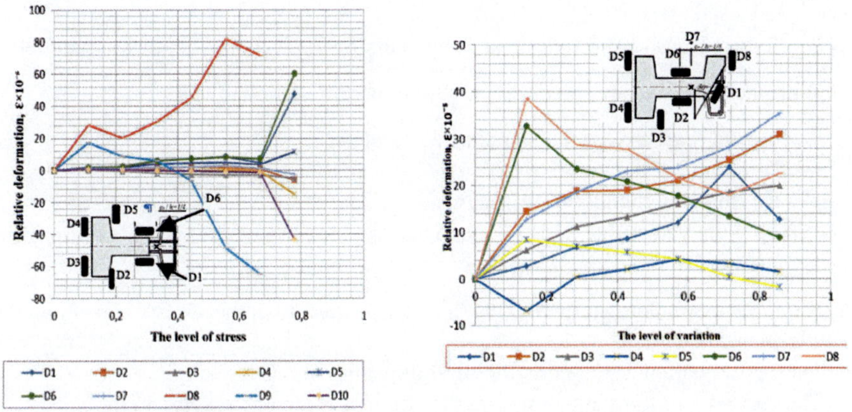

(a) (b)

Fig. 9. Relative deformations of concrete and reinforcement: (a) column C - 1 1 1; (b) coloumn C - 0 1 0. Sensors D7, D8 - on compressed fittings, D9, D10 - on stretched.

5 Research Scientific Innovations

The analyzed experimental data allowed to provide recommendations for the creation of an analytical method for calculating the residual bearing capacity of two-brand reinforced concrete columns damaged during operation.

These proposals are based on the main provisions of the current rules for the calculation of reinforced concrete structures and extend their effect in case of damage to structures for the most common case of cross section – H-shaped.

In the general case, the current DBN determines the bearing capacity taking into account the nonlinear diagram of the dependence of stresses from the deformations of reinforced concrete sections, but the use of a simplified dependence is allowed. The rationale for the use in the calculation of a simplified dependence is that the proposed method of calculation can be applied to existing damaged elements, considering it as a test. When performing test calculations, the uniform nature of the distribution of normal stresses in the compressed zone is considered.

Therefore, summarizing the above conclusions and test results, the following main prerequisites for the calculation were formulated:

1. The hypothesis of flat sections is accepted.
2. The work of bare reinforcing rods is taken into account by introducing lowering factors that take into account its flexibility.

3. Stresses in the compressed zone of concrete are evenly distributed and taken equals to fcd.
4. The forces in the stretched zone are completely perceived by the reinforcement.
5. The tensile stress in the reinforcement is taken not more than the calculated tensile strength f_t, the compression - not more than f_{yd}. Stresses in the reinforcement are determined based on the position of the neutral line and the height of the compressed zone of concrete.
6. The force planes of the outer and inner pairs of forces coincide, or are parallel.

With inclined damage (anglc $\Theta \neq 0°$) we have five unknowns:

N – bearing capacity of the sample by calculation;
x – the height of the compressed cross-sectional area;
φ – the angle of inclination of the neutral line;
δ and β – the values that must be found to describe the position of the coordinates of the center of mass of the concrete compressed zone.

We have five unknown quantities, that is, we need to compose five equations that include these quantities.

The first equation is the equation of equilibrium about the x-axis.

The second and third equations are the sums of moments relative to the x-axis and y-axis.

The fourth and fifth equations are the equations of static moments of the compressed zone of concrete, their composition is possible due to the hypothesis that the stresses are uniform in section.

Solving the system of equations, we find the residual bearing capacity of H-shaped compressed elements of the T-profile, damaged during operation.

In the case of planar damage (angle $\Theta = 0°$) the number of unknowns (and, hence, equations) decreases, and the solution of the system of equations and the determination of the residual load-bearing capacity of damaged H-shaped compressed columns is simplified.

6 Conclusions

Load-bearing capacity - one of the main indicators of serviceability of the structure, i.e., knowing the value of load-bearing capacity, we can determine the technical condition of individual structures by calculation (rather than by expert method), and hence buildings or structures as a whole. On the basis of a certain technical condition, it is possible to make reasoned (calculated) decisions on the further operation of buildings: repair, reinforcement, dismantling or even failure to take any measures.

In the future it is planned to create a system of equations that will take into account all the variety of shapes and sizes of the cross section of the element as a whole, shapes and sizes of the compressed zone of concrete. The validity of these proposals should be confirmed by comparison with experimental data and statistical processing of such comparisons.

References

1. Bobalo, T., Blikharskyy, Y., Kopiika, N., Volynets, M.: Serviceability of RC beams reinforced with high strength rebar's and steel plate. In: Lecture Notes in Civil Engineering, vol. 47, pp. 25–33 (2020). https://doi.org/10.1007/978-3-030-27011-7_4
2. Blikharskyy, Z., Brózda, K., Selejdak, J.: Effectivenes of strengthening loaded RC beams with FRCM system. Archiv. Civil Eng. **64**(3), 3–13 (2018)
3. Bobalo, T., Blikharskyy, Y., Kopiika, N., Volynets, M.: Theoretical analysis of RC beams reinforced with high strength rebar's and steel plate. In: IOP Conference Series: Materials Science and Engineering, vol. 708, no. 1, p. 012045 (2019)
4. Klymenko, Y.V.: Technical Operation and Reconstruction of Buildings and Structures. Center for Educational Literature, Kyiv (2004)
5. Lobodanov, M., Vegera, P., Blikharskyy, Z.: Influence analysis of the main types of defects and damages on bearing capacity in reinforced concrete elements and their research methods. Prod. Eng. Arch. **22**(22), 24–29 (2019). https://doi.org/10.30657/pea.2019.22.05
6. Kharchenko, Y.V., Blikharskyy, Z.Y., Vira, V.V., Vasyliv, B.D.: Study of structural changes in a nickel oxide containing anode material during reduction and oxidation at 600 °C. In: Springer Proceedings in Physics, vol. 221 (2019). https://doi.org/10.1007/978-3-030-17759-1_42
7. Blikhars'kyi, Z.Y., Obukh, Y.V.: Influence of the mechanical and corrosion defects on the strength of thermally hardened reinforcement of 35GS steel. Mater. Sci. **54**, 273–278 (2018). https://doi.org/10.1007/s11003-018-0183-2
8. Blikhars'kyi, Y.Z.: Anisotropy of the mechanical properties of thermally hardened A500s reinforcement. Mater. Sci. **55**, 175–180 (2019). https://doi.org/10.1007/s11003-019-00285-0
9. Klymenko, Y.V.: Technical condition of buildings and structures. Monograph. ODABA, Odessa (2010)
10. Blikharskyy, Y., Kopiika, N., Selejdak, J.: Non-uniform corrosion of steel rebar and its influence on reinforced concrete elements' reliability. Prod. Eng. Arch. **26**(2), 67–72 (2020). https://doi.org/10.30657/pea.2020.26.14
11. Klymenko, Y.V., Dudenko, T.A.: Calculation of damaged reinforced concrete columns. Resour. Conserv. Mater. Constr. Equipment Equipment **27**, 448–453 (2013)
12. Zhang, Q., Mol'kov, Y.V., Sobko, Y.M., et al.: Determination of the mechanical characteristics and specific fracture energy of thermally hardened reinforcement. Mater. Sci. **50**, 824–829 (2015). https://doi.org/10.1007/s11003-015-9789-9
13. Krainskyi, P., Blikharskyy, Y., Khmil, R., Blikharskyy, Z.: Experimental study of the strengthening effect of reinforced concrete columns jacketed under service load level. In: MATEC Web of Conferences, vol. 183, p. 02008. https://doi.org/10.1051/matecconf/201818302008
14. Toryanik, M.S.: Oblique Eccentric Compression and Oblique Bending in Reinforced Concrete. Gosstroyizdat, Kyiv (1961)
15. Toryanik, M.S., Vakhnenko, P.F., Faleev, L.V.: Calculation of Reinforced Concrete Structures With Complex Deformations. Stroyizdat, Moscow (1974)
16. Klymenko, Y.V., Oreshkovich, M., Zadravich, V., Kos, Z.: Structural reliability and evaluation of current state of construction. Tehnički glasnik **4**, 426–431 (2015)
17. Zatkalíková, V., Markovičová, L.: Influence of temperature on corrosion resistance of austenitic stainless steel in cl− containing solutions. Prod Eng. Arch. **25**(25), 43–46 (2019). https://doi.org/10.30657/pea.2019.25.08

Designing of Standard Cross Sections of Composite Bending Reinforced Concrete Elements by the Method of Design Resistance of Reinforced Concrete

Dmitro Kochkarev[1], Taliat Azizov[2], Anna Azizova[3], and Tatiana Galinska[3(✉)]

[1] National University of Water and Environment Engineering, Rivne, Ukraine
[2] Pavlo Tychyna Uman State Pedagogical University, Uman, Ukraine
[3] National University "Yurii Kondratyuk Poltava Polytechnic", Poltava, Ukraine
galinska@i.ua

Abstract. The principles of designing standard cross sections of composite bending reinforced concrete elements using modern deformation models have been considered in the proposed article. This paper is dedicated to composite beams with symmetrical cross sections, which are most frequently used in practical activity in construction sphere. In such elements there are neither torsion, nor tension in perpendicular plane of effect of forces, thanks to which they have the highest loading-carrying ability as compared to asymmetrical composite constructions. The authors have developed an alternative calculation method for the composite bending elements by using the method of analysis of design resistance of reinforced concrete. The given method makes it quite easy to make calculations of such elements, concurrently its precision corresponds to the accuracy of calculation of modern deformation models, taking into account nonlinear deformation curves of materials. The procedure for calculating three types of problems, which are typically used in construction practice, has been proposed in this research work. This article presents the procedure of calculation of the following problems: determination of the required area of reinforcing steel according to the known size dimensions of composite beams and defined strength grade of materials; determination of the bearing capacity of standard cross sections with known reinforcement; checking procedure of the bearing capacity of composite bending elements. An example of calculation of standard cross sections of composite bending reinforced concrete elements made of different concrete grades.

Keywords: Deformation model · Method of design resistance of reinforced concrete · Composite beams with symmetrical cross sections

Composite bending elements take their rightful place in modern design practice. Such elements are prevalently used in the reconstruction of existing beam elements. One of complicated problems that arise in the design of composite constructions is the issue of ensuring collaboration between composite constructions made of different concrete grades and, in some cases, even between different materials. Collaboration of composite elements is assured both by the adhesion between concrete of different grades,

Z. Blikharskyy (Ed.): EcoComfort 2020, LNCE 100, pp. 202–211, 2021.
https://doi.org/10.1007/978-3-030-57340-9_25

and by setting special anchors, screw bolts, drawing rods and confining elements. As these issues were properly investigated in the scientific study [1–13], we have not examined them in this paper.

1 Calculation Elements Is Performed Based on Nonlinear Calculation Models Through the Use of Factual Deformation Curves of Concrete and Reinforcing Steel

Modern deformation models of calculation make it possible to perform the calculation of such elements, but at the same time this calculation is quite complex and can be realized only with the help of purposely designed programs on electronic computers.

Calculation of such elements is performed based on nonlinear calculation models through the use of factual deformation curves of concrete and reinforcing steel. Basic prerequisites for the calculation are the following:

1) Composite cross-sections collaborate without displacing one respecting another;
2) The dependence of longitudinal deformations of concrete in the cross section of the composite elements is assumed to be linear;
3) The dependence of “tension – deformations” in the concrete of the compression area is taken as nonlinear - in the functional form, proposed in Eurocode-2 [14, 18, 20];
4) The dependence of “tension – deformations” in reinforcing steel is taken as a bilinear Prandtl diagram;
5) We do not take into account the force in the tensile zone of the concrete in composite beams.

Forces in compressed concrete can be determined in two ways. The first method involves breaking down the cross section of a composite beam into a certain number of sections within which stresses are taken to be constant [15–17]. In that case, the shape of the accepted stress function in the concrete of the compression area is not significant. The second method is to integrate the obtained equations directly. In such instance, it is advisable to use polynomial dependencies [19, 21–26] that can easily be integrated. This calculation, regardless of the means adopted, is performed by the method of iterations. At the same time, it remains quite difficult in practical application, especially in the absence of purposely designed computer programs.

2 Basis for Calculation of Standard Cross Sections of Composite Bending Reinforced Concrete Elements by the Method of Design Resistance of Reinforced Concrete

Let us consider composite bending reinforced concrete elements with symmetrical cross section. Such elements are most commonly used in the engineering practice of building composite constructions. In such elements there are neither torsion, nor tension in perpendicular plane of effect of forces, thanks to which they have the highest loading-carrying ability as compared to asymmetrical composite constructions.

We would like to offer an alternative method of calculating standard cross sections of composite bending elements. This method implies the usage of the properties of design resistance of reinforced concrete and greatly simplifies the whole process of calculation of composite elements.

All the foregoing prerequisites are the framework for the method of design resistance of reinforced concrete. However, after simple transformations and, taking into account the fracture criteria, the basic equations of equilibrium are transformed into the strength condition. The real design decisions of the master plans of grain terminals or hopper enterprises, involve a simple arrangement of silos in the form of small groups with parallel, perpendicular placement or some angular deviation. Preferably, silo parks include silos of the same diameter and height, which greatly simplifies their study.

$$M = f_z W_c, \tag{1}$$

where f_z - the design resistance of reinforced concrete in bending, MPa, W_c - moment resistance of working cross section of concrete, m^3.

The design resistance of reinforced concrete in bending can be calculated by the formula:

$$f_z = k_z f_{cd}. \tag{2}$$

The parameter k_z is functionally dependent on the mechanical percentage of reinforcement, which can be determined by the following expression:

$$\omega = \frac{\rho_f f_{yd}}{f_{cd}}, \tag{3}$$

where ρ_f - the reinforcement ratio of the cross-section with longitudinal reinforcement, f_{yd} - design resistance of the longitudinal reinforcement, f_{cd} - design resistance of compressive strength of concrete.

Functional dependency $k_z = f(\omega)$ is nonlinear in nature, so, to make its usage easier, it can be approximated by straight-line sections. In this case, the defining formulae for the parameter k_z or for the mechanical percentage of reinforcement will take the following form:

$$k_z = \alpha + \beta\omega; \tag{4}$$

$$\omega = (k_z - \alpha)/\beta. \tag{5}$$

This dependency and its approximation are presented in Table 1.

Let us consider cross section of a composite beam with rectangular cross-section (Fig. 1). The cross-section consists of three rectangles – the average rectangular cross-section has compressive strength of concrete f_{c1} and width b_1, two boundary cross sections f_{c2} and the corresponding widths $b_2/2$. In such a situation, the total width of the cross-section will be $b = b_1 + b_2/2 + b_2/2$.

Let us conditionally decompose a cross section of a composite beam in accordance with Fig. 1, in such a way, so that the strength of two cross sections from different concrete grades would correspond to the strength of the composite cross-section. Namely, upon the given geometrical parameters of two beams from different concrete grades, it is necessary to distribute the longitudinal reinforcement between them accordingly. In this case, according to the method of design resistance of reinforced concrete, the strength of the composite cross section will be equal to:

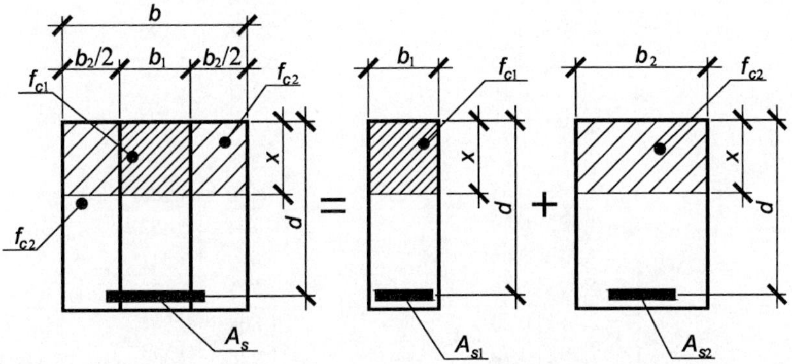

Fig. 1. To the calculation of cross sections of a composite beam by the method of design resistance of reinforced concrete.

$$M_u = M_{u1} + M_{u2};\ M_u = f_{c1}k_{z1}\frac{b_1 d^2}{6} + f_{c2}k_{z2}\frac{b_2 d^2}{6}. \tag{6}$$

As you can see from Eq. (6), it is possible to divide a composite beam into two beams only on condition of equality:

$$f_{c1}k_{z1} = f_{c2}k_{z2}. \tag{7}$$

Genuinely, if the condition of equality (7) is correct, then

$$M_u = f_{c1}k_{z1}\frac{d^2}{6}(b_1 + b_2) = f_{c1}k_{z1}\frac{bd^2}{6}. \tag{8}$$

In this case, the problem of calculating the strength of the composite beam reduces to finding the corresponding allocation of reinforcement between cross sections of regular beams.

Let us write the statement of the problem Eq. (7) with consideration to the Eq. (4):

$$f_{c1}(\alpha_1 + \beta_1 \omega_1) = f_{c2}(\alpha_2 + \beta_2 \omega_2). \tag{9}$$

In expanded form, Eq. (9) will be the following:

Table 1. Functional dependency $k_z = f(\omega)$ and its approximation

n/n	Parameters for calculating bending elements			
	k_z	Mechanical percentage ω	Approximating parameters	
			α	β
1	0,000	0,00	0,000	5,678
2	0,568	0,10	0,048	5,196
3	0,828	0,15	0,096	4,874
4	1,071	0,20	0,161	4,553
5	1,299	0,25	0,241	4,231
6	1,511	0,30	0,338	3,910
7	1,706	0,35	0,450	3,588
8	1,885	0,40	0,744	2,852
9	2,028	0,45	1,649	0,841
10	2,070	0,50	1,722	0,696
11	2,140	0,60	1,808	0,554
12	2,195	0,70	1,928	0,382
13	2,310	1,00	2,143	0,166
14	2,476	2,00	0,000	5,678
15	2,542	3,00		

$$f_{c1}\left(\alpha_1 + \beta_1 \frac{A_{s1} f_{yd}}{b_1 d}\right) = f_{c2}\left(\alpha_2 + \beta_2 \frac{A_{s2} f_{yd}}{b_2 d}\right). \tag{10}$$

Given that:

$$A_{s2} = A_s - A_{s1}, \tag{11}$$

after simple transformations of the Eq. (10) we will get:

$$A_{s1} = \frac{\frac{\beta_2 f_{yd} A_s}{b_2 d} + \alpha_2 f_{c2} - \alpha_1 f_{c1}}{\frac{\beta_1 f_{yd}}{b_1 d} + \frac{\beta_2 f_{yd}}{b_2 d}}. \tag{12}$$

For ease of use of the Eq. (12) we will introduce the notation

$$f_1 = \frac{A_s f_{yd}}{b_1 d},\ f_2 = \frac{A_s f_{yd}}{b_2 d}. \tag{13}$$

Then the Eq. (12) will be written in the following form:

$$A_{s1} = A_s \frac{\beta_2 f_2 + \alpha_2 f_{c2} - \alpha_1 f_{c1}}{\beta_1 f_1 + \beta_2 f_2}. \tag{14}$$

The abovementioned expressions allow you to solve three types of problems for calculating standard cross sections of composite beams: 1. Determination of the required area of the working reinforcement of composite beams; 2. Determination of strength of standard cross sections of composite beams; 3. Checking the strength of standard cross sections of composite beams.

The following offers the calculating procedure for the proposed problems.

The procedure for determining the required area of the working reinforcement of composite beams:

1) Initially, we define the necessary parameters:

$$k_{z1} = \frac{6M_u}{bd^2 f_{c1}},\ k_{z2} = \frac{6M_u}{bd^2 f_{c2}}. \tag{15}$$

2) Then we will determine mechanical reinforcement ratios:

$$\omega_1 = (k_{z1} - \alpha_1)/\beta_1,\ \omega_2 = (k_{z2} - \alpha_2)/\beta_2 \tag{16}$$

3) Now it is necessary to determine the required area of reinforcement of each element:

$$A_{s1} = \frac{f_{c1}}{f_{yd}} \omega_1 b_1 d,\ A_{s2} = \frac{f_{c2}}{f_{yd}} \omega_2 b_2 d. \tag{17}$$

4) The full area of the longitudinal reinforcement of a composite element is equal to:

$$A_s = A_{s1} + A_{s2}. \tag{18}$$

The procedure for determining the strength of standard cross sections of composite beams with the known longitudinal reinforcement is the following:

1) We determine the area of the longitudinal reinforcement A_{s1}. by the Eq. (17). The fault-identifying variables (auxiliary parameters) α and β are taken preliminarily and can be refined as necessary.
2) We find out the mechanical percentage of reinforcement by the following expressions:

$$\omega_1 = \frac{f_{yd}A_{s1}}{f_{c1}b_1d}, \ \omega_2 = \frac{f_{yd}(A_s - A_{s1})}{f_{c2}b_2d}.$$

3) We define the fault-identifying variables (accessory parameters) α_1, α_2, β_1, β_2 according to Table 1 and check the correctness of their assumption in example 1.
4) We define an accessory parameter $k_{z1} = \alpha_1 + \beta_1\omega_1$.
5) We define the bearing capacity of a composite bending element by the following expression:

$$M_u = f_{c1}k_{z1}\frac{bd^2}{6}. \tag{20}$$

The proposed technique is based on the universal dependence of strength characteristics on stress-related characteristics, that is why it can also be applied to different concrete grades and, even, to rock materials, for which in some cases it is necessary to limit the value of mechanical percentage of reinforcement ω.

3 Calculation Examples of Standard Cross Sections of Composite Bending Reinforced Concrete Elements

Example. It is necessary to determine the cross-sectional area of the longitudinal working reinforcement of the composite beam. A beam with cross section b × h = 400 × 400 mm (d = 35 cm) made of concrete grade C12/15 b_1 = 150 mm (f_{c1} = 8,5 MPa) and concrete C20/25 b_2 = 250 mm (f_{c2} = 14,5 MPa) should perceive the beam moment (moment of flection) M_u = 100 kN × m (see Fig. 2).
The solution to the problem.

1) We define the necessary parameters

$$k_{z1} = \frac{6M_u}{bd^2f_{c1}} = \frac{6 \times 100 \times 10^6}{400 \times 350^2 \times 8,5} = 1,441. \, k_{z2} = \frac{6M_u}{bd^2f_{c2}} = 0,844.$$

2) We set up mechanical percentages of reinforcement:

$$\omega_1 = (k_{z1} - \alpha_1)/\beta_1 = (1,441 - 0,241)/4,231 = 0,284; \ \omega_2 = 0,153.$$

3) Then we determine the required area of reinforcement of each element:

$$A_{s1} = \frac{f_{c1}}{f_{yd}}\omega_1 b_1 d = \frac{8,5}{415}0,284 \times 150 \times 350 = 305,4 \text{ mm}^2.$$

$$A_{s2} = \frac{f_{c2}}{f_{yd}}\omega_2 b_2 d = \frac{14,5}{415}0,153 \times 250 \times 350 = 467,8 \text{ mm}^2.$$

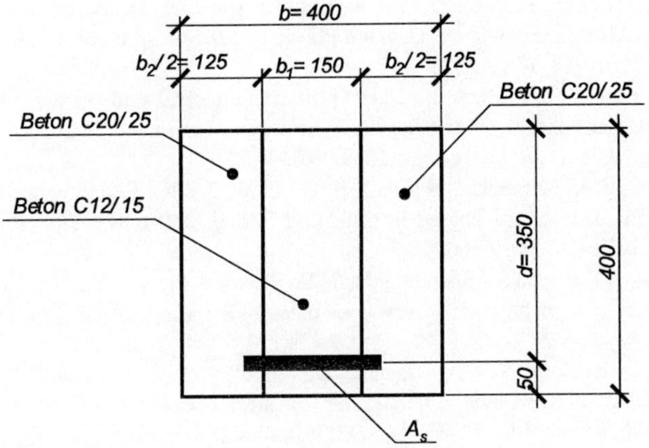

Fig. 2. To examples.

4) The full area of the longitudinal reinforcement of a composite element is equal to: $A_s = A_{s1} + A_{s2} = 305,4 + 467,8 = 773,2\ mm^2$

For reinforcement we take on 4∅16 A500 (f_{yd} = 415 MPa), A_{sl} = 804 mm^2.

4 Conclusions

The authors of the article have offered the calculation method for standard cross sections of composite bending reinforced concrete elements by using the method of analysis of design resistance of reinforced concrete. This approach makes it possible to solve three types of problems: determination of the required area of reinforcing steel according to the known size dimensions of composite beams and defined strength grade of materials; determination of the bearing capacity of standard cross sections with known reinforcement; checking procedure of the bearing capacity of composite bending elements. In the future, it is planned to make an experimental comparison of the proposed approach (method) on composite beams made of different concrete grades.

References

1. Wahab, N., Soudki, K., Topper, T.: Experimental investigation of bond fatigue behavior of concrete beams strengthened with NSM prestressed CFRP rods. J. Compos. Constr. **16**(6), 684–692 (2014)
2. Minelli, F., Plizzari, G., Cairns, J.: Flexure and shear behavior class of RC beams strengthened by external reinforcement. In: Concrete Repair, Rehabilitation And Retrofittingii, pp. 377–378 (2009)

3. Aymerich, F., Fenu, L., Loi, G.: FE Analysis of the Flexural Behavior of Cementitious Composites Using the Concrete Damage Plasticity Model. Springer, Cham (2020). https://doi.org/10.1007/978-3-030-23748-6_10
4. Wang, Y., Yu, J., Liu, J., Zhou, B., Chen, Y.F.: Experimental study on assembled monolithic steel-prestressed concrete composite beam in negative moment. J. Constr. Steel Res. **167** (2020). https://doi.org/10.1016/j.jcsr.2019.06.004
5. Thevendran, V., Shanmugam, N.E., Chen, S., Liew, J.Y.R.: Experimental study on steel-concrete composite beams curved in plan. Eng. Struct. **22**(8), 877–889 (2000). https://doi.org/10.1016/s0141-0296(99)00046-2
6. Thevendran, V., Chen, S., Shanmugam, N.E., Richard Liew, J.Y.: Nonlinear analysis of steel-concrete composite beams curved in plan. Finite Elem. Anal. Des. **32**(3), 125–139 (1999)
7. Hamoda, A., Hossain, K.M.A., Sennah, K., Shoukry, M., Mahmoud, Z.: Behaviour of composite high performance concrete slab on steel I-beams subjected to static hogging moment. Eng. Struct. **140**, 51–65 (2017). https://doi.org/10.1016/j.engstruct.2017.02.030
8. Zhang, J., Li, S., Xie, W., Guo, Y.: Experimental study on shear capacity of high strength reinforcement concrete deep beams with small shear span-depth ratio. Materials **13**(5) (2020). https://doi.org/10.3390/ma13051218
9. Ahmadi, M., Kheyroddin, A., Dalvand, A., Kioumarsi, M.: New empirical approach for determining nominal shear capacity of steel fiber reinforced concrete beams. Constr. Build. Mater. **234** (2020). https://doi.org/10.1016/j.conbuildmat.2019.117293
10. Shahnewaz, M., Rteil, A., Alam, M.S.: Shear strength of reinforced concrete deep beams – a review with improved model by genetic algorithm and reliability analysis. Structures **23**, 494–508 (2020). https://doi.org/10.1016/j.istruc.2019.09.006
11. Peng, J., Zhao, P., Wang, S., Lee, S.W., Kang, S.: Interface shear transfer in reinforced engineered cementitious composites under push-off loads. Eng. Struct. **206** (2020). https://doi.org/10.1016/j.engstruct.2019.110013
12. Wang, D., Zhang, J., Guo, J., Fan, R.: A closed-form nonlinear model for spatial Timoshenko beam flexure hinge with circular cross-section. Chin. J. Aeronaut. **32**(11), 2526–2537 (2019). https://doi.org/10.1016/j.cja.2019.01.025
13. Azizov, T., Melnik, O., Myza, O.: Strength and deformation of combined beams with side reinforced plates. In: Materials Science Forum Submitted 24 Apr 2019, vol. 968, pp. 234–239 (2019). ISSN: 1662-9752. https://doi.org/10.4028/www.scientific.net/MSF.968.234. Accepted: 29 May 2019
14. Azizov, T., Derkowski, W., Jurkowska, N.: Consideration of the torsional stiffness in hollow-core slabs' design. In: Materials Science Forum Submitted 28 May 2019, vol. 968, pp. 330–341 (2019). ISSN: 1662-9752 (2019)
15. Dem'yanov, A., Kolchunov, Vl., Iakovenko, I., Kozarez, A.: Load Bearing capacity calculation of the system "reinforced concrete beam – deformable base" under torsion with bending. In: E3S Web Conference vol. 97, XXII International Scientific Conference "Construction the Formation of Living Environment" (FORM-2019) (2019). https://doi.org/10.1051/e3sconf/20199704059
16. Kolchunov, V.I., Dem'yanov, A.I., Naumov, N.V.: The second stage of the stress-strain state of reinforced concrete constructions under the action of torsion with bending (theory). In: Paper presented at the IOP Conference Series: Materials Science and Engineering, vol. 753, no. 3 (2020). https://doi.org/10.1088/1757-899x/753/3/032056
17. Iakovenko, I.A., Kolchunov, V.I.: The development of fracture mechanics hypotheses applicable to the calculation of reinforced concrete structures for the second group of limit states. J. Appl. Eng. Sci. **15**(3), 371–380 (2017). https://doi.org/10.5937/jaes15-14662

18. Kochkarev, D., Azizov, T., Galinska, T.: Bending deflection reinforced concrete elements determination. In: Paper Presented at the MATEC Web of Conferences, vol. 230 (2018). https://doi.org/10.1051/matecconf/201823002012
19. Khmil, R., Tytarenko, R., Blikharskyy, Y., Vegera, P.: Development of the procedure for the estimation of reliability of reinforced concrete beams, strengthened by building up the stretched reinforcing bars under load. Eastern-Eur. J. Enterp. Technol. **5/7**(95) (2018)
20. Azizov, T., Jurkowska, N., Kochkarev, D.: Basis of calculation on torsion for reinforced concrete structures with normal cracks. In: Concrete Innovations in Materials, Design and Structures. Fib Symposium 2019. Cracow 27–29 May 2019. Book of Abstracts. S, pp. 489–490 (2019)
21. Kochkarev, D., Galinska, T.: Calculation methodology of reinforced concrete elements based on calculated resistance of reinforced concrete. In: Matec Web of Conferences, vol. 116, p. 02020 (2017). Materials science, engineering and chemistry, Transbud–2017, Kharkiv, Ukraine, 19–21 April (2017)
22. Lobodanov, M., Vegera, P., Blikharskyy, Z.: Planning experiment for researching reinforced concrete beams with damages. In: Blikharskyy, Z., Koszelnik, P., Mesaros, P. (eds.) Proceedings of CEE 2019. CEE 2019. Lecture Notes in Civil Engineering, vol. 47. Springer, Cham (2020)
23. Blikharskyy, Z., Vashkevych, R., Vegera, P., Blikharskyy, Y.: Crack resistance of RC beams on the shear. In: Blikharskyy, Z., Koszelnik, P., Mesaros, P. (eds.) Proceedings of CEE 2019. CEE 2019. Lecture Notes in Civil Engineering, vol. 47. Springer, Cham (2020)
24. Azizov, T., Kochkarev, D., Galinska, T.: Design of Effective Statically Indeterminate Reinforced Concrete Beams (2020). https://doi.org/10.1007/978-3-030-42939-3_10
25. Azizov, T., Kochkarev, D., Galinska, T.: Reinforced concrete rod elements stiffness considering concrete nonlinear properties. In: Blikharskyy, Z., Koszelnik, P., Mesaros, P. (eds.) Proceedings of CEE 2019. CEE 2019. Lecture Notes in Civil Engineering, vol. 47. Springer, Cham (2020). https://doi.org/10.1007/978-3-030-27011-7_1
26. Dmitriy, K., Tatyana, G.: Nonlinear calculations of the strength of cross-sections of bending reinforced concrete elements and their practical realization, cement based materials. In: El-Din, H., Saleh, M., Abdel Rahman, R.O. (eds.) IntechOpen (2018). https://doi.org/10.5772/intechopen.75122

Promising Trends in Design of LED Lighting Combined with Systems of Natural Lighting

Lidiia Koval(✉), Volodymyr Yehorchenkov, and Viacheslav Martynov

Kyiv National University of Construction and Architecture, Kyiv 03037, Ukraine
likocolor@gmail.com

Abstract. The article deals with promising trends in LED lighting design combined with daylight systems. In the process of research such methods as analysis, synthesis, generalization, visual and graphic modeling, and scientific observation were applied. The article characterizes the features of combined lighting and its main conventional schemes, as well as analyzes modern examples of combining LED lighting with the daylight systems. It has been discovered that the adjustment of natural and artificial lighting systems can be secured by the combination of their elements, by the addition of luminous flux from artificial sources to solar radiation (combined lighting), or by the combinations of these two schemes. The research has enabled to point out the absence of the modern examples of the structural combination of LED lighting with the most common systems of lateral daylight lighting used in the apartment buildings. A relevant project proposal of the autonomous lighting system, made on the basis of solar batteries and LED light sources, has been made. The study is both theoretical and applied, it may be used when preparing design proposals and for creating light environment of various premises.

Keywords: Lighting design · LED lighting · Systems of natural lighting · Combined lighting · Autonomous lighting system · Solar battery

1 Introduction

The question of combined lighting arises when natural and artificial lighting systems are used together. In general, a separate chapter of the work edited by Ju. B. Ajzenberg [1] is devoted to combined lighting. D. Carter considers technological and design features of the modern daylight transport system, calling it "passive tubular daylight guidance systems" [3]. In D. Radomtsev's article, "integrated lighting systems" means an innovative method of input, transportation and distribution of luminous flux in the premises, which combines both natural luminous flux (from the sky and the sun) during the day, and artificial, created by lighting devices at night [12]. K. I. Suvorova and L. D. Hurakova define a similar lighting transport system as "hybrid system of combined lighting" [14]. Thus, the preliminary analysis confirms the presence of rather diverse research classifications of modern combined lighting systems which need further generalization.

Z. Blikharskyy (Ed.): EcoComfort 2020, LNCE 100, pp. 212–219, 2021.
https://doi.org/10.1007/978-3-030-57340-9_26

At the same time, the popularity of LEDs has highlighted the problem of light pollution outlined by P. Boyce [2]. He notes that due to the existing technical preconditions that make it easier to control the LED light distribution, one could expect a reduction in the sky glow due to the adoption of LED sources in outdoor lighting. However, the analysis of satellite observations, conducted by Kyba et al., has demonstrated that in the period from 2012 to 2016, the area of artificially illuminated locations on the Earth has increased by at least 2.2% per year, and the total glow – by 1.8% per year [8]. M. Perry and L. Fennelly believe that this problem of outdoor artificial lighting can be solved by providing only minimal levels of illumination [11]. However, the purposeful use of lighting will also help reduce light pollution. For example, the light coming out in the evening and at night from the windows of buildings can be considered untargeted lighting. Accordingly, an interesting task is to create a system of artificial lighting, which by its design and logic of operation would play an additional role of a certain "limiter" (according to D. Norman [10]), preventing or minimizing the leakage of artificial sources light from the building to the outside.

2 Purpose

The purpose of this study is to identify promising directions of LED lighting design combined with natural lighting systems and to develop a design proposal of LED lighting, structurally combined with the lateral daylight lighting systems, most commonly used in the apartment buildings.

3 Methods

In the course of this research such widely used general scientific methods as analysis, synthesis and generalization were applied. The study was carried out in the following sequence. First, the main conventional schemes of combined lighting were described. Then, the current state of the development of daylight transport lighting systems was analyzed on the example of different manufacturers' products. The analysis was performed grounding on the illustrative, video and textual information from free access sources [13], in particular from the official websites of the manufacturers [4], as well as from operating manuals and technical characteristics of products [5], available by subscription which is provided by the manufacturers.

As a result of the analysis, the absence of examples of a structural combination of LED lighting with lateral daylight lighting systems was established. Further scientific observation of current trends in lighting has revealed the widespread use of photovoltaic cells (solar cells) to power LED sources in autonomous systems and luminaires. Therefore, the final stage of the work was the development of an appropriate project proposal for an autonomous lighting system based on a solar battery and LED light sources, structurally combined with a system of lateral daylight lighting. Computer visual and graphic modeling was used for this purpose.

4 Results

Depending on the planning concept, geometrical proportions, the purpose of premises and the systems of natural lighting applied in them, the following basic generally accepted schemes of the combined lighting [1], shown in Fig. 1, are distinguished: the configuration in Fig. 1a can be used in small rooms with the depth of 6–8 m; the configuration in Fig. 1b is typical for rooms with upper natural lighting, without lateral illumination; the configuration in Fig. 1c is suitable for rooms whose interior consists of two zones, one of which has sufficient natural light and the other is illuminated only by artificial light; the configuration in Fig. 1d is used in deep wide-area rooms with natural lateral lighting.

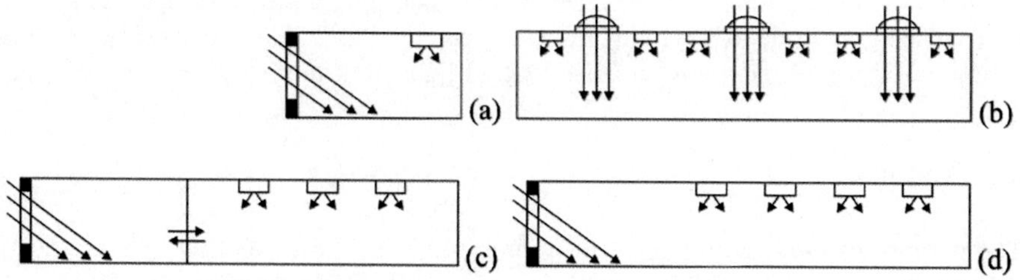

Fig. 1. The basic configurations of combining daylight lighting with artificial one inside buildings, according to [1].

The configurations in Fig. 1 demonstrate the options for combining artificial lighting systems with the main systems of daylight – top and one-sided lateral – illumination. Having considered these configurations, it becomes obvious that they demonstrate such a way of combining lighting, which provides *the addition of luminous flux from artificial sources to solar radiation sources*. It is known that one-sided lateral daylight lighting systems have a significant disadvantage which is considerably uneven luminance of the (horizontal) work surface. Modern systems of luminous flux redistribution aim at reducing this shortcoming by directing the reflected flux to a remote area of the white ceiling. However, one-sided lateral lighting remains the most commonly used daylight lighting system, especially in apartment buildings. That is why the use of energy sustainable light sources (LEDs) in order to compensate for the lack of natural light by artificial means has significantly expanded the scope of combined lighting. It helps to increase the overall energy efficiency of the building, increases the depth of the premises and the width of the house, which improves the compactness.

Further analysis of modern daylight transport systems produced by well-known manufacturers has demonstrated that most of them provide, in addition to the usual combined lighting, an additional structurally conditioned variant of combining artificial lighting with natural one. The principle of operating such lighting systems is based on maintaining the required level of illumination of the room with an integrated photosensor to control electric light, which complements the daylight one. In the evening and with a significant reduction in the sky glow, an inbuilt LED module starts working. In automatic mode, it adds the luminous flux which is necessary to provide a given level of luminance. Therefore, combined lighting based on daylight transport lighting systems is mainly characterized by a *constructive unity with artificial light sources and a simultaneous combination of artificial and natural luminous fluxes*. This is facilitated by such a system component as a diffuser placed in front of the light output, which scatters both daylight itself and combined with artificial light. This element ensures the unity of perception of both kinds of lighting. Therefore, the main feature of a *structural combination* of natural and artificial lighting systems are common light outputs.

Daylight transport systems are similar to roof lights, which by their location are not directly seen by the user and do not provide a view to the outside. Lateral lighting systems provide an external view, have a small depth and are directly seen by the user. Therefore, a combination of artificial lighting systems with natural lateral lighting ones by *a structural unity of both systems' components* is possible, but it does not provide a combination of luminous fluxes. Under such conditions, the resulting lighting cannot be called combined. However, the light output openings, which are common to both systems, indicate the *possibility of their structural combination*. Accordingly, a design proposal for LED lighting was developed, which can also be classified (according to [9]) as a system of accumulated natural lighting.

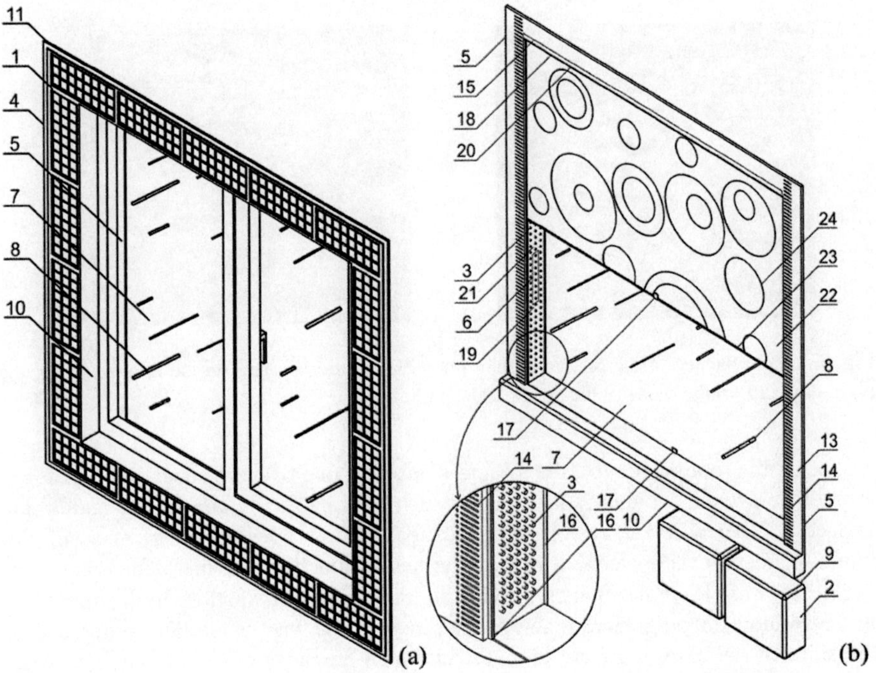

Fig. 2. General view of an autonomous lighting system based on a solar battery and LED light sources: a - outside the building; b - inside the room.

This problem can be solved with the help of an autonomous lighting system based on a solar battery and LED light sources containing: a solar battery; standalone battery; LEDs; the outer panel on which the solar battery is mounted; the case on which LEDs are mounted; control system; opaque screen, which is made in one piece and whose surface from the side of the room has a reflection coefficient of 0.7 and above. The idea of a utility model is explained in the drawings, where Fig. 2a demonstrates a general view of the proposed outdoor lighting system, and Fig. 2b shows its general view from inside the room. The legend is the following: 1 – solar battery, 2 – stand-alone battery, 3 – LEDs, 4 – outer panel, 5 – case, 6 – control system, 7 – opaque screen, 8 – cutouts on the opaque screen, 9 – batteries below the window sill, 10 – window opening, 11 – the flat surface of the outer panel, 13 – case front, 14 – holes in the case front, 15 – blinds, 16 – grooves, 17 – clasp, 18 – case top, 19 – lateral case surface, 20 – the top flat surface of the case, 21 – the lateral flat surface of the case, 22 – extra light-diffusing screen, 23 – extra light-screen with several layers, 24 – shaped slots in the light-scattering screen.

The outdoor lighting system (Fig. 2a) contains one or more solar batteries, which are mounted on an external panel having a flat (embossed or corrugated) surface. The embossed (or corrugated) surface of the outer panel allows placing a part of the photovoltaic cells of the solar battery at an angle to the incident sunlight, which is optimal for their performance in the area. The opaque screen covers the window slot and may have curly thin slots.

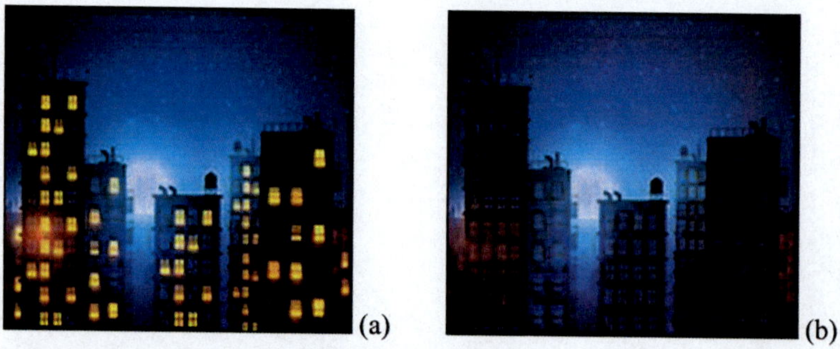

Fig. 3. The influence of the offered autonomous illumination system on the light image of the night city and on the visual indicator of light pollution.

Inside the room (Fig. 2b), the system comprises one or more stand-alone batteries located below the window sill, LEDs, which are placed on the lateral and/or top surfaces of the case. The case is built into the window opening and is made of aluminium, a composite material with high thermal conductivity, or thermally conductive plastic; its front has holes, which provide additional heat dissipation. In the upper part of the window opening, there is an opaque screen in the form of a roller shutter, and on the sides of the case, there are grooves on which the sides of the screen slide when untwisting. Close to the lower wall of the window opening (window sill), there is a clasp for fixing the lower side of the opaque screen. An additional light scattering screen can be placed parallel to it, which can have several layers of different light transmission with shaped slots.

The proposed lighting system works as follows. In daylight, solar energy is converted in the solar battery into electricity, which is delivered to a standalone battery, where it accumulates. At night or in cloudy weather, the window opening is closed by an opaque screen and an additional light-scattering screen, the control panel is used to turn on the required number of LEDs. Placing an autonomous battery (batteries) inside the room below the window sill greatly simplifies maintenance, namely replacement, and minimizes the distance from the battery to light sources, and, consequently, the loss of the wire resistance. Placing the outer panel outside the building along the perimeter of the window opening simplifies its installation and maintenance, in particular the cleaning of photovoltaic cells to prevent a decrease in efficiency.

Figure 3 provisionally demonstrates how the luminous image of the night city and the visual indicator of light pollution (Fig. 3a) may change in the case of implementation of the proposed lighting method (Fig. 3b). At the same time, thin shaped slots in the opaque screen of the lighting system can serve as marking illumination, supporting the visual perception of the building dimensions at night and aesthetically enriching the light environment of the city with minimal contribution to total light pollution.

This article presents an abridged description of the design of the developed autonomous lighting system. A complete set of design variations and drawings is available by reference [7] in the patent description. Since the lighting part of this system is a surface that emits uniformly scattered light, the illuminance from it at a room point can be calculated by means of the adaptation of the famous technique (developed by G. M. Knorring [6]) for defining the illuminance generated by a diffusely emitting surface. Besides, the total luminous flux of the illuminating surface will be defined as the product of the luminous flux of one LED and the total number of employed LEDs, taking into account the reflection coefficient of the internal surfaces of the system and the transmission coefficient of the light scattering screen.

5 Scientific Novelty

The topic of the article is determined by such modern trends in lighting design as an increase in energy efficiency in buildings by means of different options for daylighting and the use of LED sources for artificial lighting. This has raised the problem of finding new ways to combine LED lighting with natural lighting systems. Among them, the following have been singled out: a constructive combination of structural components of natural and artificial lighting systems, the addition of luminous flux of artificial sources to solar radiation (combined lighting), and a combination of these two methods. Based on this, the study proposes and substantiates the use of a constructive combination of LED lighting with natural side lighting systems.

6 Practical Significance

The main conclusions and results of the study can be applied in the design of the lighting environment of modern energy-efficient buildings, as well as in the optimization and improvement of the energy efficiency of lighting in buildings with low energy efficiency. The developed lighting system can be used in newly built buildings and integrated into existing buildings, including multistoried residential complexes. In particular, it can be used for combined lighting, supplemented by ceiling luminaires (or local luminaires), thus compensating insufficient lighting from the accumulated natural light. The light scattering screen can perform an additional sun protecting function. The opaque screen can perform the following additional functions: sun protection in hot weather, protection against heat loss at cold winter nights (if it is insulated, for example, with foam), anti-vandal protection (if it is made as a protective shutter). In the long run, the widespread implementation of the proposed lighting system can help reduce light pollution produced by large cities.

7 Conclusions

The research resulted in the outline of such promising directions of LED lighting design combined with natural lighting systems as: the combination with daylight lighting systems, the combination with daylight transport lighting systems, the combination with lateral daylight lighting systems. In the course of the study, a project proposal for LED lighting, structurally combined with natural side lighting systems, was developed. The proposed autonomous lighting system based on solar batteries and LED light sources (Patent of Ukraine 137082) has the following characteristics:

- solely constructive combination of structural elements of artificial and natural lighting systems, while the combination of luminous fluxes of natural and artificial lighting does not take place;
- the use of inexhaustible sources to power LEDs due to the accumulation of solar energy during the day;
- promoting the reduction of light pollution by preventing or minimizing the leakage of artificial light through the window openings to the outside.

References

1. Ajzenberg, Ju.B., (ed.): A Guidebook on Lighting Engineering. 3nd edn. "Znak" Publishing, Moscow (2006). (in Russian)
2. Boyce, P.: Editorial: voting with their eyes. Light. Res. Technol. **50**, 189 (2018). https://doi.org/10.1177/1477153518758610
3. Carter, D.: Tubular daylight guidance systems. Light. Res. Technol. **46**, 369–387 (2014). https://doi.org/10.1177/1477153514526081
4. Chatron Homepage. https://www.chatron.pt/en/. Accessed 20 Apr 2020
5. Ciralight Homepage. http://www.ciralight.com/. Accessed 20 Apr 2020

6. Knorring, G.M.: Lighting facilities. Energoizdat, Leningrad (1981). (in Russian)
7. Koval, L.M.: Patent of Ukraine 137082. Autonomous lighting system based on a solar battery and LED light sources. State Patent Office of Ukraine, Kyiv (2019). https://base.uipv.org/searchINV/getdocument.php?claimnumber=u201904348&doctype=ou
8. Kyba, C.C.M., Kuester, T., Miguel, A.S., Baugh, K., Jechow, A., Hölker, F., Bennie, J., Elvidge, C.D., Gaston, K.J., Guanter, L.: Artificially lit surface of Earth at night increasing in radiance and extent. Sci. Adv. **3**(11), e1701528 (2017). https://doi.org/10.1126/sciadv.1701528
9. Natural and artificial lighting, DBN V.2.5-28:2018. State Building Codes of Ukraine. Ukrarkhbudinform, Kyiv (2018). (in Ukrainian)
10. Norman, D.A.: The Design of Everyday Things. Russian language ed. Williams Publishing House, Moscow (2008). (in Russian)
11. Perry, M., Fennelly, L.: The Handbook for School Safety and Security. Butterworth-Heinemann is an Imprint of Elsevier, Waltham, USA (2014)
12. Radomtsev, D.: The improvement of energy efficiency of buildings by means of the modern integral lighting systems. Build. Struct. **77**, 274–278 (2013). (in Ukrainian)
13. Solatube Homepage. http://www.solatube.com/. Accessed 20 Apr 2020
14. Suvorova, K.I., Hurakova, L.D.: Modern lighting systems as a resource for energy saving. Electrical Engineering. Urban Community Facilities **7**(146), 121–126 (2018). (in Ukrainian)

Influence of Basalt Fiber Dispersed Reinforcement on the Work of Concrete Beams with Non-metallic Composite Reinforcement

Petro Koval[1], Maksym Koval[2], Yaroslav Balabukh[3], and Oleh Hrymak[3](✉)

[1] National Academy of Fine Arts and Architecture, Kyiv, Ukraine
[2] Scientific-Industrial Enterprise "Triada" Ltd., Co, Lviv, Ukraine
[3] Lviv Polytechnic National University, Lviv, Ukraine
grymak5oleg@gmail.com

Abstract. Recently, non-metallic composite reinforcement has been used more frequently to reinforce concrete beam structures. It has a number of advantages in comparison with steel: much greater tensile strength, high corrosion resistance, low weight, dielectric constant. However, non-metallic composite reinforcement has a lower elasticity compared to steel, and beams reinforced with non-metallic composite reinforcement have worse characteristics of deformability and crack resistance. To improve these characteristics, dispersed basalt fiber reinforcement of concrete beams with non-metallic composite reinforcement has been proposed, the effectiveness of which has been proven experimentally. The test specimens were beams with a cross section of 100 × 200 mm and a length of 2100 mm, which were made of laboratory-made concrete. The basic reinforcement of each of the beams was one rod of working basalt-plastic reinforcement type ANPB with length of 2080 mm (diameter varied in series: 4, 6, 8, 10, 12 or 13 mm). The beams were tested on a power bench according to the scheme of pure bending – two concentrated forces located in the thirds of the run, transmitting the load from the hydraulic jack.

Keywords: Basalt reinforcement · Plastic reinforcement · Basaltic fiber · Beams · Fiber-reinforced concrete

1 Introduction

Concrete has high compressive strength and much lower tensile strength. For the perception of tensile forces, concrete structures are reinforced with rod reinforcement – both metal and non-metal [1–4]. Promising in transport constructions is the use of non-metallic composite reinforcement, made of basalt roving, formed in the rod by a polymeric binder. However, such structures have some disadvantages, in particular - lower (compared to traditional reinforced) concrete crack resistance and higher deformability [5–8].

Fiber dispersed reinforcement can improve the tensile strength of concrete and increase its crack resistance [9–11]. To improve the properties of cement concretes, various fibers are used as fillers: steel, polymer, polypropylene, copolymer, basalt,

Z. Blikharskyy (Ed.): EcoComfort 2020, LNCE 100, pp. 220–226, 2021.
https://doi.org/10.1007/978-3-030-57340-9_27

glass, carbon and others. [10, 12–14]. In some cases, the use of fiber reduces the need for reinforcement. An important argument for the use of dispersed reinforcement is the ease of use [15–17].

With the widespread use of concrete, composite materials deserve special attention, in which the role of the matrix is performed by cement stone obtained on the basis of portland cement, and for microreinforcement – basalt fibers. The use of basalt fiber can largely compensate the main disadvantages of concrete - low tensile strength and brittle fracture. Usage of basalt fiber reduces shrinkage deformations, increases frost resistance, heat resistance, abrasion resistance, moisture resistance of the material, improves crack resistance, impact strength, dielectric properties, etc. [18, 19]. Basalt fiber helps to eliminate the negative impact of stress concentration in places of structural defects of cement stone, increases the deformability and durability of concrete, and is cost-effective [14]. An important condition for the use of basalt fiber for concrete reinforcement is its chemical resistance in the alkaline environment of concrete. Studies [18] show that under certain conditions, basalt fiber can be used for concrete reinforcement. Experiments [19, 20] show that reinforcement with basalt fiber up to 12 mm long does not change the brittle nature of the destruction of concrete prisms in bending and fiber content of 1–2% leads to increased tensile strength in bending concrete class C20 in 1.19–1.23 times. Concrete C20, reinforced with basalt fiber length of 24 and 50 mm with a fiber content of 1–3%, shows a more plastic nature of fracture and tensile strength of concrete in bending increases by 1.79–2.24 times.

According to the given the above facts, the use of basalt-fiber concrete is promising for the construction of bridges, roads and airfields. Therefore, the urgent task is to study the characteristics of concretes reinforced with basalt fiber for use in transport constructions.

2 Method

In order to experimentally determine the effect of fiber on the work of bent basalt concrete structures at the National University "Lviv Polytechnic" tests were performed on 12 concrete (series C) and 12 fiber concrete (series FC) beams reinforced with non-metallic composite reinforcement with different reinforcement coefficients. In the experimental studies, basalt fiber made of basalt roving with a diameter of 16 microns and a length of 24 mm was used. The fiber content was taken by weight from the weight of cement in the dry state and ranged from 0% (control samples without microreinforcement) to 6%. The test specimens were beams with a cross section of 100 × 200 mm and a length of 2100 mm (Fig. 1), which were made of laboratory-made concrete. The basic reinforcement of each of the beams was one rod of working basalt-plastic reinforcement type ANPB with length of 2080 mm (diameter varied in series: 4, 6, 8, 10, 12 or 13 mm). In the extreme thirds of the span, transverse reinforcement was provided with rods of Ø6 mm class A-I with a length of 180 mm. The pitch of the transverse rods was 100 mm, the total number of rods of the transverse reinforcement - 16 pcs. The upper reinforcement is made of rods Ø6 mm class A-I with a length of 730 mm in the extreme thirds of the span. The reinforcement coefficient of the cross section of the structure $\rho_{f,tot}$ is 0.00073 at Ø4 mm; 0.00158 at Ø6 mm; 0.00286 at Ø8 mm; 0.00446 at Ø10 mm; 0.00649 at Ø12 mm; 0.0077 at Ø13 mm.

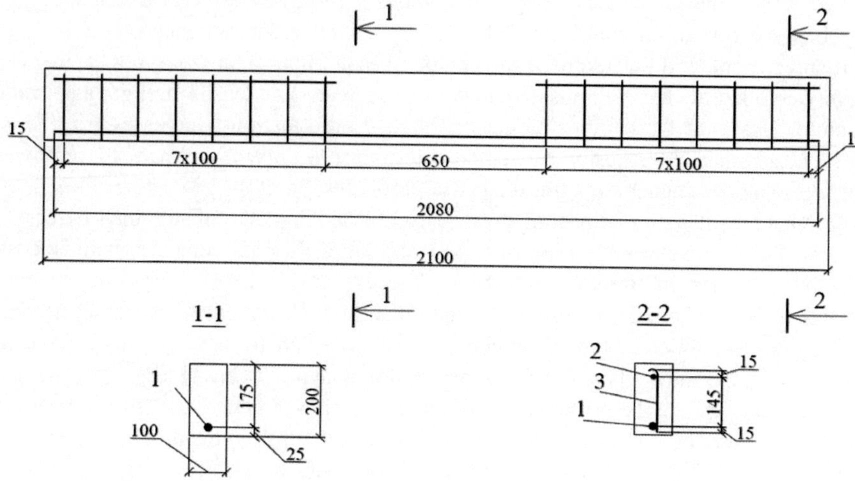

Fig. 1. Construction of basalt concrete beam. 1 – rod of working basalt-plastic reinforcement, 2 – structural rods Ø6 A-I; 3 – structural rods Ø6 A-I.

The prototypes were divided into 6 series (I, II, III, IV, V, VI) depending on the reinforcement coefficient $\rho_{f,tot}$.

The beams were tested on a power bench according to the scheme of pure bending – two concentrated forces located in the thirds of the run, transmitting the load from the hydraulic jack. The samples were loaded in steps of 5… 10% of the expected destructive load. The load endurance at each stage was at least 15 min, after which the results were taken from the devices: the deflections of the beams were measured with 6-PAO deflectometers with a division price of 0.01 mm; fiber deformations of concrete along the height of the beam in the middle of the run were measured with clock-type indicators with a division price of 0.001 mm on the basis of 200 mm and strain gages with a base of 50 mm; the displacement of the ends of the basalt-plastic reinforcement at the ends of the beams was measured by indicators with a division price of 0.001 mm; the crack opening width was determined using a MPB-2 microscope with a partition price of 0.05 mm.

3 Results

The results of studies of basalt concrete beams under static loads – the value of displacements (deflections) immediately before the failure, the nature of the destruction, the value of the destructive load and the value of the load at which cracks began to appear, are given in Table 1. Similar results of researches of basalt-fiber concrete beams are given in Table 2.

Table 1. The results of experimental studies of bent basalt concrete beams under static load.

Series	Sample names	Bending moment of crack formation Mcrc, kNm	Destructive bending moment Mu, kNm	Deflection f at Mu, mm	The nature of the destruction
I	I-C1	2,67	2,67	1,39	Rupture of working reinforcement
	I-C2	2,67	2,67	1,27	
II	II-C1	2,67	4,00	12,82	
	II-C2	3,33	3,67	16,50	
III	III-C1	3,00	9,30	35,80	Rupture of working reinforcement and crushing of concrete in compressed zone
	III-C2	3,33	8,00	37,80	
IV	IV-C1	2,33	12,00	47,82	
	IV-C2	3,00	11,00	42,01	
V	V-C1	2,33	13,00	33,44	Crushing of concrete in compressed zone
	V-C2	2,67	14,00	33,92	
VI	VI-C1	2,67	14,67	37,53	
	VI-C2	2,67	13,30	41,13	

Table 2. The results of experimental studies of bent basalt-fiber concrete beams under static load.

Series	Sample names	Bending moment of crack formation Mcrc, kNm	Destructive bending moment Mu, kNm	Deflection f at Mu, mm	The nature of the destruction
I	I-FC1	2,83	2,83	0,68	Rupture of working reinforcement
	I-FC2	2,67	2,67	0,72	
II	II-FC1	3,33	5,00	4,60	
	II-FC2	3,00	5,00	6,39	
III	III-FC1	3,17	8,00	38,75	Rupture of working reinforcement and crushing of concrete in compressed zone
	III-FC2	3,67	9,30	32,60	
IV	IV-FC1	3,83	12,00	36,70	
	IV-FC2	3,00	13,00	39,26	
V	V-FC1	3,50	14,67	33,44	Crushing of concrete in compressed zone
	V-FC2	3,67	13,30	28,07	
VI	VI-FC1	3,50	15,00	33,63	
	VI-FC2	3,33	14,00	30,27	

Comparison of test results of basalt concrete and basalt fiber concrete beams (Table 3) shows that fiber microreinforcement increases the bearing capacity of beams by 3%... 8.7%. The exception is series II, where the carrying capacity is greater by 30.37%. It is interesting to compare the results of crack resistance of beams: in series I and II (non-reinforced beams) the moment of crack formation increased slightly - by 3%... 5.5%. But in the beams of series III-VI a significant increase in the moment of crack formation was recorded: from 29.5% to 54.25%.

Table 3. Comparison of the average moments of crack formation M_{crc} and destructive moments M_u of basalt concrete (C) and basalt-fiber concrete (FC) beams.

Series	Bending moment of crack formation Mcrc, kNm		Difference, %	Destructive bending moment Mu, kNm		Difference, %
	C	FC		C	FC	
I	2.67	2.75	−3.0	2.67	2.75	−3.0
II	3	3.165	−5.5	3.835	5	−30.37
III	2.83	3.42	−29.5	8.65	8.65	0
IV	2.665	3.415	−37.5	11.5	12.5	−8.70
V	2.5	3.585	−54.25	13.5	13.985	−3.59
VI	2.67	3.415	−37.25	13.985	14.5	−3.68

4 Scientific Novelty

As part of the experimental studies new experimental data were obtained on the stress-strain state, the nature of fracture, strength and deflection of bent basalt concrete and basalt fiber concrete beam elements. For the first time the experimental data on the effect on the work of beam bent basalt concrete elements with basalt fiber micro-reinforcement were obtained.

5 Practical Significance

The practical significance consist in the experimentally established possibility of improving the deformability and crack resistance characteristics of concrete beams with non-metallic composite reinforcement by usage of dispersed reinforcement with basalt fiber.

6 Conclusions

Experimental studies have shown that the additional disperse reinforcement with basalt fiber of concrete beams, reinforced with non-metallic composite reinforcement, improves strength and deformation characteristics. Disperse reinforcement of concrete with basalt fiber increases the crack resistance of beams: in basalt-fiber concrete beams the moment of crack formation was greater compared to the moment of crack formation in basalt concrete beams from 29.5% to 54.25%. Disperse reinforcement of concrete with basalt fiber also increased the bearing capacity of beams in bending moment: the strength of basalt-fiber concrete beams was 3%–8.7% greater than basalt-concrete.

The obtained results allow to recommend the use of basalt fiber for dispersed reinforcement of concrete elements of transport structures.

References

1. Blikharskyy, Y., Khmil, R., Blikharskyy, Z.: Research of RC columns strengthened by carbon FRP under loading. In: Matec Web of Conferences, vol. 174, p. 04017 (2018). https://doi.org/10.1051/matecconf/201817404017
2. Blikharskyy, Z., Vashkevych, R., Vegera, P., Blikharskyy, Y.: Crack resistance of RC beams on the shear. In: Lecture Notes in Civil Engineering, vol. 47, pp. 17–24 (2020). https://doi.org/10.1007/978-3-030-27011-7_3
3. Bobalo, T., Blikharskyy, Y., Kopiika, N., Volynets, M.: Serviceability of RC beams reinforced with high strength rebar's and steel plate. In: Lecture Notes in Civil Engineering, vol. 47, pp. 25–33 (2020). https://doi.org/10.1007/978-3-030-27011-7_4
4. Selejdak, J., Blikharskyy Y., Khmil R., Blikharskyy Z.: Calculation of reinforced concrete columns strengthened by CFRP. In: Lecture Notes in Civil Engineering, vol. 47, pp. 400–410 (2020). https://doi.org/10.1007/978-3-030-27011-7_51
5. Blikhars'kyi, Y.Z.: Anisotropy of the mechanical properties of thermally hardened A500s reinforcement. Mater. Sci. **55**, 175–180 (2019). https://doi.org/10.1007/s11003-019-00285-0
6. Blikharskyy, Y., Kopiika, N., Selejdak, J.: Non-uniform corrosion of steel rebar and its influence on reinforced concrete elements' reliability. Prod. Eng. Arch. **26**(2), 67–72 (2020). https://doi.org/10.30657/pea.2020.26.14
7. Bobalo, T., Blikharskyy, Y., Kopiika, N., Volynets, M.: Theoretical analysis of RC beams reinforced with high strength rebar's and steel plate. In: IOP Conference Series: Materials Science and Engineering, vol. 708, no. 1, p. 012045 (2019)
8. Khmil, R., Tytarenko, R., Blikharskyy, Y., Vashkevych, R.: Influence of load level during strengthening of reinforced concrete beams on their reliability. In: IOP Conference Series: Materials Science and Engineering, vol. 708, no. 1, p. 012054 (2019)
9. Jiang, C., Fan, K., Wu, F., Chen, D.: Experimental study on the mechanical properties and microstructure of chopped basalt fibre reinforced concrete. Mater. Des. **58**, 187–193 (2014)
10. High, C., Seliem, H.M., El-Safty, A., Rizkalla, S.H.: Use of basalt fibers for concrete structures. Constr. Build. Mater. **96**, 37–46 (2015)
11. El Refai, A., Ammar, M.A., Masmoudi, R.: Bond performance of basalt fiber-reinforced polymer bars to concrete. J. Compos. Constr. **19**(3), 04014050 (2015)
12. Kabay, N.: Abrasion resistance and fracture energy of concretes with basalt fiber. Constr. Build. Mater. **50**, 95–101 (2014)
13. Tomlinson, D., Fam, A.: Performance of concrete beams reinforced with basalt FRP for flexure and shear. J. Compos. Constr. **19**(2), 04014036 (2015)
14. Atutis, M., Valivonis, J., Atutis, E.: Experimental study of concrete beams prestressed with basalt fiber reinforced polymers. Part I: flexural behavior and serviceability. Compos. Struct. **183**, 114–123 (2018)
15. Kumbhar, V.P.: An overview: basalt rock fibers-new construction material. Acta Eng. Int. **2**(1), 11–18 (2014)
16. Afroz, M., Patnaikuni, I., Venkatesan, S.: Chemical durability and performance of modified basalt fiber in concrete medium. Constr. Build. Mater. **154**, 191–203 (2017)

17. Kizilkanat, A.B., Kabay, N., Akyüncü, V., Chowdhury, S., Akça, A.H.: Mechanical properties and fracture behavior of basalt and glass fiber reinforced concrete: an experimental study. Constr. Build. Mater. **100**, 218–224 (2015)
18. Ouyang, L.J., Lu, Z.D., Chen, W.Z.: Flexural experimental study on continuous reinforced concrete beams strengthened with basalt fiber reinforced polymer/plastic. J. Shanghai Jiaotong Univ. (Sci.) **17**(5), 613–618 (2012)
19. Lypez-Buendha, A., Romero-Sanchez, M., Climent, V., Guillem, C.: Surface treated polypropylene (PP) fibres for reinforced concrete. Cement Concrete Res. **54**, 29–35 (2013)
20. Ghanbari, A., Karihaloo, B.: Prediction of the plastic viscosity of self-compacting steel fibre reinforced concrete. Cement Concrete Res. **39**, 1209–1216 (2009)

Strength of Reinforced Concrete Beams Strengthened Under Loading with Additional Reinforcement with Different Levels of its Pre-tension

Bogdan Kovalchuk[1], Yaroslav Blikharskyy[1](✉), Jacek Selejdak[2], and Zinoviy Blikharskyy[1]

[1] Lviv Polytechnic National University, Lviv 79013, Ukraine
yaroslav.z.blikharskyy@lpnu.ua
[2] Czestochowa University of Technology, 42200 Czestochowa, Poland

Abstract. This article describes the strengthening of normal sections of reinforced concrete beams by pre-stressed reinforcement under load. Experimental samples had the length of 2100 mm, the width of 100 mm and the height of 200 mm. Strengthening of experimental beams is designed from flat frames and isolated rods, which were combined into a spatial frame by transverse rods. In all the be the longitudinal working reinforcement were steel bars of 2 2 Ø 12 A500C, constructive and transverse reinforcement was represented by– Ø 8 A240C bars with spacing of 50…100 mm. The load was applied by means of a hydraulic jack in thirds of the span. The beams were tested for bending. The beams were reinforced with the use of two unstressed or pre-stressed reinforcement rods of Ø12 A500C mm.

Their connection to the main beam reinforcement by welding through Ø28 mm reinforcement rods. Three series of beams were tested: the first one consisted of two beams of the VO series, the second –of two beams reinforced with unstressed reinforcement at the level of 0.7Mcr (beams of the BR series), the third - beams reinforced with pre-stressed reinforcement at the level of 0.7Mcr (beams of the BRR series) with additional reinforcement tension level of 70% and 90%. The effect of beam reinforcement, at the level of 0.7Mcr, by additional reinforcement without its pre-stressing was 49%, and with the use of pre-stressing of additional reinforcement - 108% and 118%, respectively.

Keywords: Pre-stressing · Reinforced concrete beams · Real-size constructions · Strength · Reinforcement

1 Introduction

Today the most widely used material for industrial and civil construction is reinforced concrete constructions [1–6]. Due to the long-term operation of reinforced concrete structures of buildings and structures, the limit state occurs when they do not meet the operational requirements and as a result are damaged. Therefore, strengthening of certain elements is required [7–10].

Z. Blikharskyy (Ed.): EcoComfort 2020, LNCE 100, pp. 227–236, 2021.
https://doi.org/10.1007/978-3-030-57340-9_28

The effectiveness of strengthening building structures consists in eliminating of existing defects, increasing of the load-bearing capacity of this structure [11–14]. Nowadays the issue of effective strengthening of reinforced concrete structures, which are under a certain residual load, is insufficiently studied and therefore still remains relevant and requires detailed research.

Studies on the strengthened reinforced concrete structures with the help of pre-stressing of strengthening materials (in particular, steel bars) were conducted in many countries [15–19], but showed a significant variation in strength, deformability and crack resistance. However, insufficient research is devoted to the strengthening of such structures if they are subjected to the previous load level. This additionally confirmed the fact that the real efficiency of the pre-stressing of the additional strengthening reinforcement has not been studied yet.

2 Review of Scientific Sources and Publications

Nowadays there are many ways for strengthening reinforced concrete structures. In particular, one of the most popular is the use of traditional reinforced concrete jacketing and external steel plates [20, 21]. The use of composite materials for reinforcement is becoming increasingly popular. One of the methods of reinforcement, which requires additional research, is the cross-section increasing with additional pre-stressed reinforcement, which would be extremely effective due to increasing both normal sections' strength of the reinforcement structure, as well as its rigidity (limiting deflections and crack widths) which is achieved by the pre-tension of the additional reinforcement.

Recently wide spread have become methods of reinforcement based on the inclusion of additional pre-stressed steel bars in the reinforced concrete element, which changes its stress state and cross section [22, 23]. They have the number of valuable advantages over previously used methods: significant reduction in the amount of site work, minimizing of structural disturbances in the reinforcement process, eliminating of "wet" processes, minimum time and possibility of strengthening without removing the existing load. All this is especially valuable when it is necessary to strengthen structures without stopping of production.

Rather simple is the method of strengthening of reinforced concrete beams by means of additional pre-stressed reinforcement. It can be both horizontal and sprung. Due to installation of the additional reinforcement, with its preliminary tension, the stress-strain state of the strengthened beam changes. The pre-stress reliably and opportunely includes additional reinforcement in the work, and the beam could be further considered as a reinforced concrete bending structure with an increased reinforcement area and changed effective height.

The work of the following scientists [24–26] is devoted to the study of bending elements reinforced in the stretched zone by additional pre-stressed and non-prestressed reinforcement. The researches of the strengthened bending elements presented in works are more concerned with studying of strengthening ways and technology of its performance. However, all scientists emphasize that for the economic solution of strengthening measures and predicted assessment of the bending elements' stress state after strengthening it is necessary to determine the elements' stress state before strengthening and ensure the joint work of the initial and additional reinforcement. This should be especially taken into account when pre-stressing of additional rebar (heating and cooling processes that directly affect the stress level and the beginning of joint work) mostly of structures with large spans. Even in the work [3] attention was paid to the necessity of the residual load-bearing capacity usage and taking it into account with additional reinforcement, after strengthening as one of the main factors of structure reliable operation after reconstruction.

In Lviv Polytechnic National University have been previously studied reinforced concrete beams strengthened by increasing the cross section of reinforcement under load in a stretched zone [21] or using different composite material for it [16]. During the studies the effect of strengthening at increase in reinforcement section area in a reinforced concrete beam design was defined. The strengthening effect is higher if lower the stresses in the working rebar's at the strengthening moment take place.

Due to the fact that the strengthening in practical situations is carried out under the certain load (in the conditions of re-equipment of production line or reconstruction), the design strength level is first reached by the main rebar's. It is believed that the stress in the additional rebar's should be limited depending on the level of initial stresses in the existing rebar. This is necessary for the correct joint operation of the main rand additional strengthening reinforcement during the further structure exploitation.

In this case the purpose and objectives of the study are to obtain data on the operation of reinforced concrete beams strengthened under load in a stretched area with pre-stressed reinforcement.

3 Research Methodology

In order to achieve this goal, series of reinforced concrete beams and corresponding concrete prisms and cubes were made. According to the type of tests, all beams were divided into three series, the characteristics of which are given in Table 1.

All the reinforced concrete beams had the rectangular cross-section of 2100 mm length, 100 mm weight and 200 mm height. The concrete used for beams – C32/40. In all the beams the longitudinal working reinforcement was 2Ø12 mm A500C steel bars with the yield strength equal to f_{yd} = 540 MPa; constructive and transverse – Ø8 mm A240C with 75…100 mm spacing. Connection of the rebars in the spatial frame was industrial by welding.

Three series of beams were tested: the first one consisted of two beams of the BO series, the second –of two beams reinforced with unstressed reinforcement at the level of 0.7Mcr, the third - beams reinforced with pre-stressed reinforcement at the level of 0.7Mcr. During the experiment the beams were marked according to the following: BO-the usual beam, BR- the beam, strengthened by unstressed reinforcement, BRR-the beam, strengthened by pre-stressed reinforcement. The first group of numbers includes the numbers, which: first- indicates the beam series, second-the sequence number of the beam in this series. The second group of numbers indicates up to which level from the maximal the beam was loaded, depending on the reinforcement yield strength of the usual RC beams at short-term experiment (0.7) and up to which level the additional reinforcement was pre-stressed (0, 0.7, 0.9). For each experiment type two types of twin-samples were used.

Table 1. Program of experimental research.

Series №	№ of beam	Beam marking	Beam cross-section, mm	Reinforcement characteristics		Stress level in the additional reinforcement
				Main	Additional	
1	1	BO-1.1	100 × 200	2Ø12 A500C	2Ø12 A500C	–
	2	BO-1.2				–
2	3	BR-2.1-0.5/0				0
	4	BR-2.2-0.5/0				0
3	5	BRR-3.1-0.7/0.7				0.7
	6	BRR -3.2-0.7/0.7				0.7
	7	BRR -3.3-0.7/0.9				0.9
	8	BRR -3.4-0.7/0.9				0.9

Thus, for example, the marking "BRR-3.2-0.7/0.9" shows, that the second beam from third series was strengthened by pre-stressed reinforcement at the load level equal to 0.7 from the maximum level of the non-strengthened sample and stress level of the additional reinforcement equal to 0.9 with further testing of the beam up to failure.

Beams were tested for the pure bending. Load was applied with the use of hydraulic jack in the thirds of the span. Beams were strengthened by two pre-stressed A500C Ø12 mm bars. Their connection was performed with the use of welding through Ø28 mm steel bars. During the experiment with the use of clock-type micro-indicators deformations of concrete and reinforcement were measured.

Additional strengthening reinforcement was installed at the design length in accordance with the internal forces' diagram. Its tension was performed electro-thermally. Electro-thermal pre-stressing of additional steel bars must be performed with the use of transformer and clock-type gauges (in order to control the rebar's strain), additionally the heating temperature of the additional rebar is also controlled.

The beam was loaded by stages $\Delta P = 0.05P$ with endurance at each stage during 5 min (where P–destructive load of non-strengthened beam). After reaching of corresponding load level for the beam, additional reinforcement is included in the joint work with the use of electrothermal pre-stressing. After this the strengthening reinforcement is fixed by welding of short bars, which is connected with the main working reinforcement of the beam (Fig. 1).

Fig. 1. The general view of strengthened sample of RC beam

4 Results of the Experimental Study

On the first stage was provided testing of strengthened RC beams of BO series. According to experimental results the destructive value was obtained for bending moment M_c^{exp}. Moreover, on the special holders, fixed on the stretched reinforcement the strain was measured on the each stage. Additionally was indicated the value of load, which corresponds to its yield strength M_s^{exp}. According to strain of the stretched rebar and compressed concrete of tested samples graphs were obtained, which correspond to averaged strain values of the first series of tested samples BO-1.1 and BO-1.2 (Fig. 2).

After stresses in stretched rebar reach the yield strength further loading of beams is accompanied with significant deformation increase in reinforcement and concrete, increase in deflections and cracks' opening with further parceling of the compressed concrete. Such destruction is typical for unstrengthened beams.

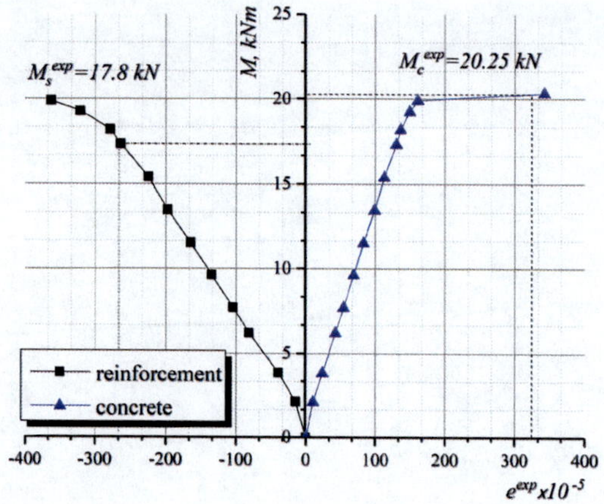

Fig. 2. Averaged strain graphs for stretched reinforcement (left) and compressed concrete (right) for unstrengthened beams of the first series BO-1.1 and BO-1.2.

Deformations in unstrengthened beams BO-1.1 and BO-1.2 were increasing smoothly; after the load has reached the value of 17.8 kNm the increase in rebar strain was indicated, after that stresses in main reinforcement reached the yield strength level under loading. Samples' failure was indicated at the bending moment equal to 20.25 kNm, which corresponds to the level, at which concrete deformation in the most compressed fiber has reached the limit value. Next, the fracture destruction takes place (physical destruction) of the compressed zone of concrete.

During BR-2.1-0.5/0 and BR-2.2-0.5/0 samples' testing additional reinforcement starting from the first stages was included in the work and perceived the part of tensile deformations (Fig. 3). After the main reinforcement of the beam reaches the yield strength, tensile stress increase is fully additional rebar. This could be seen by typical strain increase in additional reinforcement after reaching the load level, which corresponds yield strain of the main rebar at M_s^{exp} = 25.7 kNm. Next, at the load level of M_c^{exp} = 30.15 kNm concrete deformations of the most compressed fiber has reached the value of ε_{cu1} = 0.00325 and the depletion of bearing capacity of samples' has place. Stress in additional reinforcement was equal to 47% of the yield strength.

Deformations during the experiment testing of BRR-3.1-0.7/0.7 and BRR-3.2-0.7/0.7 were increasing smoothly (Fig. 4). As could be seen from graphs, obtained from strengthened samples' testing additional reinforcement starting from the first stages was included in the work. During the cooling of the reinforcement to the room temperature the unloading of the beam took place. It could be confirmed by decrease in deformations of the main reinforcement and concrete. After the main reinforcement reaches the yield strength, the tensile stress increase is perceived by additional rebar.

This could be confirmed by the typical strain increase in additional reinforcement after reaching the load level, which corresponds to yield strength of the main rebar at M_s^{exp} = 41.0 kNm. After that additional reinforcement reaches its limit values at $M_{s,add}^{exp}$ = 42.0 kNm. Next, at the same load level concrete deformation in the most compressed fiber reaches the value of ε_{cu1} = 0.00325 and the bearing capacity of the sample is depleted.

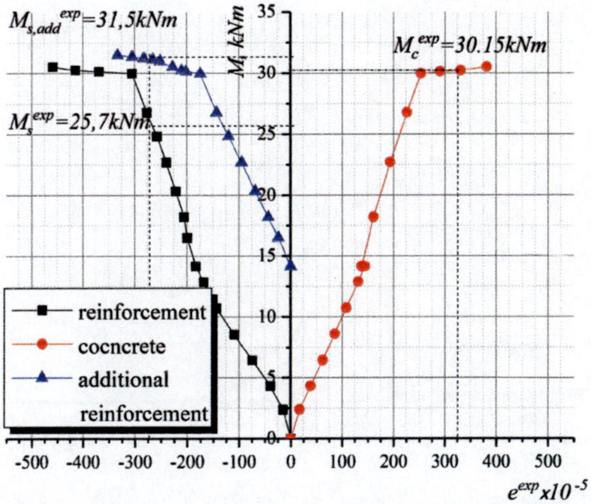

Fig. 3. Averaged strain graphs of the stretched reinforcement (left) and compressed concrete (right) for strengthened beams of the second series BR-2.1-0.5/0 and BR-2.2-0.5/0.

Samples BRR-3.3-0.7/0.9 and BRR-3.4-0.7/0.9 (Fig. 5) were destructed after reaching the yield strength of the main and additional reinforcement (which occurred almost simultaneously) at M_s^{exp} = 42.6 kNm with further destruction of the compressed concrete fiber at the load level of M_c^{exp} = 44.2 kNm concrete deformations of the most compressed fiber has reached the value of ε_{cu1} = 0.00325.

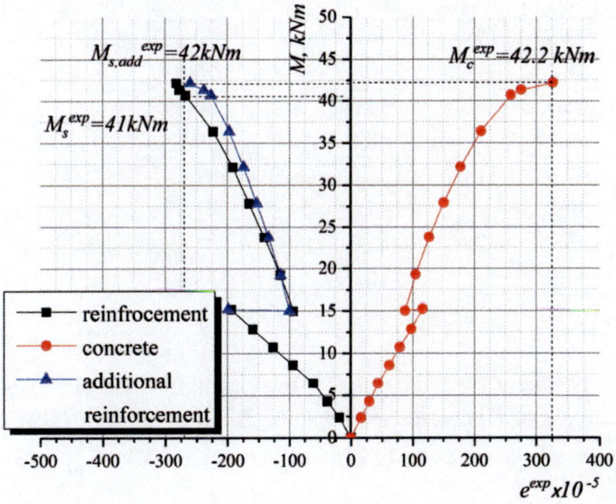

Fig. 4. Averaged strain graphs for the stretched reinforcement (left) and compressed concrete (right) for strengthened beams of the second series BRR-3.1-0.7/0.7 and BRR-3.2-0.7/0.7.

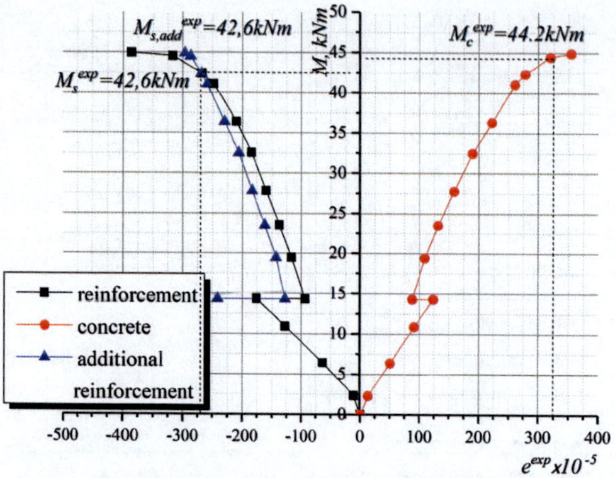

Fig. 5. Averaged strain graphs for the stretched reinforcement (left) and compressed concrete (right) for strengthened beams of the second series BRR-3.3-0.7/0.9 and BRR-3.4-0.7/0.9.

Obtained results of the experiment are given in the Table 2.

Table 2. Results of experimental research

Sample marking	Yield strength of the main reinforcement, M_s^{exp}, kNm		Strengthening effect, %		Depletion of the bearing capacity, M_c^{exp}, kNm		Strengthening effect, %	
	Sample	Average	Sample	Average	Sample	Average	Sample	Average
BO-1.1	17.3	17.8	–	–	20.5	20.25	–	–
BO-1.2	18.2		–		20.0		–	
BR-2.1-0.7	25.4	25.7	54.4	50.1	30.55	30.15	50.9	49.0
BR-2.2-0.7	26.07		46.6		29.75		47.0	
BRR-3.1-0.7/0,7	42.0	41.0	136.0	130.0	42.7	42.2	111.0	108.0
BRR-3.1-0.7/0,7	40.0		125.0		41.7		106.0	
BRR-3.1-0.7/0,9	43.2	42.6	143.0	140.0	44.6	44.2	120.0	118.0
BRR-3.2-0.7/0,9	42.0		136.0		43.8		116.0	

After the experiment was conducted the strength increase was indicated for strengthened samples of BR series for 49%, of BRR series-for 108% and 118%.

5 Conclusions

1. Strengthening of bended elements with pre-stressed reinforcement is simple in usage and does not include high costs;
2. Strengthening effect at level of $0.7M_{cr}$, by additional reinforcement without its pre-stressing was 49%; if the pre-stressing of the additional reinforcement is used – 108% and 118% respectively.
3. Strengthening efficiency is higher, if the pre-stressing level of additional reinforcement is higher. It enables to instantly include additional rebar in the work.

References

1. Blikharskyy, Z., Vashkevych, R., Vegera, P., Blikharskyy, Y.: Crack resistance of RC beams on the shear. In: Lecture Notes in Civil Engineering, vol. 47, pp. 17–24 (2020). https://doi.org/10.1007/978-3-030-27011-7_3
2. Karpiuk, V., Somina, Y., Maistrenko, O.: Engineering method of calculation of beam structures inclined sections based on the fatigue fracture model. In: Lecture Notes in Civil Engineering, vol. 47, pp. 135–144 (2019). https://doi.org/10.1007/978-3-030-27011-7_17
3. Khmil, R., Tytarenko, R., Blikharskyy, Y., Vashkevych, R.: Influence of load level during strengthening of reinforced concrete beams on their reliability. In: IOP Conference Series: Materials Science and Engineering, vol. 708, no. 1, p. 012054 (2019)
4. Kos, Ž., Klimenko, Y.: The development of prediction model for failure force of damaged reinforced-concrete slender columns. Tehnički vjesnik **26**(6), 1635–1641 (2019). https://doi.org/10.17559/TV-20181219093612
5. Krainskyi, P., Vegera, P., Khmil, R., Blikharskyy, Z.: Theoretical calculation method for crack resistance of jacketed RC columns. In: IOP Conference Series: Materials Science and Engineering, vol. 708, no. 1, p. 012059 (2019)
6. Vatulia, G., Lobiak, A., Chernogil, V., Novikova, M.: Simulation of performance of CFST elements containing differentiated profile tubes filled with reinforced concrete. In: Materials Science Forum, vol. 968, pp. 281–287 (2019). Trans Tech Publications Ltd. https://doi.org/10.4028/www.scientific.net/MSF.968.281
7. Blikhars'kyi, Y.Z.: Anisotropy of the mechanical properties of thermally hardened A500s reinforcement. Mater. Sci. **55**, 175–180 (2019). https://doi.org/10.1007/s11003-019-00285-0
8. Kharchenko, Y.V., Blikharskyy, Z.Y., Vira, V.V., Vasyliv, B.D.: Study of structural changes in a nickel oxide containing anode material during reduction and oxidation at 600 °C. In: Fesenko, O., Yatsenko, L. (eds.) Nanocomposites, Nanostructures, and their Applications. NANO 2018. Springer Proceedings in Physics, vol. 221. Springer, Cham (2019). https://doi.org/10.1007/978-3-030-17759-1_42
9. Kharchenko, Y., Blikharskyy, Z., Vira, V., Vasyliv, B., Podhurska, V.: Study of nanostructural changes in a Ni-containing cermet material during reduction and oxidation at 600 °C. Appl. Nanosci. (2020). https://doi.org/10.1007/s13204-020-01391-1
10. Lobodanov, M., Vegera, P., Blikharskyy, Z.: Planning experiment for researching reinforced concrete beams with damages. In: Lecture Notes in Civil Engineering, vol. 47, pp. 243–250 (2020). https://doi.org/10.1007/978-3-030-27011-7_31
11. Blikharskyy, Y., Kopiika, N., Selejdak, J.: Non-uniform corrosion of steel rebar and its influence on reinforced concrete elements' reliability. Prod. Eng. Arch. **26**(2), 67–72 (2020). https://doi.org/10.30657/pea.2020.26.14

12. Lipiński, T.: Roughness of 1.0721 steel after corrosion tests in 20% NaCl. Prod. Eng. Arch. **15**(15), 27–30 (2017). https://doi.org/10.30657/pea.2017.15.07
13. Lobodanov, M., Vegera, P., Blikharskyy, Z.: Influence analysis of the main types of defects and damages on bearing capacity in reinforced concrete elements and their research methods. Prod. Eng. Arch. **22**(22), 24–29 (2019). https://doi.org/10.30657/pea.2019.22.05
14. Zatkalíková, V., Markovičová, L.: Influence of temperature on corrosion resistance of austenitic stainless steel in cl− containing solutions. Prod. Eng. Arch. **25**(25), 43–46 (2019). https://doi.org/10.30657/pea.2019.25.08
15. Azizov, T.N., Kochkarev, D.V., Galinska, T.A.: New design concepts for strengthening of continuous reinforced-concrete beams. In: IOP Conference Series: Materials Science and Engineering, vol. 708, no. 1, p. 012040. IOP Publishing, December 2019. https://doi.org/10.1088/1757-899X/708/1/012040
16. Blikharskyy, Z., Brózda, K., Selejdak, J.: Effectivenes of strengthening loaded RC beams with FRCM system. Arch. Civil Eng. **64**(3), 3–13 (2018)
17. Bobalo, T., Blikharskyy, Y., Kopiika, N., Volynets, M.: Theoretical analysis of RC beams reinforced with high strength rebar's and steel plate. In: IOP Conference Series: Materials Science and Engineering, vol. 708, no. 1, p. 012045 (2019)
18. Wight, R.G., Green, M.F., Erki, M.A.: Prestressed FRP sheets for poststrengthening reinforced concrete beams. J. Compos. Constr. **5**(4), 214–220 (2001)
19. Pan, L., Chen, S., Zhao, S., Li, C.: Experimental study of cyclic loading behaviors ofimpaired reinforced concrete beams strengthening with prestressed CFRP sheets. Appl. Mech. Mate. **201**, 483–486 (2012)
20. Bobalo, T., Blikharskyy, Y., Kopiika, N., Volynets, M.: Serviceability of RC beams reinforced with high strength rebar's and steel plate. In: Lecture Notes in Civil Engineering, vol. 47, pp. 25–33 (2020). https://doi.org/10.1007/978-3-030-27011-7_4
21. Krainskyi, P., Blikharskyy, Y., Khmil, R., Vegera, P.: Crack resistance of RC columns strengthened by jacketing. In: Lecture Notes in Civil Engineering, vol. 47, pp. 195–201 (2020). https://doi.org/10.1007/978-3-030-27011-7_25
22. Barros, J.A., Taheri, M., Salehian, H., Mendes, P.J.: A design model for fibre reinforced concrete beams pre-stressed with steel and FRP bars. Compos. Struct. **94**(8), 2494–2512 (2012)
23. Tan, K.H., Tjandra, R.A.: Strengthening of RC continuous beams by external prestressing. J. Struct. Eng. **133**(2), 195–204 (2007)
24. Du, J., Liu, X.: Experimental study of RC continuous beams strengthened by external prestressing. In: Innovation & Sustainability of Modern Railway Proceedings of ISMR, vol. 2, pp. 221–227 (2008)
25. Kim, S., Yang, K., Byun, H., Ashour, A.: Tests of reinforced concrete beams strengthened with wire rope units. Eng. Struct. **29**(10), 2711–2722 (2007)
26. Minelli, F., Plizzari, G., Cairns, J.: Flexure and shear behavior class of RC beams strengthened by external reinforcement. In: Concrete Repair, Rehabilitation and Retrofitting, pp. 377–378 (2009)

Effects of Nano-liquids on the Durability of Brick Constructions for External Walls

Tetiana Kropyvnytska[1(✉)], Roksolana Semeniv[2], Roman Kotiv[1], and Yurii Novytskyi[1]

[1] Lviv Polytechnic National University, Lviv 79013, Ukraine
tkropyvnytska@ukr.net
[2] Ivano Frankivsk National Technical University of Oil and Gas, Ivano-Frankivsk 76000, Ukraine

Abstract. The effectiveness of the influence of modifying substances on the properties of ceramic facing brick and masonry surface for decorative external walls is considered in the article. X-ray diffraction and spectral analysis showed that sodium sulfate, potassium sulfate and syngenite are the salt effloresces from the ceramic brick surface. It is established that polymeric one-component coatings based on siloxanes are not effective, especially under the action of alternating temperatures. Studies have shown that the improvement of the performance properties of ceramic facing brick increase due to the surface coating by nano-liquids with the addition of nano-Al_2O_3. The surface modification of the ceramic facing brick by hydrophobic nano-liquids allows sealing the structure due to the colmatation of pores, which reduces the capillary tightening of the masonry. It also leads to increased weather resistance and frost resistance of masonry. It is established that the surface protection of brick structures with hydrophobic nano substances provides a decrease in the porosity, increase the water resistance and durability. This allows to confirm the relevance of the revealed influence of the modifiers on the formation of properties and the practical attractiveness of the proposed technological solutions.

Keywords: Brick constructions · Ceramic facing brick · Salt effloresces · Polymeric substances · Nano-liquid · Water absorption

1 Introduction

The durability of structures depends on their composition and the stability of the physicochemical properties depending on the level of influence of environmental factors [1–3]. During the operation of brick structures, changes in temperature and humidity cause defects, which significantly impair the architectural expressiveness and adversely affect the physical and technical properties of the masonry facades. The durability of bricks in different conditions, especially in the presence of moisture, promotes progressive changes in ceramic material. Expansive corrosion products can form effloresces in the ceramic body. This causes the stresses and cracking of in the material [4–6]. Brick structures are exposed to aggressive environment during their service life. This leads to formation of crystallization pressure as a result of the soluble salt evaporation in clay masonry structures [7].

Z. Blikharskyy (Ed.): EcoComfort 2020, LNCE 100, pp. 237–244, 2021.
https://doi.org/10.1007/978-3-030-57340-9_29

Special impregnating materials are used to protect the surface of the decorative external walls of the of brick structures facade [8–11]. Silicone impregnating substances are widely used for surface treatment. Technical characteristics of modified ceramic brick with hydrophobic substances based on acrylic polymers are increased; in particular, the water absorption and capillary diffusion rates are reduced, while the consumption properties are reduced. Studies showed that siloxane admixture has negative impact on the frost resistance, compression strength and adhesion of cement paste of mortars with basalt fibres [12]. However, the substances with fixed indicators of chemical and physical properties are not effective enough, especially under the influence of alternating temperatures. In addition, covered facing brick is aging during its service life, which is accompanied by irreversible chemical and physical processes under the influence of external and internal factors [13]. Signs of the coating destruction are their cracking, peeling, loss of mass and color.

Durability of building constructions plays a crucial role in the modern construction [14–18]. Nowadays, to increase the functional properties of building materials and products, a new promising direction in science is the use of nanotechnology, which involves the creation of new materials and systems at the molecular, nanoscale and microlevels [19–22]. Nanosized particles and components of a nanoscale system have new properties that are important for the development of new products and applications. Studies [23, 24] have proven to be more effective in protecting surfaces with nano-liquids, which reduces the water absorption, the permeability coefficient, and increases the water resistance of surface structures. G. Kamal's [25] investigations showed that the nano-liquid does not affect the strength when applied to the surface of hardened concrete. At the same time, the coefficient of permeability decreases by 40–65%, water absorption – by 10%, and also improves wear resistance. Investigations showed that penetration of a nanopolymer into the brick surface occurs through mechanical adhesion, which is affected by roughness and structure of the hydrophobised surface, the thickness of the formed layer [26, 27].

Studies by A. Nazari and S. Riahi [28] showed that the use of Al_2O_3 nanoparticles plays a crucial role in the production of ceramic materials and in the use as filler in nano-composites to improve their properties. In addition, Al_2O_3 nanoparticles are able to act as nanofillers and restore pore structure by reducing the number of larger pores. This improves the mechanical and physical properties of the materials. Researches showed [29], that coatings made on the basis of silicate substances with the Al_2O_3 and TiO_2 nanoadditives are similar. However, according to the X-ray diffraction, the reflex intensity increases for Al_2O_3, indicating that Al_2O_3 nanoparticles have a higher permeability. This determines the possibility of obtaining a durable construction with increased properties. That is why the investigations of the nano-liquids on the durability (porosity, water absorption, frost resistance) of brick structures for external walls are relevant.

2 Aim

The aim of the article is to increase the durability and operational reliability of the external walls of brick structures by their surface treatment with water-repellent nano-liquids of the new generation.

3 Method

The research was carried out using ceramic facing bricks and ceramic clinker bricks (Ukraine). Hydrophobizing substances – polymethylphenylsiloxane substances (PMPhS), acrylic polymers (AP), modifiers of permeable complex action (nano-Al_2O_3) are used for surface treatment of ceramic facing bricks. To increase the durability of brick masonry, a hydrophobic nano-liquid consisting of PMPhS, alumina and iron oxide powder, and alumina nanopowder was developed. Nanodispersed powder of alumina (Sigma-Aldrich Chemie GmbH, Germany) contains particles with a size of 30–40 nm, the true density is 0.12 g/cm^3, the specific surface area is 70 m^2/g. A protective coating was obtained by dispersing the components automatically. Clinker-efficient mortars based on low-energy modified multicomponent cement for masonry MC 22.5 (clinker factor – 0.40) of PJSC Ivano-Frankivskcement (Ukraine) were used to obtain brick masonry.

The chemical compositions of salts from the brick masonry surface were determined using an X-ray spectrometer ARL 9800 XP. The microstructure of the efflorescences from the ceramic brick surface was examined using a Philips XL30 ESEM-FEG scanning electron microscope. Non-destructive analysis by the Carsten method (RILEM Test Method II.4) according to ASTM E 514 and according to DSTU B V.2.7-126:2011, was used to determine the water absorption index. The weather resistance of the ceramic facing brick and the brick masonry sample was determined by alternating hydration and drying. The resistance of ceramic facing bricks to salt efflorescences was tested according to DSTU B V.2.7-171:2008. The surface of the samples, located above the water, was blown with air with a temperature of (20 ± 2) °C, for at least 3 h a day. After 7 days of testing, the presence of efflorescences on the brick surface was determined visually and fixed salts by physico-chemical methods of analysis. Evaluation of frost resistance of ceramic brick and masonry sample was performed according to the degree of damage and loss of weight and strength according to DSTU B V.2.7-42-97. The compressive strength of brick masonry was determined according to DSTU B EN 1052-1:2011.

4 Results

It is known that ceramic clinker bricks are mainly used for finishing decorative protective layer of structures. At the same time, currently used cheaper building material – ceramic facing bricks. The comparative researches of a ceramic clinker and facing brick are carried out to analyze and substantiate the expediency of ceramic facing brick application for masonry of an external finishing walls. It was found that ceramic clinker brick is characterized by a dense structure using the scanning electron microscope method. At the same time, ceramic facing brick is characterized by a heterogeneous microstructure with an increased number of pores (21.5%). The water absorption by weight for clinker bricks is 4.7%, the water absorption index during capillary tightening is $0.5\,\mathrm{kg/m^2} \cdot \mathrm{h}^{0.5}$. The water absorption increases to 15.8% for ceramic facing bricks and the water absorption during capillary tightening reaches the value of $2.2\,\mathrm{kg/m^2} \cdot \mathrm{h}^{0.5}$, which is 4.4 times more compared to clinker bricks.

Increased porosity of the facing brick promotes the penetration of water into the capillary-porous structure of the material. It leads to cracking and peeling of plasters and paints, as well as the destruction of bricks and mortar in the joints of the masonry. The changes in the texture are the effect of the process of salt crystallisation. Salt efflorescence accumulates in the wall material [8]. According to chemical analysis, the efflorescence are characterized by a high content of SO_3 (49.5 wt%), alkaline oxides – Na_2O (31.8 wt%) and K_2O (9.2 wt%). Minor inclusions of CaO (2.6 wt%) and SiO_2 (1.8 wt%) are recorded. The lines of sodium sulfate Na_2SO_4 (d/n = 0.467; 0.318; 0.278; 0.232 nm), potassium sulfate K_2SO_4 (d/n = 0.288; 0.208 nm) and syngenite $K_2Ca(SO_4)_2{\cdot}H_2O$ (d/n = 0.951) were established by X-ray phase analysis. As can be seen from the photomicrograph (Fig. 1a), alkaline salts are formed on the ceramic bricks surface: sodium sulfate – in the form of small crystals (crystal size 1–2 μm) and potassium sulfate, which crystallize more locally plate-like layering aggregates (crystal size – 15…20 μm), (Fig. 1b). The presence of salts phases Na_2SO_4 and K_2SO_4 is confirmed by X-ray spectral analysis (Fig. 1c, d).

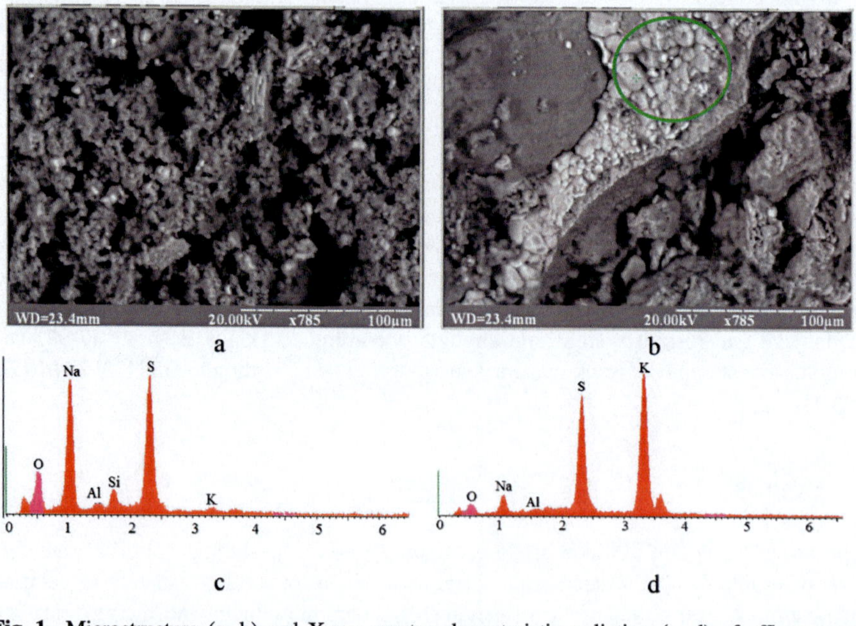

Fig. 1. Microstructure (a, b) and X-ray spectra characteristic radiation (c, d) of efflorescence from ceramic facing brick.

At the same time, the formation of efflorescence on the ceramic clinker bricks surface is not observed. Therefore, the method of surface modification with water-repellent substances was used to prevent salting and to improve the operational properties of ceramic facing bricks. Thus, water absorption decreases by 3.3 wt%, water absorption by capillary tightening – by $0.22\,kg/m^2 \cdot h^{0.5}$ by applying the water repellent PMPhS. Surface treatment with a protective substance AP leads to a decrease in water absorption from 16.5 to 8.7 wt%, water absorption during capillary tightening – from 2.2 to $1.08\,kg/m^2 \cdot h^{0.5}$.

Electron microscopy showed that there is an intensive formation of microcracks on the surface of the brick treated with a water repellent based on PMPhS after alternating freezing and thawing. This leads to an increase in water absorption from 13.2 to 18.7 wt%. For AP-modified bricks, cracks on the sample surface were formed locally with less opening, which led to an increase in water absorption from 8.7 to 11.14 wt%. Therefore, the research of nano-liquids influence is carried out in the further work. was studied to improve the performance properties of ceramic facing bricks.

Compositions of nano-liquids containing alumina oxide powder, optimal ratio of PMPhS and nano-Al_2O_3 was determined due to plan of two-factor three-level experiment. The amount of PMPhS (X1 = 30; 35; 40 wt%) and nano-Al_2O_3 (X2 = 0; 0.6; 1.2 wt%) was chosen as variables. According to test results obtained by the regression equation and constructed isoparameter diagram (Fig. 2) was established the optimal balance between additives (nano-Al_2O_3 is 0.9…1.2 wt%) that ensure modified surface with water absorption of 1.2–0.8% and water absorption with capillary tightening – up to $0.8{-}0.06\,kg/m^2 \cdot h^{0.5}$.

Defectoscopy using a Carsten tube showed that the lowest water absorption (0.004 ml/cm^2) after 2 h of exposure-drop is characterized by the surface of ceramic bricks modified with nano-liquid, while for uncoated bricks – 0.15 ml/cm^2. In particular, electron microscopy revealed that the surface of the uncoated ceramic brick sample is inhomogeneous with protrusions and capillary micropores (size – 10–15 μm).

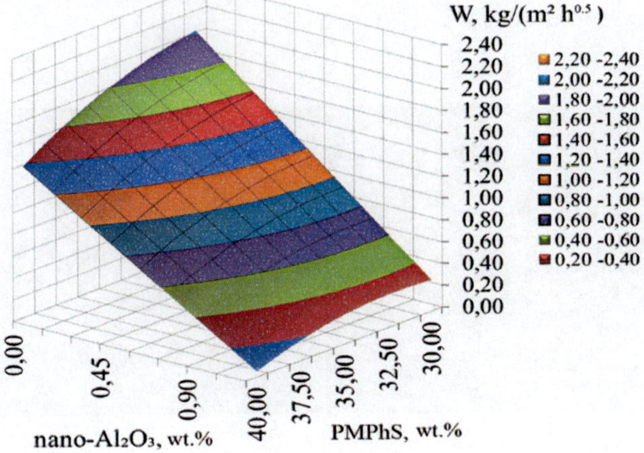

Fig. 2. Isoparametric diagram of the change in capillary suction of ceramic facing brick modified with nano-liquid.

The microstructure is leveled and compacted due to the penetration of nanoparticles into the pore structure of the material. The efficiency of nano-liquid application is due to the free energy of the surface, as well as the surface colmatation at the submicro level by nano-Al_2O_3 particles in the composition of the hydrophobic substance. At the same time pulling up of water-soluble salts from masonry is blocked.

Studies of weather resistance found that the loss of strength was 16.7% for uncoated ceramic facing bricks after 100 cycles of alternating drying and hydration. Herewith, cracks with a width of 0.4–0.8 mm were observed on the samples. At the same time, the loss of strength was 1.6% for ceramic facing bricks, the surface of which was modified with nano-liquid, without the formation of cracks. It should be noted that frost resistance increased for ceramic facing brick modified with nano-liquid. Thus, the surface treatment with nano-liquid provides protection of the brick building structure and increase its durability.

5 Scientific Novelty and Practical Significance

The article experimentally confirms the possibility of physical and chemical modification of the ceramic facing brick surface with hydrophobic protective substances based on nano-liquids of permeable and colmatative action using nano-Al_2O_3. The developed nano-liquids give the surface structure a uniform and denser character at the submicrolevel with the formation of nanoreinforcing crosslinked structure, which provides reduced water absorption and increased atmospheric and frost resistance. The authors obtained an isoparametric diagram of different amounts of nano-Al_2O_3 in combination with PMPhS on the capillary suction and water absorption of the surface sample. The effect of the optimized nano-liquid composition was investigated under the action of alternating freezing and thawing.

The enterprise PE "Termit" (Ukraine) made weather-resistant protective nano-liquid. Nano-liquid (PMPhS, alumina oxide, nano-Al_2O_3) was applied on the surface of a ceramic brick structure with a thickness of 0.4–0.6 mm using an atomizer. The nano-liquid was used to protect the surface of brick fencing structures during repair and restoration works in Lviv (Ukraine) (Fig. 3a, b).

a b

Fig. 3. Fencing construction: a – before applying nano-liquid; b – after applying nano-liquid

6 Conclusions

1. It is established that efflorescence based on sodium sulfate, potassium sulfate and syngenite appear under the action of moisture on the brick structure, which is constructed with the use of ceramic facing bricks.
2. It is shown that the use of hydrophobic coatings based on acrylic polymers and organosilicon compounds provides the formation of a waterproof film. In this case, the effect of variable temperatures (cycles of freezing and thawing) leads to the formation of local defects and cracks on the treated ceramic bricks surface.
3. It is established that indicators of water absorption and capillary tightening are significantly reduced (to $W_m = 0.8$ wt% and $W = 0.06\,\mathrm{kg/m^2} \cdot h^{0.5}$), frost-resistance increases in 2 times without formation of defects and cracks with the surface treatment containing nano-Al_2O_3. The surface covered by nano-Al_2O_3 becomes denser and smoother, the number of pores decreases. The surface treatment with nano-liquid provides an increase in weather resistance and efflorescence resistance. This leads to increased durability of the finishing wall of brick structures that are exposed to aggressive environmental influences, in particular alternating temperatures.

References

1. Krivenko, P., Kovalchuk, O., Boiko, O.: Practical experience of construction of concrete pavement using non-conditional aggregates. Mater. Sci. Eng. **708**(1), 012089 (2019)
2. Hradil, P., Toratti, T., Vesikari, E., Ferreira, M., Häkkinen, T.: Durability considerations of refurbished external walls. Constr. Build. Mater. **53**, 162–172 (2014)
3. Krivenko, P.V., Rudenko, I.I., Petropavlovskyi, O.M., Konstantynovskyi, O.P., Kovalchuk, A.V.: Alkali-activated Portland cement with adjustable proper deformations for anchoring application. In: IOP Conference Series: Materials Science and Engineering, vol. 708, no. 1, p. 8 (2019)
4. Van Hess, R., Brocken, H.: Damage development to treated brick masonry in a long-term salt crystallisation test. Constr. Build. Mater. **18**, 333–338 (2004)
5. Benavente, D.: Influence of microstructure on the resistance to salt crystallisation damage in brick. Mater. Struct. **39**, 105–113 (2006)
6. Stryszewska, T., Kaeka, S.: The effects of salt crystallization in ceramic bricks in terms of line deformations. Procedia Eng. **193**, 120–127 (2017)
7. Abu Bakar, B.H., Wan Ibrahim, M.H., Megat Johari, M.A.: A review: durability of fired masonry wall due to salt attack. Int. J. Integr. Eng. (Issue on Civil and Environmental Engineering) **1**, 111–127 (2011)
8. Varshavets, P., Svidersky, V., Chernyak, L.: Peculiarities of the structure and hydrophysical properties of face brick. Eur. Appl. Sci. **1**, 106–110 (2014)
9. Šadauskienė, J., Ramanauskas, J., Stankevičius, V.: Effect of hydrophobic materials on water impermeability and drying of finish brick masonry. Mater. Sci. (Medžiagotyra) **9**, 94–98 (2003)
10. Novák, V., Zach, J.: Study of hydrophobic modification of ceramic elements. Key Eng. Mater. **776**, 121–126 (2018)

11. Ginchitskaia, I., Yakovlev, G., Kizinievich, O., Polyanskikh, I., Pervushin, G., Taybakhina, P., Balobanova, I.: Damage to polymer coating on facing brick surface in operated buildings. Procedia Eng. **195**, 189–196 (2017)
12. Barnat-Hunek, D., Łagód, G., Fic, S., Jarosz-Hadam, M.: Effect of polysiloxanes on roughness and durability of basalt fibres-reinforced cement mortar. Polymers **10**(420), 20 (2018)
13. Kropyvnytska, T., Semeniv, R., Kotiv, R., Kaminskyy, A., Gots, V.: Studying the effect of nano-liquids on the operational properties of brick building structures. Eastern-Eur. J. Enterp. Technol. **5/6**(95), 27–32 (2018)
14. Runova, R.F., Gots, V.I., Rudenko, I.I., Konstantynovskyi, O.P., Lastivka, O.V.: The efficiency of plasticizing surfactants in alkali-activated cement mortars and concretes. In: MATEC Web of Conferences, vol. 230, p. 03016 (2018)
15. Krainskyi, P., Blikharskyy, Y., Khmil, R., Blikharskyy, Z.: Experimental study of the strengthening effect of reinforced concrete columns jacketed under service load level. In: MATEC Web of Conferences, vol. 183, p. 5 (2018)
16. Savchuk, Y., Plugin, A., Lyuty, V., Pluhin, O., Borziak, O.: Study of influence of the alkaline component on the physico-mechanical properties of the low clinker and clinkerless waterproof compositions. In: MATEC Web of Conferences, vol. 230, p. 03018 (2018)
17. Solodkyy, S., Markiv, T., Sobol, K., Hunyak, O.: Fracture properties of high-strength concrete obtained by direct modification of structure. In: MATEC, vol. 116, p. 7 (2017)
18. Kropyvnytska, T., Rucinska, T., Ivashchyshyn, H., Kotiv R.: Development of eco-efficient composite cements with high early strength. In: International Conference Current Issues of Civil and Environmental Engineering, pp. 211–218 (2020)
19. Sikora, P., Elrahman, M., Dietmar, S.: The influence of nanomaterials on the thermal resistance of cement-based composites – a review. Nanomaterials **8**(465), 33 (2018)
20. Krivenko, P., Sanytsky, M., Kropyvnytska, T.: The effect of nanosilica on the early strength of alkali-activated portland composite cements. Solid State Phenom. **296**, 21–26 (2018)
21. Kropyvnytska, T., Sanytsky, M., Rucinska, T., Rykhlitska, O.: Development of nanomodified rapid hardening clinker-efficient concretes based on Portland composite cements. Eastern-Eur. J. Enterpr. Technol. **6**(6), 38–48 (2019)
22. Sanytsky, M., Marushchak, U., Olevych, Y., Novytskyi, Y.: Nano-modified ultra-rapid hardening portland cement compositions for high strength. In: Lecture Notes in Civil Engineering, vol. 47, pp. 392–399 (2020)
23. Sharobim, K., Mohammedin, H.: The effect of Nano-liquid on the properties of hardened concrete. HBRC J. **9**, 210–215 (2013)
24. Fic, S., Szewczak, A., Barnat-Hunek, D., Łagód, G.: Processes of fatigue destruction in nanopolymer-hydrophobised ceramic bricks. Mater. (Basel) **10**(1), 16 (2017)
25. Kamal, G.S., Mohammedin, H.A.: The effect of nano-liquid on the properties of hardened concrete. Hous. Build. Res. Centre J. **9**, 210–215 (2013)
26. Baldan, A.: Adhesion phenomena in bonded joints. Int. J. Adhes. **38**, 95–116 (2012)
27. Rudawska, A.: Selected Issues in Constituting Homogeneous and Hybrid Adhesive Joints. Lublin University of Technology, Lublin (2013)
28. Nazari, A., Riahi, S.: Al_2O_3 nanoparticles in concrete and different curing media. Energy Build. **43**, 1480–1488 (2011)
29. Demirba, Ç., Ayday, A.: Effects of an Al_2O_3 nano-additive on the performance of ceramic coatings prepared with micro-arc oxidation on a titanium alloy. Mater. Technol. **51**(4), 613–616 (2017)

Monitoring of Dynamic Loads on Steel Headframes

Volodymyr Kushchenko[1(✉)] and Oleksandr Nechytailo[2]

[1] Lviv Polytechnic National University, Lviv, Ukraine
volodymyr.m.kushchenko@lpnu.ua
[2] NTU "Dnipro Polytechnic", Dnipro, Ukraine

Abstract. According to the regulatory requirements of Ukraine, structures of headframes belong to CC-3 importance class. Such a regulatory requirement as systems for safety monitoring of a technical state is available for the constructions of CC-3 importance class. Dynamic strain loads resulting from the hoist rope tension is the basic type of technological loads; their monitoring is of principal importance to evaluate their operational safety. Following theoretical and experimental methods have been applied to substantiate a system to monitor loads on a headframe structure: finite-element method; tensometric method for strain measurement; and method of experiment planning. Relying upon the generalized results of theoretical and experimental research concerning regularities of distribution of local stresses within the bearing support pulleys, the paper proposes a principle of tensiometer allocation agreed with the methods determining hoist rope tension based upon a mathematical model developed by the authors. The proposed method to monitor dynamic loads resulting from tension of hoist ropes makes it possible to record intermittent (emergency) loads as well as load cycles of normal service. The abovementioned helps evaluate structural fatigue strength of a mine headframe structure as well as the mine hoist components.

Keywords: Headframes · Hoist ropes · Monitoring of loads

1 Introduction

Structures of steel headframes are a part of structures of the mine hoists (Fig. 1). According to the regulatory requirements of Ukraine, they belong to CC-3 importance class. Such a regulatory requirement as systems for safety monitoring of a technical state is available for them [1–3]. In their operational process, structures of headframes resist cyclical dynamic loads resulting from the hoist rope tension [2, 4, 5]. As a consequence of the dynamic load action, fatigue structural damages and mechanical ones may accumulate within the structures of mine headframes [5]. Hence, monitoring of dynamic loads by the mine hoist tension is the important component to provide their technological safety since it helps record emergency loads and control structural fatigue strength of a headframe structure. The paper considers structures of steel headframes (Table 1) equipped with friction-pulley hoists (Fig. 1 a, c) and drum-type machines (Fig. 1 b, d).

Z. Blikharskyy (Ed.): EcoComfort 2020, LNCE 100, pp. 245–252, 2021.
https://doi.org/10.1007/978-3-030-57340-9_30

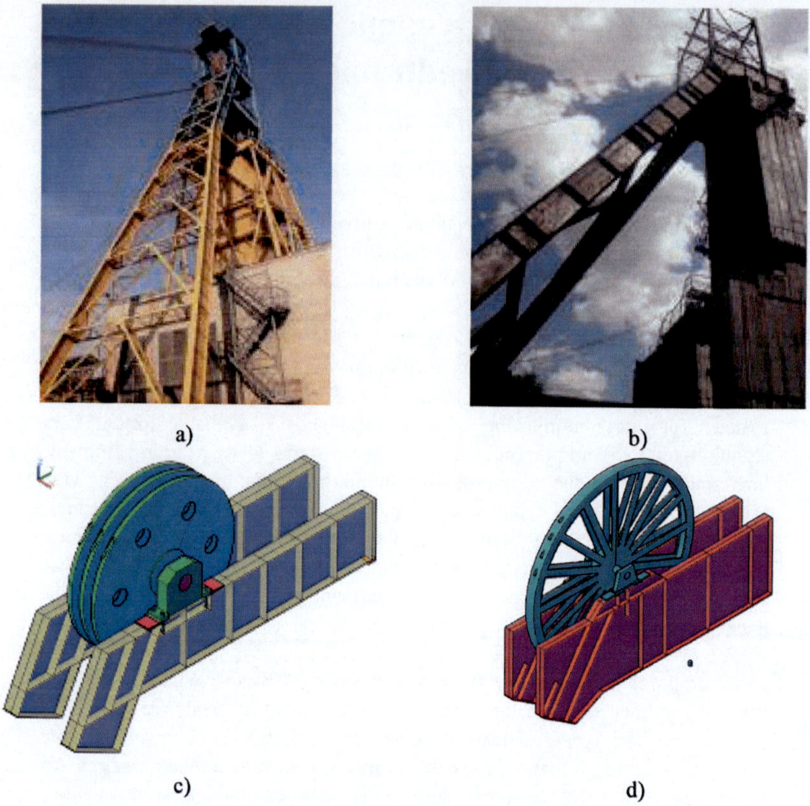

Fig. 1. Subject of the research – steel headframes equipped with surface-based hoists: a) a headframe of a multirope hoist with a friction pulley; b) a headframe of a hoist with a drum-type machine; c) sub-pulley structures of a headframe shown in Fig. 1 a; d) sub-pulley structures of a headframe shown in Fig. 1 b.

Paper [5] mentions dynamic character of loads by hoist tension on the structures of headframes being of cyclic or impulsive nature. Paper [6] represents the results of vibration measurement of a headframe structure in terms of normal operation loads and proposes a method of their technical state monitoring on the basis of the comparison of the measured vibration velocity amplitudes within the control points of the structure with the critical values being determined using statistical analysis of preliminary measurement data. However, the method prevents from the identification of effect of cyclic and impulsive natures of loads by hoist tension on the strength of components of mine hoist structure. Papers [7–10] describe different techniques to measure tension within hoists using load gages mounted either on a hoist near a pull-type device of a conveyance or on the pull-type device itself. However, the techniques are not applicable for continuous monitoring of headframes due to insufficient durability of the facilities in aggressive environments of a hoisting shaft. Papers [11, 12] represent the research results of stress state of bearing support pulleys being of interest for monitoring of loads by hoist tension since the papers carried out factorial parameter analysis of stress state of bearing support pulleys and experiments concerning local stress distribution within the nodes of sub-pulley structures.

2 Purpose of the Research

Purpose of the research is to substantiate a method of monitoring of loads by hoist tension while measuring local stresses within the bearing support pulleys (see Fig. 1c, d).

3 The Research Procedure

The research considers local stresses within the bearing support pulleys to sub-pulley headframe structures, measured by means of tensometric method, as the parameter determining time variations in a hoist tension (see Fig. 1c, d). The polynomial dependence has been obtained as a result of mathematical modeling of tension of bearing support pulleys using a software complex Ansys Workbench 14.0 (see Fig. 2). In the context of the analytical model, developed by means of the software complex Ansys Workbench 14.0, a node of geometrically constant sub-pulley structure of a headframe was simulated together with a guiding pulley which had only one kinematic degree of freedom (i.e. rotation). That helped model its interaction with a hoist when load was transferred to the headframe. Diagrams of force monitoring within the hoists of an operating winder have been a result of tensometric measurements of local stresses within the bearing support pulleys in terms of two actual objects (see Fig. 3).

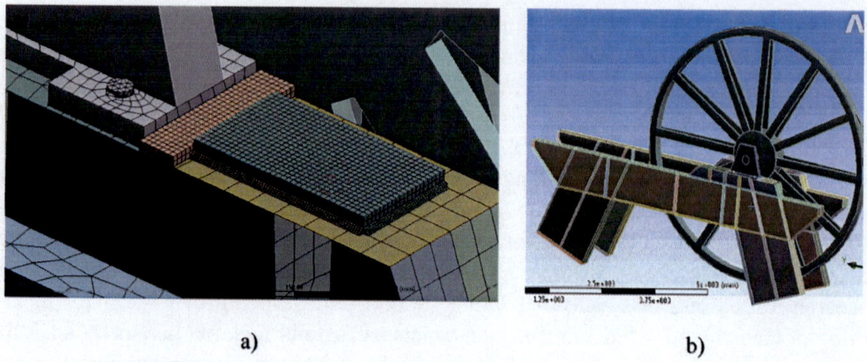

Fig. 2. Analytical model of a sub-pulley structure of a headframe in terms of Ansys Workbrench 14.0 SC: a) fragment of approximation of the bearing support pulley by means of finite elements; b) general view of the analytical model.

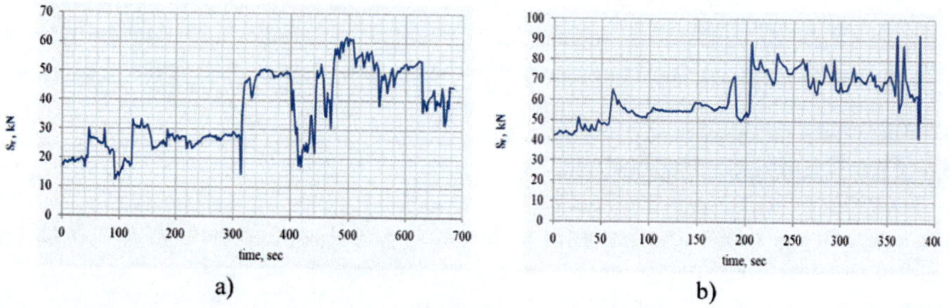

Fig. 3. Monitoring diagram of hoist tension during the parts of operation schedule of mine hoisting facilities: a) a headframe with a drum-type hoisting machine (Table 1, Fig. 1b); b) a headframe with a multirope hoisting machine with a friction pulley (Table 1, Fig. 1a).

4 Research Results of a Stress-Strain State of the Bearing Nodes of the Guiding Pulleys

Following regularities of local stress distribution within the bearing nodes of the guiding pulleys have been identified relying upon the generalization of papers [11, 12]: a) distribution of local stresses under the main bearing of a pulley at 0.4b distance from the pulley axis has the localized zone ("A" zone in Fig. 4a) with the expressed maximum of the major principal stresses; linear type of the stress state is typical for the zone (see Fig. 4).

The factorial experiment has helped determine polynomial dependence of the main stresses within the localized maximum upon following factors: S_r – force within the hoisting rope; t_w, – thickness of a wall of the sub-pulley structure; α – inclination of the hoist cord:

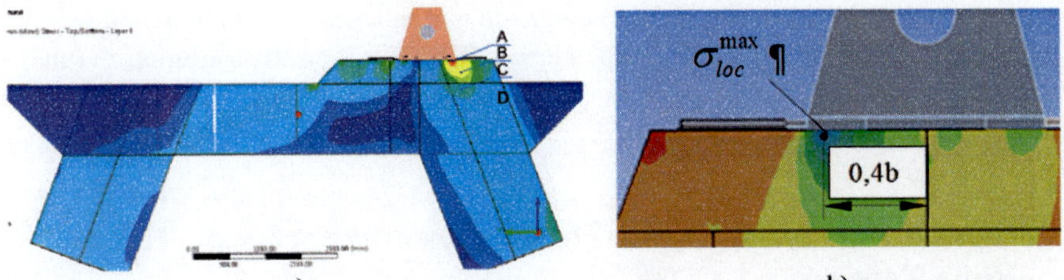

a) b)

Fig. 4. Results of mathematical modeling of a stress state of the sub-pulley structures: a) distribution of isofields of the main stresses within a sub-pulley structure A-zone of the localized maximum of the main compressive stresses; b) regularity of location of A-zone centre of the localized maximum (σ_{loc}^{max}) of the main stresses under supporting bearings of guiding pulleys; b is the length of a supporting share of the pulley bearing.

$$\sigma_{loc}^{max}(x_1, x_2, x_3) = 268.5 + 207.3\tfrac{x_1-2653625}{2041015} + 12.5\tfrac{x_2-50}{15} - 41.1\tfrac{x_3-16}{4} + 7.2\left(\tfrac{x_1-2653625}{2041015}\right)^2 + 6.6\left(\tfrac{x_2-50}{15}\right)^2 \\ -14.01\left(\tfrac{x_3-16}{4}\right)^2 + 6.6\left(\tfrac{x_1-2653625}{2041015}\right)\left(\tfrac{x_2-50}{15}\right) - 27.4\left(\tfrac{x_1-2653625}{2041015}\right)\left(\tfrac{x_3-16}{4}\right) - 16.4\left(\tfrac{x_2-50}{15}\right)\left(\tfrac{x_3-16}{4}\right)[M\Pi a] \quad (1)$$

where x_1 – a rope tension force (S_r), H; x_2 – inclination of the hoist cord to the horizon (α), degrees; and x_3 – the wall thickness (t_w), mm.

A-zone area (Fig. 4b) with the localized maximum of the main compressive stresses is 50–100 cm^2 for different nodes. Inside the area, the main stresses are of minor gradient which makes it possible to arrange strain gauge sensors within it to measure values of the mentioned stresses (Fig. 5a); data of the measurements concerning stress state parameters in the nodes of pulley supporting, performed for objects in Fig. 1 a, b and in Table 1 (Fig. 5b), verified the results of numerical experiments obtained using analytical model with the help of software complex Ansys Workbrench 14.0 (Fig. 6).

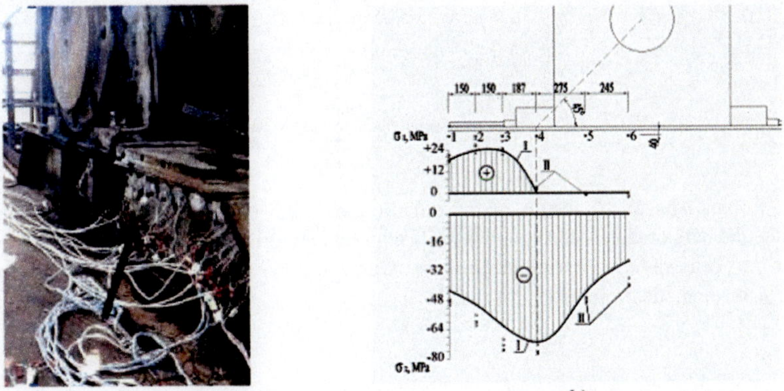

a) b)

Fig. 5. Measurement of the parameters of a stress state of the sub-pulley structures within a zone of the localized maximum of the main compressive stresses (A-zone in Fig. 4a): a) arrangement of tensoresistors; b) experimentally recorded distribution of the main stresses under the supporting pulley bearing as a result of tensometric measurements.

The graphs, shown in Fig. 6, help conclude that in terms of the fixed values of the geometrical parameters t_w and α, dependence of a hoist rope tension upon the maximums of local stresses $\sigma_{loc}^{\max}$ is of linear nature being determined with the help of a polynomial model (1).

5 Monitoring of Headframe Stresses Resulting from the Hoist Rope Tension

Experimental data were applied for two typical steel headframes (Table 1) shown in Fig. 1 to identify tension of hoist ropes relying upon the strain gauge measurements of the maximums of local stresses within the supporting nodes of the guiding pulleys.

Strain-gauge sensors were used to determine the main compressive stresses in the process of technological cycles of a mine hoist within the supporting nodes of the guiding pulleys. The measurements were performed inside a zone of the localized maximum of the main stresses *A* (see Fig. 4). Data by the strain-gauge sensors were recorded online with the help of a digital multichannel tensometric system on the basis of instrumental modules OVEN (device type is MB-110-224.4TД).

According to the identified dependences between the hoist tension and the measured main stresses within the points of their localized maximum (i.e. within A-point, Fig. 4a) represented in Fig. 6, relevant rope tension diagrams were calculated (Fig. 3). Essentially, the diagrams result from the monitoring of the dynamic stress impact of the headframe structures.

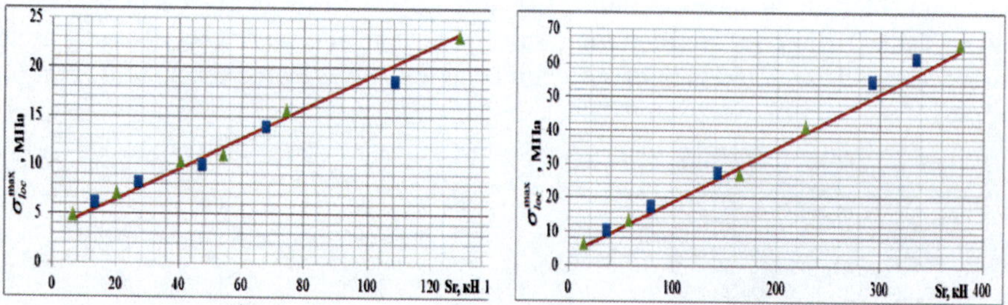

a) b)

Fig. 6. Comparison of the results of tensometric measurements with the results of numerical modeling and data obtained during a full-scale experiment: a) headframe with a drum-type hoist (Fig. 1b); b) headframe with a multirope hoist with a friction pulley (Fig. 1c); ■ experimental results; ▲ mathematical modeling results.

Table 1. Technical specifications of back-lag headframes.

Specification	Object in Fig. 1b	Object in Fig. 1a
Headframe height, m	36.0	40.0
Inclination of a back-lag to the horizon, degree	59° 15′	63° 58′
Hoisting height, m	1206	1213
Hoist type	БЦК-9/5 × 2.5	МПМН 6.2 × 3
Putting into service, year	1988	2005

6 Originality

Of the results is in the formulation of innovative method to monitor stresses by the rope tension. The method is based upon the regularities of stress distribution within the supporting nodes of the guiding pulleys. It helps record dynamics within the hoist ropes online.

7 Practical Implications

The proposed method to monitor operating loads of headframes makes it possible to get information concerning dynamic stress cycles. The information is required to evaluate fatigue strength of structural components of headframes which is the important components to provide safety of mine hoists.

8 Conclusions

1. Structures of headframes belong to CC-3importance class. According to Ukrainian regulatory requirements, they need monitoring of their technical state.
2. Monitoring of the cyclical dynamic loads resulting from the hoist rope tension is the important component of methods evaluating technical state of steel headframes since it is the basis to evaluate structural fatigue strength of steel headframe structures.
3. Analysis of stress state of the supporting nodes of the guiding pulleys has helped identify regularities of local stress distribution and define location of the localized maximum zone of the main compressive stresses within a wall of the sub-pulley structures. Moreover, polynomial dependence between the maximum local stresses and hoist tension has been determined.
4. Tensometric measurements within a zone of the localized maximum of the main compressive stresses under the supporting bearings of the guiding pulleys have made it possible to derive diagrams of temporal changes in hoist tension during operation.
5. The results of monitoring of stresses by hoist rope tension involve critical information concerning structural fatigue strength of steel headframe structures.

References

1. DBN Buildings. General principles of ensuring the reliability and constructive safety of buildings, structures, constructions and foundations, vol. 30. Minregionbud Ukraine, Kyiv (2018). (in Ukrainian)
2. RD 12.005_94. Metal structures of mine headframes. Operation requirements. Gosugleprom of Ukraine, Kyiv 1994, vol. 68 (1994). (in Russian)
3. DSTU-NB.1.2-18: 2016. Guidelints for the inspection of building and structures for the determination. DP "UkrDPC", Kyiv, vol. 44 (2017). (in Ukrainian)
4. ISO 19426-2 – part 2: Headframe structures, vol. 22 (2018)
5. Kushchenko, V.N.: Organizations of safety of building constructions of cut_sample mine headframe. Monograph. Makiivka, vol. 203 (2006). (in Russian)
6. Kushchenco, V.M., Khomitskyi, D.O.: Vibration monitoring of steel shaft headgears. In: Advances in Resource-saving Technologies and Environmental Engineering, Lecture Notes in Civil Engineering, vol. 47, pp. 227–234. Springer, Cham (2019)
7. Bezhok, V.R., Dvornikov, V.I., Manets, I.G., Pristrom, V.A.: Shaft lift, OOO "Yugo-Vostok, ATMD" Donetsk, vol. 624 (2007). (in Russian)
8. Wolny, S.: Dynamic loading of the pulley block in a hoisting installation in normal operation conditions. Arch. Min. Sci. **54**(2), 261–284 (2009)
9. Wolny, S., Matuchowsky, F.: Analysis of loads and stresses in structural elements of hoisting installation in mines. Eng. Trans. **58**(3–4), 153–174 (2010)
10. Jian, L.: Multi-rope hoist steel rope tension on line monitoring system. In: Conference on Information Technology and Computer Science (SITES 2012), pp. 229–232. Atlantis Press (2012)
11. Kuschenko, V., Nechitaylo, A.: Analysis of the mode of deformation of the joints of guide pulley resting on shaft sloping headgear structures. Metal Constr. **18**(2), 97–109 (2012)
12. Kuschenko, V., Nechitaylo, A.: Experimental research of the mode of deformation of sub-pulley structures of shaft frame-type sloping headgear. Metal Constr. **19**(3), 143–154 (2013)

Dependence of Evaporation Temperature and Exergetic Efficiency of Air Split-Conditioners Heat Pumps from the External Air Temperature

Volodymyr Labay(✉), Vitaliy Yaroslav, Oleksandr Dovbush, and Bohdan Piznak

Lviv Polytechnic National University, Lviv 79013, Ukraine
wlabay@i.ua

Abstract. In the face of growing shortages and rising prices for fuel and energy resources, the problem of energy conservation and the use of alternative energy sources to solve the problem of reducing energy consumption for the Ukrainian economy becomes very important. Today, the use of air split-conditioner heat pumps in buildings' heating systems is becoming more common. Therefore, the improvement of the design and operation of power equipment to which air split-conditioner heat pumps ("air-air") are related, is related to a detailed study of their operation and an objective assessment of their degree of energy perfection, which can only be determined on the basis of analysis their exergy efficiency. This made it possible to substantiate the relevance of such a research task due to the lack of information on the operating modes and the exergy efficiency of the use of air split-conditioner heat pumps. It was used the author's innovation, the mathematical model to analysis of the operation of one-step freon heat pumps, which are used in air split-conditioners according to the exergetic method. The dependence of the evaporation temperature and the exergetic output-input ratio (OIR) of the air split-conditioner heat pump by "Mitsubishi Electric" firm with nominal heating capacity of 3067 W in the standard external temperature conditions on the refrigerant R32, was determined from the out of doors temperature.

Keywords: Heat pump · Air Split-Conditioner · Evaporation temperature · Exergetic efficiency · External air temperature

1 Introduction

In the face of growing shortages and rising prices for fuel and energy resources, the problem of energy saving and the use of alternative energy sources to solve the problem of reducing energy consumption for the Ukrainian economy becomes very actually [1].

A significant reduction in the use of traditional organic energy sources for building heating is possible thanks to, for example, heat pumps (HP) ("air-to-air") of air split-conditioners, that use renewable energy. Today, the use of air split-conditioners heat

Z. Blikharskyy (Ed.): EcoComfort 2020, LNCE 100, pp. 253–259, 2021.
https://doi.org/10.1007/978-3-030-57340-9_31

pumps in building's heat supply systems is becoming more common [2–4, 6–8, 14]. This is due to the fact that, by consuming 1 kW of electricity, HP air split-conditioners can carry up to 5 kW of energy for heating the room air. The use of HP provides not only energy-saving but also environmental impact.

According to the World Energy Committee, by 2020, 75% of heat supply (municipal and industrial) in developed countries will be supplied by heat pumps. In the USA, more than 30% of residential homes are equipped with HP.

Currently for Ukraine the problem of saving energy resources is particularly relevant in conditions of market economy, limited resources of primary energy commodities – oil and gas.

In recent decades both abroad and in Ukraine with the aim of saving energy resources, fundamental research in a number of industries and technologies from the standpoint of exergetic methodology have been conducted [7–18].

Exergy not only quantifies energy of any kind, but also allows estimate its quality. It defines the convergence, suitability of energy for its technical use in any conditions.

Since exergy is the only measure of the work ability, that is, suitability of energy resources for use, it enables to objectively evaluate the energy resources of any kind. Consequently, exergy is some universal measure of the suitability of energy resources. And the exergetic balance, on the basis of which the volume of energy resources is set, indicates the possibility of increasing the OIR of the process.

In some leading European countries and in USA exergy analysis has been introduced as a mandatory component in development of projects and plans of modernization of manufacturing.

2 Analysis of Recent Studies and Publications

Successful application of the exergetic method of analysis of different technical systems, in particular of refrigerating machine of air split-conditioners, have been grounded in the works of R. K. Clausius, John V. Gibbs, G. Gouy, A. Stodola, J. Szargut, R. Petela and V. M. Brodyansky for their technical and economic optimization and our works [7–18]. We have applied this method to the analysis of air split-conditioners heat pumps and in this article are used.

Modern air split-conditioners, which are used to create a suitable microclimate in small rooms, have achieved some definite technical improvement. As it is known, the energy efficiency of heat pumps of these air split-conditioners depends on parameters of both external and internal temperature conditions of their operation and type of refrigerant [8]. Therefore, to further improving of efficiency of operation of air split-conditioners heat pumps a detailed analysis of their functioning is required.

For this purpose, based on the work [8], the innovative mathematical model of exergy analysis of the operation of air split-conditioners heat pumps has been developed by the authors, adapted for different refrigerants and manufacturers. This mathematical model makes it possible to carry out exergy investigations of heat pumps as a whole one and of its individual parts, for obtaining full information about the processes of energy transformation that have place in such systems [16, 17]. The result of the analysis is to finding the temperature mode of operation and the exergy OIR of process in overall and the exergy losses in the individual elements of the air split-conditioner heat pump with the aim of its optimization.

So, based on analysis of available literature data, decreasing the energy costs that used by air split-conditioners heat pumps can be the most complete achieved on the basis of exergy analysis, which takes into account not only the quantity but also the quality of spent energy [16, 17].

For perform research, the air split-conditioner heat pump scheme, which is shown in Fig. 1, and respectively the construction of the processes of its operation on the (p, i)-diagram – in Fig. 2, and work agent refrigerant-32 (R32) [18] is using.

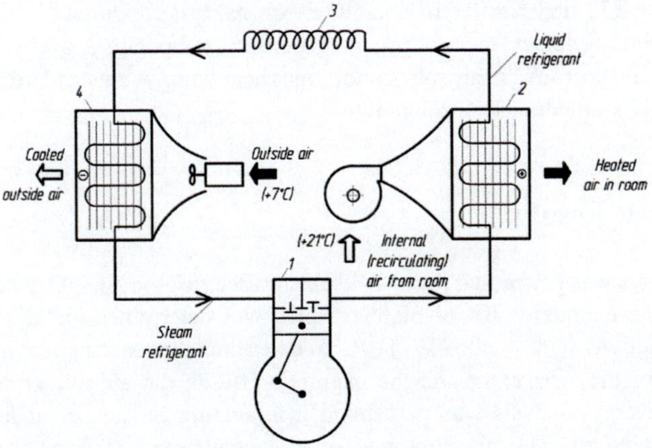

Fig. 1. Scheme of air split-conditioner heat pump: 1 is compressor; 2 is condenser; 3 is capillary tube; 4 is evaporator

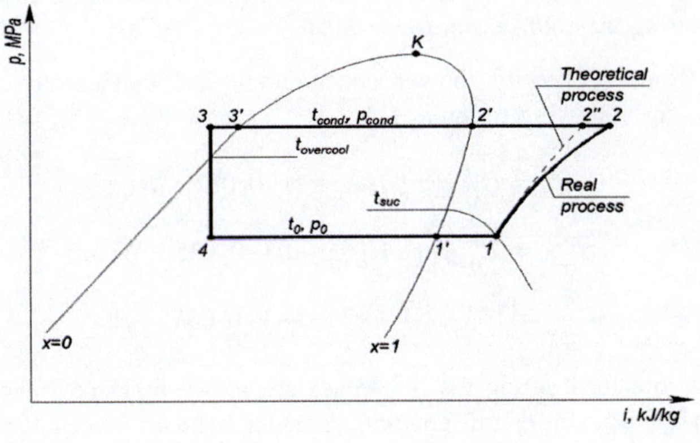

Fig. 2. Construction of the processes of work on *(p,i)*-diagram for air split-conditioner heat pump: 1, 2, 3, 4 are characteristic points of the thermodynamic cycle

3 The Purpose of the Research

The purpose of this work is to determine the evaporation temperature and exergy efficiency of the air split-conditioner heat pump on the one-component refrigerant R32.

To achieve this purpose, the following main tasks were formulated:

- determine the evaporation temperature for the air split-conditioner heat pump, for example, "Mitsubishi Electric", with the standard heat capacity of 3067 W on refrigerant R32 under different operating external (out door) test conditions;
- to set the exergy efficiency on the example of the air split-conditioner heat pump with the rated heat capacity of 3067 W at standard temperature conditions on refrigerant R32 under different operating external test conditions;
- to set analytical dependences between evaporation temperature, exergy efficiency of "Mitsubishi Electric" the air split-conditioner heat pump with standard heat capacity of 3067 W and external air temperature.

4 Research Results

Exergy analysis was performed for a "Mitsubishi Electric" the air split-conditioner heat pump with the exergetic OIR of 38.0%, which was determined under standard temperature conditions ($Q_H^{st} = 3067$ W) [19]. To determine the working temperature of the evaporation of the refrigerant and the exergetic OIR in the air split-conditioner heat pump, this exergy analysis was performed in operating mode, i.e. under conditions other than standard. The air flow rate on the condenser (1507 m^3/h) and on the evaporator (614 m^3/h) was kept constant during these tests.

The following initial data were taken for the calculation:

- external (out door) temperature $t_{C1} = -15\ldots20\,°C$;
- temperature of indoor (recirculating) air in the air-conditioned room, which is constant during the cold season $t_{H1} = 21\,°C$.

Working heating capacity, power consumption and condensate amount were determined by the following formulas [5]:

$$Q_H^w = Q_H^{st} \cdot [1 + (t_{C1} - 7) \cdot 0,035], \text{ W}; \quad (1)$$

$$N_{cons}^w = N_{cons}^{st} \cdot [1 + (t_{C1} - 7) \cdot 0,035], \text{ W}; \quad (2)$$

$$W_{cond}^w = W_{cond}^{st} \cdot [1 + (7 - t_{C1}) \cdot 0,035], \text{ 1/h}. \quad (3)$$

The results obtained during the researches are shown in Table 1 (boldly marked technical characteristics of the air conditioner under standard temperature conditions) and are shown graphically in Fig. 3 and 4.

Table 1. Results of studies of evaporation temperature and exergetic OIR of "Mitsubishi Electric" air split-conditioner heat pump with heat capacity of 3067 W, depending on external air temperature

$t_{out} = t_{C_1}$, °C	Q_H^w, W	N_{cons}^w, W	W_{cond}^w, l/h	$t_0 = t_{ev}$, °C	η_e, %
–15	705	179	1,68	–21,0	58,1
–11	1135	289	1,55	–17,1	54,4
–7	1564	398	1,42	–13,1	50,9
0	2316	589	1,18	–6,2	44,7
+7	**3067**	**780**	**0,95**	**+0,7**	**38,0**
+10	3389	862	0,85	+3,6	34,8
+15	3926	998	0,68	+8,7	28,6
+20	4462	1135	0,52	+13,5	20,7

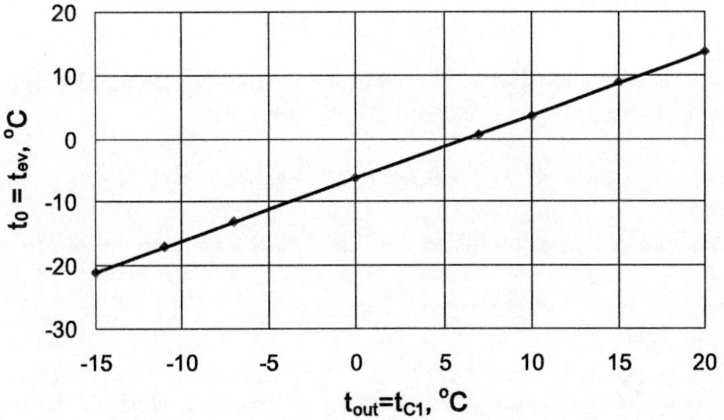

Fig. 3. Dependence of the evaporation temperature of the "Mitsubishi Electric" air split-conditioner heat pump with a heat capacity of 3067 W under standard conditions, depending on the external air temperature

Under other conditions: Q_H^{st} = 3067 W; N_{cons}^{st} = 780 W; W_{cond}^{st} = 0,95 l/h; L_{cond} = 614 m^3/h; L_{ev} = 1507 m^3/h; Q_C^{st} = 2567 W; $t_{in} = t_{H_1} = +21\,°C$; $t_{cond} = +39,9\,°C$.

The dependence of the evaporation temperature of the "Mitsubishi Electric" air split-conditioner heat pump of 3067 W under standard conditions on the external temperature (Fig. 3) is approximated by the formula:

$$t_0 = t_{ev} = 0,9873 \cdot t_{out} - 6,2073, ^\circ C, \quad (4)$$

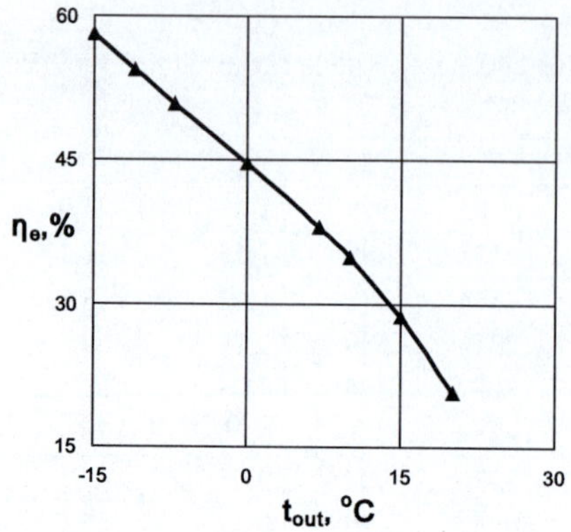

Fig. 4. Dependence of the exergetic OIR of the "Mitsubishi Electric" air split-conditioner heat pump 3067 W with a heat capacity of 3067 W under standard conditions, depending on the external air temperature

and the dependence of the exergetic OIR of the air split-conditioner heat pump on the external temperature (Fig. 4) – according to the formula:

$$\eta_e = 45{,}122 - 0{,}9811 \cdot t_{out} - 0{,}0104 \cdot t_{out}^2, \ \%. \tag{5}$$

The maximum error according to the Eq. (4) is 1.94%, and the Eq. (5) is 3.09%.

5 Conclusions

The author's innovative mathematical model was used to analysis of the operation of one-step freon heat pumps used in local autonomous air-conditioners to create comfortable conditions in buildings during the cold season. The researches have been conducted on the basis of the mathematical model have been established the evaporation temperature and the exergetic output-input ratio (OIR) of the air split-conditioner heat pump on the example of the air conditioner with a nominal heat capacity of 3067 W by "Mitsubishi Electric" firm under standard external conditions on refrigerant R32.

Analyzing the data obtained in the Table 1 and in Fig. 3 and 4, one can come to the following conclusions. An increase in external air temperature by (20 − (−15)) · 100/15 = 233% leads to an increase in the evaporation temperature of the air conditioner by (13,5 − (−21,0)) · 100/21,0 = 164%, and the exergetic OIR decreases by (58,1 − 20,7) · 100/58,1 = 64% with a significant increase in the heat capacity of the air conditioner by (4462 − 705) · 100/705 = 533%.

The mathematical model of operation of air split-conditioners heat pumps used for this article can be applied to different types of refrigerants and models of split-conditioners, provided that the thermodynamic properties of the refrigerant and the characteristics of the heat-pump of the air split-conditioner are known.

References

1. Energy Strategy of Ukraine until 2030 [Electronic resource]. http://www.ukrenergo.energy.gov.ua. (in Ukrainian)
2. Heat Pump, Types and Applications of Heat Pumps [Electronic resource]. http://www.ecosvit.net/ua/teplovij-nasos-vidi-tazastosuvannya. (in Ukrainian)
3. Heat pumps [Electronic resource]. http://www.npblog.com.ua/index.php/hi-tech/teplovi-nasosi.html. (in Ukrainian)
4. Bezrodniy, M.N., Dranik, T.V.: Thermodynamic efficiency of heat pump application for providing comfortable conditions in indoor pools. Eastern Eur. J. Enterp. Technol. **3**(8), 25–30 (2013). http://nbuv.gov.ua/UJRN/Veipt_2013_3_8_8. (in Ukrainian)
5. Bogoslovskiy V.N., Kokorin O.Ya., Petrov, L.V.: Air Conditioning and Cold Supply. Stroyizdat, Moscow (1985). (in Russian)
6. Zalewski, P.K.: Pompy ciepła. Podstawy teoretyczne i przykłady zastosowania. I.P.P.U. MASTA Sp. z o.o., Kraków (2001). (in Polish)
7. Shargut, E., Petela, R.: Exergy. Energy, Moscow (1968). (in Russian)
8. Sokolov, E.Ya., Brodiansky, V.M.: Energy bases of heat transformation and cooling processes. Energoizdat, Moscow (1981). (in Russian)
9. de Oliveira Junior, S.: Exergy. Production, Cost and Renewability. Springer (2013)
10. Sazhin, B.S., Bulekov, A.P., Sazhin, V.B.: Exergetic Analysis of Industrial Plants. Chemistry, Moscow (2000). (in Russian)
11. Bejan, A.: Advanced Engineering Thermodynamics. Wiley, New York (1988)
12. Bejan, A., Tsatsaronis, G., Moran, M.: Thermal Design and Optimization. Wiley, New York (1996)
13. Morosuk, T., Nikulshin, R., Morosuk, L.: Entropy-cycle method for analysis of refrigeration machine and heat pump cycles. Therm. Sci. **10**(1), 111–124 (2006)
14. Morozyuk, T.V.: The theory of refrigeration machines and heat pumps. Studio "Negotsiant", Odessa (2006). (in Russian)
15. Tsatsaronis, J.: The interaction of thermodynamics and economics to minimize the cost of an energy conversion system. Studio "Negotsiant", Odessa (2002) (in Russian)
16. Labay, V.Yo., Khanyk, Ya.M.: Energy saving ratio between the air flows at the evaporator and condenser air split-conditioners. Sci. Tech. J. Refrig. Eng. Technol. **6**(116), 28–31 (2008). (in Ukrainian)
17. Labay, V.Yo., Mysak, Yo.S.: Adduction of work of refrigeration's machines of air split-conditioners to the identical internal temperature condition. Sci. Techn. J. Refrig. Eng. Technol. **4**(126), 19–22 (2010). (in Ukrainian)
18. Jakobsen, A., Rassmussen, B.-D., Skovrup, M.-J., Andersen, S.-E.: CoolPack – a collection of simulation tools for refrigeration – Tutorial – Version 1.46. – Department of Energy Engineering Technical University of Denmark (2001)
19. Mitsubishi Electric Catalogo Split (2019). (in English)

Influence of Damages in the Compressed Zone on Bearing Capacity of Reinforced Concrete Beams

Maxim Lobodanov, Pavlo Vegera(✉), Roman Khmil, and Zinoviy Blikharskyy

Lviv Polytechnic National University, Lviv 79013, Ukraine
Pavlo.I.Vehera@lpnu.ua

Abstract. The paper present the results of reinforce concrete beams damaged at compressed fiber. Study of damaged reinforced concrete elements is associated with certain complexity in both theoretical and practical issues. In such cases the number of factors take place which influence on elements' performance, which is caused by composite features of reinforced concrete elements. In addition certain factors occur due to damages. In the article experimental data is analyzed concerning concrete damages' influence in the compressed zone on the bearing capacity of the reinforced concrete elements. Results of 4 reinforced concrete beams' testing are proposed one of which was the control one (tested without damages) and three,- typically damaged in compressed zone at different load levels. As the result the effect of damage and load influence on bearing capacity of beams was indicated, as well as effect of neutral axis position change. Research results demonstrate increase of 3.8% in reinforced concrete beams bearing capacity, if they are damaged under the load, comparing with the unloaded damaged reinforced concrete beams.

Keywords: Reinforced concrete beam · Damages · Defects · Corrosion · Bearing capacity

1 Introduction

Increase in tendency of minimizing of reconstruction costs encourages research dedicated to indicating of residual bearing capacity reinforced concrete elements [1, 2]. It is especially relevant when conducting technical examination of reinforced concrete elements at functionality changes in the building in general [3, 4]. Thus increases the necessity in retrofitting of existing buildings and structures and renovating of non-operating buildings [5–7].

Research tendencies in reconstruction field, formed by market needs are the basis for development of construction, as well as related industries [8–10]. Therefore, research in the field of residual bearing capacity indication are highly topical for Ukraine as well as for other countries [11–15].

Most of studies are dedicated to restoration of damaged structures with the use of RC casing [16, 17]. Especially attention should be paid to strengthening of reinforced

Z. Blikharskyy (Ed.): EcoComfort 2020, LNCE 100, pp. 260–267, 2021.
https://doi.org/10.1007/978-3-030-57340-9_32

concrete structures with modern composite materials [18, 19], which have high strength parameters and low weight.

The basis for damaged reinforced concrete beams' study is understanding of stress-strain state of structures, which work without damages. Especially attention should be paid to materials' characteristics, as they are the main components, which determine the stress-strain state of constructions [20–25].

2 Purpose of the Research

The main purpose of the research is to compare the results of three samples' testing, including the following: control, damaged before load application and during the load application. According to obtained results, analysis of load influence on bearing capacity was conducted.

3 Methodology of the Research

Research was conducted for test samples- one-span reinforced concrete beams. Reinforcement was performed by working stretched rebar of Ø14 mm, compressed rebar in zone of maximum shear force action– Ø10 mm. Transverse reinforcement was introduced by smooth rebar of Ø8 mm, located in supporting zones. During the experiment, physic-mechanical characteristics of materials were indicated: concrete - C35/45; reinforcement introduced by A500C class. The samples' construction was described in more detail in the article [24].

For concrete deformation measurements clock-type indicators I1…I8 were used with basis of 20 cm in the zone between the loads applied and spacing of 40 mm (see Figs. 1 and 2). Indicators I11…I14 were located in the zone of concrete stresses' concentration. These indicators indicate the highest strains and include in the basis the damaged concrete zone. On the stages, previous to load application and after, maximum compression deformation are identified by I15 and I16. Strain measures for reinforcement were made with the use of indicators I9 and I10. Aistov deflection-meters P2…P4 measured beams' deflection (Figs. 1 and 2) and subsidence's on supports were indicated by P1 i P5.

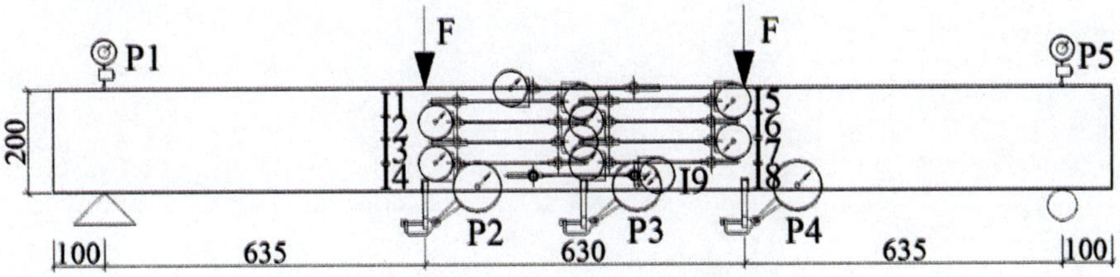

Fig. 1. Scheme of devices' location.

Fig. 2. General view of devices location on the research sample

Experiment was conducted in stages with loading increases of 0.05 $P_{max.c}$ (where $P_{max.c}$ – is the loading at which the bearing capacity of control sample was depleted). This increase was used until the loading has reached the value of 0.3 $P_{max.c}$ or until the first cracks appear. Further increases of loading were equal to 0.1 $P_{max.c}$. At the load level equal to 0.8–0.9 of the limit strength of the sample, the increase was equal again to 0.05 $P_{max.c}$. Observation of the cracks formation and development was conducted with the use of microscope MPB-2 with the division price of 0.05 mm.

Damages were performed with the use of device-strobe-cutter. Depending on the damage type strobe-cutter was customized according to following parameters: damage height, disks 'spacing, disks' number.

For reinforced concrete beams, damaged before the load application, the damages were made without additional stages at impairments of 10, 20, 30 mm. For experimental samples BD 2.6.1-0-20 and BD 1.1.2-0-20 damages were made in one stage with the use of two disks (Fig. 3).

Fig. 3. General view of the damages in the compressed zone.

Experiment was conducted according to developed program, given in Table 1. Each beam had its individual marking. For example BD 1.4.1-0.7-20, is the beam of first series; first testing sample of fourth research at loading level equal to 70% of the control sample strength with damage weight of 20 mm.

Table 1. Program of the experiment.

№	Marking of the experiment	Marking of the testing sample	Type of the experiment
1	B 1.1	B 1.1.1-0	Without damages (control sample)
	BD 1.2	BD 1.1.2-0-20	With damages of 20 × 30 mm in the center of pure bending zone at load level of 0% from bearing capacity of control sample
2	BD 2.1	BD 1.2.1-0.3-20	With damages of 20 × 30 mm in the center of pure bending zone at load level of 30% from bearing capacity of control sample
3	BD 3.1	BD 1.3.1-0.5-80	With damages of 80 × 30 mm in the center of pure bending zone at load level of 50% from bearing capacity of control sample

4 Analysis of Experiment Results

According to the developed research program, control samples were tested for obtaining of initial data for further study. Experimental samples BD 1.1.1-0 of the first series were tested without damages and impairments. During the analysis, all further research data was compared with the control sample parameters (Fig. 4).

For comparison and analysis of deformation distribution in concrete three sections of measures were considered. First section- measures, obtained with the use of indicators I1-4, second - from indicators I11-14 and third – from indicators I5-8 (Fig. 1).

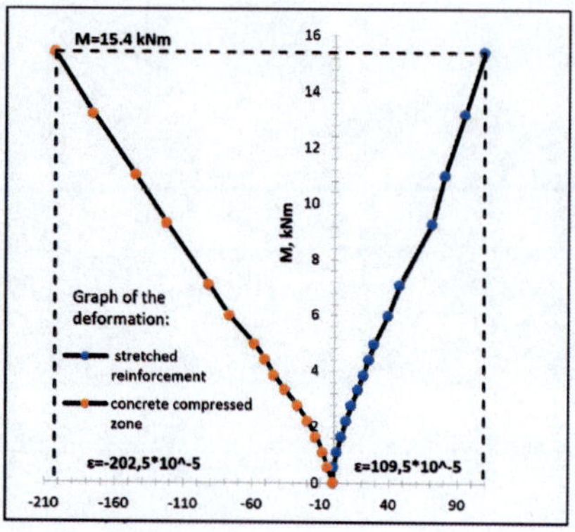

Fig. 4. The sample of strain graph for reinforcement (left) and compressed concrete (right) of undamaged beams BD 1.1.1-0 of the first series

Obtained results demonstrate typical changes of compressed concrete zone height for undamaged reinforced damaged concrete beams. Additionally these results demonstrate certain symmetricity in measures in first and third sections (Fig. 5). Such the difference in measures in stretched zone was caused by cracks' opening, at the end of first stress-strain state of the element. Obtained results are given in the Table 2.

Table 2. Program of the experiment

Testing samples' marking	Sample bearing capacity, kNm	Deviation in bearing capacity according to B 1.1.1-0	Concrete compressed zone height at M = 4,9 kNm, mm	Concrete compressed zone height at M = 15,4 kNm, mm	Deviation in changes of concrete compressed zone height at M = 15,4к kNm according to M = 4,9 kNm, %
B 1.1.1-0	25.703	–	66.71	65	2.563
BD 1.1.2-0-20	20.869	−18.807	67.446	67.759	−0.464
BD 1.2.1-0,3-20	21.655	−15.749	67.12	75.711	−12.8
BD 1.3.1-0.5-80	21.02	−18.298	75.83	80.21	−5.776

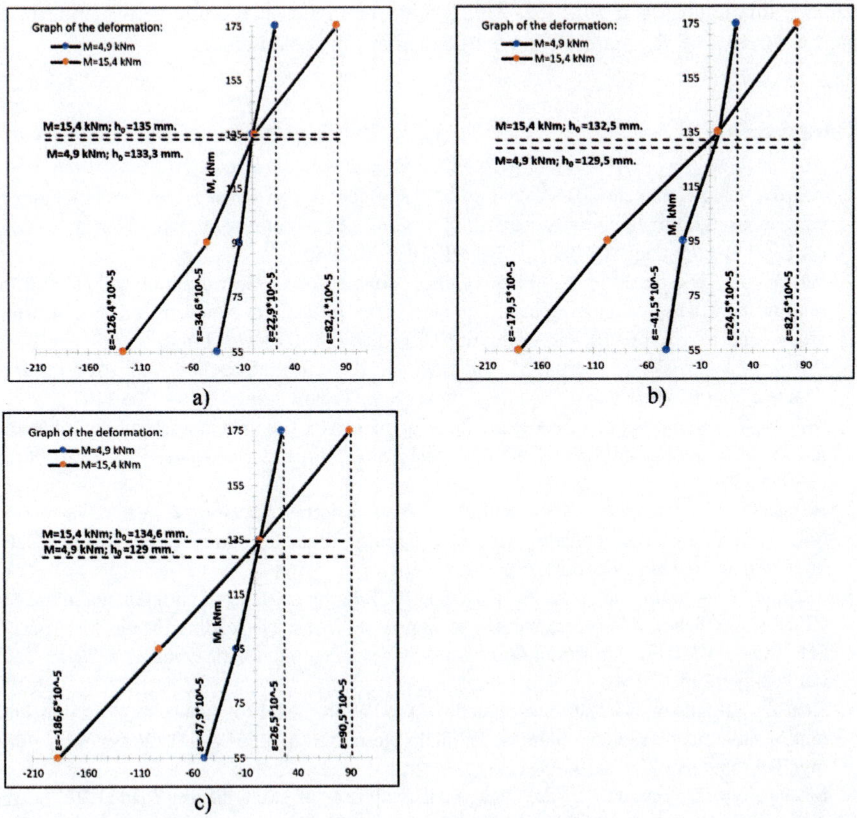

Fig. 5. Deformation graph for concrete in undamaged beams BD 1.1.1-of the first series: a) first section; b) second section; c) third section.

Analyzing the results, it was indicated that bearing capacity could be strongly influenced by damages, which appear during the loading. Damages at 0.3 and 0.5 M from the limit strength for the sample demonstrate distribution of stresses and increase of bearing capacity, comparing with the sample, damaged before load application. Additionally changes in neutral axis position was fixed, where the transition from the stretched zone to the compressed zone takes place.

5 Conclusions

Research results demonstrate increase of 3.8% in reinforced concrete beams bearing capacity, if they are damaged under the load, comparing with the unloaded damaged reinforced concrete beams. Additionally was fixed the effect of neutral axis position change: if the sample is damaged the stretched concrete zone is decreased (in the damaged zone), the compressed zone is increased. If the damages occur during the loading, this tendency is fulfilled. This occurrence takes place due to distribution of internal stresses in the spatial changes of element in dynamics.

References

1. Azizov, T.N., Kochkarev, D.V., Galinska, T.A.: New design concepts for strengthening of continuous reinforced-concrete beams. IOP Conf. Ser. Mater. Sci. Eng. **708**(1), 012040 (2019). https://doi.org/10.1088/1757-899X/708/1/012040
2. Bobalo, T., Blikharskyy, Y., Kopiika, N., Volynets, M.: Serviceability of RC beams reinforced with high strength rebar's and steel plate. In: Lecture Notes in Civil Engineering, vol. 47, pp. 25–33 (2020). https://doi.org/10.1007/978-3-030-27011-7_4
3. Pavlikov, A., Kosior-Kazberuk, M., Harkava, O.: Experimental testing results of reinforced concrete beams under biaxial bending. Int. J. Eng. Technol. **7**(3.2), 299–305 (2018)
4. Bobalo, T., Blikharskyy, Y., Kopiika, N., Volynets, M.: Theoretical analysis of RC beams reinforced with high strength rebar's and steel plate. IOP Conf. Ser. Mater. Sci. Eng. **708**(1), 012045 (2019)
5. Blikharskyy, Y., Kopiika, N., Selejdak, J.: Non-uniform corrosion of steel rebar and its influence on reinforced concrete elements' reliability. Prod. Eng. Arch. **26**(2), 67–72 (2020). https://doi.org/10.30657/pea.2020.26.14
6. Karpiuk, V.M., Syomina, Y.A., Antonova, D.V.: Bearing capacity of common and damaged CFRP-strengthened RC beams subject to high-level low-cycle loading. Mater. Sci. Forum **968** 185–199 (2019). https://doi.org/10.4028/www.scientific.net/MSF.968.185. Trans Tech Publications Ltd
7. Kos, Ž., Klimenko, Y.: The development of prediction model for failure force of damaged reinforced-concrete slender columns. Tehnički vjesnik **26**(6), 1635–1641 (2019). https://doi.org/10.17559/TV-20181219093612
8. Blikharskyy, Z., Brózda, K., Selejdak, J.: Effectivenes of strengthening loaded RC beams with FRCM system. Arch. Civ. Eng. **64**(3), 3–13 (2018)
9. Blikharskyy, Z., Vashkevych, R., Vegera, P., Blikharskyy, Y.: Crack resistance of RC beams on the shear. In: Lecture Notes in Civil Engineering, vol. 47, pp. 17–24 (2020). https://doi.org/10.1007/978-3-030-27011-7_3
10. Selejdak, J., Blikharskyy Y., Khmil R., Blikharskyy Z.: Calculation of reinforced concrete columns strengthened by CFRP. In: Lecture Notes in Civil Engineering, vol. 47, pp. 400–410 (2020). https://doi.org/10.1007/978-3-030-27011-7_51
11. Blikhars'kyi, Y.Z.: Anisotropy of the mechanical properties of thermally hardened A500s reinforcement. Mater. Sci. **55**, 175–180 (2019). https://doi.org/10.1007/s11003-019-00285-0
12. Lipiński, T.: Roughness of 1.0721 steel after corrosion tests in 20% NaCl, Production Engineering Archives, **15**(15), 27–30 (2017). https://doi.org/10.30657/pea.2017.15.07
13. Sykora, M., Holicky, M., Prieto, M., Tanner, P.: Uncertainties in resistance models for sound and corrosion-damaged RC structures according to EN 1992-1-1. Mater. Struct. **48**(10), 3415–3430 (2015)
14. Tigeli, M., Moyo, P., Beushausen, H.: Behaviour of corrosion damaged reinforced concrete beams strengthened using CFRP laminates. In: Nondestructive Testing of Materials and Structures, pp. 1079–1085. Springer, Dordrecht (2013)
15. Wang, X.H., Li, B., Gao, X.H., Liu, X.L.: Shear behaviour of RC beams with corrosion damaged partial length. Mater. Struct. **45**(3), 351–379 (2012)

16. Khmil, R., Tytarenko, R., Blikharskyy, Y., Vashkevych, R.: Influence of load level during strengthening of reinforced concrete beams on their reliability. IOP Conf. Ser. Mater. Sci. Eng. **708**(1), 012054 (2019)
17. Krainskyi, P., Vegera, P., Khmil, R., Blikharskyy, Z.: Theoretical calculation method for crack resistance of jacketed RC columns. IOP Conf. Ser. Mater. Sci. Eng. **708**(1), 012059 (2019)
18. Al-Saidy, A.H., Saadatmanesh, H., El-Gamal, S., Al-Jabri, K.S., Waris, B.M.: Structural behavior of corroded RC beams with/without stirrups repaired with CFRP sheets. Mater. Struct. **49**(9), 3733–3747 (2016)
19. Benjeddou, O., Ouezdou, M.B., Bedday, A.: Damaged RC beams repaired by bonding of CFRP laminates. Constr. Build. Mater. **21**(6), 1301–1310 (2007)
20. François, R., Khan, I., Dang, V.H.: Impact of corrosion on mechanical properties of steel embedded in 27-year-old corroded reinforced concrete beams. Mater. Struct. **46**(6), 899–910 (2013)
21. Imam, A., Anifowose, F., Azad, A.K.: Residual strength of corroded reinforced concrete beams using an adaptive model based on ANN. Int. J. Concr. Struct. Mater. **9**(2), 159–172 (2015)
22. Kharchenko, Y.V., Blikharskyy, Z.Y., Vira, V.V., Vasyliv, B.D.: Study of structural changes in a nickel oxide containing anode material during reduction and oxidation at 600 °C. In: Fesenko, O., Yatsenko, L. (eds.) Nanocomposites, Nanostructures, and Their Applications, NANO 2018. Springer Proceedings in Physics, vol. 221. Springer, Cham (2019). https://doi.org/10.1007/978-3-030-17759-1_42
23. Kharchenko, Y., Blikharskyy, Z., Vira, V., Vasyliv, B.: Vasyliv, B: study of nanostructural changes in a Ni-containing cermet material during reduction and oxidation at 600 °C. Appl. Nanosci. (2020). https://doi.org/10.1007/s13204-020-01391-1
24. Lobodanov, M., Vegera, P., Blikharskyy, Z.: Planning experiment for researching reinforced concrete beams with damages. In: Lecture Notes in Civil Engineering, vol. 47, pp. 243–250 (2020). https://doi.org/10.1007/978-3-030-27011-7_31
25. Prem, P.R., Murthy, A.R., Ramesh, G., Bharatkumar, B.H., Iyer, N.R.: Flexural behaviour of damaged RC beams strengthened with ultra high performance concrete. In: Advances in Structural Engineering, pp. 2057–2069. Springer, New Delhi (2015)

Influence of Humidity of Wood Fuel on the Gasification Process in a Continuous Layer

Stepan Lys(✉), Oksana Yurasova, and Yuriy Vashkurak

Lviv Polytechnic National University, Lviv 79013, Ukraine
stepan.s.lys@lpnu.ua

Abstract. In the world has been stepped up the work to develop technologies for biomass thermal conversion into gaseous fuels that can be used for combustion in different types of boilers, in internal combustion engines, or for liquefaction. The main advantage of the biomass thermal conversion technology for synthesis gas is its low environmental impact compared to fossil fuels. Therefore, gasification is one of the most promising ways of converting wood fuel into energy. According to the results of experimental research, the regression dependence of the lower calorific heat of synthesis gas during thermal processing of wood into gaseous fuel was obtained from the relative humidity and fractional composition of wood, the amount of air fed into the gasification chamber. Were found the rational values of humidity and fractional composition of the wood, as well as the amount of air supplied to the gasification chamber, at which the lower calorific heat of the synthesis gas reaches the maximum.

Keywords: Lower calorific heat · Synthesis gas · Thermal recycling of wood fuel · Wood moisture · Regression dependencies

1 Introduction

In the world has been stepped up the work to develop technologies for biomass thermal conversion into gaseous fuels that can be used for combustion in different types of boilers, in internal combustion engines, or for liquefaction. The main advantage of the biomass thermal conversion technology for synthesis gas is its low environmental impact compared to fossil fuels. One of the advantages of the gasification process over direct combustion is that the underfuel is much less, since there is an almost complete conversion of carbon when it is converted from solid to gaseous. As a result of the almost complete combustion of the synthesis gas, a significantly lower amount of harmful substances in the flue gas is produced. Therefore, gasification is one of the most promising ways of recycling wood fuel into energy.

There is sufficient supply for raw materials for this purpose. Timber wastes and logging residues are a significant reserve for raw materials for further thermal processing, since they are practically not used. Most wood wastes from woodworking enterprises are not used for industrial production. In the woodworking complex,

Z. Blikharskyy (Ed.): EcoComfort 2020, LNCE 100, pp. 268–276, 2021.
https://doi.org/10.1007/978-3-030-57340-9_33

the largest amount of waste that can be used for the thermal recycling of wet wood into gaseous fuel is in sawmilling and the production of rough and blanks.

Another source of wood biomass production that is provided for thermal recycling into gaseous fuels is “energy plantations” - the cultivation of fast-growing energy crops (willow, poplar and aspen).

Therefore, increasing the efficiency of the process of thermal conversion of wood fuels into gaseous fuels is an urgent task to solve which will create an environmentally friendly source of energy, which is an alternative to natural gas and gasification of coal.

2 Analysis of Literature and Problem Statement

The composition and calorific value of the synthesis gas resulting from the gasification of solid fuels may vary depending on various factors [1]. The main factors affecting the gasification process of wood fuels are humidity, particle sizes, the amount of air supplied to the gasification chamber and a number of other factors dependent on the gasification fuel and gasifier parameters. The output and composition of gasification products depend not only on the properties of input raw materials [2], but also on the process mode and the design of the gas generator.

The gas-fired plants used for gasification used timber in the form of meter firewood, which were air-dried. The negative impact of increasing humidity on the yield of liquid products is affected by the simultaneous increase in the particle size of the wood. Modern gas-generating installations [3, 4] use crushed wood. The use of crushed wood allows to increase the heat of combustion of synthesis gas, which leads to an increase in the productivity of gas consumers and, at the same time, to the reduction of gas consumption rates per unit of production, resulting in increased productivity of the gas generator.

During gasification of wetter wood, the yield of liquid products decreases significantly and the amount of gas, on the contrary, increases [5]. Gasification of wood with increasing humidity leads to a decrease in the quality of synthesis gas. The optimal relative humidity of crushed wood, which is gasified, will be about 25% [6, 7]. This is of great practical importance because the relative humidity of 25% is close to critical. The humidity of crushed wood, when it is lower than critical, reduces the drying rate. Higher synthesis gas combustion heat decreases with increasing raw material humidity.

Gasification of crushed wood with high humidity up to 60% is advisable. During the processing of wet crushed wood W = 57.2–60.7%, the intensity of gasification decreases in terms of unit area of the reactor core cross section compared to the gasification of dry wood [8, 9]. Synthesis gas, which is formed in the gasification zone during thermal processing of wet wood, contains a significant amount of H_2 and is unstable in composition: CO_2 = 8–12%; CO = 17–23%; CH_4 = 0.5–2%; C_nH_m = 0.2–0.6%; H_2 = 10–17%; O_2 = 0.2–0.4%; N_2 = 52–55%. As the temperature increases, the amount of CO in the synthesis gas increases, but the amount of H_2 decreases, these two gases being the main combustion gases of the synthesis gas, so it is important that the content of these gases increases.

Development of a rational design of gas generators for crushed wood is to create such an unit that would provide high productivity per unit area of the reactor core area [10], as well as ease of maintenance of this unit and the possibility of automation of the gasification process.

Therefore, the task was to develop a design of the gas generator, which would be able to obtain synthesis gas with greater calorific value, gasify the fuel with greater humidity and was easy to operate.

3 The Aim and Objectives of the Research

The aim is to research the effect of humidity of wood fuel on the gasification process in a continuous layer.

To achieve this aim it was necessary to perform the following tasks:

- to carry out the analysis of theoretical provisions and experimental researches of the process of gasification of wood fuel with different humidity in a continuous layer;
- to develop a schematic diagram of the design of a gas generator with a continuous layer for gasification of wood with a large range of changes in humidity;
- determine the rational parameters of the process of gasification of wood fuel in a gas generator with a continuous layer.

4 Materials and Methods of Research of the Influence Input Factors on the Process of Thermal Recycling of Wood into Gaseous Fuels and Calorific Value of Synthesis Gas

4.1 Materials and Equipment Used in Experimental Research

The analysis of the influence of the input parameters on the calorific value of the synthesis gas will allow to find the optimal mode parameters of the gasification process and the operation of the gas-generating unit. This will allow to develop the gasification process technology and the design scheme of the industrial gas generator.

The following materials were used for the experimental studies: a mixture of wood of different species (willow (Salix alba L.) - 1/3 mass of the mixture, pine (Pinus sylvestris) - 1/3; birch (Betula pendula Roth. - 1/3).

The task was to determine the dependence of the lower combustion heat of the synthesis gas on the relative humidity of the wood, the particle size of the wood and the amount of air.

For the purpose of experimental research and development of technological process of synthesis gas production, a gas generator with a continuous layer was developed (Fig. 1).

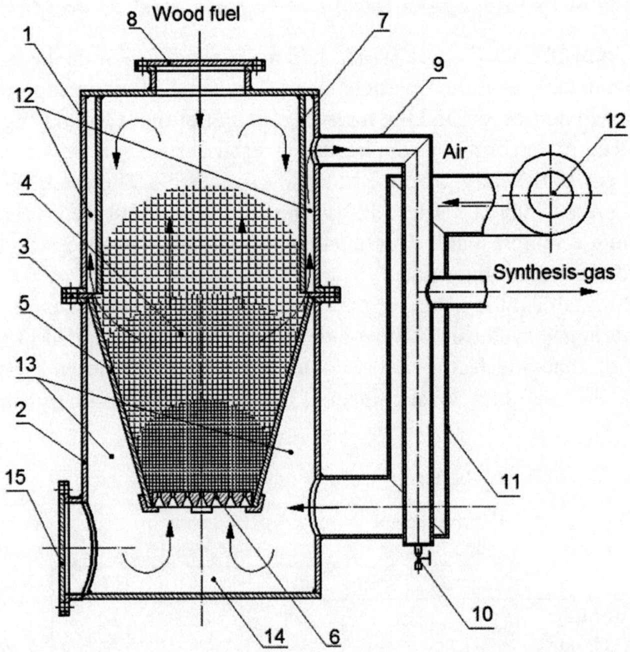

Fig. 1. Scheme of the gas generator unit with a continuous layer: 1, 2 - the housing; 3 - bolt connection; 4 - gasification chamber; 5 - lower cut cone; 6 - grate; 7 - upper cut cone; 8 - hatch; 9 - pipe for drainage of synthesis gas; 10 - condenser sump; 11 - a casing for cooling the synthesis gas; 12 - air blower; 13 - interconical space; 14 - ash-trapping device; 15 - a hatch for removal of ashes and ignition of fuel.

Modern solid-state gas generators allow to obtain synthesis gas with not high calorific value (about 7 MJ/nm^3) and demanding for moisture and fuel quality, the task was to develop a gas generator design that would allow to obtain synthesis gas with higher calorific value. and gasify with higher humidity.

On the developed gas-fired unit with a continuous layer, this is achieved due to the fact that the gases formed during gasification are re-passed through the layer of heated fuel in the oxidation and recovery zone [11, 12]. Where, at high temperatures, a heterogeneous carbon dioxide reduction reaction takes place, that is, $C + CO_2 \rightarrow 2CO$ and a combustible component of synthesis gas, carbon monoxide CO, is formed. If there is water vapor in the recovery zone, the reaction of water vapor conversion, i.e. $C + H_2O \rightarrow CO + H_2$ and $CO + H_2O \rightarrow CO_2 + H_2$, takes place at high temperature. In this case, the second combustible component of synthesis gas - hydrogen H_2, is formed. Thus, due to the high content of carbon monoxide CO and H_2 in the synthesis gas, the lower combustion heat is relatively high.

The use of gas generators of the proposed design, as shown by research, allows to increase the efficiency of work by increasing the speed and intensity of the process of gasification of wet wood.

4.2 Methods of Processing the Results of Experimental Research

In a series of experiments, the task was to find the dependence of the lower combustion heat of the synthesis gas on the particle size of the crushed wood supplied to the gas generator, the amount of air, and the moisture content of the wood during gasification.

The number of experiments duplicated in each series, n = 6. The data of each experiment were statistically processed to find gross errors, and doubtful results were verified using the Student's t-test [13]. The doubtful result (y_i) was temporarily excluded from the sample and the arithmetic mean ($\bar{y}$) and variance estimate (S^2) were calculated from the remaining data.

In order to establish the nature of the influence of the variable factors on the lower heat of synthesis gas synthesis, a three-level B-plan (B_3) was applied [13]. The levels and intervals of changing factors are given in Table 1. In the chosen experiment plan, each factor changes at three levels: upper (+), lower (−), and primary (0).

Table 1. The levels and intervals of the varying factors.

The name of the factor	The symbol of the factor		The levels of the factor changing			The interval of the factor changing
	Natural	Normalized	(−1)	(0)	(+1)	
The dimensions of the particles of wood, mm	l	x_1	10	30	50	20
The amount of air, Nm^3/h	G	x_2	40	65	90	25
Wood moisture, %	W	x_3	10	30	50	20

For the transition from natural to normalized notation factors used the ratio:

$$x_1 = \frac{l - 30}{20}; \; x_2 = \frac{G - 65}{25}; \text{ and } x_3 = \frac{W - 30}{20}; \tag{1}$$

The regression equation that can be obtained as a result of the implementation of second-order plans, that is, a plan by which you can obtain a mathematical description of objects in the form of second-order polynomials, is:

$$y = b_0 + \sum_{i=1}^{k} b_i x_i + \sum_{i=1}^{k} b_{ii} x_i^2 + \sum_{i,j=1}^{k} b_{ij} x_i x_j, \tag{2}$$

where x denotes the variable factors; b stands for the regression coefficients; k denotes the number of the variables.

The matrix of planning in the natural and normalized values for pine wood is given in Table 2.

After processing the experimental data and obtaining the regression equation, their statistical analysis was performed. Two problems were solved: the significance of the regression coefficients was evaluated and the adequacy of the mathematical model was checked using the Fisher F_p.

Table 2. Results of the B-plan calculation

The matrix of the experimental design							The experiment results		
No	x_1	x_2	x_3	The size of the wood particles, mm	The amount of air, Nm^3/h	Wood moisture W, %	$\bar{y}$	S^2	y^p
1	−1	−1	−1	10	40	10	6,925	0,02683	6,8750
2	1	−1	−1	50	40	10	8,674	0,02621	8,6638
3	−1	1	−1	10	90	10	7,147	0,02993	7,1619
4	1	1	−1	50	90	10	8,866	0,02007	8,8632
5	−1	−1	1	10	40	50	7,041	0,02294	7,0536
6	1	−1	1	50	40	50	8,821	0,03605	8,8159
7	−1	1	1	10	90	50	7,415	0,03279	7,4350
8	1	1	1	50	90	50	9,050	0,01788	9,1098
9	−1	0	0	10	65	30	7,893	0,03265	7,9175
10	1	0	0	50	65	30	9,669	0,02294	9,6493
11	0	−1	0	30	40	30	9,545	0,02449	9,6197
12	0	1	0	30	90	30	9,980	0,02949	9,9101
13	0	0	−1	30	65	10	9,838	0,02177	9,9081
14	0	0	1	30	65	50	10,186	0,03277	10,1207

According to the requirements of regression analysis [14], the correct processing and use of research results are only possible if the variances of response measurements at each experiment point are the same (homogeneity of variances). Verification of the homogeneity of the experiments was carried out according to the Kohren criterion G_p.

5 Results of Researches Influence of Wood Fuel Humidity on Process of Gasification in a Continuous Layer

It is established that in the developed experimental gas-generating unit it is possible to gasify the wood, both with high relative humidity higher than W = 50% and low below W = 10% with a sufficiently high value of lower combustion heat of synthesis gas.

This is due to the fact that the gases formed during gasification are repeatedly passed through a layer of hot fuel in the recovery zone, where the reaction of water vapor conversion proceeds with a high temperature.

As a result of the B_3-plan, a mathematical description of the object in the form of a second-order polynomial is obtained, which looks like:

This is due to the fact that the gases formed during gasification are repeatedly passed through a layer of hot fuel in the recovery zone, where the reaction of water vapor conversion proceeds with a high temperature.

As a result of the B_3-plan, a mathematical description of the object in the form of a second-order polynomial is obtained, which looks like:

$$Q = 0.8998 + 0.27185 \cdot l + 0.11416 \cdot G + 0.04415 \cdot W - 0.00375 \cdot l^2 - 0.000832 \cdot G^2 - 0.000675 \cdot W^2 - 0.00004 \cdot l \cdot G - 0.000025 \cdot l \cdot W + 0.00004 \cdot G \cdot W. \quad (3)$$

The transfer from the normalized to the natural factors is based on the given correlation:

$$l = 20 \cdot x_1 + 30; G = 25 \cdot x_2 + 65; W = 20 \cdot x_3 + 30, \quad (4)$$

where x_i is the natural value of the factor.

The obtained dependence was rationalized in order to determine the values of the factors that provide the maximum value of the function [14].

After rationalizing the regression equation for the wood mixture, we obtain the values of the input parameters at which the lower calorific heat reaches the maximum (see Table 3).

Table 3. The regression equation rationalization for the wood mixture

The coded values of the factors		The natural values of the factors	
x_1	0,287303	1	35,746 mm
x_2	0,138703	G	68,467 Nm^3/h
x_3	0,20062	W	34,012%
Q_l	10,427 MJ/nm^3		

Therefore, as a result of the B_3-plan, a mathematical description of the object in the form of a second-order polynomial for wood fuel was obtained. After streamlining the process, the values of the input parameters were obtained (the sizes of the particles of wood 1 = 36 mm, the amount of air G = 68,5 nm^3/h, the humidity of the wood W = 34%), at which the calorific heat reaches a maximum of Q = 10,4 MJ/nm^3.

6 Conclusions

The analysis of the technology of thermal processing of wood biomass into gaseous fuel, which can be used for combustion in boilers of various types, in internal combustion engines, or for liquefaction. It has been shown that there is sufficient wood and waste stock for this purpose. The analysis is carried out and it is shown that the most promising way is gasification in a continuous layer, followed by the use of synthesis gas for energy purposes. It is shown that gasification of wood biomass with increasing its humidity leads to a decrease in the quality of synthesis gas.

A gas-fired unit with a solid layer for the thermal processing of wood into gaseous fuel was developed. The use of gas generators of the proposed design makes it possible to increase the efficiency of the process of thermal conversion of wet wood biomass into gaseous fuel by increasing the speed and intensity of the gasification process of the fuel. Due to the fact that the gases formed during the gasification, again pass through the layer of hot fuel in the recovery zone, where the reaction of water vapor conversion takes place with high temperature. It allows to solve problems of ecological utilization of industrial and household waste of wood and to obtain cheap energy.

The influence of wood moisture in the process of gasification on the lower heat of synthesis gas combustion is determined. It is established that the experimental gasifier allows gasification of wood, both with low below W = 10% and high humidity more than W = 50% with high value of lower heat of synthesis gas combustion. As a result of the B_3-plan, a mathematical description of the object in the form of a second-order polynomial is obtained. The rationalization of the obtained results is performed, the values of the input parameters are obtained, under which the lower calorific heat of the synthesis gas reaches a maximum of Q = 10.4 MJ/nm^3. The average value of the rational input parameters of the gasification process in the continuous layer are: the size of the particles of wood l = 36 mm, the amount of air G = 68,5 nm^3/h, the humidity of the wood W = 34%.

References

1. Subbotin, D., Kazakov, A.: Gasification of wood under different conditions of the pyrolysis process. In: Intelligent systems: Proceedings of the International Youth Forum, no. 1, pp. 276–279 (2015). (in Russian)
2. Batenin, V., Kovbasyuk, V.: On the technology for effective utilization of humid fuels. TVT, **53**(3), 475–477 (2015). https://doi.org/10.7868/s0040364415030023. (in Russian)
3. Donskoy, I., Kozlov, A., Svishcheva, D., Shamanskiy, V.: Numerical investigation of the staged gasification of wet wood. Therm. Eng. 64(4), 21–29 (2017). https://doi.org/10.1134/S0040363617040026. (in Russian)
4. Batenin, V., Ivanov, P., Kovbasyuk, V.: Improvement of thermodynamic efficiency of the humid biofuel application in the distributed generation power suppliers. TVT **55**(1), 76–80 (2017). https://doi.org/10.7868/S0040364417010033 (in Russian)
5. Leonovich, O.: Influence of chip moisture on its thermal efficiency during combustion. Bryansk State Acad. Eng. Technol. **43**, 184–187 (2015). (in Russian)
6. Lys, S.: Review of wood gasification technology. Sci. Bull. UNFU **19**(12), 101–105 (2009). (in Ukrainian)
7. Hamad, M., Radwan, A., Heggo, D., Moustafa, T.: Hydrogen rich gas production from catalytic gasification of biomass. Renew. Energy **85**, 1290–1300 (2016). https://doi.org/10.1016/j.renene.2015.07.082
8. Lopez, G., Alvarez, J., Amutio, M., Arregi, A., Bilbao, J., Olazar, M.: Assessment of steam gasification kinetics of the char from lignocellulosic biomass in a conical spouted bed reactor. Energy, no. 107, 493–501 (2016). https://doi.org/10.1016/j.energy.2016.04.040
9. Mysak, Y., Lys, S., Martynyak-Andrushko, M.: Research on gasification of low-grade fuels in a continuous layer. Eastern Eur. J. Enterpr. Technol. **2**(8)(86), 16–23 (2017). https://doi.org/10.15587/1729-4061.2017.96995
10. Lys, S., Mysak, Y.: Thermal recycling of wood by method of a continuous layer in the gaseous fuel. Eastern Eur. J. Enterpr. Technol. **3**(8(57)), 47–49 (2012). http://doi.org/10.15587/1729-4061.2012.4076

11. Zhang, Y., Yang, M., Song, Y.: Effect of fuel origin on synergy during co-gasification of biomass and coal in CO_2. Bioresour. Technol. **200**, 789–794 (2016). https://doi.org/10.1016/j.biortech.2015.10.076
12. Lys, S.: Analysis of experimental studies of the process of gasification of low-grade fuels. Sci. Bull. UNFU, **27**(1), 154–156 (2017). (in Ukrainian)
13. Pylypchuk, M., Hryhoryev, A., Shostak, V.: Basics of the Scientific Research. Znannya, Lviv (2007). (in Ukrainian)

Experimental Study of Crack Resistance and Shear Strength of Single-Span Reinforced Concrete Beams Under a Concentrated Load at a/d = 1

Solomiya Maksymovych(✉), Olha Krochak, Ihor Karkhut, and Rostyslav Vashkevych

Lviv Polytechnic National University, Lviv 79013, Ukraine
solo4max@gmail.com

Abstract. At a stage following the crack development, a reinforced concrete beam has a complex structure that does not correspond to such simple systems as the truss, arch or strutted frame. Therefore, to establish a possible design model for calculating the shear strength, it is necessary to know not only the failure pattern, but also the order of crack nucleation and propagation in the loaded structures. The pattern of nucleation and propagation of diagonal cracks differs significantly from that of vertical cracks, which are the reason for the stretching stress or bending moment. Diagonal cracks can be far more dangerous, as the time interval between the crack nucleation and failure is considerably shorter than for pure bending. The paper discusses the experimental study of diagonal cross-sections of single-span reinforced concrete beams under a concentrated load for a/d = 1. The tested beams had identical characteristics and differed only in the type of transverse reinforcement. The analysis of the experimental data revealed the specifics of nucleation and propagation of diagonal cracks in the zone of impact of the shear force and bending moment. The dynamics of propagation and width of diagonal cracks opening was recorded depending on the load. The effect of different types of transverse reinforcement on the bearing capacity of the beams was established.

Keywords: Reinforced concrete beams · Crack resistance · Shear strength · Mechanisms of shear failure · Transverse reinforcement

1 Introduction

The RC constructions are mostly useful at the world [1–5]. However, there are some disadvantages of this constructions [6–12]. In this case, are a lot of researching for strengthening this constructions [13], also for improving shear strength. The study of the issue of diagonal cross-section strength or shear strength, as written in the effective regulations [14] and EU standards [15], preserves its topical importance. We agree with the high-profile researchers from École Polytechnique Fédérale de Lausanne (Switzerland), who in their recent paper [16] wrote, 'The shear strength of beams and one-way slabs has been acknowledged for more than one century as one of the most

Z. Blikharskyy (Ed.): EcoComfort 2020, LNCE 100, pp. 277–285, 2021.
https://doi.org/10.1007/978-3-030-57340-9_34

complex, yet fundamental, topics to be addressed in structural concrete design.' The authors also substantiated the need for further study of crack resistance and shear strength, using modern approaches and methods of measuring deformations.

The results of the experimental studies obtained by this school of thought and presented in [17] include the testing of 20 reinforced concrete beams without transverse reinforcement with a detailed description and analysis of crack growth and shear failure of the beams. The digital image correlation was used for deformation measurement. The beams with the cross section dimension of 250 × 600 mm were tested using different procedures (single-span, continuous, cantilever beams) under a concentrated or distributed load. The authors mentioned the strong effect the shear slenderness ratios a/d and l/d have on the development of the critical crack and on the shear strength.

After testing 200 × 400 mm reinforced beams, the researchers concluded [18], that 'the mechanism of 'shear' failure is closely associated with the loss of the dowel force'.

The total of 381 samples of deep beams with and without transverse reinforcement was analysed [19]. The analysis showed that the shear span-to-depth ratio (a/d), horizontal and vertical shear reinforcement ratio are the most important parameters affecting the shear strength. Smith, K.N., Vantsiotis, A.S. [20] found out that the contribution of the vertical reinforcement decreases as a/d drops (a/d < 1.0). In [21] Ashour, A.F. reported that horizontal transverse reinforcement is more effective than vertical transverse reinforcement for a/d < 0.75.

The program of our experimental studies included the tests of four series of beams (21 reinforced beams under a concentrated load with positive (negative) and positive/negative bending moment diagrams). The results of testing the series 1 beams (console beams) for a/d = 1 were presented in the paper [22], and series 2 single-span beams for a/d = 0.5 in [23]. The present paper discusses the experimental studies of the series 3 single-span beams under a concentrated load for a/d = 1.

2 Aim

The aim of the research is further study of crack resistance and shear strength of reinforced concrete beams. The tests of the series 3 beams makes it possible to obtain the required experimental data on the strain/stress state of the reinforcement and concrete in the zone of the simultaneous impact of the moment and shear force for a/d = 1 and to compare it with the results of the series 1 tests (console beams with a positive/negative bending moment diagram). The various loading patterns used for the series of experimental samples allow revealing the specific pattern of the influence of positive (or negative) and positive/negative bending moment diagrams on the shear strength.

The various types of shear reinforcement for the same amount of transverse reinforcement in the beams 6, 7 of the series 1, 2, 3 enables finding an effective method of shear reinforcement.

3 Testing Method

The series 3 included testing three single-span reinforced concrete beams. The experimental beam samples were designed to be 120 × 400 × 1150 mm. The height of the experimental samples was assumed at 400 mm, as for h ≥ 400 mm the relative shear bearing capacity remains almost constant, and the authors in [24] recommend conducting experimental studies on shear fracture using beams with the height no less than 400 mm.

The samples were tested as single-span beams with the span l = 700 mm under a concentrated load applied at middle span with a/d = 1.

The schematic diagram of beams testing and diagrams of the M and Q moments are shown in Fig. 1. Table 1 summarized the characteristics of the experimental beam samples.

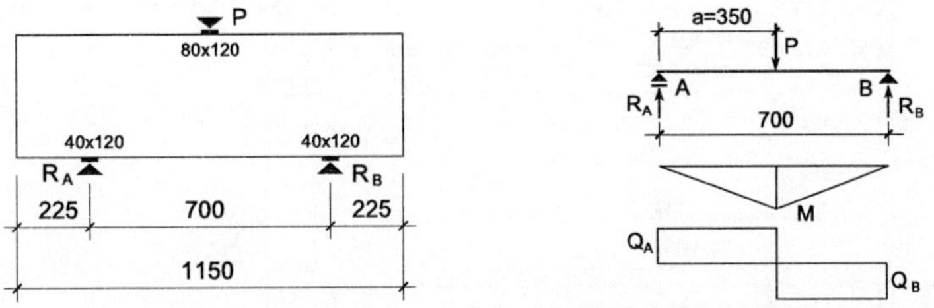

Fig. 1. Schematic diagram of testing the series 3 beams and M and Q moment diagram.

Table 1. Characteristics of the series 3 experimental beams.

Beam type	*l*, mm	*b*, mm	*h*, mm	*d*, mm	f_c, MPa	Longitudinal reinforcement: Reinforcement (top and bottom)	A_s, cm^2	σ_T, MPa	ρ_f, % (bottom)	Transverse reinforcement: Reinforcement	s_w, mm	A_{sw}, cm^2	σ_T, MPa
B-3-1	700	123	399	363	47.5	4Ø12 A400C	4.52	410.4	1.01	–	–	–	–
B-3-6	700	123	401	371	47.5	4Ø12 A400C	4.52	410.4	0.99	Ø6.5 A240	–	0.664	262
B-3-7	700	124	401	356	47.5	4Ø12 A400C	4.52	410.4	1.02	Ø6.5 A240	50	0.664	262

The beams differ in the type of shear reinforcement, but their reinforcement is identical to that of the corresponding series 1 beams [22] and series 2 beams [23]. B-3-1 beam has no transverse reinforcement (Fig. 2). In B-3-6 beam, the shear reinforcement is implemented using horizontal bars 4Ø6.5 A240 located at the top and at the bottom of the diagonal cross-sections of the beam (Fig. 2). In B-3-7 beam, horizontal bars 8Ø6.5 A240 located along the whole depth of the cross-section every 50 mm were used for shear reinforcement (Fig. 2). The deformations were measured using a strain gauge.

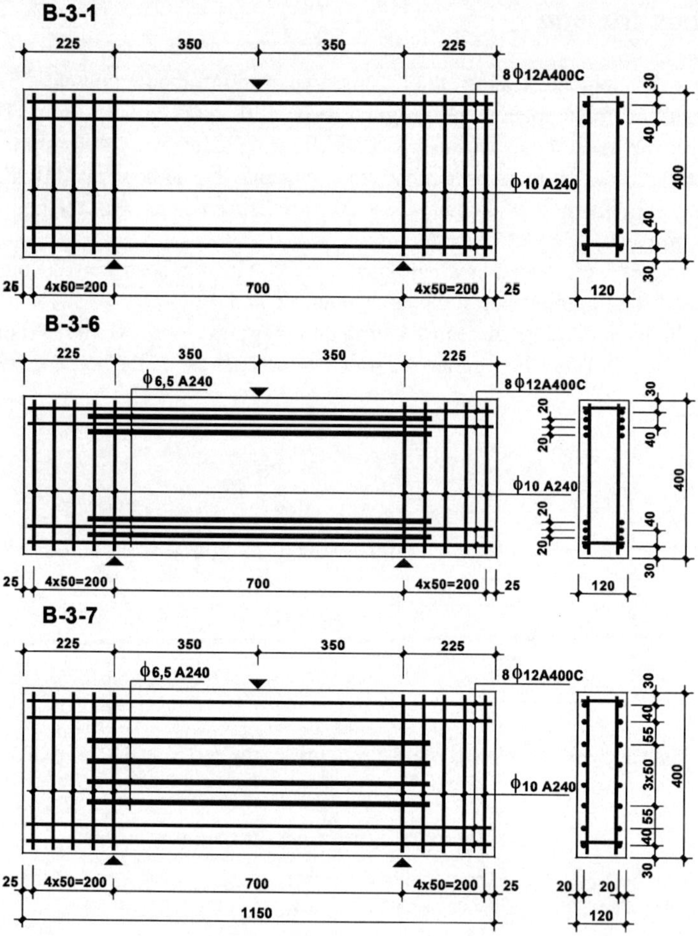

Fig. 2. Structures of the experimental series 3 beam samples.

4 Experimental Results

The first vertical cracks in the series 3 beams appeared at the load P = 200 kN in the middle of the span earlier than in the series 1 and series 2 beams. At P = 300 kN, closer to the supports, new vertical cracks occurred and started developing into diagonal cracks, deviating from the applied force; diagonal cracks also started nucleating in the middle of the beam at approximately 1/3 h off the bottom side. As the load rose to P = 400 kN, the existing diagonal and vertical cracks propagated, and new diagonal cracks nucleated in the middle of the beam (almost in parallel to the existing ones), which were located on the direct line connecting the support with the applied force (B-3-6, B-3-7) and along which later the failure occurred. In the B-3-1 beam, a new diagonal crack appeared at P = 500 kN in the mid-height of the beam along the direct line between the support and applied force, along which the beam failed at P = 590 kN. The failure of all the series 3 beams occurred along the diagonal cracks.

The results of the experimental study of the series 3 beams are presented in Table 2, using the schematic symbols assumed in Fig. 1.

Table 2. Results of the experimental study of the series 3 beams.

Beam type	*a*, *mm*	*a/d*	P, kN	$R_A = R_B$, kN	M, kNm	$V_A = V_B$, kN	Failure site	Failure type	Crack number
B-3-1	350	0.96	530	265.0	92.8	265.0	Near support B	Along the diagonal crack	17
B-3-6	350	0.94	575	287.5	100.6	287.5	Near support A		6
B-3-7	350	0.98	590	295.0	103.3	295.0	Near support A		9

The pattern of nucleation and propagation of the cracks and failure of the series 3 beams are shown in photos in Fig. 3.

As seen from Fig. 3, the diagonal cracks in the series 3 beams started to nucleate at the bottom side and in the middle of the beams, but the failure proceeded only along the diagonal cracks formed in the mid-beam almost immediately before the failure.

The above-presented analysis of the series 3 beams showed that the type of shear reinforcement does not have a significant effect on the moment of diagonal cracks formation, i.e. all of them appeared at the same load P = 200 kN.

In the series 3 beams, the diagonal cracks formed at a lower load, and the difference between the diagonal crack nucleation load and beam failure load is smaller than for the series 2 beams [23]. At the load 500 kN, the maximum opening width of the diagonal cracks along which the failure occurred was as follows:

B-3-1 Crack 17 – 0.7 mm;
B-3-6 Crack 6 – 0.8 mm;
B-3-7 Crack 9 – 0.4 mm.

This data show that the B-3-7 beam has the smallest opening width (almost one-half as large).

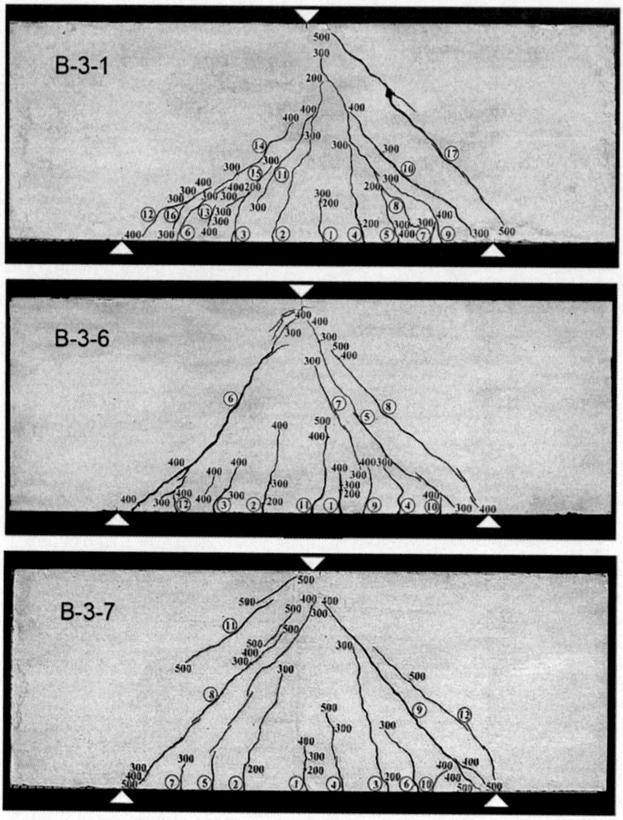

Fig. 3. Patterns of nucleation and propagation of the cracks and failure of the series 3 beams.

The parameters at which the series 3 beams' failure occurred are listed in Table 2. Comparing the series 3 beams in terms of the bearing capacity, one can determine the effect of the transverse reinforcement type on the shear strength of the beams with positive (or negative) moments diagram for a/d = 1.

The series 3 beams failed at the following loads:

B-3-1 − P = 530 kN;
B-3-6 − P = 575 kN;
B-3-7 − P = 590 kN.

Let us calculate the difference in the shear strength between the beams with transverse reinforcement and those without it:

B-3-1 – no transverse reinforcement;
B-3-6 (575 − 530)/530 · 100% = 8.49%;
B-3-7 (590 − 530)/530 · 100% = 11.32%.

The analysis reveals that the type of reinforcement has somewhat stronger effect on the shear strength of the series 3 beams than on the series 1 and series 2 beams (columns 10, 11, 12, 13 Table 2 [22]; columns 6, 7 Table 2 [23]).

5 Scientific Novelty

The tests of the series 3 beams revealed that the type of transverse reinforcement affects the crack resistance and shear strength of single-span beams if the force is applied near a support (a/d = 1).

The comparison of the results of the experimental tests of the series 3 beams presented herein with the series 1 tests [22] demonstrated the differences in the patterns of cracks development and failure of single-span and console beams for a/d = 1, the moment being positive (negative) and positive/negative. According to the effective design regulations, diagonal cross-sections are calculated based on the shear force that is constant in the segment from the intermediate support to the force, whereas the moment changes the sign.

The analysis of the results of both the series 3 beams tests and all the four series in a bulk enables a better understanding of the effect that the transverse reinforcement type and shear span have on the crack resistance and shear strength.

6 Practical Value

The detailed design of the tests for the four series of beams made it possible to increase the experimental basis for further analysis and study of the shear strength. The obtained results are consistent and provide the rationale for understanding the shear phenomenon in reinforced concrete beams both with transverse reinforcement and without it. The beam height of 400 mm was assumed based on the real size of the constructions, so as to eliminate the effect of the scale factor on the results of the experiment.

An efficacious method of shear reinforcement of reinforced concrete beams with positive (or negative) and positive/negative bending moment diagrams under a concentrated load for $a/d \leq 1$ was proposed.

7 Conclusions

1. The type of shear reinforcement in beams at a/d = 1 does not have a significant effect on the development of diagonal cracks, i.e. they develop at approximately the same total load (200 kN).
2. The nucleation of diagonal cracks in beams with positive (or negative) bending moment diagram at a/d = 1 occurs on the bottom side of the beam from vertical cracks that developed earlier.
3. The maximum opening width of diagonal cracks, starting from their nucleation and until the beam failure, is observed at the sites of their nucleation.

4. At a/d = 1, in the beams with positive (negative) and positive/negative bending moment diagrams the most effective type of transverse reinforcement (from the point of view of the shear bearing capacity and width of diagonal cracks opening) is the horizontal transverse reinforcement.
5. Longitudinal reinforcement has a detectable effect on the shear strength of reinforced concrete beams both with positive (negative) bending moment diagram and with positive/negative one.

References

1. Bobalo, T., Blikharskyy, Y., Kopiika, N., Volynets, M.: Theoretical analysis of RC beams reinforced with high strength rebar's and steel plate. IOP Conf. Ser. Mater. Sci. Eng. **708**(1), 012045 (2019)
2. Lobodanov, M., Vegera, P., Blikharskyy, Z.: Planning experiment for researching reinforced concrete beams with damages. In: Lecture Notes in Civil Engineering, vol. 47, pp. 243–250 (2020). https://doi.org/10.1007/978-3-030-27011-7_31
3. Khmil, R., Tytarenko, R., Blikharskyy, Y., Vashkevych, R.: Influence of load level during strengthening of reinforced concrete beams on their reliability. IOP Conf. Ser. Mater. Sci. Eng. **708**(1), 012054 (2019)
4. Krainskyi, P., Vegera, P., Khmil, R., Blikharskyy, Z.: Theoretical calculation method for crack resistance of jacketed RC columns. IOP Conf. Ser. Mater. Sci. Eng. **708**(1), 012059 (2019)
5. Blikharskyy, Z., Vashkevych, R., Vegera, P., Blikharskyy, Y.: Crack resistance of RC beams on the shear. In: Lecture Notes in Civil Engineering, vol. 47, pp. 17–24 (2020). https://doi.org/10.1007/978-3-030-27011-7_3
6. Blikhars'kyi, Y.Z.: Anisotropy of the mechanical properties of thermally hardened A500s reinforcement. Mater. Sci. **55**, 175–180 (2019). https://doi.org/10.1007/s11003-019-00285-0
7. Kharchenko, Y., Blikharskyy, Z., Vira, V., Vasyliv, B.: Vasyliv, B: study of nanostructural changes in a Ni-containing cermet material during reduction and oxidation at 600 C. Appl. Nanosci. (2020). https://doi.org/10.1007/s13204-020-01391-1
8. Kharchenko, Y.V., Blikharskyy, Z.Y., Vira, V.V., Vasyliv, B.D.: Study of structural changes in a nickel oxide containing anode material during reduction and oxidation at 600 °C. In: Fesenko, O., Yatsenko, L. (eds.) Nanocomposites, Nanostructures, and Their Applications, NANO 2018. Springer Proceedings in Physics, vol 221. Springer, Cham (2019). https://doi.org/10.1007/978-3-030-17759-1_42
9. Bobalo, T., Blikharskyy, Y., Vashkevich R., Volynets M.: Bearing capacity of RC beams reinforced with high strength rebars and steel plate. In: MATEC Web of Conferences, vol. 230, p. 02003 (2018). https://doi.org/10.1051/matecconf/201823002003
10. Blikharskyy, Z., Brózda, K., Selejdak, J.: Effectivenes of strengthening loaded RC beams with FRCM system. Arch. Civ. Eng. **64**(3), 3–13 (2018)
11. Selejdak, J., Blikharskyy Y., Khmil R., Blikharskyy Z.: Calculation of reinforced concrete columns strengthened by CFRP. In: Lecture Notes in Civil Engineering, vol. 47, pp. 400–410 (2020). https://doi.org/10.1007/978-3-030-27011-7_51
12. Blikharskyy, Y., Khmil, R., Blikharskyy, Z.: Research of RC columns strengthened by carbon FRP under loading. Matec Web of Conferences **174**, 04017 (2018). https://doi.org/10.1051/matecconf/201817404017

13. Blikharskyy, Z., Khmil, R., Vegera, P.: Shear strength of reinforced concrete beams strengthened by PBO fiber mesh under loading. In: MATEC Web of Conferences, vol. 116, p. 02006 (2017). https://doi.org/10.1051/matecconf/201711602006
14. DSTU B V.2.6-156:2010. Konstruktsii budynkiv i sporud. Betonni ta zalizobetonni konstruktsii z vazhkoho betonu. Pravyla proektuvannia. – K.: Minrehionbud Ukrainy. – 118 s (2011)
15. Eurocode 2: Design of Concrete Structures. EN 1992 – 1.1: General Rules and Rules for buildings. – Brussels: CEN, 2004 (2004). 226 p.
16. Cavagnis, F., Simões, J.T., Fernández Ruiz, M., Muttoni, A.: Shear strength of members without transverse reinforcement based on development of critical shear crack. ACI Struct. J. **117**(1), 103–118 (2020)
17. Cavagnis, F., Fernández Ruiz, M., Muttoni, A.: An analysis of the shear-transfer actions in reinforced concrete members without transverse reinforcement based on refined experimental measurements. Struct. Concr. **19**(1), 1–16 (2017)
18. Chana, P.S.: Investigation of the mechanism of shear failure of reinforced concrete beams. Mag. Concr. Res. **39**(141), 196–204 (1987)
19. Shahnewaz, M., Rteil, A., Alam, M.S.: Shear strength of reinforced concrete deep beams – a review with improved model by genetic algorithm and reliability analysis. Structures **23**, 494–508 (2020)
20. Smith, K.N., Vantsiotis, A.S.: Shear strength of deep beams. J. Am. Concr. Inst. **79**(3), 201–213 (1982)
21. Ashour, A.F.: Shear capacity of reinforced concrete deep beams. J. Struct. Eng. **126**(9), 1045–1052 (2000)
22. Maksymovych, S.B.: Experimental study of crack resistance and strength of oblique planes of reinforced concrete point-force loaded beams with positive and negative bending moment diagram. J. Theory Pract. Build. no. 823, pp. 334–342 (2015)
23. Maksymovych, S.B., Krochak, O.V.: Eksperementalni doslidzhennia trishchynostiikosti i mitsnosti pokhylykh pereriziv zalizobetonnykh balok, zavantazhenykh zoseredzhenymy sylamy. J. Resour. Saving Mater. Build. Struct. **37**, 164–174 (2019). Rivne
24. Laupa, A., Siess, C.P., Newmark, N.M.: The shear strength of simple-span reinforced concrete beams without web reinforcement. Research Series No. 52, University of Illinois at Urbana-Champaign, USA (1953)

Effect of Plasticizing and Retarding Admixtures on the Properties of High Strength Concrete

Taras Markiv(✉), Sergii Solodkyy, Khrystyna Sobol, and Djire Rachidi

Lviv Polytechnic National University, Lviv 79013, Ukraine
taras.y.markiv@lpnu.ua

Abstract. From year to year the Earth's temperature increases due to global warming resulting in new challenges in the concrete industry. It still remains very important branch of the modern economy. Concrete is the material which is used widely. Due to the temperature rising, the transportation of fresh concrete in hot weather becomes complicated, especially high strength concrete, which contains higher amount of cement in comparison with traditional concrete. In spite of this the fresh concrete should be delivered in plastic state to provide proper placing and compaction. That's why it is compulsory to use retarders and plasticizers if high strength concrete is designed. In this article the influence of plasticizers and retarder on properties of cement paste, fresh and hardened concrete was studied, because it is important to enhance the workability and slump retention of such concrete under hot weather. The setting time and compressive strength of cement paste with different dosages of polycarboxylate and lignosulphonate based plasticizers with retarding effect and retarder on the basis of sodium gluconate (SG) were studied. Experimental researches were carried out to determine the optimum dosage of plasticizing and retarding admixtures and their effect on compressive strength of concretes. The influence of optimal amount of polycarboxylate based plasticizer and SG on slump loss and compressive strength of concrete was established as well. The obtained results show that correlation between setting time of cement paste and slump loss isn't observed, but rational technical decisions allow to obtain designed properties both fresh and hardened concretes.

Keywords: High strength concrete · Fresh concrete · Retarder · Plasticizer · Slump loss · Compressive strength

1 Introduction

The significant changes have taken place since the beginning of the industrial revolution. The global warming causes the permanent temperature increase and results in new challenges in the concrete industry, which remains very important branch of the modern economy. Concrete is still the material, which is used widely to satisfy the human needs. The new types of concretes such as high strength concrete (HSC), ultra high strength concrete (UHSC), self-compacting concrete (SCC), high performance

Z. Blikharskyy (Ed.): EcoComfort 2020, LNCE 100, pp. 286–293, 2021.
https://doi.org/10.1007/978-3-030-57340-9_35

concrete (HPC) become very popular nowadays and are more often used in construction industry to achieve sustainability [1–6]. HSC is used to put the concrete into service earlier (for example opening the road traffic at 3 days), to build structures by reducing cross section area (high-rise buildings), to build superstructures such as long-span bridges and bridge desk with enhanced durability. It can be also used to satisfy the specific needs of special application such as flexural strength, modulus of elasticity and durability. Portland cements with higher compressive strength class is used as a rule to manufacture HSC. Due to the higher temperature caused by global warming, the problems related with the transportation of fresh high strength concrete in hot weather are more often risen. The transportation of high strength concrete becomes more and more complicated in such conditions, because it usually contains higher amount of cement with the higher activity in comparison with traditional concrete and as a result more heat is released when fresh concrete is delivered due to cement hydration causing its quick stiffening. Though, it is well known that ready-mixed concrete should be delivered to the construction site in plastic state to provide proper placing and compaction. That's why the effective retarders, air entraining agent, different modifiers and both plasticizer or superplasticizer are used in contemporary concrete technologies, including HSC production, to improve the technological properties of fresh and mechanical and durability characteristics of hardened concretes [7–12]. As a result, it is compulsory to use plasticizers and retarders when such concretes are designed. There are several types of commercially available superplasticizers such as lignosulphonates, naphthalene-based, melamine-based, and modified polycarboxylates and different retarding admixtures [13].

Runova et al. [14] revealed the slump loss problem in hot weather if SNF-based superplasticizer is used for ready-mixed concrete production. Multifunctional chemical admixture was developed consisting of sulfonated naphthalene formaldehyde condensate in combination with the lignosulphonate based retarder and sodium borate to prevent rapid slump loss. Many researchers also study the efficiency of such organic retarder as sodium gluconate [15–17]. SG is commonly used in real concrete production resulting in the significant retarding effect and rather good compatibility with different superplasticizers. Ma et al. [15] concluded that SG results in the formation of AFt at early stage of hydration if the dosage is less than 0.03 mass%. On the contrary, SG slows down the formation of AFt if the dosage is more than 0.05 mass%. Most researchers study the setting time of cements containing retarders at normal temperature, but the influence of SG on setting time of cement at higher temperature is rarely discussed [16]. Lv et al. [17] shows that sodium gluconate has positive influence on the compressive strength at the dosage which is less than 0.15 mass% and if it is exceeded, the negative effects are observed. That's why the aim of this researches is to study the effect of dosage different types of plasticizers and retarder on properties of fresh and hardened concrete.

2 Materials and Methods

Commercially available Portland cement CEM II/A-S 42,5R was used in this study.

The physical and mechanical properties of Portland cement are presented in Table 1.

Table 1. Physical and mechanical properties of Portland cement.

Specific surface, m^2/kg	Residue on sieve 008, %	Water demand, %	Setting time, min		Compressive strength, MPa	
			Initial	Final	2 days	28 days
380	0.4	29.5	180	290	31.8	52.7

The tests of both Portland cement and aggregates properties were carried out according to Ukrainian standards [18–22]. The results of aggregates' investigations are shown in Table 2.

Table 2. Aggregates' properties.

Aggregate type	Density, (g/cm^3)	Bulk density, (kg/m^3)	Voidage, (%)	Dust and clay particles, (%)	Fineness modulus
Fine	2.64	1321	50.0	1.5	1.49
Coarse (5–20 mm)	2.70	1408	43.7	0.5	–

Commercially available polycarboxylate (PCE-RE) and lignosulfonate (LS-RE) based superplasticizers with retarding effect as well as traditional retarder on the basis of sodium gluconate (R-SG) were used in researches.

The concrete mix designed was carried out according to DSTU B V.2.7-214:2009 [23]. The mathematical planning of experiments was carried out to determine the optimal amount of plasticizing and retarding admixtures [24]. Compressive strength of concrete was determined according to DSTU B V.2.7-214:2009 [25].

3 Results and Discussion

At first stage the researches were focused on the study of the influence of plasticizing and retarding admixtures on the properties of cement paste and both fresh and hardened concretes. As seen from the Fig. 1, cement paste containing plasticizer with retarding effect and retarder results in extending the hydration induction period and thereby lengthening the setting times. It should be noted that the use of plasticizers with secondary retarding effect is one of the most effective method to improve fresh and hardened concrete properties. The addition of LS-RE and PCE-RE delay the initial and final setting time of cement pastes by approximately 100 and 140 min respectively compared to control cement paste without admixtures. The setting time of cement paste incorporating R-SG shows the tendency of gradual significant increasing with the dosage growth from 0.2 to 0.3 mass%. Thus, the initial and final setting time for cement containing 0.3 mass% of SG extend by 380 min compared to control cement paste. However, the interval time between initial and final time is rather short and ranges between 80–110 min.

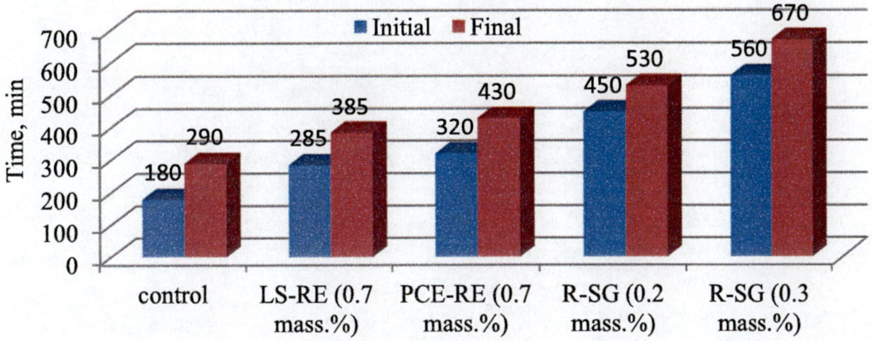

Fig. 1. Setting time of cement pastes.

The compressive strength of cement paste (paste 1:0, samples-cubes 2 × 2 × 2 cm) containing plasticizing and retarding admixtures is discussed in this part and the results are given in Table 3. The water demand of cement pastes added with LS-RE and PCE-RE decreases by 8 and 9% respectively, but cement paste containing different dosages of SG present the tendency of increasing by 3%. The compressive strength of cement pastes incorporating LS-RE and different dosages of R-SG decreases at early age (after 1 and 7 days) of cement paste hardening in comparison with control paste without admixtures. The highest decrease (16 and 25.6%) is observed for cement paste with 0.3 mass% of R-SG after 1 and 7 days of hardening respectively. It should be noted that compressive strength of cement pastes incorporating LS-RE, R-SG (0.2 mass%) exceeds the compressive strength of control paste by 8 and 4% respectively, except cement paste containing 0.3 mass% of R-SG where 8% decrease is observed after 28 days of hardening.

Table 3. Effect of admixtures on compressive strength of cement paste.

Age, days	Compressive strength, MPa				
	Control	LS-RE (0.7 mass%)	PCE-RE (0.7 mass%)	R-SG (0.2 mass%)	R-SG (0.3 mass%)
1	16.4	15.5	16.6	14.8	13.8
7	36.3	35.0	37.0	27.5	27.0
28	60.0	65.0	67.5	62.5	55.0

The compressive strength of cement paste containing PCE-RE has shown permanent increase for 28 days of hardening. The PCE-RE and R-SG were used for further researches such as these admixtures show the best plasticizing and retarding effect respectively.

The optimization of the content of PCE-RE and R-SG was carried out using the mathematical planning of experiments. The results are presented in Fig. 2.

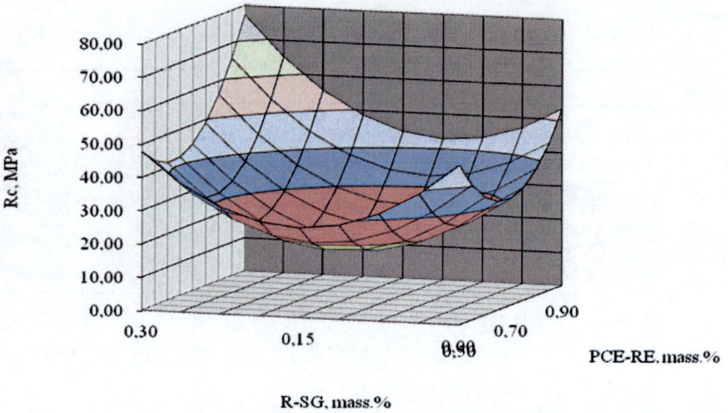

Fig. 2. Compressive strength of concretes after 1 (a) and 28 (b) days of hardening.

The optimum dosage of PCE-RE and R-SG are found on the basis on the highest ultimate compressive strength of fine-grained concrete (Cement:Sand = 1:2). The results are presented at the age of 28 days. Dosage with lower content of PCE-RE than this optimum value reduces the compressive strength and higher is not recommended by producer, because can cause negative effect related with the over dosage of plasticizer. The content of R-SG will depend on the workability retention of fresh concrete and compressive strength of hardened concrete.

The concrete mix design was done according to Ukrainian standard to study the influence of optimal amount on plasticizing and retarding admixtures on slump loss and compressive strength of concrete. The following mix-proportion has been obtained: C = 408 kg/m^3, S = 600 kg/m^3, G = 1138 kg/m^3, W = 246 kg/m^3, PCE-RE (0,9 mass%), R-SG (0,3 mass%). Consistency class of concrete mixes was S4.

The results of slump loss determination are shown in Fig. 3. The data show the relation between dosages of R-SG and slump loss. The results show that slump reduces with time. More R-SG (0.9 mass%) than designed (0.3 mass%) was added to retain the concrete in liquid state for a longer time and, as a result, it would reduce the slump loss during the transportation of concrete to the construction site. According to Fig. 3, the designed consistency class S4 is retained during the test and the conclusion can be made that retarder R-SG (0.9 mass%) is more effective in comparison with plasticizer with retarding effect PCE-RE in retaining the slump of the high strength concrete.

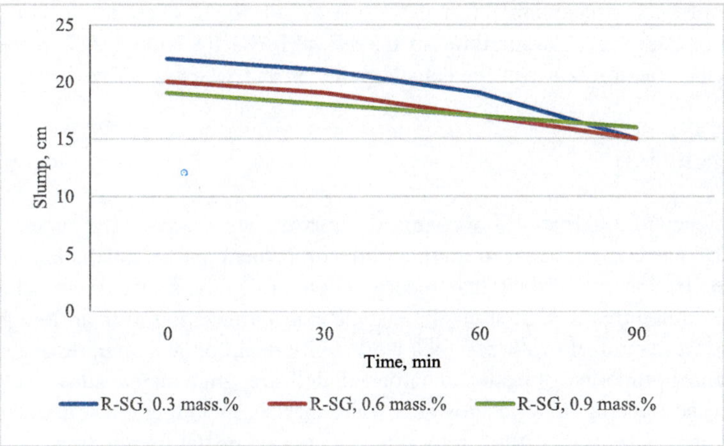

Fig. 3. Slump loss of fresh concretes.

The dosage of R-SG presents different behavior on the compressive strength of modified high strength concrete (Fig. 4). At early age, the addition of extra amount (0.6 and 0.9 mass%) of R-SG is not able to increase the compressive strength of concrete. On the contrary, the strength significantly reduces from 43.5 MPa (0.3 mass% of R-SG) to 18,6 MPa (0.9 mass% of R-SG), because the addition of the extra amount of retarder to the concrete delays the reaction of C3S and C3A and, as a result, the strength development is low. The situation changes only after 7 days from casting, compressive strength of high strength concrete containing R-SG slightly improves and exceeds the compressive strength of concrete incorporating 0.3 mass% of R-SG. As seen from the graph, continuous strength gain is observed with age and compressive

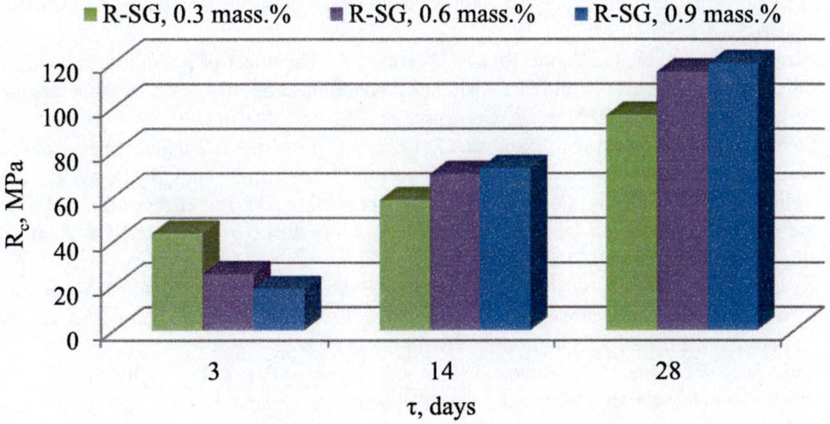

Fig. 4. Compressive strength of concretes.

strength of HSC containing 0.6 and 0.9 mass% of R-SG exceeds the compressive strength of concrete incorporating 0.3 mass% of R-SG by 20 and 24% respectively, because the reaction between the cement particles and water is active.

4 Conclusion

The efficiency of different type of chemical admixture was studied. The obtained results show that correlation between setting time of cement paste and slump loss isn't observed. It was established that incorporation rationally selected plasticizing and retarding admixtures in high strength concrete and optimization of their dosage allows obtaining fresh concrete with designed workability retention and compressive strength. The obtained properties of fresh and hardened high strength concrete allow transporting it longer and opening up a gate towards wider market, including construction of high-rise buildings. Use of this structurally safe and environmental friendly material enables to realize concept and idea of vertical cities.

References

1. Sanytsky, M., Marushchak, U., Olevych, Y., Novytskyi, Y.: Nano-modified ultra-rapid hardening Portland cement compositions for high strength. In: Lecture Notes in Civil Engineering, vol. 47, pp. 392–399 (2020)
2. Solodkyy, S., Markiv, T., Sobol, K., Hunyak, O.: Fracture properties of high-strength concrete obtained by direct modification of structure. In: MATEC Web of Conferences, vol. 116, p. 01016 (2017)
3. Kew, H., Donchev, T., Petkova, D., Iliadis, I.: Behaviour of high strength concrete (HSC) under high temperatures. In: 5th International Conference on Concrete Repair Concrete Solutions Proceedings, Belfast, Northern Ireland, pp. 493–497 (2014)
4. Dvorkin, L., Bezusyak, A., Lushnikova, N., Ribakov, Y.: Using mathematical modeling for design of self compacting high strength concrete with metakaolin admixture. Constr. Build. Mater. **37**, 851–864 (2012)
5. Markiv, T., Hunyak, O., Sobol, Kh., Blikharskyy, Z.: The effect of active mineral additives on properties of HSC in different hardening conditions. In: IBAUSIL. 20 Internationale Baustofftagung, Band 2, Weimar, Germany, pp. 851–857 (2018)
6. Stechshyn, M., Sanytsky, M., Poznyak, O.: Durability properties of high volume fly ash self-compacting fiber reinforced concretes. Eastern Eur. J. Enterp. Technol. **3**(11), 49–53 (2015)
7. Tolmachov, S., Brazhnik, G., Belichenko, O., Tolmachov, D.: The effect of the mobility of the concrete mixture on the air content and frost resistance of concrete. IOP Conf. Ser. Mater. Sci. Eng. **708**, 012109 (2019)
8. Kropyvnytska, T., Sanytsky, M., Rucinska, T., Rykhlitska, O.: Development of nanomodified rapid hardening clinker-efficient concretes based on Portland composite cements. Eastern Eur. J. Enterp. Technol. **6**(6), 38–48 (2019)
9. Runova, R.F., Gots, V.I., Rudenko, I.I., Konstantynovskyi, O.P., Lastivka, O.V.: The efficiency of plasticizing surfactants in alkali-activated cement mortars and concretes. In: MATEC Web of Conferences, vol. 230, p. 03016 (2018)

10. Markiv, T., Sobol, K., Petrovska, N., Hunyak, O.: The effect of porous pozzolanic polydisperse mineral components on properties of concrete. In: Blikharskyy, Z., Koszelnik, P., Mesaros, P. (eds.) Advances in Resource-Saving Technologies and Materials in Civil and Environmental Engineering 2019. LNCE, vol. 47, pp. 275–282. Springer, Heidelberg (2019)
11. Turba, Y., Solodkyy, S., Markiv, T.: Strength and fracture toughness of cement concrete, dispersedly reinforced by combination of polypropylene fibers of two types. In: Blikharskyy, Z., Koszelnik, P., Mesaros, P. (eds.) Advances in Resource-Saving Technologies and Materials in Civil and Environmental Engineering 2019. LNCE, vol. 47, pp. 488–494. Springer, Heidelberg (2019)
12. Kroviakov, S., Mishutin, A., Pishev, O.: Management of the properties of shipbuilding expanded clay lightweight concrete. Int. J. Eng. Technol. **7**(3.2), 245–249 (2018)
13. Alsadey, S.: Effects of super plasticizing and retarding admixtures on properties of concrete. In: International Conference on Innovations in Engineering and Technology, pp. 271–274 (2013)
14. Runova, R.F., Kochevyh, M.O., Rudenko, I.I.: On the slump loss problem of super plasticised concrete mixes. In: Admixtures - Enhancing Concrete Performance, pp. 149–156 (2005)
15. Ma, S., Li, W., Zhang, S., Ge, D., Yu, J., Shen, X.: Influence of sodium gluconate on the performance and hydration of Portland cement. Constr. Build. Mater. **5**(91), 138–144 (2015)
16. Li, B., Lv, X., Dong, Y., Zhou, S., Zhang, J.: Comparison of the retarding mechanisms of sodium gluconate and amino trimethylene phosphonic acid on cement hydration and the influence on cement performance. Constr. Build. Mater. **168**, 958–965 (2018)
17. Lv, X., Li, B., Shi, Y., Yang, H.: Comparison of the influence of amino trimethylene phosphonic acid and sodium gluconate on the performance of concrete. In: Proceedings of International Conference on Architectural, Civil and Hydraulics Engineering, pp. 1–8. Atlantis Press, Guangzhou (2015)
18. DSTU B V.2.7-188:2009: Building materials. Cements. Methods of determination of fineness. Ukrarkhbudinform, Kyiv, Ukraine (2010)
19. DSTU B V.2.7-185:2009: Building materials. Cements. Methods of determination of normal thickness, setting time and soundness. Ukrarkhbudinform, Kyiv, Ukraine (2010)
20. DSTU B V.2.7-187:2009: Building materials. Cements. Methods of determination of bending and compression strength. Ukrarkhbudinform, Kyiv, Ukraine (2010)
21. DSTU B V.2.7-71-98: Building materials. Mauntainous rock road-metal and gravel, industrial waste products for construction works. Methods of physical and mechanical tests, Kyiv, Ukraine (1998)
22. DSTU B V.2.7-232:2010: Building materials. Sand for construction work testing methods, Kyiv, Ukraine (2010)
23. DSTU-N B V.2.7-299:2013: Guidelines for appointments of the heavy concrete. Ukrarkhbudinform, Kyiv, Ukraine (2014)
24. Dvorkin, L., Dvorkin, O., Ribakov, Y.: Mathematical experiments planning in concrete technology. Nova Science Publishers, New York (2012)
25. DSTU B V.2.7-214:2009: Building materials. Concrete. Methods of determining the strength of control samples. Ukrarkhbudinform, Kyiv (2010)

Analysis of the Water Consumption in the Apartment House – Case Study

Oksana Matsiyevska[1(✉)], Peter Kapalo[2], Jakub Vrana[3], and Cristina Iacob[4]

[1] Lviv Polytechnic National University, Lviv, Ukraine
oksanamatsiyevska@gmail.com
[2] Technical University of Kosice, Kosice, Slovak Republic
[3] Brno University of Technology, Brno, Czech Republic
[4] Technical University of Cluj-Napoca, Cluj-Napoca, Romania

Abstract. The aim of the research is to analyze the water demand in apartment building calculated according to the state regulations of Poland, Czech Republic, Romania, Slovak Republic and Ukraine and to compare them with the results of experimental measurements in apartment building. Poland and the Czech Republic are characterized by the minimum average coefficient of water demand for residents of the houses with local hot water preparation. For Romania, Slovak Republic and Ukraine this value is higher by 20, 35 and 110–135% respectively. Czech Republic has the lowest values of theoretically calculated daily (minimum, average and maximum) water demand. These values are higher for Romania – by 38.9, 15.3 and 9.7%; Slovak Republic – by 35.2, 34.7 and 34.9%; Ukraine – by 114.8, 101.4 and 109.6% respectively, as compared with water demand in Czech Republic. The average daily water consumption in an apartment building, measured experimentally, corresponds to the normative value of the specific water demand adopted in the Czech Republic. Maximum daily water demand exceeds the estimated value and corresponds to the maximum coefficient of the daily variation in water demand, which equals 1.63.

Keywords: Water demand · Water consumption · Variability of water consumption

1 Introduction

In the last 100 years, water consumption in the world has increased almost 6-fold [1]. In recent decades, global demand for water has increased at a rate of about 1% per year, depending on population growth and economic development. In the future, demand for water will continue increase. Water demand for industry and every-day purposes will increase much faster than in agriculture sector. However, on the whole, agriculture will remain the largest consumer of water [2].

The share of water used for domestic purposes is about 12% of total water consumption [3]. Specific consumption of water for domestic use is increasing rapidly in low-income countries and decreasing in high-income countries. The general trend

Z. Blikharskyy (Ed.): EcoComfort 2020, LNCE 100, pp. 294–302, 2021.
https://doi.org/10.1007/978-3-030-57340-9_36

towards water demand reduction is related to the increase in water use efficiency with the help of modern water intake fittings and sanitary appliances [1, 4, 5].

Improving the accuracy of forecasting changes in water demand is crucial in the design of new water supply systems and reconstruction of existing ones. The operation of high volume systems leads to increase in the additional costs of transporting water and, consequently, increase in water tariffs. In such systems, the duration of water transportation to consumers is increased, which can lead to a deterioration of its quality. Conversely, underestimation of the future water demand will lead to the hydraulic overload of the system, water shortages and so on [6, 7].

With the help of a computer hydraulic model of the settlement's water supply system, performance of all its elements can be optimized. However, the creation and correct performance of such a model is possible only on the basis of accurate values of the estimated water consumption, including the maximum daily allowance, and the coefficients of daily uneven water demand [8, 9]. The normative documents of different countries give indicative values of water consumption (average per year) and coefficients of uneven water consumption. The results of the experience of operating water systems in different countries indicate that these indicators need to be adjusted/require adjustments [10–13]. Therefore, in the context of globalization, it is important to compare and unify the above indicators [14–16].

2 Aim

The aim of the research is to analyze the water demand in apartment building calculated according to the state regulations of Poland, Czech Republic, Romania, Slovak Republic and Ukraine and to compare them with the results of experimental measurements in apartment building.

3 Method

The research methodology consisted of comparative analysis of the water demand for categories of selected apartment building in Poland, Czech Republic, Romania, Slovak Republic and Ukraine. The calculated results were compared with results from experimental measurements water consumption for selected apartment building.

For the purposes of the experimental measurements of water consumptions has been selected apartment building with three flats. The apartment building is located in a town with around 380,000 inhabitants. There are six people living together in an apartment building. Hot water preparation is local in every apartment. On the main supply pipe in the house behind the water meter assembly on the water connection was installed the flowmeter. A measuring center was located near the flowmeter. The number of sampling points was as follows: 3 toilet bowl with cistern; 3 kitchen sinks; 3 washbasins; 3 showers; 1 bath; 2 automatic washing machines.

Measuring device: the flowmeter was used Ahlborn Almemo FVA915VTH25 with Ahlborn Almemo 5690-2 measuring unit for measuring water flow. The measurement accuracy is ±5% of the measured value. Measured was the flow rate of water entering in the building into liters per second. Measurement took several weeks. The article presents two days, Tuesday 15.1.2019 and Tuesday 22.1.2019.

4 Results

4.1 Analysis of Legislative Requirements

The values of the estimated daily water demand (average per year) for consumers of the five countries given in Table 1 and Fig. 1. Average specific water demand for apartment buildings are accepted according to national regulations and standards.

In the Poland such a document is Regulation of the Minister of Infrastructure as of 14 January 2002 on determining the average standards of water consumption (Journal of Laws as of 31 January 2002) (in Polish). They are of less importance for residential buildings in non-canalized areas of settlements, while the standards for houses/buildings in the sewer areas are higher.

In the Czech Republic such a document is Decree of the Ministry of Agriculture of the Czech Republic No. Amending Decree no. 48/2014 Coll., Implementing Act no. 274/2001 Coll., On water supply and sewerage systems for public use and amending certain acts (Water Supply and Sewerage Act), as amended (in Czech). Decree does not divide flats into types listed in categories. For all types of flats in which hot water flows 100 l/person per day is considered.

Romanian standard – SR 1343-1:2006. Water supply: Calculation of drinking water supply quantities in urban and rural sites (in Romanian). The Standard does not take into account the bathroom equipment (showers/bathtubs). In the case of local hot water preparation, Romanian values of the daily water demand are close to Slovakian ones, but in the case of centralized hot water preparation, Romanian values are slightly higher (up to 24%).

In the Slovak Republic such a document is Decree no. 684/2006 Coll. (2008) Decree of the Ministry of Environment of the Slovak Republic, which lays down details on technical requirements for the design, project documentation and construction of public water supply and public sewerage systems (in Slovak). In Slovak Republic the water demand reduced can be by 25% if the consumer lives in house, which on public sewerage is not connected. For apartments in a house with over-standard sanitary equipment, the water demand is increased by 15%.

In Ukraine, this value is regulated by SBN V.2.5-64-2012. Internal plumbing and sewage system. Part I. Designing. Part II. Building and SBN V.2.5-74-2013. Water supply. External networks and constructions. Basic principles designing's (in Ukrainian). The daily drinking water demand is indicated in Table 1 in a certain range of values because the territory of Ukraine according to architectural and construction climatic conditions is divided into four districts (according to SSTU-N B V.1.1-27-2010. Protection from hazardous geological processes, harmful exploitative influences, from fire. Construction Climatology (in Ukrainian). For further calculations, the northwestern architectural and construction climatic region of Ukraine (district I) has been selected – for town Lviv.

The analysis of Table 1 shows that Poland and the Czech Republic are characterized by the minimum average specific water demand. For Romania and Slovak Republic, this value is higher than the minimum by 20 and 35% respectively. For Ukraine, (taking into account the architectural, structural and climatic regionalization of the territory) this value is more than the minimum by 110–135%.

4.2 Theoretical Comparison of the Water Demand

For the comparison was the selected an apartment building with six occupants, situated in a town with 380,000 inhabitants. The daily water demand for the apartment building with local prepare hot water in flat was calculated. In Poland, the coefficients of the daily variations in water demand are adopted on the basis of technical guidelines, databases, etc. individually for specific groups of consumers. Therefore, further calculations are carried out for four countries. The results of calculating the daily demand of drinking water for apartment building in Table 2.

Table 1. Drinking water daily demand (average per year), l/day per inhabitant.

Apartment building	Daily water demand for one person [L/(day · pers.)]				
	Poland	Czech Republic	Romania	Slovak Republic	Ukraine
The apartment with local hot water preparation, bath and with gas water heaters	80–100	100	100–120	135	210–235

Table 2. Daily water consumption in apartment building.

Number of persons in house	Water demand [L/(day. pers.)]	Unaccounted consumption		Coefficient of daily non-uniformly consumption		Daily water consumption [m^3/day]		
		%	γ_i	K_{min}	K_{max}	Minimum	Average	Maximum
Czech Republic								
6	100	20	1.20	0.75	1.15	0.54	0.72	0.83
Romania								
6	120	15	1.15	0.90	1.10	0.75	0.83	0.91
Slovak Republic								
6	135	20	1.20	0.75	1.15	0.73	0.97	1.12
Ukraine								
6	210	15	1.15	0.80	1.20	1.16	1.45	1.74

The analysis of Table 2 shows the significant fluctuation of daily water consumption by the residents of the selected house for different countries. The lowest values are observed in the Czech Republic. For Romania, the minimum, average and maximum daily water consumption is higher by 38.9, 15.3 and 9.7% respectively than in the Czech Republic. For Slovak Republic – by 35.2, 34.7 and 34.9% respectively. The largest theoretical values of daily water consumption are common Ukraine. These values are higher by 114.8, 101.4 and 109.6% respectively, as compared to the Czech Republic. Figure 1 shows the excess of theoretically calculated daily water consumption for different countries compared to the Czech Republic.

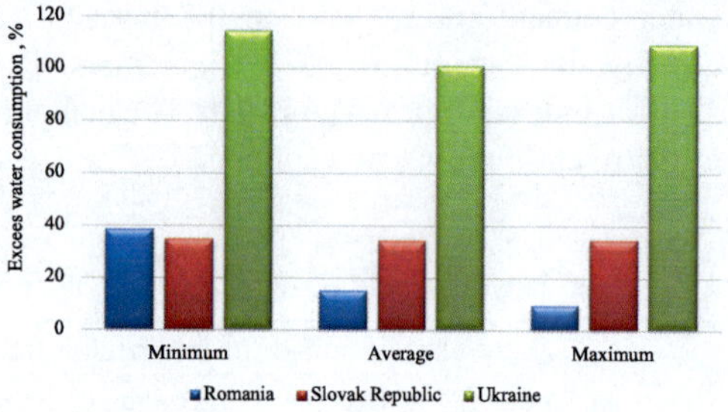

Fig. 1. Excess of estimated daily water consumption for different countries compared to the Czech Republic.

4.3 Experimental Measurement of Water Consumption

The experimental measurements in an apartment house was in January 2019. A parts of the water flow of measurement record are in Fig. 2 and Fig. 3.

The maximum flow rates measured on individual days ranged from 0.21 to 0.42 L/s. The flow rate of 0.42 L/s (maximum during the measurement period) it was measured on Tuesday, January 15, 2019. These days probably the largest number of inhabitants stayed in the house. A similar flow rate (0.40 L/s) it was measured on Tuesday, January 22, 2019. The graphs in Fig. 2 and Fig. 3 show the flow rates during January 15 and January 22, 2019 and show that peak flow rates were in the morning around 8 am.

It was calculated average and maximum of water consumption per person and per day from measured of water consumption. The average water consumption is 101 L/(day · pers.) According to the normative value of the specific water consumption of 100 L/(day · pers.), adopted in the Czech Republic. However, there is a significant unevenness of water consumption during the measurement day. The maximum value of daily water consumption is 165 L/(day · pers.) It is much higher than the estimated one and corresponds to the maximum coefficient of the daily irregularity of water consumption 1.63. Figure 4 and Fig. 5 shows the water consumption in each hour on selected days (Sunday - Weekend and Tuesday - Weekdays).

In the figures shown are peaks of consumption in the morning, respectively in the mornings and evening hours. However, on a working day - Tuesday there is a peak of water consumption from 3 p.m. to 4 p.m. Is the assumption that the significant variance the consumption of water is the caused by a small population of the building.

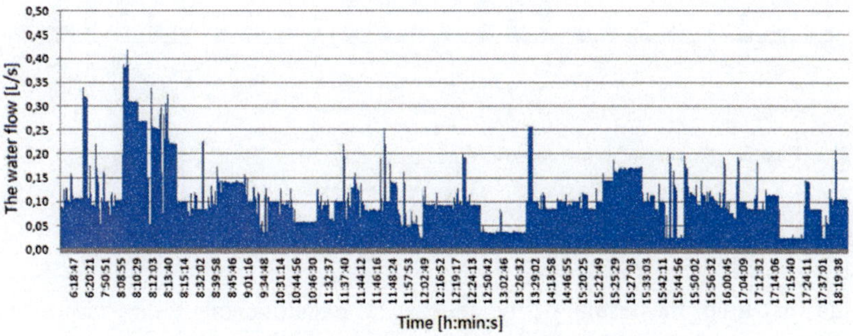

Fig. 2. The water flow during the day 2019-01-15 (Tuesday) [16].

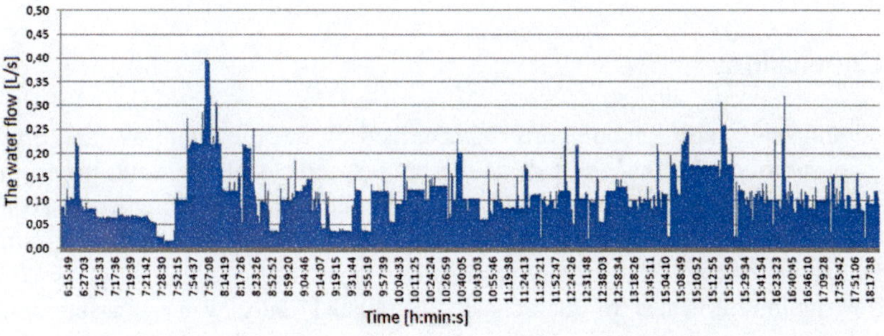

Fig. 3. The water flow during the day 2019-01-22 (Tuesday) [16].

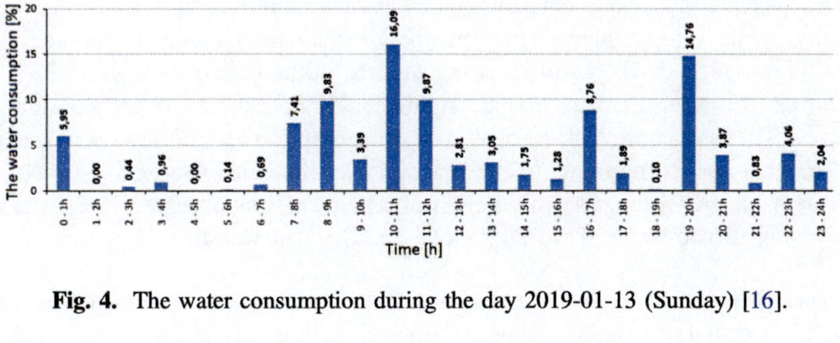

Fig. 4. The water consumption during the day 2019-01-13 (Sunday) [16].

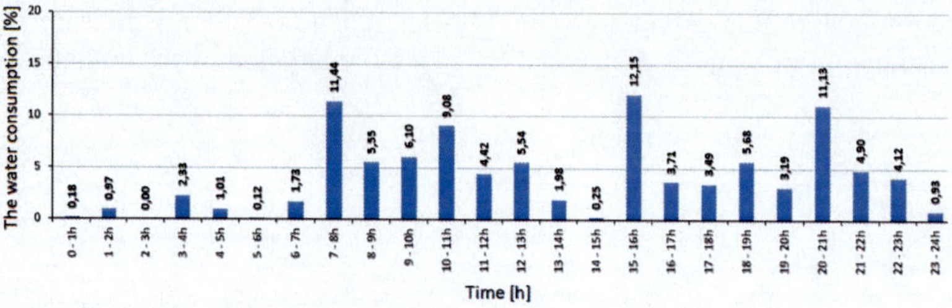

Fig. 5. The water consumption during thc day 2019-01-22 (Tuesday) [16].

5 Scientific Novelty and Practical Significance

For the first time, the results of theoretical and experimental studies of water consumption in an apartment building on the example of different European countries are analyzed and compared. The studies were conducted with a small number of residents and sanitary appliances. However, even such partial results are useful for practicing more accurate design of water supply networks, especially in Slovakia and Ukraine.

6 Conclusions

Poland and the Czech Republic are characterized by the minimum average drinking water consumption for the residents of the houses with local hot water production – 100 L/(day · pers.). In Romania and Slovak Republic this value is higher by 20% and 35%. In Ukraine (taking into account the architectural, structural and climatic regionalization of the territory) this value is higher by 110–135%.

The minimum values of theoretically calculated daily water consumption are common in the Czech Republic. Minimum, average and maximum water consumption are respectively 0.54, 0.72 and 0.83 m^3/day. These values are higher in Romania – by 38.9, 15.3 and 9.7%; Slovak Republic – by 35.2, 34.7 and 34.9%; Ukraine – by 114.8, 101.4 and 109.6% respectively, as compared with water demand in Czech Republic.

The average daily water consumption in an apartment building, measured experimentally, is 101 L/(day · pers.). This corresponds to the normative value of the specific water consumption of 100 L/(day · pers.) adopted in the Czech Republic.

During the measurement period, significant variations in water demand was observed. The maximum value of daily water consumption 165 L/(day · pers.) exceeds the estimated and corresponds to the maximum coefficient of the daily variation in water demand, which equals 1.63. Peak water demand was observed in the morning and evening. However, on a working day - Tuesday – it was from 3 p.m. to 4 p.m.

Acknowledgments. This article was elaborated in the framework of the project VEGA 1/0697/17 "Proposal of technical platform of hygienic audit for elimination of microbiological pollution in water distribution and ventilation in hospitals" and FAST-S-19-5863 "Analysis of the internal environment of buildings and buildings with almost zero energy consumption".

References

1. Wada, Y., Flörke, M., Hanasaki, N., Eisner, S., Fischer, G., Tramberend, S., Satoh, Y., van Vliet, M.T.H., Yillia, P., Ringler, C., Burek, P., Wiberg, D.: Modeling global water use for the 21st century: the Water Futures and Solutions (WFaS) initiative and its approaches. Geosci. Model Dev. **9**, 175–222 (2016). https://doi.org/10.5194/gmd-9-175-2016
2. Boretti, A., Rosa, L.: Reassessing the projections of the world water development report. npj Clean Water **2**, 15 (2019). https://doi.org/10.1038/s41545-019-0039-9
3. Flörke, M., Kynast, E., Bärlund, I., Eisner, S., Wimmer, F., Alcamo, J.: Domestic and industrial water uses of the past 60 years as a mirror of socio-economic development: a global simulation study. Glob. Environ. Change **23**(1), 144–156 (2013). https://doi.org/10.1016/j.gloenvcha.2012.10.018
4. Gorączko, M., Pasela, R.: Causes and effects of water consumption drop by the population of cities in Poland – selected aspects. Bull. Geogr. Socio-econ. Ser. **27**(27), 67–79 (2015). https://doi.org/10.1515/bog-2015-0005
5. Gutierrez-Escolar, A., Castillo-Martinez, A., Gomez-Pulido, J.M., Gutierrez-Martinez, J.-M., Garcia-Lopez, E.: A new system for households in Spain to evaluate and reduce their water consumption. Water **6**, 181–195 (2014). https://doi.org/10.3390/w6010181
6. Matsiyevska, O.O.: Study of water quality in the distribution network of the centralized water supply system in the city of Lviv. East Eur. J. Enterp. Technol. **6**(6), 62–70 (2015). https://doi.org/10.15587/1729-4061.2015.56225. (in Ukrainian)
7. Rinaudo, J.D.: Long-term water demand forecasting. In: Grafton, Q., Daniell, K., Nauges, C., Rinaudo, J.D., Chan, N. (eds). Understanding and Managing Urban Water in Transition. Global Issues in Water Policy, vol. 15, pp. 239–268 (2015). Springer, Dordrecht. https://doi.org/10.1007/978-94-017-9801-3_11
8. Georgescu, A.-M., Perju, S., Georgescu, S.-C., Anton, A.: Numerical model of a district water distribution system in Bucharest. Procedia Eng. **70**, 707–714 (2014). https://doi.org/10.1016/j.proeng.2014.02.077
9. Gwoździej-Mazur, J., Świętochowski, K.: Non-uniformity of water demands in a rural water supply system. J. Ecol. Eng. **20**(8), 245–251 (2019). https://doi.org/10.12911/22998993/111716
10. Bergel, T., Kotowski, T., Woyciechowska, O.: Daily water consumption for household purposes and its variability in a rural household. J. Ecol. Eng. **17**(3), 47–52 (2016). https://doi.org/10.12911/22998993/63312
11. Piasecki, A., Górski, Ł.: Analysis of water consumption in 2014–2017 in Toruń. Infrastrukture Ekology Rural Areas **IV/1**, 973–984 (2018). https://doi.org/10.14597/INFRAECO.2018.4.1.067
12. Pietrucha-Urbanik, K., Szeligowski, A.: Analysis of water consumption changeability in the exemplary water system. Czasopismo Inżynierii Lądowej, Środowiska i Architektury, **63** (2/I), 231–240 (2016). https://doi.org/10.7862/rb.2016.125
13. Wichowski, P.P., Rutkowska, G., Kamiński, N., Trach, Y.: Analysis of water consumption in the campus of Warsaw University of life sciences – SGGW in years 2012–2016. J. Ecol. Eng. **20**(5), 193–202 (2019). https://doi.org/10.12911/22998993/105473
14. Kapalo, P., Matsiyevska, O., Adamski, M.: Water demand for apartment buildings in Slovakia, Ukraine, and Poland (Potreba vody pre bytové domy na Slovensku, Ukrajine, a v Pol'sku). In: Peráčková, J. (ed.). Proceedings of SANHYGA 2019, vol. 24, pp. 45–48. SSTP Bratislava, Bratislava (2019). (in Slovak)

15. Matsiyevska, O., Kapalo, P., Vovk, L.: Analysis of the water demand in the Slovak Republic and Ukraine. In: Blikharskyy, Z., Koszelnik, P., Mesaros, P. (eds.). Proceedings of CEE 2019, CEE 2019. Lecture Notes in Civil Engineering, vol. 47, pp. 299–306. Springer, Cham (2020). https://doi.org/10.1007/978-3-030-27011-7_38
16. Vrána, J., Moštěk, J.: Measurement of water flows and water consumption in apartment buildings (Měření průtoků vody a spotřeby vody v bytových domech). In: Peráčková, J. (ed.). Proceedings of SANHYGA 2019, vol. 24, pp. 49–58. SSTP Bratislava, Bratislava (2019). (in Slovak)

Precise Explicit Approximations of the Colebrook-White Equation for Engineering Systems

Viktor Mileikovskyi(✉) and Tetiana Tkachenko

Kyiv National University of Construction and Architecture, Kyiv, Ukraine
v_mil@ukr.net

Abstract. Modern automated engineering systems have variable hydraulic/aerodynamic conditions with Reynolds number from zero to hundred thousand with a wide range of roughness. Simple approximations of the Colebrook-White equation cannot give enough precision. The aim of the work is a universal simple precise approximation of the Colebrook-White equation. The methods are selected by the analysis of the equation curve. In the whole range of turbulent flow, it is near to linear. Thus, Newton's method is very effective. The algorithm is proposed for getting high-precision approximations. The results are two simple explicit ones for rough and careful calculations with a deviation of 5.36% and 0.00072% in a wide range of parameters. It is shown on the examples of the objects: the highest building "Biotecton" and researches of "green roofs" in a wind tunnel. The scientific novelty is that we scientifically grounded the effective usage of Newton's method, which provides new universal, precise and simple explicit approximations of Colebrook-White equation. The practical value is that the approximations are covered different practical tasks of hydraulic and aerodynamic calculations in the whole range of turbulent flow.

Keywords: Colebrook-White equation · Hydraulic calculation · Aerodynamic calculation · Microclimate system · Turbulent flow · Approximation

1 Introduction

Quality and reliability of engineering systems in buildings depend on the correctness of hydraulic and aerodynamic calculations of pipelines or air ducts. Friction losses in most cases are significant in the total pressure losses. Most flows in the systems are turbulent, and the friction losses correspond to the accurate but implicit Colebrook-White equation [1]:

$$EQ = f^{-1/2} + \frac{2}{\ln(10)}\ln\left(\frac{2.51}{Re}f^{-1/2} + \frac{\Delta_e/D}{3.71}\right) = \frac{2}{\ln(10)}\ln\left(\frac{a}{3.71Re}\right) + f^{-1/2} = 0, \tag{1}$$

where Δ_e – equivalent roughness of the pipeline, m; Re – Reynolds number; D – hydraulic (equivalent) diameter of the pipeline, m; a – parameter:

Z. Blikharskyy (Ed.): EcoComfort 2020, LNCE 100, pp. 303–310, 2021.
https://doi.org/10.1007/978-3-030-57340-9_37

$$a = Re(\Delta_e/D) + 9.3121 \times f^{-1/2}. \tag{2}$$

In older systems with an approximately constant hydraulic or aerodynamic regime, most flows have a developed turbulent regime, and the Reynolds number is more than 10,000. In these conditions very simple logarithmic (after conversion to natural one) and power approximations by A. Altshul [2] were used:

$$f^{-1/2} = -(1.8/\ln(10))\ln(((\Delta_e/D)/10) + 7/Re); \tag{3}$$

$$f = 0.11((\Delta_e/D) + (68/Re))^{1/4}. \tag{4}$$

Due to increasing energy efficiency, modern systems are highly automated. A significant amount of research is devoted to the appropriate air distributors [3–6] or optimal control strategies [7–9]. However, the simulation of hydraulic and aerodynamic conditions [10] is no less important. The Reynolds number varies from zero to hundreds of thousands. The range of equivalent roughness, m, in modern pipelines and air ducts has also expanded. Plastic technologies decrease it by order. Flexible corrugated pipelines rise it up to commensurable to the diameter [m]. Therefore, universal, simple and accurate explicit approximations of the Colebrook-White equation are becoming especially relevant today. Due to the forced use of iterative procedures, the modelling of the variable thermal-hydraulic regime of a renovated one-pipe heating system [10] of an apartment building took more than a day. Through the efforts of the English Wikipedia Society, a table of historically significant approximations was created, which at May 2020 had 26 positions [11]. At May 2020, the last entry corresponds to 2018 [12]. Dejan Brkić and Pavel Prax [1] obtained the most accurate modern approximations in 2019 by one and two steps of Padé approximation. It has 0.0259% of deviation, but it is too bulky. If we pick a simple fast-converging method, a compact approximation will be found. For example, Newton's method has been effectively used for computer calculation of Darcy coefficient [13, 14].

2 Aim

The aim of the work is a simple precise approximation of the Colebrook-White equation, which is applicable for manual calculation and complex simulations.

3 Method

To choose a method for solving Eq. (1), let us analyse its properties. Partial derivatives of the solution according to the parameters $Re \geq 2320$ and $\Delta_e/D \geq 0$ (to save time taken in the Maxima system) $\partial(f^{-1/2})/\partial Re > 0$ and $\partial(f^{-1/2})/\partial(\Delta_e/D) < 0$. Therefore, the solution $f^{-1/2}$ monotonically increases with increasing Reynolds number and decreases with increasing relative roughness Δ_e/D. The range of the solution is determined from Eq. (1). For $Re = \infty$ and $\Delta_e/D = 0$, $f = 0$. At $Re = 2320$ and $\Delta_e/D = 1$ Eq. (1) has been solved numerically. Thus

$$\left(f^{-1/2}\right)_{min} = 1.1348004636910905664 \leq f^{-1/2} \leq \infty; \quad (5)$$

$$0 \leq f \leq 0.7765349209745279322\,0 = f_{max}; \quad (6)$$

$$a_{min} = 9.3121 \cdot 1.1348004636910905664 = 10.567375397937803782 < a < \infty. \quad (7)$$

The first derivative of the EQ function of Eq. (1) by the unknown parameter $f^{-1/2}$ monotonically decreases with increasing $f^{-1/2}$ and $Re\ \Delta_e/D$:

$$1 < \partial \mathrm{EQ}/\partial(f^{-1/2}) = 1 + ((18.6242/\ln(10))/a) < 1.7654111816109958780. \quad (8)$$

The narrow range of change of the derivative (8) indicates the closeness of the equation function EQ to the linear one. The second derivative is always negative and changes from minus 0.67448966236883048814 to zero. The function is convex with the curvature

$$\kappa = (173.43041282/\ln(10))/\left(\left((a + (18.6242/\ln(10)))/a^{1/3}\right)^2 + a^{4/3}\right)^{3/2}. \quad (9)$$

The derivative of (9) by a has one positive root $a_{extr} = 2.328025\ (17^{1/2} - 1)/\ln(10) \approx$ $\approx 3.1576109808930112860$, at which $\kappa_{extr} = 4(17^{1/2} - 1)\ln(10)/(3(17^{1/2} + 7))^{3/2} \approx$ $\approx 0.14922492819316431408$. For $a \to \infty$, the curvature (9) goes to zero. Therefore, the root corresponds to the maximum curvature (8). In the range (5), the curvature monotonically decreases: $\kappa_{max} = 0.080752349164383260460 \leq \kappa \leq 0$. Therefore, the curve should be close enough to the line.

Let us test the maximum nonlinear curve EQ. We will assume that all approximations in the process of solving Eq. (1) are close enough to the root (otherwise, these parameters in the process of solving should be considered as an independent). As it is shown above, the solution $f^{-1/2}$ increases by Re and decreases by Δ_e/D. To clarify the total effect on the curvature, let us express the parameter a from Eq. (1):

$$a = 3.71\,Re/10^{(1/2)f^{-1/2}}. \quad (10)$$

The parameter a by Eq. (10) increases with Re and decreases with $f^{-1/2}$, which, in turn, decreases (as it is shown above) with Δ_e/D. Therefore, taking into account the decrease in curvature (9) by a, the largest value of κ corresponds to $Re = 2320$ and $\Delta_e/D = 0$ (Fig. 2). The obtained result (Fig. 1) indicates the practical linearity of the curve EQ by the solution $f^{-1/2}$. Thus, the methods of linear approximation are the most suitable for solving the Eq. (1). Preference should be given to a method that requires only one approximation, i.e. the Newton method. At manual calculations, the first approximation can be taken not less than $(f^{-1/2})_{min}$.

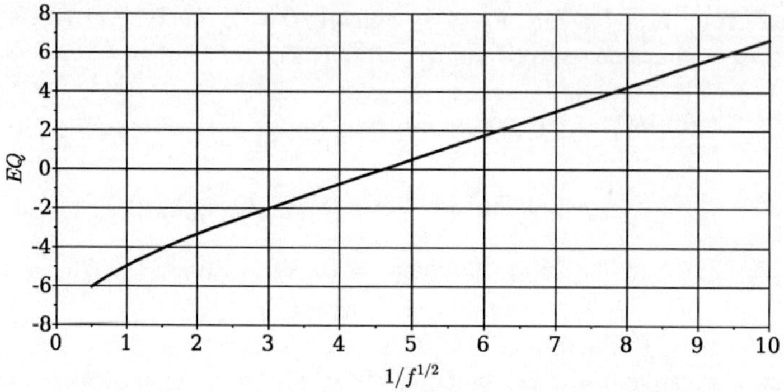

Fig. 1. The graph of the function EQ in Eq. (1) depending on the approximation of $f^{-1/2}$

In machine calculation, excessive verification reduces the performance of the algorithm. However, Fig. 1 in the critical case of the greatest curvature shows that the proximity to linearity is maintained when the approximation deviates more than two times in the smaller direction.

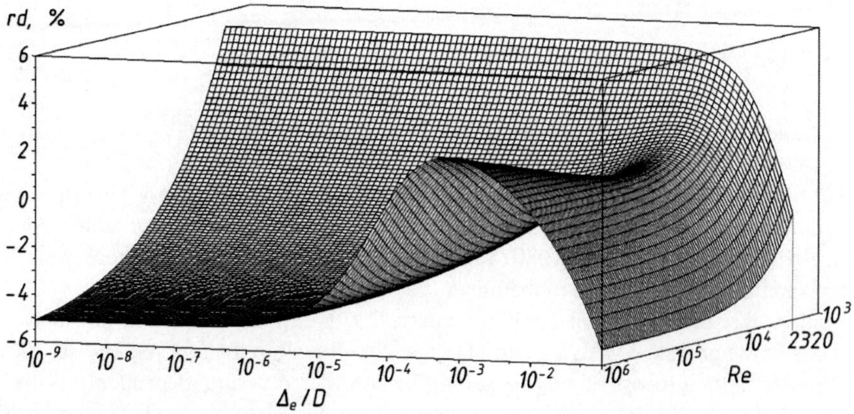

Fig. 2. Deviation of *rd*,%, approximation (13) from the Colebrook-White Eq. (1) depending on the Reynolds number *Re* and the relative equivalent roughness Δ_e/D.

According to Eqs. (1) and (8), the formula of Newton's method is simple and can be used for computer and manual calculations:

$$f_{i+1}^{-1/2} = f_i^{-1/2} - \frac{EQ\left(f_i^{-1/2}\right)}{\left(dEQ/d\left(f^{-1/2}\right)\right)_{f_i^{-1/2}}} = \frac{\frac{18{,}6242}{\ln(10)} f_i^{-1/2} - \frac{2}{\ln(10)} a_i \ln\left(\frac{a_i}{3{,}71 Re}\right)}{\left(18{,}6242/\ln(10)\right) + a_i}, \quad (11)$$

where i is the iteration number, and $i = 0$ corresponds to the first approximation. For the efficient calculation of *Eq.* (11), it is recommended to calculate the constants with the maximum accuracy of the corresponding computer system (or calculator) and enter or store them digitally. Very high accuracy is not required for engineering calculations. Therefore, instead of performing iterations (11), another approach is proposed to achieve high accuracy. We assume a sufficiently dense grid of Reynolds numbers $Re = 2320\ldots10^9$ with a variable step ΔRe: $\Delta Re = 20$ at $Re \leq 10000$, $\Delta Re = 200$ at $10000 < Re \leq 100000$, $\Delta Re = 2000$ at $100000 < Re \leq 1000000$, and so on. Similarly, we take a grid of relative equivalent roughness $\Delta_e/D = 0\ldots0.1$ with a variable step $\Delta(\Delta_e/D)$: $\Delta(\Delta_e/D) = 2\cdot10^{-8}$ for $\Delta_e/D \leq 10^{-5}$, $\Delta(\Delta_e/D) = 2\cdot10^{-7}$ for $10^{-5} < \Delta_e/D \leq 10^{-4}$, $\Delta(\Delta_e/D) = 2\cdot10^{-6}$ for $10^{-4} < \Delta_e/D \leq 10^{-3}$, etc. We accept the first approximation, in this paper – (3). To increase accuracy, all numerical coefficients are considered as unknowns C_1, C_2 and C_3:

$$f^{-1/2} = -C_1 \ln(((\Delta_e/D)/C_2) + (C_3/Re)). \tag{12}$$

Their values in Eq. (3) are considered as the first approximation. In the system of computer algebra (in this work – SciLab), a program has been created that single time numerically solves Eq. (1) with the maximum possible accuracy in each node of the grid. After that, a function multiple times returns the maximum relative deviation on the grid of the value according to Eq. (12) from the obtained solution of Eq. (1) for any values of the coefficients C_1, C_2 and C_3. The obtained result is optimized (in this work by the fsolve function according to the Nelder-Mead method) to minimize the maximum deviation. During the process, accuracy increases by an order in comparison with Eq. (3). Next, the grid should be extended to test the possibility of expanding the range of application of the formula without reducing accuracy.

After that, a single iteration by Eqs. (2) and (11) is applied to the first approximation (12). In this case, again, all numerical coefficients are replaced by unknowns C_j. The obtained coefficients C_1, C_2 and C_3 are also not considered as constants but are used for the first approximation for optimization, which is performed similarly. This allows significant refining of the result obtained by Newton's method.

Similarly, two iterations of Newton's methods with more unknown coefficients can be performed. Since Newton's iterative process was broken at the first iteration because of optimization, there is no guarantee that the previously obtained coefficients C_j are a better second approximation than the standard coefficients in Eqs. (2) and (11). Thus, standard coefficients were used to ensure the convergence of the process.

4 Results

As a result, after a few days of machine calculations we have (equations are given in the order of calculation in a form that provides a minimum of operations):

- with a deviation up to 5.36% (Fig. 2) within $Re = 2320\ldots10^9$, $\Delta_e/D = 0\ldots0.65$ for rough engineering calculations:

$$f = \left(0.8284\ln\left(\frac{\Delta_e/D}{4.913} + \frac{10,31}{Re}\right)\right)^{-2}; \quad (13)$$

- with a deviation up to 0.00072% (Fig. 3) at Re = 2320…10⁹, $\Delta_e/D = 0\ldots0.65$ for the most application in engineering and science (further refinement is impractical):

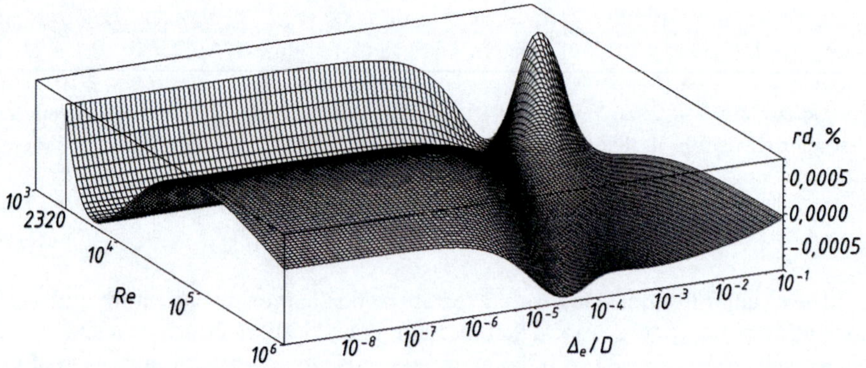

Fig. 3. Deviation of *rd*,%, approximation (14) from the Colebrook-White Eq. (1) depending on the Reynolds number *Re* and the relative equivalent roughness Δ_e/D.

$$\begin{cases} f_0^{-1/2} = -0.79638\ln\left(\frac{\Delta_e/D}{8.208} + \frac{7.3357}{Re}\right), \\ a_1 = (Re\Delta_e/D) + 9,3120665 f_0^{-1/2}, \\ f = \left(\frac{8.128943 + a_1}{8.128943 f_0^{-1/2} - 0.86859209 a_1 \ln(a_1/3.7099535Re)}\right)^2. \end{cases} \quad (14)$$

Large Reynolds numbers (over 10^6) are used in special technologies and constructions. An example is the project of the highest skyscraper "Biotecton" [15] of 1 km high for cities with polluted air. Its ventilation takes very clean air at the level of 1 km and supplies it to through an air-duct of 10 m in diameter. It is thermally insulated and soundproofed. Only technical and economic indicators limit the air velocity. The Reynolds number exceeds 10 million. The Eqs. (14) are the best universal ones for calculations of all microclimate systems in the object.

An example of pipelines with high equivalent roughness is the study of "Green structures" in a straight wind tunnel. The first author's research was carried out [16] in the Eiffel chamber (Fig. 4a) with unlimited internal height [m]. To increase the length of the models it is planned to use a wind tunnel (Fig. 4b) with a straight duct of a cross-section of 1 × 1 m (hydraulic diameter D_h = 1 m) and a flow velocity up to u = 10 m/s (Re = 666667). The grass on the "green roof" model grows up to (averaged) 450 mm (Fig. 4c). Such a model with a width of 1000 mm is planned to be installed at the bottom of the duct. The equivalent roughness of other walls can be considered 0.1 mm [17]. The average equivalent channel roughness will be (450 + 0.1 + 0.1 + 0.1)/4 = 112.6 mm. The relative equivalent roughness is $\Delta_e/D = 0.1126$. According to the Eq. (14) $f = 0.1085$. For comparison, by Eq. (4) $f = 0.06373$, i.e. 1.7 times less. Using air density $\rho = 1.2$ kg/m^3 (at temperature of $T = 293.15$ K), pressure loss per meter according to Darcy's formula [2] $\Delta p_\ell = (f/D_h)(\rho u^2/2) = 6.51\,\text{Pa/m}$.

Fig. 4. Wind tunnels: a – Eiffel chamber; b – straight wind tunnel at Czestochowa Polytechnic (Czestochowa, Poland); c – the model with grass of 300…500 mm high.

This pressure gradient at a model length of 1…2 m is insignificant and it is within the pressure gradient that occurs when the wind goes around the roofs of real buildings.

5 Scientific Novelty and Practical Significance

The scientific novelty is that we scientifically grounded the effective usage of Newton's method, which provides new universal, precise and simple explicit approximations of Colebrook-White equation. The practical value is that the approximations are covered different practical tasks of hydraulic and aerodynamic calculations, and allow simulation of variable hydraulic conditions in microclimate systems with adequate computational resources.

6 Conclusions

The obtained approximations of the solution of the Colebrook-White equation allow solving effectively a wide range of problems – engineering calculations and scientific researches. The effectiveness of the use of approximations is confirmed by the example of studies of green roofs in the wind tunnel in the form of a straight channel.

References

1. Brkić, D., Praks, P.: Colebrook's flow friction explicit approximations based on fixed-point iterative cycles and symbolic regression. Computation **7**(3), Article no. 48 (2019)
2. Altshul, A., Kiselev, P.: Hydraulics and Aerodynamics (Fundamentals of Fluid Mechanics). Stroyizdat, Moscow (1975)
3. Korbut, V., Voznyak, O., Myroniuk, Kh., Sukholova, I., Kapalo, P.: Examining a device for air distribution by the interaction of counter non-coaxial jets under alternating mode. Eastern-Eur. J. Enterp. Technol. **2**(8), 30–38 (2017)
4. Voznyak, O., Korbut, V., Davydenko, B., Sukholova, I.: Air distribution efficiency in a room by a two-flow device. In: Blikharskyy, Z., Koszelnik, P., Mesaros, P. (eds.) Proceedings of CEE 2019, CEE 2019. Lecture Notes in Civil Engineering, vol. 47, pp. 526–533. Springer, Cham (2020)
5. Vozniak, O., Dovhaliuk, V., Sukholova, I., Dovbush, O.: Mathematical Simulation of a Twisted Inlet Jet at Variable Mode with Using Various Turbulence Models. Ventyliatsiia, Osvitlennia ta Teplohazopostachannia **31**, 6–15 (2019)
6. Kapalo, P., Voznyak, O., Yurkevych, Y., Myroniuk, K., Sukholova, I.: Ensuring comfort microclimate in the classrooms under condition of the required air exchange. Eastern-Eur. J. Enterp. Technol. **5**(10), 6–14 (2018)
7. Seong, N.-C., Kim, J.-H., Choi, W.: Optimal control strategy for variable air volume air-conditioning systems using genetic algorithms. Sustainability **11**(18), Article no. 5122 (2019)
8. Rismanchi, B., Zambrano, J.M., Saxby, B., Tuck, R., Stenning, M.: Control strategies in multi-zone air conditioning systems. Energies **3**(12), Article no. 347 (2019)
9. Gładyszewska-Fiedoruk, K., Zhelykh, V., Pushchinskyi, A.: Simulation and analysis of various ventilation systems given in an example in the same school of indoor air quality. Energies **15**(12), Article no. 2845 (2019)
10. Mileikovsky, V.: Mathematical simulation of the variable hydraulic regime of one-pipe vertical water heating systems. Danfoss INFO 3-4, pp. 25–30 (2011)
11. Darcy friction factor formulae. https://en.wikipedia.org/wiki/Darcy_friction_factor_formulae#Colebrook–White_equation. Accessed 06 May 2020
12. Bellos, V., Nalbantis, I., Tsakiris, G.: Friction modeling of flood flow simulations. J. Hydraul. Eng. **144**(12), Article no. 04018073 (2018)
13. Brkić, D., Praks, P.: Advanced iterative procedures for solving the implicit Colebrook equation for fluid flow friction. Adv. Civil Eng. **2018**, Article no. 5451034 (2018)
14. Moreno, E.O.L., Ubaque, C.A.G., Vaca, M.C.G.: Darcy-Weisbach resistance coefficient determination using Newton-Raphson approach for android 4.0. Tecnura **23**(60), 52–58 (2019)
15. Krivenko, O., Mileikovskyi, V., Tkachenko, T.: The principles of energy efficient microclimate provision in the skyscraper "Biotecton" of 1 km height. Eur. J. Formal Sci. Eng. **1**(3), 8–17 (2018)
16. Tkachenko, T., Mileikovskyi, V.: Methodology of thermal resistance and cooling effect testing of green roofs. Songklanakarin J. Sci. Technol. **1**(42), 50–56 (2020)
17. Barkalov, B., Pavlov, N., Amirjanov, S., et al.: Internal sanitary facilities. In: 3 Hours, Part III. Ventilation and Air Conditioning. Prince 2, 4th edn. Stroyizdat, Moscow (1992)

Technical and Economic Efficiency After the Boiler Room Renewal

Khrystyna Myroniuk, Orest Voznyak, Yuriy Yurkevych, and Bohdan Gulay(✉)

Lviv Polytechnic National University, Lviv 79013, Ukraine
08bogdan1986@gmail.com

Abstract. The article is devoted to decision of actual task of energy saving. As a fact, a large-scale energy saving policy is being implemented in Ukraine, and energy efficiency objectives are comprehensive and cover both the legislative framework and technical innovations. It is consumed in the boiler rooms too much amount of gas. One of the effective ways to reduce energy consumption for boiler gas supply is to carry out the thermal modernization of the boiler rooms. This article describes the economic indicators of thermal renovation measures during the reconstruction of the boiler rooms. During the reconstruction of boiler rooms the following thermal modernization measures were taken for comparison: equipping the existing boiler with a modern domestic gas burner, economizer or condensing economizer; replacement of the existing boiler with a modern Ukrainian-made boiler with a modern domestic burner with the possibility of its additional equipping with economizer or condensing economizer; replacement of the existing boiler by Viessmann boiler. Conducted energy audit of the boiler room during its reconstruction has showed that the profit from the introduction of energy-saving technologies during their operation is the maximum, if it is provided the modernization of the existing boiler by its equipment at the same time modern burner domestic production and condensing economizer.

Keywords: Boiler room · Energy saving · Energy audit · Thermal renewal · Heat utilization · Condensing economizer

1 Introduction

The problem of significantly reducing of hydrocarbon energy sources consumption is very acute. It has become a priority in most countries of the world, particularly in Ukraine, both for economic and environmental reasons. In the utility sector the issue of energy savings is solved by accounting of energy resources [4], their consumption managing, by increasing of thermal protection level of systems for providing microclimate of premises [5], by reducing of heat losses in thermal networks [1] and by efficiency increasing of the heating boilers [2]. Only with complex solution of these problems, it can be achieved significant savings of a fuel and energy resources. Among of these measures the thermal renewal of the boiler rooms is the most important, as it requires great funds to be implemented.

Z. Blikharskyy (Ed.): EcoComfort 2020, LNCE 100, pp. 311–318, 2021.
https://doi.org/10.1007/978-3-030-57340-9_38

Ukraine still operates a large number of gas and solid fuel boilers installed in the second half of the twentieth century, which are outdated and do not meet current energy efficiency criteria. In the most cases, the life of these boilers exceeds 40 years, the equipment of these boilers has exhausted its physical and moral resources and is inefficient. The total number of heating boilers exceeds 35 thousand, with the boiler capacity up to 3.5 MW accounting for over 88% [9].

The average efficiency of boilers equipped with NIISTU-5, KV-300, KCHM, "Fakel", "Universal" boilers often does not exceed 60%. Much of these boilers have been used up and need replacement.

But in many cases, the physical condition of such boilers (especially cast iron ones) allows them to operate for more than one year. In addition, the replacement of the boiler, in addition to the significant costs of a new heat generator, is associated with the need for simultaneous reconstruction of the heating system (especially in the case of low-temperature and condensing boilers).

In this regard, two directions of modernization of the following boiler-houses were formed [10]: replacement of old boilers with modern ones; modernization of small and medium capacity boilers. Such tasks require a preliminary (basic) technical and economic analysis.

2 Analysis of Current Research and Problem Statement

An important priority for European economic policy is careful using of energy. This problem is complex and covers both legislative framework and technical innovation. A large amount of thermal energy is consumed by indoor climate systems [3, 6]. There is an objective need to reduce them due to energy saving [7, 8] and energy audit [15]. However, the issue of energy audit of heat supply systems and boiler rooms was not considered. It is in this aspect that the energy efficiency of the system as a whole could be achieved by reducing of the gas demand at its combustion in boilers.

The main disadvantage of existing boilers is their excessive energy consumption due to the low efficiency of burners, the lack of economizers (conventional and condensing) and other factors [12]. There is an objective need to address these disadvantages and achieve of the gas savings. Furthermore, thermal modernization of boiler rooms is an effective way of reducing energy consumption for heat supply needs. An important role is played by the technical and economic evaluation of the measures effectiveness [11]. Using of the modern methods of the economic effectiveness evaluating of thermal modernization is taken into account in the latest concept of economic calculations, particular of UNIDO (United Nations Industrial Development Organization) recommendation [14]. Therefore, it is advisable to use terminology, symbols and basic economic characteristics according to paper [14]. Nevertheless, these recommendations considered only such measures that can be applied at the same time.

3 The Purpose and Objectives of the Study

The purpose of the work is to increase the energy efficiency of boiler rooms and to improve the energy audit methodology, taking into account measures that can not be applied at the same time.

To achieve this goal, the task was set to carry out an energy audit of the boiler room during its reconstruction, taking into account measures that can not be applied at the same time.

4 Methods, Materials and Research Results

Increasing of the boiler efficiency can be achieved due to a number of technical measures. One of them is deep cooling of combustion products. As a consequence to do this, an economizer must be installed behind the boiler. Cooling of combustion products by 10–12 °C allows to increase the boiler efficiency by 1%. If the products of combustion are cooled below the dew point temperature, then the condensation of the part of the steam contained in the flue gas will condense. This allows not only to increase the amount of apparent heat used, but also to use the latent heat of water vapor condensation. A condensing economizer could be used to implement this measure, which in-creases the boiler efficiency by 12–15% [13].

Another effective measure is to reduce the heat loss from the chemical incompleteness of combustion, reduce the coefficient of excess air and automatically maintain its value throughout the range of changes in the productivity of the boiler. To do this substitution of the old gas burners should be realized, in which the level of the combustion process automation is limited only by the protection functions.

The most effective are modular burners among modern burners. They can operate in the wide power range - from 10% to 100% and the effect of their using can be 5–15% of gas consumption economy [13].

A number of calculations were carried out to determine the feasibility of comparing various options for the reconstruction of the boiler room. They consisted of the determined gas consumption for the heating period by the boiler capacity of 3 MW in different variants of boiler equipment layout in process of thermal modernization.

For the main calculation, a boiler house with three boilers with the capacity of 1 MW each was selected. As the basic variant of the boiler room, equipped with NIISTU-5 boilers, gas in which is burned in the draft burners was considered.

During the reconstruction of the boiler room, the following options were considered: modernization of existing boilers, replacement of boilers by domestic production, and installation of boilers by a leading foreign manufacturer. As an option for the modernization of the outdated equipment of the boilers are the possibility of replacing the hearth burner with the modern burner of domestic production type PGS-BM, installing the traditional economizer (Ateplo Wichlocz WE) or the condensing economizer (La Term) was considered.

As a next variant, the replacement of existing boilers by modern domestic production boilers (CSWa-1) with the standard for this type of boilers burner PGS-BM was considered. Additionally, the possibility of bundling CSWa-1 boilers with

Weishaupt modular burners and their additional equipment with the traditional or condensing economizer is considered. Installation of three Viessmann Vitoplex 100 boilers is considered separately.

Gas consumption for each variant was determined taking into account change of the boiler efficiency due to the technical decisions made, and due to change of the magnitude of the relative load of the boiler during the heating period. In the course of technical and economic calculations, the estimated cost of equipment and installation work has been determined on the basis of current regulatory documents. The final gas price, taking into account the cost of delivery of gas and other mark-ups and surcharges in Ukraine differs depending on the tariffs of regional gas and is in the range of 183,3–233,3 EUR per 1000 m^3. For calculation it was accepted 200 EUR per 1 thousand m^3.

Let the boiler heat load Q (kW), the gas combustion heat q (J/m^3) and the burner efficiency $\eta_0 = 0.88$. To supply this amount of heat, gas must be supplied in the amount of V (m^3/hour). Its cost is C_0 (EUR). Replacing the burner with the modern efficiency factor $\eta_1 = 0.99$ results in an efficiency increase $\alpha = \eta_1/\eta_0$ times. As a result, it is gas saving by α times. In monetary terms it is C_1 EUR.

Using of an economizer (conventional or condensing) saves Q_{ec} of heat, and therefore – gas amount V by cost ΔC_{ec}.

The installation of Viessmann boiler with the condensing economizer takes all these aspects into account.

Energy audit of boiler room reconstruction was carried out with the following energy saving measures:

A - installation of a modern burner in the existing boiler;
B - installation of the economizer in the existing boiler;
C - installation of the condensing economizer in the existing boiler;
D - installation of Viessmann boiler;
E - installation of a boiler of Ukrainian production with a Weishaupt burner;
F - installation of the boiler of the Ukrainian production with the economizer;
G - installation of the boiler of the Ukrainian production with the condensing economizer;
H - installation of a Ukrainian-made boiler with a modern Weishaupt burner.

The initial data for the energy audit of the boiler room are: annual gas consumption in the base variant V_0, m^3/year, cost of gas P_g, EUR/thousand m^3, data for calculating the estimated cost of thermal modernization works I_i, EUR, discount rate r (economic analysis is carried out at the condition of constant time prices and investment consideration time t = 15 years).

The following solution algorithm is proposed:

1. Determination of annual energy and gas consumption Q_0, MJ/year for the base variant.
2. Selection of a list of simple thermal renewal measures for this system.
   The cost of gas (P_g value) was taken according to the data of the National Commission for State Regulation of Energy and Utilities [11] from the calculation of the average in Ukraine 6.9 EUR per 1 thousand m^3, i.e. P_g = 0,0069 th.EUR/th.m^3.

3. Determination of the gas savings ΔV_i of each measure as $\Delta V_i = V_o - V_i$ and, consequently, the annual savings of K_i, EUR/year.
The results of the algorithm implementation are given in Table 1.

Table 1. Characteristics of energy saving measures

	Measurements	Annual gas consumption under the basic option V_o, th.m^3/year	After the change V_i, th.m^3/year	Gas savings ΔV_i $\Delta V_i = V_o - V_i$, th.m^3/year	Cost savings K_i $K_i = \Delta V_i\ P_g$, th.EUR/year
1.	A	866,1	805,5	60,6	11,28
2.	B	866,1	814,1	52,0	9,66
3.	C	866,1	762,2	103,9	19,33
4.	D	866,1	679,1	187,0	34,77
5.	E	866,1	750,1	116,0	21,57
6.	F	866,1	726,9	139,2	25,89
7.	G	866,1	634,1	232,0	43,15
8.	H	866,1	773,3	92,8	17,26

4. Determination of each thermal renewal measure indexes (Table 2) at condition of the different discount rates in Ukraine: $r = 0{,}18$ (2019) and $r = 0{,}11$ (2020). Table 2 has 2 recent columns at the different discount rates (0,18 and 0,11) for comparison and conclusions.

Table 2. Economic indicators of thermal renewal measures

№	Measurements	Ii	K_i'	SPBTi	NPVRi r = 0,18	NPVRi r = 0,11
		th.EUR	th.EUR/Year	Year	th.EUR	th.EUR
1	A	11,2	11,28	1,0	+2,81	+24,39
2	B	15	9,66	1,5	−2,92	+15,35
3	C	16,2	19,33	0,8	+7,97	+44,51
4	D	239,85	34,77	7,0	−196,52	−131,03
5	E	193,55	21,57	9,0	−166,47	−92,18
6	F	180,4	25,89	7,0	−147,85	−98,78
7	G	181,6	43,15	4,1	−127,46	−45,55
8	H	165,4	17,26	9,5	−143,76	−111,01

5. Optimization of aggregate thermal renewal options (Table 3).

In order to improve the quality of the energy audit, optimization was carried out to ensure the best possible result. In its implementation there was taken into account fact that some certain energy saving measures cannot be applied in the same time.

The dynamics of funds by years have been determined (Fig. 1).

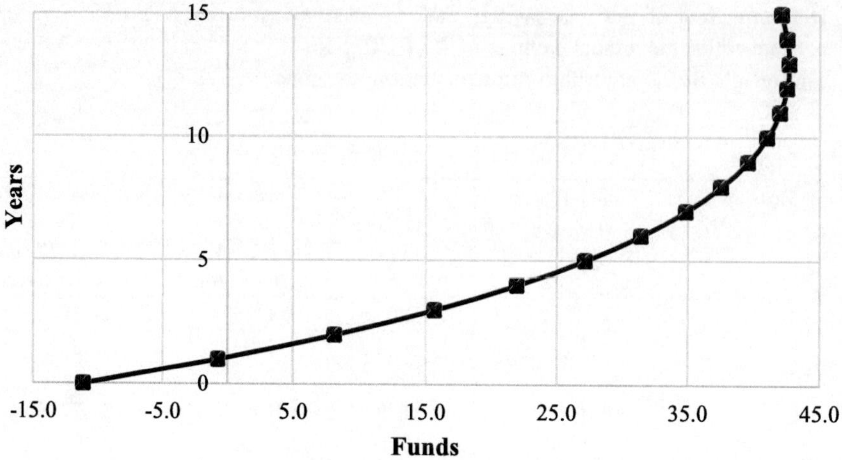

Fig. 1. Dynamics of funds by years.

Table 3. Optimization of cumulative thermal innovation options

№	Measurements	Variants							
		I	II	III	IV	V	VI	VII	VIII
1	A	+	+	–	–	–	–	–	–
2	B	–	+	–	–	–	–	–	–
3	C	–	–	–	–	–	–	–	–
4	D	–	–	–	+	–	–	–	–
5	E	–	–	–	–	+	+	+	–
6	F	–	–	–	–	–	+	–	–
7	G	–	–	–	–	–	–	+	–
8	H	–	–	–	–	–	–	–	+
Indexes									
1	*I*, th.EUR	11,2	26,2	27,4	239,83	193,53	373,93	375,13	165,4
2	*K*, th.EUR	11,27	20,93	30,6	34,77	21,57	47,47	64,7	17,27
3	*SPBT*, year	1,0	1,3	0,9	7,0	9,0	8,0	6,0	9,5
4	*NPVR, (r = 0,18)* th.EUR	+2,81	+0,05	+10,1	−191,7	−166,47	−311,43	−291,7	−143,73
5	*NPVR, (r = 0,11)* th.EUR	+24,4	+39,73	**+68,9**	−13517	−125,5	−227,4	−165,8	-111

The variants have been optimized taking into account the data of claim 5 and the optimal variant **III** with the maximum $NPVR_j$. has been determined. This means that the maximum economic effect will be in the case of application in the same time of the following measures: installation of a modern burner in the existing boiler and a condensing economizer. The profit under this option is 2067 th.EUR, that is, approximately EUR 2 million.

5 Conclusions

1. Energy audit methodology is improved taking into account measures that cannot be implemented in the same time.
2. The energy audit of the boiler room during its reconstruction has showed that using of Ukrainian production boilers allows to save energy costs for the boiler room and has the lowest payback time.
3. The maximum profit from the introduction of energy-saving technologies during their operation is EUR 2 million at condition of the existing boiler modernization due to its equipment at the same time by modern burners of domestic production and condensing economizer.

References

1. Basok, B.I., Davydenko, B.V., Farenuyk, G.G., Goncharuk, S.M.: Computational modeling of the temperature regime in a room with a two-panel radiator. J. Eng. Phys. Thermophys. **87** (6), 1433–1437 (2014)
2. Basok, B., Davydenko, B., Isaev, S., Goncharuk, S., Kuzhel', L.: Numerical modeling of heat transfer through a triple-pane window. J. Eng. Phys. Thermophys. **89**(5, 1), 1277–1283 (2016)
3. Bilous, I., Deshko, V., Sukhodub, I.: Parametric analysis of external and internal factors influence on building energy performance using non-linear multivariate regression models. J. Build. Eng. **20**, 327–336 (2018)
4. Bilous, I., Deshko, V., Sukhodub, I.: Building inside air temperature parametric study. Mag. Civ. Eng. **68**(8), 65–75 (2016)
5. Buyak, N., Deshko, V., Sukhodub, I.: Buildings energy use and human thermal comfort according to energy and energy approach. Energy Build. **146**(1), 172–181 (2017)
6. Deshko, V., Buyak, N.: A model of human thermal comfort for analysing the energy performance of buildings. East.-Eur. J. Enterp. Technol. **4**(8–82), 42–48 (2016)
7. Kapalo, P., Sedláková, A., Košicanová, D., Voznyak, O., Lojkovics, J., Siroczki, P.: Effect of ventilation on indoor environmental quality in buildings. In: The 9th International Conference "Environmental Engineering", Vilnius, Lithuania (2014). Selected Papers, eISSN 2029-7092/eISBN 978-609-457-640-9
8. Kapalo, P., Voznyak, O., Yurkevych, Yu., Myroniuk, Kh., Sukholova, I.: Ensuring comfort microclimate in the classrooms under condition of the required air exchange. East. Eur. J. Enterp. Technol. **5/10**(95), 6–14 (2018)
9. Klymchuk, O., Denysova, A., Shramenko, A., Borysenko, K., Ivanova, L.: Theoretical and experimental investigation of the efficiency of the use of heat-accumulating material for heat supply systems. EUREKA Phys. Eng. (3), 32–40 (2019)
10. Klymchuk, O., Denysova, A., Mazurenko, A., Balasanian, G., Tsurkan, A.: Construction of methods to improve operational efficiency of an intermittent heat supply system by determining conditions to employ a standby heating mode. East.-Eur. J. Enterp. Technol. **6**(8–96), 25–31 (2018)
11. National Commission for State Regulation of Energy and Utilities. http://www.nerc.gov.ua
12. Redko, A., Dzhyoiev, R., Davidenko, A., Pavlovskaya, A., Pavlovskiy, S., Redko, I., Kulikova, N., Redko, O.: Aerodynamic processes and heat exchange in the furnace of a steam boiler with a secondary emitter. Alex. Eng. J. **58**(1), 89–101 (2019)

13. Redko, A., Kulikova, N., Pavlovskiiy, S., Redko, A.: Simulation and optimization of heat-exchanger parameters of heat pipes by changes of entropy. Heat Transf. Res. **49**(16), 1545–1557 (2018)
14. United Nations Industrial Development Organization. https://www.unido.org
15. Voznyak, O., Korbut, V., Davydenko, B., Sukholova, I.: Air distribution efficiency in a room by a two-flow device. In: Blikharskyy, Z., Koszelnik, P., Mesaros, P. (eds.) Proceedings of CEE 2019, CEE 2019. Lecture Notes in Civil Engineering, vol. 47. Springer, Cham (2020)

The Calculation of Indoor Air Forecast Temperature of a Space with the Replaceable Thermotechnical Characteristics of the Enclosure Structures While in Operation

Viktor Petrenko(✉), Konstantin Dikarev, Anatolii Petrenko, and Ruslan Papirnyk

Prydniprovs'ka State Academy of Civil Engineering and Architecture, Dnipro 49600, Ukraine
petrenko@pgasa.dp.ua

Abstract. In this article there are given the calculations of indoor air forecast temperature of a space, having the outer enclosure structures with the replaceable thermotechnical characteristics of structural layers while in operation. The indoor temperature forecasting is an interesting and actual task, which makes it possible to solve problems related to the process of ensuring the parameters of the indoor microclimate, as well as social. The solving of this task depends on many factors influencing the maintenance of a stable indoor air temperature. The one-dimensional thermal field in a flat wall with a fixed heat transfer process between warm and cold air, which have constant temperatures and processes of internal destruction of building materials is considered as a basis for solving this task. The solution of the problem involves determining the coefficient of heat transfer, or the thermal resistance of the outer wall under conditions of their variation in operation and the final determination of the indoor air forecasting temperature. The derivation of an equation of determination of the indoor air forecasting temperature is based on the assumption that the heating losses of the room are equal to heat inputs of the heating appliances/unit heaters of the heating system. As a result, the equation of determination of the indoor air forecasting temperature subject to alterations of thermotechnical characteristics of structural layers and the enclosing parts is obtained.

Keywords: Thermal comfort · Forecasting temperature · Heat transfer coefficient · Thermal resistance · Material aging

1 Introduction

The creation of comfortable conditions of a microclimate in multifunctional rooms is important and justified. Both the social and the economic determinants of human activity depend on the micro-climate that has been established. Therefore, the question of organizing such conditions of the microclimate, that would be stable in time and space during whole time period of building operation, remains a pressing issue.

Z. Blikharskyy (Ed.): EcoComfort 2020, LNCE 100, pp. 319–327, 2021.
https://doi.org/10.1007/978-3-030-57340-9_39

The research works by A. Fanger [1], V.N. Theologian [2] and other authors [3, 4] are devoted to research and forecasting of the microclimate on the premises.

These studies focused on such tasks as the creation of microclimate conditions and its impact on human health, forecasting human condition through a combination of thermal balances.

The authors [5, 6] give the directions of tasks for forecasting of microclimate in the premises due to the influence of factors of wear and tear and obsolescence, both systems for ensuring the microclimate, and heat protection of a structure. In this article, the authors set a goal to reveal the dependence of changes in the parameters of the microclimate on changes in burning characteristics of the structural layers of the enclosing parts of the building.

Taking into account the literature sources [7, 8], it can be seen that there is a need to solve the tasks aimed at determining the projected indoor air temperature related to the processes of changing the thermal characteristics of the enclosing parts.

Following the defined problem, we highlight the areas that need to be addressed in this article, namely:

- to formulate the initial data for the task of determining the projected indoor air temperature related to the processes of changing the thermal characteristics of the enclosing parts;
- to determine the force deformation relationship between the projected indoor air operation coming from the processes of changes in the thermotechnical characteristics of the enclosing parts;
- to calculate and analyze the results of the solution of the forecast indoor air temperature formula related to the processes of changing the thermal characteristics of the enclosing parts.

2 Initial Data

The current requirements for the thermotechnical characteristics of the enclosing parts in Ukraine [DBN V.2.6-31:2016] are those for which it is necessary to design the sandwich structures including various building materials. These materials can be both natural and polymeric.

The current requirements for the thermotechnical characteristics of the enclosing parts in Ukraine [DBN V.2.6-31:2016] are those for which it is necessary to design the sandwich structures including various building materials. These materials can be both natural and polymeric. In the design of building constructions, an empiric method of limit states is used. During operation, the building materials operate over time under various internal and external factors. Internal factors include natural destructive changes coming from the material properties. Major external effects include stresses, stress concentrators, temperature gradients, moisture gradients and wind pressure.

Recalculation of heat insulation values of building materials from one condition (λ_1, R_1) in other conditions (λ_2, R_2) is realized by formula [prDSTU EN 10456:201X]:

– for heat conductivity:

$$\lambda_2 = \lambda_1 \cdot F_T \cdot F_m \cdot F_a \tag{1}$$

– for thermal resistance:

$$R_2 = \frac{R_1}{F_T \cdot F_m \cdot F_a} \tag{2}$$

In Eqs. (1) and (2), the following notations are used:

F_T – is a correcting factor for the action of temperature on the thermal conductivity and resistance of the building material or article;
F_m – is a correcting factor for the effect of moisture on the thermal conductivity and resistance;
F_a – is a correcting factor for the ageing effect of the building material on the thermal conductivity and resistance.

The standard [prDSTU EN 10456:201X] does not specify the influence coefficients to obtain the correction of the thermal conductivity in dependence to the temperature related to ageing material or product F_a. If an ageing correction of the thermal conductivity F_a is applied, the thermotechnical characteristics will be calculated, taking into account the ageing over time, to be at least half of the working life of the construction product or material.

The calculation values of the thermal conductivity shall be established by means of a formula (1) or of a formula (2), the recalculation from the declared thermal conductivity is in reference to the specified design of the external barrier and to the specific actual conditions of its operation, which are determined by calculation at the specified parameters of the microclimate in the premises and the ambient operating conditions (parameters of the atmosphere).

On the basis of claims that the changes in the thermotechnical characteristics of the materials of the layers of enclosing part shall be changed as well as the heat loss of the room to be used. The change in the heating loss of the room should lead to a change in the indoor air temperature [9, 10].

Hereinafter, the authors provide the calculations by which the processes of indoor air temperature changes are confirmed, depending on the changes in the thermotechnical characteristics of the enclosing structures.

3 Derivation of a Formula

Considering the above calculations, we accept that heating losses through the enclosing structures for the first conditions are determined by the formula:

$$Q_1 = k_1 \cdot F \cdot (t_{in1} - t_{on}) \tag{3}$$

where k_1 – is a thermal transmission coefficient of the enclosing structure for the first conditions, $\frac{Wt}{m^2 \cdot {}^\circ C}$, that is defined by the formula:

$$k_1 = \frac{1}{\frac{1}{\alpha_{in}} + \sum \frac{\delta_i}{\lambda_i} + \frac{1}{\alpha_{on}}} \tag{4}$$

F – is an area of the enclosing structure, m^2;
t_{in1} – is the design temperature of indoor air for the first conditions, °C;
t_{on} – is the design temperature of parameters of the atmosphere, °C.

Since the thermal transmission coefficient depends on the change in the coefficient of the thermal conductivity of the materials of the enclosing structure and this variation depends on Eq. (1) or Eq. (2), we introduce such an indicator as the fraction of a decrease in the temperature conductivity coefficient or thermal resistance of materials in the enclosing structures:

$$\zeta_i = \frac{\lambda_{2i}}{\lambda_{1i}} = F_{Ti} \cdot F_{mi} \cdot F_{ai} \text{ or } \zeta_i = \frac{R_{2i}}{R_{1i}} = \frac{1}{F_{Ti} \cdot F_{mi} \cdot F_{ai}} \tag{5}$$

where ζ_i can vary from 0 to 1.

With Eqs. (4) and (5) we obtain the formula for determining the conductivity coefficient on change of the temperature conductivity coefficient of each building material of the enclosing structures:

$$k_2^m = \frac{1}{\frac{1}{\alpha_{in}} + \sum \frac{\delta_i}{\lambda_i \cdot \zeta_i} + \frac{1}{\alpha_{on}}} \tag{6}$$

In general, the heat transfer coefficient of building materials is changed in different ways over the whole standard operating period. For this reason, it is possible to introduce the indicator of reduction ratio of thermotechnical characteristics, in general, for an external enclosure structure:

$$\zeta_K = \frac{\sum \frac{\delta_i}{\lambda_i \cdot F_{Ti} \cdot F_{mi} \cdot F_{ai}}}{\sum \frac{\delta_i}{\lambda_i}} \tag{7}$$

where ζ_i can vary from 0 to 1.

In general, the temperature conductivity coefficient of building materials will change in different ways during the regulatory operational phase. Therefore, it is possible to introduce a percentage of the reduction in thermotechnical characteristics in general for the enclosing structures:

$$k_2^c = \frac{1}{\frac{1}{\alpha_{in}} + \zeta_c \cdot \sum \frac{\delta_i}{\lambda_i} + \frac{1}{\alpha_{on}}} \quad (8)$$

Taking into account the change in the value of the thermal transmission coefficient for the second conditions, the heating losses through the enclosing structures are defined by formula:

$$Q_2 = k_2^c \cdot F \cdot (t_{in2} - t_{on}) \quad (9)$$

where k_2^c – is the thermal transmission coefficient through the enclosing structure for the second conditions, $\frac{Wt}{m^2 \cdot {}^\circ C}$;
t_{in2} – the predicted design indoor air temperature for the second conditions, °C.

It is expected that the heating appliances of the heating system, which are installed in the premises, give the calculated amount of heat Q_1, Wt, and if there is a reduction in the thermotechnical characteristics the enclosing structure i.e. $Q_1 = Q_2$, then with Eqs. (3), (8) and (9) we get a formula for obtaining the predicted indoor air temperature for the second conditions:

$$t_{in2} = \frac{k_1 \cdot (t_{in1} - t_{on})}{k_2^c} + t_{on} \quad (10)$$

4 Analysis

For practical calculations we take the thermotechnical characteristics the enclosing structures are given below (see Table 1).

Table 1. Designs of enclosing parts

Enclosing structure	Material name	Density, $\rho_i, \frac{kg}{m^3}$	Depth, δ_i, m
Wall	Sand-cement bonding plaster	1800	0,02
	Insulant - the mineral wool mats	125	0,1
	Brick walls	1800	0,51
	Sand-cement bonding plaster	1800	0,03

(continued)

Table 1. (*continued*)

Enclosing structure	Material name	Density, $\rho_i, \frac{kg}{m^3}$	Depth, δ_i, *m*
Floor	Reinforced concrete hollow slab	2500	0,22
	Vapour barrier- the asphalt coating compound	1400	0,003
	Insulant - the mineral wool mats	125	0,25
	Floating coat - the sand-cement mortar	1800	0,05
	Oak parquet	700	0,025
Ceiling	Reinforced concrete hollow slab	2500	0,22
	Vapour barrier- the asphalt coating compound	1400	0,003
	Insulant - the perlite-plastoconcrete	100	0,26
	Floating coat - the sand-cement mortar	1800	0,05
	Waterproofing - three layers of ruberoid	600	0,009

Table 2. The data of reduction of thermal transmission coefficients of enclosure parts depending on the ratio of their reduction of thermotechnical characteristics

Part	Ratio									
	1	0,95	0,9	0,85	0,8	0,75	0,7	0,65	0,6	0,55
Wall	0,31	0,32	0,34	0,36	0,38	0,4	0,43	0,46	0,49	0,53
Floor	0,23	0,24	0,25	0,26	0,28	0,3	0,32	0,34	0,37	0,4
Ceiling	0,18	0,19	0,2	0,21	0,22	0,23	0,25	0,27	0,29	0,31

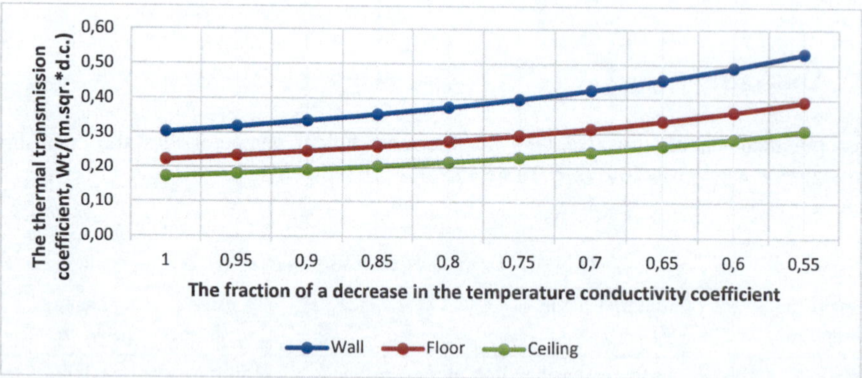

Fig. 1. The relation of thermal transmission coefficients of enclosure parts depending on the ratio of their reduction of thermotechnical characteristics

By Eq. (8) and the data of Table 1 we calculate the values of thermal transmission coefficients of enclosure parts that depend on the ratio of their reduction of thermotechnical characteristics (see Eq. 5). The calculations are shown in Fig. 1 and in Table 2.

Table 3. The data of the indoor air forecast temperature of a space, with the external walls, having the replaceable thermotechnical characteristics of structural layers in operation

Indicators	Ratio									
	1	0,95	0,9	0,85	0,8	0,75	0,7	0,65	0,6	0,55
$k_1, \frac{Wt}{m^2 \cdot °C}$	0,31	0,32	0,34	0,36	0,38	0,4	0,43	0,46	0,49	0,53
t_{in2}, °C	20	17,9	15,8	13,7	11,6	9,5	7,4	5,3	3,3	1,2

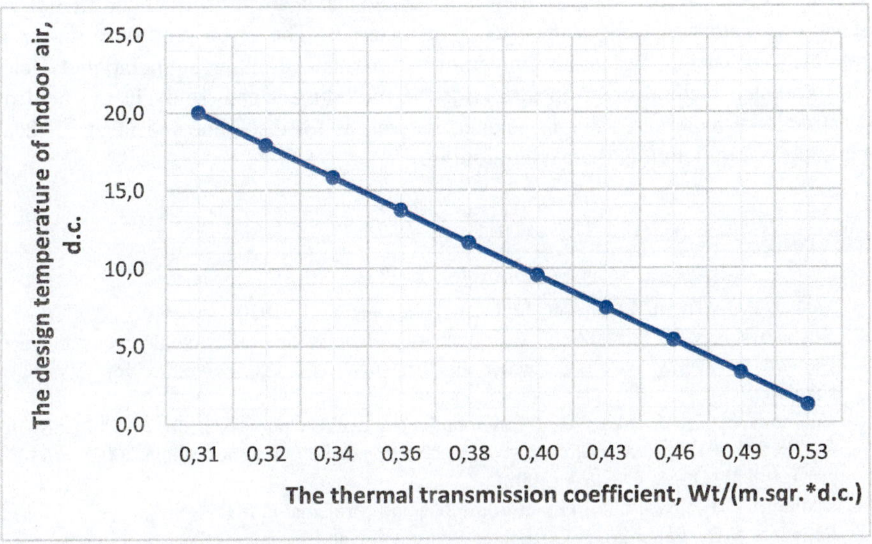

Fig. 2. The relation of the indoor air forecast temperature of a space, with the external walls, having the replaceable thermotechnical characteristics of structural layers in operation

According to the obtained data, the value of the indoor air forecast temperature of a space, with the external walls, having the replaceable thermotechnical characteristics of structural layers in operation, changes linearly and is a temperature drop of 2.1 for every 0,05% of the change of the thermotechnical characteristics between 1 and 0.55 (Table 3 and Fig. 2).

5 Conclusions

This article is devoted to the analytic survey of the temperature changes of indoor air of a space during the cold period of the year for conditions when the enclosure parts had the replaceable thermotechnical characteristics of structural layers in operation. It was discovered that when changing thermal transmission coefficients of enclosure parts (wall, floor, ceiling) they changed nonlinearly and the deviation rate was between 3 and 7%, while the reduced ratio of thermotechnical characteristics was from 1 to 0,55. It was also discovered, that the value of indoor air forecast temperature of a space with the external walls, having the replaceable thermotechnical characteristics of structural layers in operation, changes linearly and is a temperature drop of 2.1 for every 0,05% of the change of the thermotechnical characteristics between 1 and 0.55.

The analytically reported values of indoor air forecast temperature show that there is a problem of micro-climatic parameters in the building, which is operated and in which the enclosure parts change their thermotechnical characteristics under the impact of various factors as well as the ageing of materials. The given procedure makes it possible to predict the indoor air temperature of rooms with different operational cycle. The presented method open up space for the multidimensional study in the field of constructional materials, sanitary science, as well as for discipline specialists of heat gas supply, ventilation and air conditioning.

References

1. Fanger, P.O.: Thermal Comfort. McGrow Hill, New York (1970)
2. Bogoslovskij, V.N.: Stroitel'naya teplofizika (teplofizicheskie osnovy otopleniya, ventilyacii i kondicionirovaniya vozduha): uchebnik dlya vuzov, 2-e izd., pererab. i dop. Vysshaya shkola (1982)
3. Chesanov, L.G., Petrenko, V.O.: Teploobmen cheloveka v pomeshchenii. In: Chesanov, L. G., Petrenko, V.O. (eds.) Stroitel'stvo. Materialovedenie. Mashinostroenie 2002, vyp. 15, pp. 169–171. Dn-sk, PGASA (2002)
4. Banhidi, L.: Teplovoj mikroklimat pomeshchenij. Strojizdat (1981)
5. Petrenko, V.O., Petrenko, A.O.: Faktory, yaki vplyvaiut na mikroklimat v prymishchenni, shcho maie defekty ohorodzhuiuchykh konstruktsii i system OVK. In: Petrenko, V.O., Petrenko, A.O. (eds.) Budivnytstvo. Materialovedennia. Mashynobuduvannia, vyp. 93, pp. 286–291. Dnipropetrovsk, PHASA (2016)
6. Kapalo, P., Zhelykh, V.: Investigation of indoor air quality in the selected Ukraine classroom–case study. In: Kapalo, P., Zhelykh, V. (eds.) International Conference Current Issues of Civil and Environmental Engineering, Lviv-Košice–Rzeszów, pp. 168–173. Springer, Cham (2019)
7. Vorob'eva, YU.A.: Vliyanie processa stareniya ograzhdayushchih konstrukcij i inzhenernyh sistem zhilyh zdanij na mikroklimat pomeshchenij: dis. k-ta tekhn. nauk: 05.23.03. Voronezhskij gos. arh.-str. un-t. Voronezh (2006)
8. Li, A.V.: Dolgovechnost' energoeffektivnyh polimersoderzhashchih ograzhdayushchih konstrukcij: dis. kand. tekhn. nauk: 05.23.01. Habarovsk (2003)

9. Petrenko, V., Dykarev, K.: Estimation of indoor temperatures on condition that building envelope is damaged. In: Petrenko, V., Dykarev, K. (eds.) CONFERENCE, vol. 1, pp. 36–44. Revista Romana de Inginerie Civila (2017)
10. Petrenko, V., Dykarev, K.: Evaluation of indoor temperature for various building envelopes damaged. In: Petrenko, V., Dykarev, K. (eds.) E3S Web of Conferences, vol. 32, p. 01019. EDP Sciences (2018)

Hydration Products that Provide Water-Repellency for Portland Cement-Based Waterproofing Compositions and Their Identification by Physical and Chemical Methods

Andrii A. Plugin[1], Olga S. Borziak[1(✉)], Oleksii A. Pluhin[1], Tatiana A. Kostuk[2], and Dmytro A. Plugin[1]

[1] Ukrainian State University of Railway Transport, Feuerbach sq. 7, Kharkiv 61050, Ukraine
borziak.olga@gmail.com

[2] Kharkiv National University of Civil, Engineering and Architecture, Sumska st., 40, Kharkiv 61002, Ukraine

Abstract. The coatings formed by water-proofing compositions of a penetrating action are characterized by high physical and mechanical properties and good adhesion to the substrates used for their deposition. The experimental investigations of the influence multicomponent chemical additive that provide waterproofness on the formation of the hydration products of Portland cement have been carried out. The colmatation of the pores and capillaries by the hydration products of Portland cement with the complex chemical additive were simulated. Such pores were simulated by the formation of the cement-perlite mortar and the capillaries were simulated by the introduction of fiberglass to the cement paste. The data of X-ray phase analysis, infrared spectroscopy, and thermal analysis show the formation of carbonate and chloride *AFm*-phases and possibly the formation of the nitrate *AFm*-phase and calcite in the cement stone with the complex chemical additive. The data of electron microscopy research also showed that the hydration of Portland cement with complex chemical additive results in the formation of crystallohydrates that according to their morphological features can be identified as carbonate-, chloride-, and nitrate *AFm*-phases, portlandite, and calcite that grow on silicate surfaces with a negative surface charge.

Keywords: Waterproofing composition · Colmatation · Porosity of concrete · X-ray phase analysis · Infrared spectroscopy · Thermal analysis

1 Introduction

In recent years the cement-based waterproofing compositions of a penetrative action with the multicomponent complex chemical additive are increasingly used for the repair, protection and hydraulic insulation of concrete, reinforced concrete and stone structures [1]. Such compositions are produced in the form of dry mixtures [1] or

Z. Blikharskyy (Ed.): EcoComfort 2020, LNCE 100, pp. 328–335, 2021.
https://doi.org/10.1007/978-3-030-57340-9_40

multicomponent cements [2] that are applied onto the surface using plastering methods. The structure of the detachment is identical to that of the substrates onto which these are applied; these compositions are characterized by a high adhesion degree and high physical and mechanical characteristics and survive a hydrostatic pressure of any sign both in the case of compression and avulsion. In addition, these compositions compact the surface layer of concrete by the colmatation of its porous space with the products of joint hydration of cement and additive components providing thus a high water-proofness for the coating on the whole [3]. However, the compositions available on the market have very often the unbalanced composition of chemical additives and it results in the formation of the efflorescences on the surface and the formation of the cracks in the coating and insufficient water tightness. It requires a very careful approach to the selection of the composition and the dose for complex chemical additive in order to control the composition and the structure of the hydration products of Portland cement containing it. Hence, the in-depth studies of the hydration products of Portland cement with the multicomponent chemical additive that provide an appropriate waterproofness are of great importance.

The use of mineral [4] and chemical [5] additives allows adjusting the structure [6] and properties [7, 8] of silicate composite materials. The developers and producers of waterproofing compositions usually do not disclose the complex chemical additive composition. In [1, 3], sodium nitrate $NaNO_3$, sodium sulfate Na_2SO_4, sodium carbonate Na_2CO_3, calcium nitrate $Ca(NO_3)_2$, calcium chloride $CaCl_2$, calcium hydroxide $Ca(OH)_2$, fine sulfonaphthalene formaldehyde SP-1, the steel corrosion inhibitor and the rust converter in the total amount of at least 5% of the cement mass were used as the components of such additive.

It is shown in [9] that the products of hydration C_3A (and C_4AF) of Portland cement in the presence of sulfates, chlorides and nitrates are the double salts of $C_3A \times CaX \times (10\text{-}12)H_2O$ and $C_3A \times 3CaX \times (14\text{-}32)H_2O$ types, where X is either a two-charge anion SO_4^{2-} or two singly-charged anions Cl^- or NO_3^-. In [9–11], these compounds were studied and these were defined as Friedel's salts in which Cl^- according to [11] can be substituted by other anions and in [12–14] these were defined as *AFm*- and *AFt*-phases, accordingly. In [9–14], the conditions for the formation of above compounds and their stability were defined using also the thermodynamic method [14]. Based mainly on the theoretical research data, it was shown in [1] that particularly *AFm*- и *AFt*-phases serve as the main crystallohydrates in Portland cement-based waterproofing compositions that provide water resistance. However, their quantity among other hydration products is insignificant and it complicates their detection using traditional physical and chemical methods. Therefore, the experimental investigations of these phases for the given waterproofing compositions and the improvement of the techniques for their detection still remain to be vital scientific problems.

2 Methods

Portland cements analogous to CEM I 42.5 N, CEM II/B-S 32.5 were used for the investigations. The effect of the complex chemical additive on the composition and structure of the products of their hydration was defined using physical and chemical research methods, in particular the X-ray phase analysis, infrared spectroscopy (IR-spectra), differential thermal analysis (DTA), thermal gravimetry (TG) and the electron microscopy. The changes in the intensity of diffraction maxima, absorption bands, the endo- and exoeffects of the minerals were analyzed on X-ray patterns, infrared absorption spectra, and the thermograms of the cement stone with the additive in comparison with the X-ray patterns, IR-spectra and the thermograms of the cement stone with no additives. To carry out the electron microscopy investigations we simulated the colmatation of the pores and capillaries that have the walls with a negative surface charge by the hydration products of Portland cement with the complex chemical additive. Such pores were simulated by the formation of the cement-perlite mortar and the capillaries were simulated by the introduction of fiberglass to the cement paste. The electron microscopy studies were done for the surface of cement stone using the scanning electron microscope JEOL JSM-6390/6390LV with the accelerating voltages of 5 to 21 kV and the magnification of $^{\times}100$ to 5000.

3 Results and Discussion

Figure 1 gives the X-ray patterns for the cement stone with no additives and with the complex chemical additive. The addition of complex chemical additive resulted in the following changes:

- judging mainly by the intensity of individual lines 2.60, 2.73, 2.77 Å, the hydration degree of silicate phases C_3S and C_2S in the cement stone with the additive dropped down in comparison with the reference composition with no additive;
- the amount of the portlandite decreased in proportion to silicate phases;
- the X-ray pattern of the cement stone with the complex chemical additive shows that the intensity of diffraction maxima of highly basic calcium hydrosilicates of a $C_2SH(A)$ type has decreased and it is indicative of a decrease in their amount;
- the tricalcium aluminate C_3A and the tetracalcium alumoferrite C_4AF were not detected in the cement stone with the complex chemical additive;
- the diffraction maxima appeared and it is indicative of the formation of calcite $CaCO_3$ that is conditioned by the interaction of $Ca(OH)_2$ with the sodium carbonate Na_2CO_3 and carbon dioxide CO_2 contained in the air;
- the diffraction maxima appear and it is indicative of the formation of calcium hydrochloraluminate $C_3A{\cdot}CaCl_2{\cdot}10H_2O$ and calcium hydrocarboaluminate $C_3A{\cdot}CaCO_3{\cdot}12H_2O$. The formation of calcium hydrochloraluminate $C_3A{\cdot}CaCl_2{\cdot}10H_2O$ is conditioned by the availability of calcium chloride $CaCl_2$ in the complex chemical additive and it correlates with the data [10, 11, 14] on its formation due to the interaction of C_3A and $CaCl_2$. The formation of calcium hydrocarboaluminate $C_3A{\cdot}CaCO_3{\cdot}12H_2O$ is conditioned by the availability of sodium carbonate Na_2CO_3 and calcium hydroxide $Ca(OH)_2$ in the complex chemical additive and it is correlated with the data [12, 13] on its formation due to the interaction of C_3A with the given compounds and carbon dioxide CO_2 contained in the air.

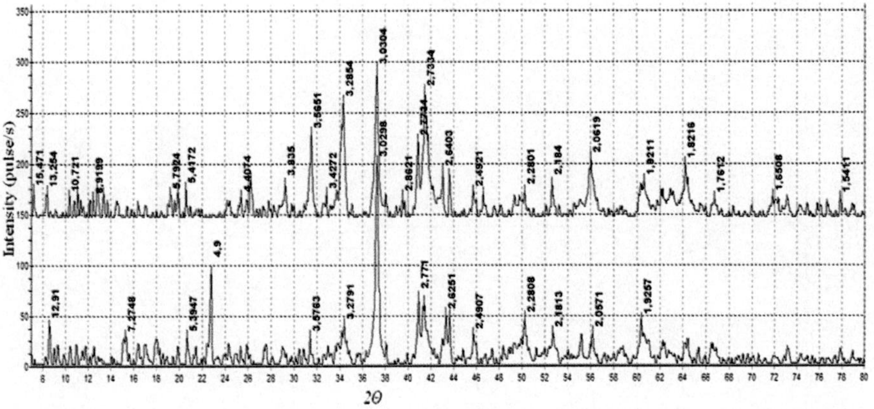

Fig. 1. X-ray patterns of the cement stone with the complex chemical additive (upper) and with no additives (lower)

Hence, the X-ray phase analysis data show that the addition of the complex chemical additive to Portland cement that includes carbonates, chlorides, nitrides, calcium sulfates and sodium sulfates provides the formation of calcium hydrocarboaluminate $C_3A{\cdot}CaCO_3{\cdot}12H_2O$ and calcium hydrochloraluminate $C_3A{\cdot}CaCl_2{\cdot}10H_2O$ and possibly calcite $CaCO_3$.

Figure 2 shows the infrared spectra of the absorption of the Portland cement-based compositions with the complex chemical additive and with no additives. The IR-spectra show the absorption bands with wave numbers, cm^{-1} for the bonds:

- the 3441, 3425, 2920, 2852 cm^{-1} bands are peculiar for the valent vibrations of the bond *OH*-groups for all hydrates;
- the 1630 cm^{-1} band are peculiar for the valent vibrations of H_2O molecules;
- the 1440 cm^{-1} band (intensive, superimposed on 1477–1479 cm^{-1}) and 876 cm^{-1} (narrow intensive) are peculiar for the valent vibrations of carbonates;
- the 1385 cm^{-1} band (narrow intensive) – are peculiar for the valent vibrations of the *N–O* bonds of nitrates;
- the 1082 cm^{-1}, 1084 cm^{-1} bands (intensive) and 462 cm^{-1} – valent and deformation vibrations of the *Si–O* bond of silicates independently of the type of crystalline lattice (and quartz as well).

We can see that IR-spectra show the availability of the absorption bands for the bonds and groups peculiar for Portland cement hydration products. The main distinctive feature is the availability of the band in the absorption spectrum of the cement stone with the complex chemical additive that is peculiar for the nitrates and a change in the shape and intensity of the bands typical for the silicates. A rather intensive band of 1082 cm^{-1} in the spectrum of the reference specimen of the cement stone acquires a doublet shape with the bands of 1083 and 1035 cm^{-1} of an average intensity in the spectrum of the specimen with the additive. This circumstance confirms a decrease in the amount of highly basic calcium hydrosilicates and an increase in the amount of low-basic calcium hydrosilicates.

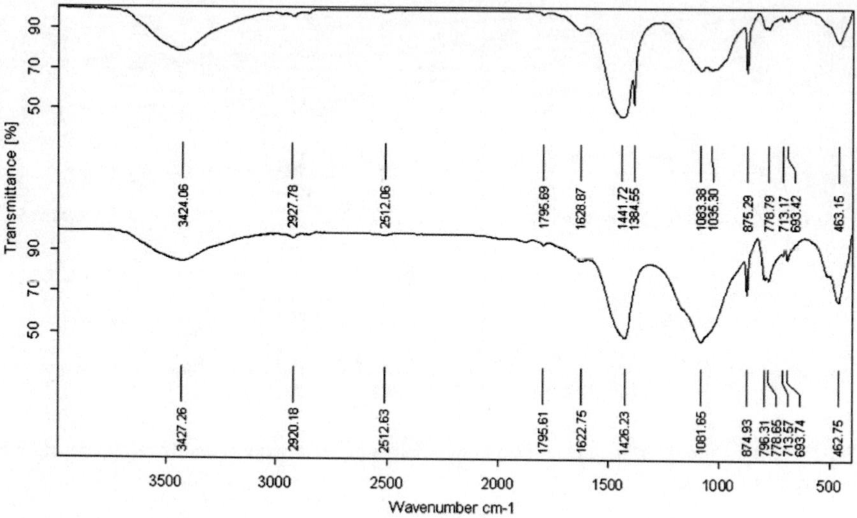

Fig. 2. Infrared absorption spectra for the cement stone with the complex chemical additive (upper) and with no additives (lower)

Hence, the analysis data of infrared absorption spectra can be indicative of the formation of nitrate *AFm*-phase. The most informative is the narrow intensive absorption band of 1385 cm^{-1} that can be indicative of the formation of such nitrate *AFm*-phase.

Figure 3 gives the thermograms of the cement stone with no additive and with the complex chemical additive.

The addition of complex chemical additive results in such changes:

- the intensity of the endoeffect at 160–170 °C and the mass loss at 20–500 °C decreased significantly, which may indicate a decrease in the amount of highly basic calcium hydrosilicates $C_2SH(A)$ with the basicity of C/S 1,5 and higher;
- a smooth endoeffect appeared at 180–230 °C that is indicative of the formation of $C_3A{\cdot}CaCO_3{\cdot}12H_2O$ (the endoeffect is typical for the temperature range of 190 to 210 °C) and $C_3A{\cdot}CaCl_2{\cdot}10H_2O$ (the endoeffect is typical for the temperature range of 180 to 230 °C);
- the presence of endoeffects at 620 °C and 835 °C may be due to the oxidation of sulfur-, carbon-containing components and the dehydration of *CASH* structures;
- the endoeffect is available in the cement stone with no additive and with the complex chemical additive at 490 °C and it confirms the availability of portlandite $Ca(OH)_2$ both in the cement stone with no additives and with the complex chemical additive;
- the exoeffect in the temperature range of 800 to 835 °C can be indicative of the formation of the to bermorite-like low-base calcium hydrosilicates of a *CSH(B)* type with the basicity of $C/S = 0.8 - 1.25$;
- the endoeffect intensity at 860 °C increased significantly that is indicative of the availability of calcite $CaCO_3$ possibly after decomposition of $C_3A{\cdot}CaCO_3{\cdot}12H_2O$.

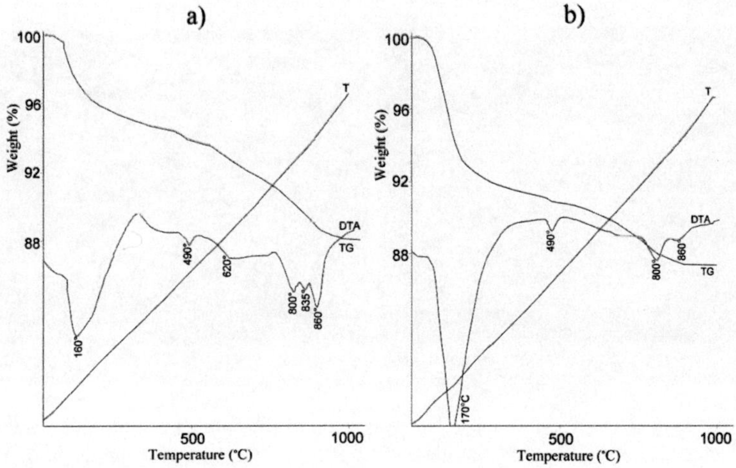

Fig. 3. Cement stone thermograms: a – with the complex chemical additive; b – with no additives

Hence, the thermal analysis data proved that the addition of complex chemical additive to Portland cement that includes carbonates, chlorides, nitrates, calcium and sodium sulfates provides the formation of carbonate and chloride *AFm*-phases and a decrease in the basicity of calcium hydrosilicates in the cement stone.

The electron microscopy snapshots of the cement-perlite mortar and the cement stone with the fiberglass are given in Fig. 4. The expanded perlite mainly contains the open pores of 25 to 100 μm in its surface layer and the cement stone with no additives occupies mainly the place around its grains. The hydration products of Portland cement are only observed in some perlite pores. Figure 4a shows that the hydration products of Portland cement with complex chemical additive occupy also the perlite pores. The hydration products have a granular structure formed by the particles with the shape similar to cubic particles, apparently calcites with the size of 0.2 to 2 μm and the aggregates formed by them. Figure 4a also shows the plates similar to hexagonal that form layers on each other. These hexagonal plates with the size of 0.4 to 3.5 μm, mostly of 0.8 to 1.5 μm, and the thickness of 0.1 to 0.15 μm together with spherolites and needles are peculiar for the calcium hydrocarboaluminates, calcium hydrochloraluminates and calcium hydrosulfoaluminates.

The Fig. 4b shows that the gaps between the fiberglasses are overgrown with hexagonal crystals peculiar for portlandite and calcium hydromonosulfoaluminate and with cubic crystals with the size of up to 2 μm peculiar for calcium hydroaluminates and calcite. At the age of 28 days the gaps are totally overgrown with hydration products. Crystalline phases have completely been formed mainly by hexagonal crystals peculiar for portlandite and calcium hydromonosulfoaluminate, by prismatic dendrites of calcium hydroaluminates with a spatial structure and the crystals of rhombohedral calcite prisms with the size of up to 10 μm.

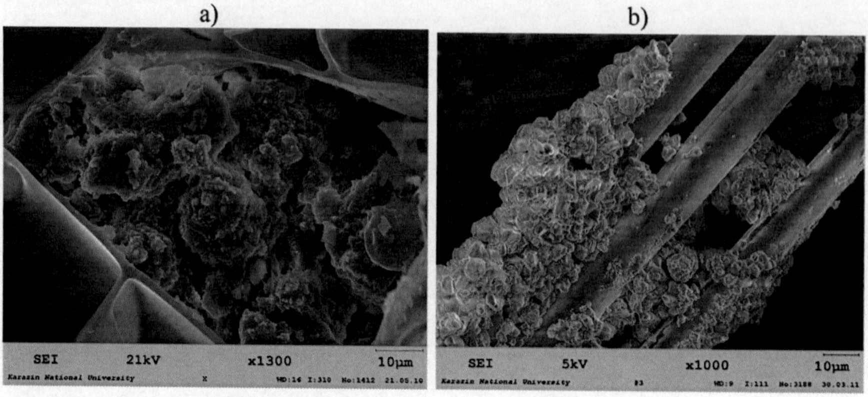

Fig. 4. Scanning electron microscopy. The simulation of the colmatation of open pores (a) and the capillaries (the gaps between the glass fibers) (b) by the hydration products of Portland cement with complex chemical additive.

Hence, the electron microscopy investigations showed that crystalline cement hydration products with the complex chemical additive are calcium hydrocarboaluminates, calcium hydrochloraluminates, calcium hydronitroaluminates, portlandite, and calcite that have a positive surface charge and due to this fact these grow on the negatively charged silicate substrates, i.e. on the pore walls of expanded perlite, and fiberglass; in the hardened cement-perlite solution the perlite pores remain unfilled, while in the composite with complex chemical additive the perlite pores are overgrown by dense crystalline structures. Hence, the hydration products that provide water-tightness can be identified by their morphological features. These are hexagonal crystals that are peculiar for calcium hydromonosulfoaluminate and portlandite and cubic crystals with the size of up to 2 μm that are peculiar for calcium hydroaluminates and calcite; hexagonal plates with the size of 0.4 to 3.5 μm and 0.1–0.15 μm thick that form layers on each other, spherolites and needles peculiar for calcium hydrocarboaluminates, calcium hydrochloraluminates and calcium hydrosulfoaluminates.

4 Conclusions

The data of X-ray phase analysis, infrared spectroscopy, and thermal analysis show the formation of carbonate and chloride *AFm*-phases and possibly the formation of the nitrate *AFm*-phase and calcite in the cement stone with the complex chemical additive.

The data of X-ray studies of the cement stone with the complex chemical additive confirm the formation of calcium hydrocarboaluminate $C_3A \cdot CaCO_3 \cdot 12H_2O$, calcium hydrochloraluminate $C_3A \cdot CaCl_2 \cdot 10H_2O$ and possibly calcite $CaCO_3$. The thermal analysis and infrared spectroscopy data proved that the addition of complex chemical additive to Portland cement provides the formation of carbonate and chloride *AFm*-phases and a decrease in the basicity of calcium hydrosilicates.

The electron microscopy research data showed that the hydration of Portland cement with complex chemical additive results in the formation of crystallohydrates that according to their morphological features can be identified as carbonate-, chloride-, and nitrate *AFm*-phases, portlandite, and calcite that grow on silicate surfaces with a negative surface charge.

References

1. Plugin, A.A., Kostuk, T.O., Proshhy`n, O.Yu., Bondarenko, D.O., Pluhin, O.A., Borziak, O.S., Arutyunov, V.A.: Gidroizolyacijni cementni kompozy`ty` prony`knoyi diyi (Waterproofing cement composites). Kolegium, Kharkiv (2018)
2. Ivashchyshyn, H., Sanytsky, M., Kropyvnytska, T., Rusyn, B.: Study of low-emission multicomponent cements with a high content of supplementary cementitious materials. East.-Eur. J. Enterp. Technol. **4/6**(100), 39–47 (2019)
3. Plugin, A.A., Pluhin, O.A., Borziak, O.S., Kaliuzhna, O.V.: The mechanism of a penetrative action for portland cement-based waterproofing compositions. In: Proceedings of CEE 2019. Lecture Notes in Civil Engineering, vol. 47, pp. 34–41 (2020)
4. Krivenko, P., Sanytsky, M., Kropyvnytska, T.: The effect of nanosilica on the early strength of alkali-activated Portland composite cements. Solid State Phenom. **296**, 21–26 (2019)
5. Krivenko, P., Petropavlovskyi, O., Rudenko, I., Konstantynovskyi, O.: The influence of complex additive on strength and proper deformations of alkali-activated slag cements. In: Materials Science Forum, vol. 968, pp. 13–19. Trans Tech Publications Ltd., Switzerland (2019)
6. Dhandapani, Y., Santhanam, M.: Investigation on the microstructure-related characteristics to elucidate performance of composite cement with limestone-calcined clay combination. Cement and Concrete Research 129, art. 105959 (2020)
7. Kharitonov, A., Smirnova, O.: Optimization of repair mortar used in masonry restoration. Spatium **42**, 8–15 (2019)
8. Krivenko, P., Gots, V., Petropavlovskyi, O., Rudenko, I., Konstantynovskyi, O., Kovalchuk, A.: Development of decisions for alkali-activated cements proper deformations control. East.-Eur. J. Enterp. Technol. **5–6**, 24–32 (2019)
9. Taylor, H.F.W.: Cement Chemistry. Academic Press, London (1990)
10. Rapin, J.P., Elkaim, E., Francois, M., Renaudin, G.: Structual transition of Friedel's salt $3CaO \cdot Al_2O_3 \cdot CaCl_2 \cdot 10H_2O$ studied by synchrotron powder diffraction. Cem. Concr. Res. **32**, 513–519 (2002)
11. Andersen, M.D., Jakobsen, H.J., Skibsted, J.: Characterization of the α-β phase transition in Friedels salt ($Ca_2Al(OH)_6Cl \cdot 2H_2O$) by variable-temperature ^{27}Al MAS NMR spectroscopy. J. Phys. Chem. A **106**(28), 6676–6682 (2002)
12. Matschei, T., Lothenbach, B., Glasser, F.P.: The AFm phase in Portland cement. Cem. Concr. Res. **37**(2), 118–130 (2007)
13. Borziak, O.S., Plugin, A.A., Chepurna, S.M., Zavalniy, O.V., Dudin O.A.: The effect of added finely dispersed calcite on the corrosion resistance of cement compositions. In: IOP Conference on Series: Materials Science and Engineering, vol. 708, p. 012080. IOP Publishing, Bristol (2019)
14. Balonis, M., Medala, M., Glasser, F.P.: Influence of calcium nitrate and nitrite on the constitution of the AFm and AFt cement hydrates - experiments and thermodynamic modelling. Adv. Cem. Res. **23**(3), 129–143 (2011)

Humidity, Air Temperature, CO_2 and Well-Being of People with and Without Green Wall

Zuzana Poorova and Zuzana Vranayova(✉)

Technical University of Kosice, 04002 Kosice, Slovakia
zuzana.vranayova@tuke.sk

Abstract. The paper presents an experiment on green wall performed to apprehend its thermal and hydrological behavior and its impact inside building. The experiment is based on a living wall set up in a classroom of the faculty in TUKE campus in Košice where the interior green wall is situated. Monitoring of temperature, humidity and CO_2 variations within the living wall and a reference case enable us to analyze effects of green walls. During the measurements, set of questions were answered. The data of respondents are used for gaining the goal of this interdisciplinary research, the effect of green wall on the well-being of people. The measurements were carried out in the classroom between January 04, 2018 and February 08, 2018. It can be stated that women are more sensitive to changes than men. Following the measurements, the green wall is very favorable for the indoor environment.

Keywords: Green wall · Experiment · Questionnaire · Case study · Well-being · Indoor

1 Introduction

The benefit of living green dividers is a surefire approach to upgrade a building's visuals, enhance air quality and also representative sharpness and vitality levels. Over the past 50 years, a remarkable increment of urban-living searchers has prompted an extensive uptick in air contamination and loss of green spaces. Living green dividers (additionally usually alluded to as vertical gardens or living dividers) are a superb answer for any property keen on enhancing their space with characteristic advantages of nature. They offer a moving and tastefully captivating characteristic lift to worker resolve [1]. Regardless of whether they are introduced on the outside or inside of a building, the structures of absolutely real vegetation make the "wow factor" [2] such a significant number of inside architects look for while championing manageability.

The paper describes the experiment in the classroom of the faculty in TUKE campus in Košice where the interior green wall was situated. The measurements were carried out in the chosen classroom between January 04, 2018 and February 08, 2018 between 08:00 and 09:15 am. This paper presents findings on effects of green on people and environment [3].

Z. Blikharskyy (Ed.): EcoComfort 2020, LNCE 100, pp. 336–346, 2021.
https://doi.org/10.1007/978-3-030-57340-9_41

2 Experiment Site

Košice city (Fig. 1), is at an altitude of 206 m above sea level and covers an area of 242.77 km^2. It is located in eastern Slovakia. Košice has a humid continental climate, as the city lies in the North Temperate Zone. The city has four distinct seasons. Precipitation varies little throughout the year with abundance precipitation that falls during summer and only few during winter. The coldest month is January, with an average temperature of −2.6 °C and the hottest month is July, with an average temperature of 19.3 °C (Figs. 2, 3, 4, 5 and 6) [4].

Fig. 1. Green wall experiment location

The months May, June, July, August and September have a nice average temperature. On average, the warmest month is July. On average, the coolest month is January. The average annual maximum temperature is: 13.0 °C. The average annual minimum temperature is: 3.0 °C.

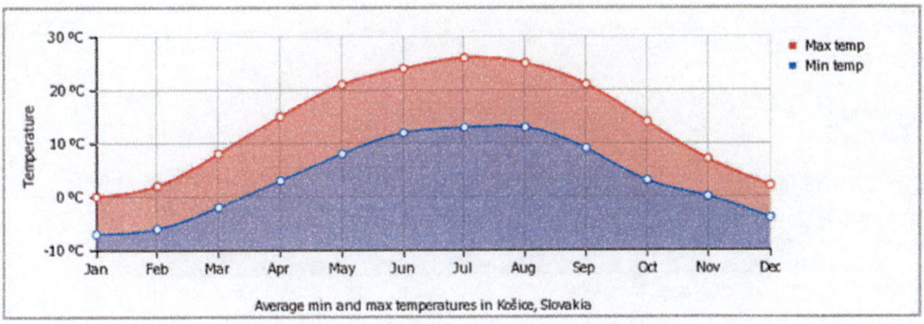

Fig. 2. Average minimum and maximum temperature over the year [4]

On average, July is the sunniest. On average, December has the lowest amount of sunshine.

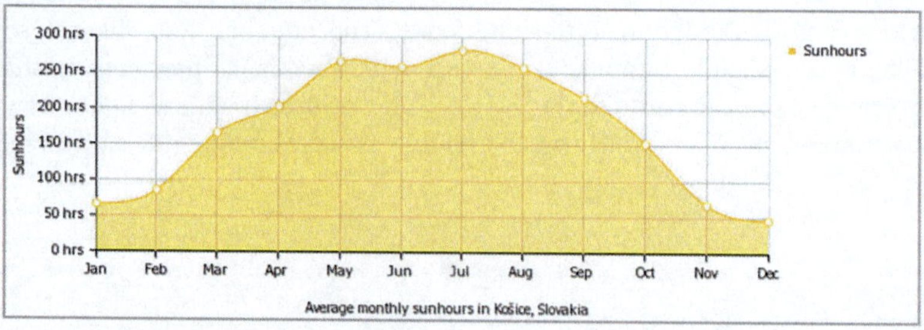

Fig. 3. Average monthly hours of sunshine over the year [4]

On average, July is the wettest month. On average, January is the driest month. The average amount of annual precipitation is: 619.0 mm.

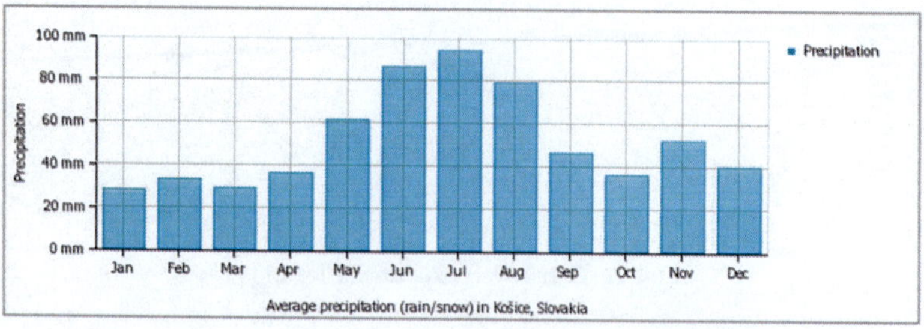

Fig. 4. Average monthly precipitation over the year (rainfall, snow) [4]

Most rainy days are in December. On average, December is the rainiest. On average, September has the least rainy days. The average annual amount of rainy days is: 147.0 days.

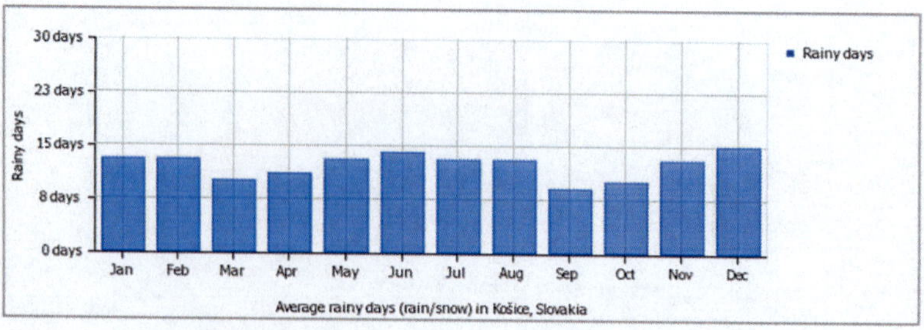

Fig. 5. Average monthly rainy days over the year [4]

On average, December is the most humid. On average, May is the least humid month. The average annual percentage of humidity is: 62.0%.

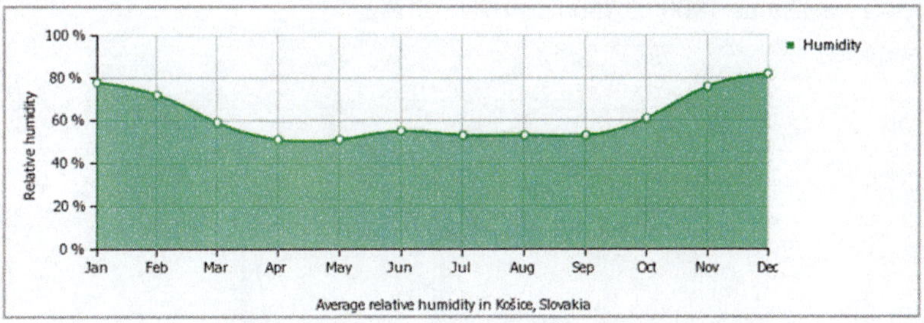

Fig. 6. Average humidity over the year [4]

3 Methodology

3.1 Classroom

The aim of the research is to analyze impact of green wall on its environment. The quality may be affected by materials of the research area. The list of materials, area and percentage of each material is listed in Table 1.

Table 1. Classroom material characteristics

Material	Object	Area	Real area m^2	Total area m^2	Percentage %
Engineered wood	OSB board	Floor	59.2	59.2	24.6
Plaster	Wall	Ceiling	61.2	106.0	44.0
		North wall	17.5		
		East wall	9.8		
		South wall	17.5		
Glass	Glass wall + Entrance door	West wall	36.2	56,9	23.6
	Windows	East wall	20,7		
Steel	Heaters	East wall	4.5	4.5	1.8
Wood	Glass wall + Entrance door	West wall	4.1	6.4	2.7
	Windows	East wall	2.3		
Plants	Green wall	North wall	3.0	7.9*	3.3
TOTAL			236.0	240.9	100

Note: *real area after calculating leaf areas.

3.2 Green Wall Construction

Size of the one green wall bearing construction is 1000 × 750 × 1750 mm. On this bearing construction made of iron jackels; diameter 50 × 50 mm; OSB board is attached; dimensions 1000 × 1500 × 10 mm (Fig. 7).

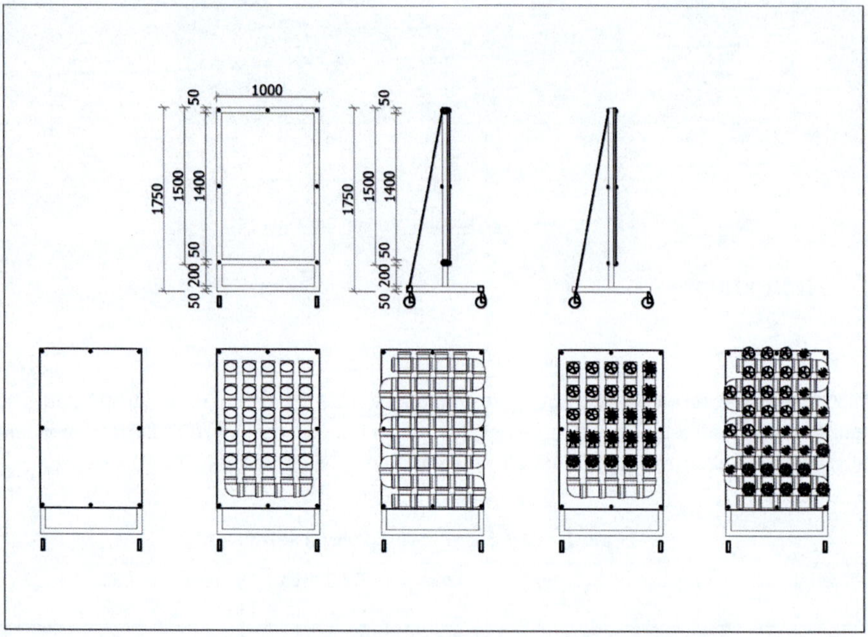

Fig. 7. TUKE self made green wall – front and right side view on both green wall designs

On these OSB boards plastic drainage pipes; diameter 110 mm; are attached. Total number of the plastic drainage pipes is 58, 30 on one board and 28 on the second (Fig. 8). These plastic drainage pipes were pre-joined into specific design (Fig. 8) and then attached to the board using straps usually used for attaching the downspout to the building facade. The idea of these green walls was to use specific material – plastic, which can be re-used.

Used flowers were pre-grown. Totally 6 types of flowers were used. Number of each species is: Dryopteris Erythrosora 'Brilliance' 19 pieces [5], Scindapsus Aureus 9 pieces [6], Aglaonema 'Silver Queen' 10 pieces [7], Philodendron Hederaceum 8 pieces [8], Chlorophytum Comosum 'Variegatum' 4 pieces [9], Anthurium Andraeanum 8 pieces [10].

These pre-grown flowers were in separate plastic pots. Each pot with its filtration layer, substrate, ceramsite layer and flower was placed in the plastic drainage pipe following the desired design of the green wall.

Fig. 8. Plastic pipes used for creating the experiment green wall

The big change in the same environment is noticeable on Fig. 9. The students were not informed about installing these green walls in the classroom. After a week of the presence of these green walls in the classroom, the questionnaire was distributed among students and teachers of the faculty.

Fig. 9. Classroom with a) and without the green wall b)

4 Results

4.1 Experiment with 20 Respondents

The aim of this measurement was to find out how the green wall affects the microclimate, environment and the well-being (Merriam-Webster 2017; Chaumillon et al. 2017) [11, 12]. The experiment was performed during days when people were in the room from 8:00 am to 9:15 am. The respondents were students and teachers. The number of participants was 185, 76 were female and 109 male. First question in the questionnaire was the perception of the temperature in the room where the answer was: cold, moderately cold, appropriate temperature, moderate warm, warm. Second was about odor of air in the room respondents perceive no odor, moderate weak odor, appropriate odor, moderate strong odor, strong odor. And the last question was to find out whether the given air quality in the room was satisfactory or unsatisfactory (Francis and Lorimer 2011 [3]; Malys et al. 2014 [13]; Otteléa et al. 2011 [14]; Medl et al. 2017. [15])

4.2 Effect on Well-Being

Effect on Respondents, Change in Temperature

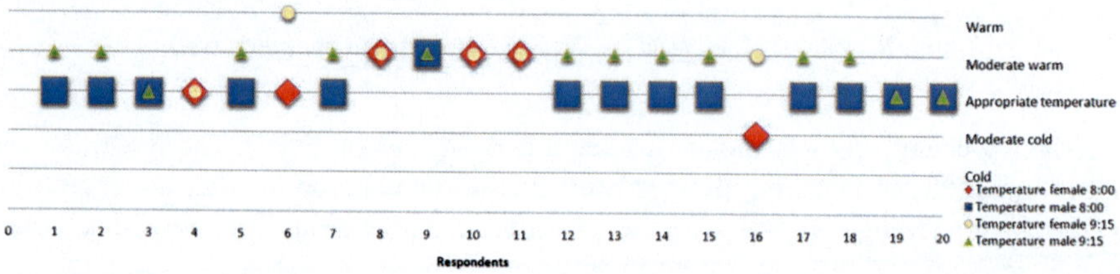

Fig. 10. Change in temperature, classroom **with** green wall January 04, 2018

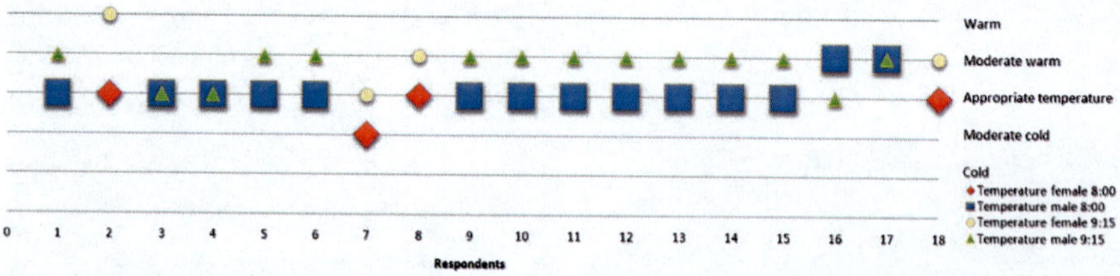

Fig. 11. Change in temperature, classroom **without** green wall January 18, 2018

We can see that men perceived temperature increase at the end of the measurement in both cases, whereas for women it is more individual, but also perceived a higher temperature at the end of the measurement. The effect of flowers in this measurement did not significantly change the temperature change.

Effect on Respondents, Change in Odor

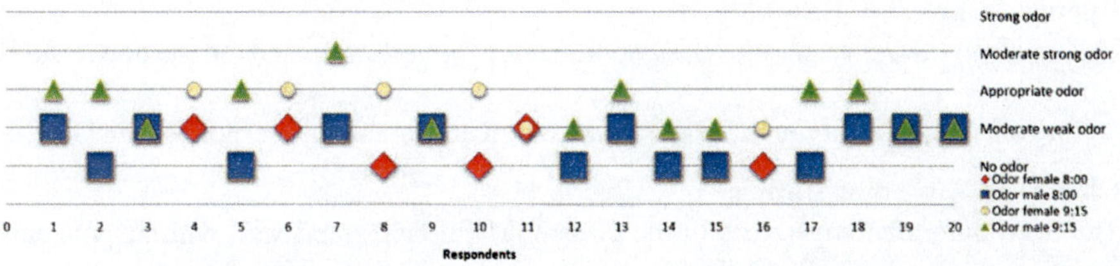

Fig. 12. Change in odor, classroom **with** green wall January 04, 2018

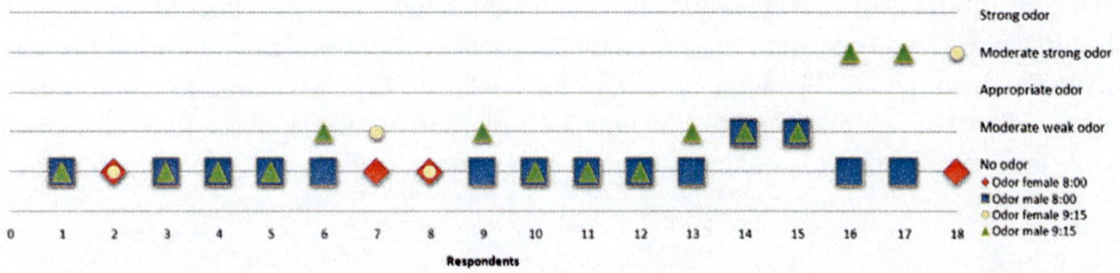

Fig. 13. Change in odor, classroom **without** green wall January 18, 2018

While observing the odor change, we can notice that in the room where there was a green wall with flowers, the respondents perceived the odor change more. Men and women perceived the odor change as well.

Effect on Respondents, Change in Air Quality

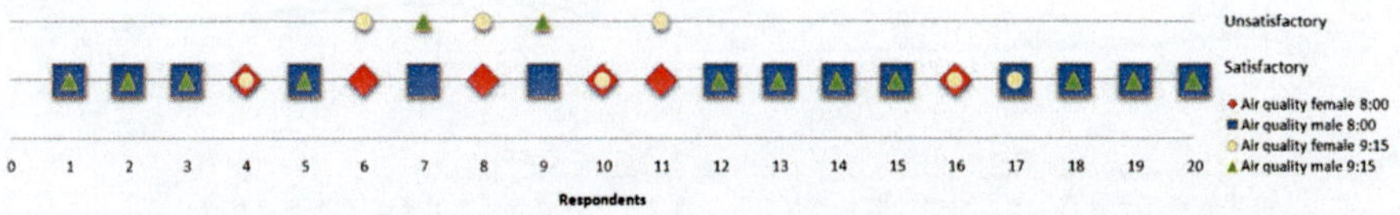

Fig. 14. Change in air quality, classroom **with** green wall January 04, 2018

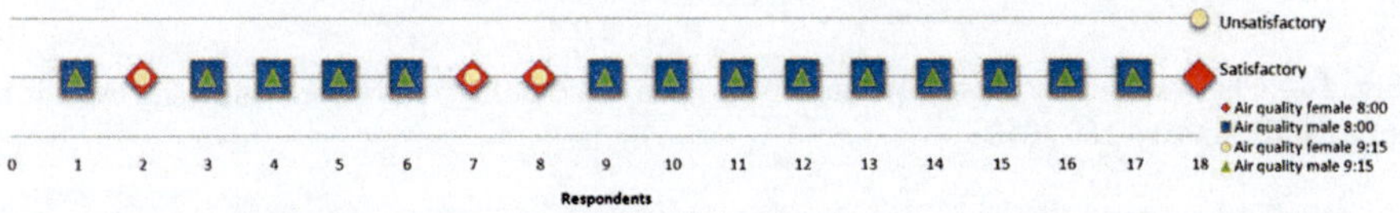

Fig. 15. Change in air quality, classroom **without** green wall January 18, 2018

With the green wall 5 respondents rated the air quality as unsatisfactory and without the green wall only one respondent rated air quality as unsatisfactory.

Change in Relative Humidity

The relative humidity in the classroom without green wall and no respondents is 28.7%.

The relative humidity in the classroom with green wall and no respondents is 30.1%.

Change in Air Temperature

The maximum air temperature in the classroom without green wall with respondents during the stay was 37.2 °C. The maximum air temperature reached in the classroom with green wall with respondents during the stay was 35.5 °C.

Change in CO_2

The CO_2 production of respondents has been calculated - and the effect of green wall: The increase in CO_2 concentration in the classroom without green wall during the stay of respondents was 8.29 mg/person. The increase in CO_2 concentration in the class-room with green wall during the stay was 7.10 mg/person. In the classroom with green wall, the increase in CO_2 concentration is 14% lower.

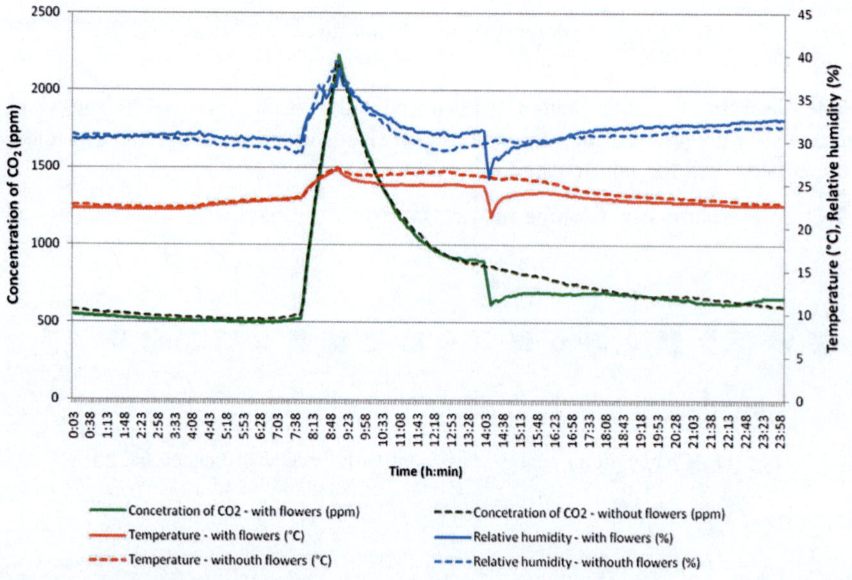

Fig. 16. Change in CO_2, air temperature and relative humidity, classroom **with** and **without** green wall January 18, 2018

5 Conclusions

The measurement on effects on well-being was carried out using a subjective method - a questionnaire, where respondents evaluated the internal microclimate at the beginning of the stay in the room and at the end of the stay in the room. It can be stated that women are more sensitive to changes than men. Following the measurements, it can be stated that the green wall is very favorable for the indoor environment. According to The Decree of the Ministry of Health of the Slovak Republic No. 210/2016 Coll. on the details of the requirements for the indoor environment of buildings and the minimum requirements for apartments of lower standard and accommodation facilities requires the optimum relative air humidity in rooms such as classrooms, hotels, theaters, from 30% to 70%. In our case, relative air humidity was measured 28.7% in the room without the green wall, which is just above the minimum required. In a room with a green wall, the relative air humidity was measured 30.1%. From a relative humidity perspective, green walls appear to be beneficial in building indoor buildings.

Acknowledgements. This work was supported by: APVV-18-0360 ACHIEve Active hybrid infrastructure towards to sponge city and SWAMP - Zodpovědný management vody v intravilánu obce ve vztahu k okolní krajině (č. CZ.02.1.01/0.0/0.0/16_026/0008403).

References

1. Benvenuti, S., Malandrin, V., Pardossi, A.: Germination ecology of wild living walls for sustainable vertical garden in urban environment. Sci. Hortic. **203**, 185–191 (2016)
2. Hoyle, H., Hitchmough, J., Jorgensen, A.: All about the 'wow factor'? The relationships between aesthetics, restorative effect and perceived biodiversity in designed urban planting. Landscape Urban Plan. **164**, 109–123 (2017). ISSN 0169-2046
3. Francis, R.A., Lorimer, J.: Urban reconciliation ecology: the potential of living roofs and walls. J. Environ. Manag. **92**, 1429–1437 (2011)
4. Climate Homepage: average monthly weather in Kosice, Slovakia online. www.weather-and-climate.com/average-monthly-Rainfall-Temperature-Sunshine. Accessed 01 Jan 2020
5. Homepage Missouri Botanical Garden: Dryopteris erythrosora 'Brilliance' online. www.missouribotanicalgarden.org/PlantFinder/PlantFinderDetails.aspx?kempercode=e149. Accessed 01 Jan 2020
6. Homepage Missouri Botanical Garden: Chlorophytum comosum 'Variegatum' online. www.missouribotanicalgarden.org/PlantFinder/PlantFinderDetails.aspx?kempercode=b547. Accessed 01 Jan 2020
7. Homepage Missouri Botanical Garden: Aglaonema 'Silver Queen' online. www.missouribotanicalgarden.org/PlantFinder/PlantFinderDetails.aspx?taxonid=243522&isprofile=0&hf=1. Accessed 01 Jan 2020
8. Homepage Missouri Botanical Garden: Scindapsus Aureus online. www.missouribotanicalgarden.org/PlantFinder/PlantFinderDetails.aspx?kempercode=b594. Accessed 01 Jan 2020
9. Homepage Missouri Botanical Garden: Philodendron hederaceum online. www.missouribotanicalgarden.org/PlantFinder/PlantFinderDetails.aspx?kempercode=b611. Accessed 01 Jan 2020

10. Homepage Missouri Botanical Garden: Anthurium andraeanum online. www.missouribotanicalgarden.org/PlantFinder/PlantFinderDetails.aspx?taxonid=276219&isprofile=0&hf=1. Accessed 01 Jan 2020
11. Homepage Merriam-Webster Questionnaire online. www.merriam-webster.com/dictionary/questionnaire. Accessed 01 Jan 2020
12. Chaumillon, R., Romeas, T., Paillard, C., Bernardin, D., Giraudet, G., Bouchard, J.F., Fauberta, J.: Enhancing data visualisation to capture the simulator sickness phenomenon: on the usefulness of radar charts. Data Brief **13**, 301–305 (2017). ISSN 2352-3409
13. Malys, L., Musya, M., Inard, C.: A hydrothermal model to assess the impact of green walls on urban microclimate and building energy consumption. Build. Environ. **73**, 187–197 (2014)
14. Otteléa, M., Perini, K., Fraaij, A.L.A., Haas, E.M., Raiteri, R.: Comparative life cycle analysis for green facades and living wall systems. Energy Build. **43**, 3419–3429 (2011)
15. Medl, A., Mayr, S., Rauch, H.P., Weihs, P., Florinetha, F.: Microclimatic conditions of 'green walls', a new restoration technique for steep slopes based on a steel grid construction. Ecol. Eng. **101**, 39–45 (2017)

Production of Fly Ash Aerated Concrete and Efficiency of Its Application

Oksana Pozniak(✉), Volodymyr Melnyk, Igor Margal, and Petro Novosad

Lviv Polytechnic National University, Lviv 79013, Ukraine
oksana.r.pozniak@lpnu.ua

Abstract. The literature sources on the production and energy efficiency of non-autoclaved aerated concrete application are analyzed. The efficiency of use in the composition of aerated concrete as a filler of fly ash and light microsphere (cenosphere) is compared. It is shown that at the same content of source materials the use of fly ash as a filler allows us to obtain aerated concrete with an average density of 969 kg/m^3 and a strength of 3.5 MPa. Substituting the fly ash in aerated concrete composition makes it possible to reduce its average density to 440 kg/m^3 with a strength of 1.14 MPa. The pore structure of the aerated concrete samples is investigated. The complex estimation of the efficiency using the developed fly ash aerated concrete in the construction of the enclosing structures of residential buildings is made. The efficiency is evaluated taking into account the costs (including the energy component) at different stages of the life cycle of the building: production, construction, operation. Selection of individual elements and the optimal design of the external enclosure of the building, based on the proposed indicator of the cost of unit saving energy, and with account of the cost of construction and installation works and operational characteristics of individual elements, is performed.

Keywords: Aerated concrete · Light microsphere · Average density · Saving energy · Strength · Energy efficiency

Improvement of the energy efficiency of buildings is now considered to be the most important task of preserving the environment and reducing energy consumption. The major concern of scientists from all over the world is creating an environment-friendly and low-energy production of composite materials. That is why there is a growing tendency to reduce energy consumption when designing energy efficient porous materials and using them in construction. Production of non-autoclaved aerated concrete from the energy industry waste is a progressive technology of obtaining new building material with good properties. The advantage of the technology of non-autoclaved cement concrete is its low-cost and low-energy consumption, environmentally friendly production, lower costs for stable quality of production.

Z. Blikharskyy (Ed.): EcoComfort 2020, LNCE 100, pp. 347–352, 2021.
https://doi.org/10.1007/978-3-030-57340-9_42

1 Introduction

One of the most important directions in the modern building material science is development and introduction of new effective heat insulating materials, which is mainly due to the growth of electricity rate and cost of energy needed for heating of buildings. The non-autoclaved aerated concrete (NAAC) has the potential of being an alternative to ceramic brick or ordinary concrete, as it reduces loads on the structure and foundation, contributes to energy conservation, and lowers the cost of production and labor cost during the construction and transportation. Lower density of aerated concrete has a good impact on environment, because lower weight provides reduction of waste from buildings made from aerated concrete in the future. The aerated concrete is the best construction material, which has a very low λ value [W/mK] [1–3].

The analysis of the scientific literature shows that the properties of aerated concretes are defined by the character of binders and fillers properties [4, 5]. The non-autoclaved aerated concrete with a modified solid component contains a lime-carbonate additive with calcium carbonate (calcite), calcium hydroxide (portlandite) and the additive with a plasticizing and accelerating effect, having a density of 500 kg/m^3 and a maximum compressive strength of 3.53 MPa, corresponding to concrete of class C2 in line with current standard, is obtained [5]. The possibility of production of the aerated concrete of a density of 350 [kg/m^3] by the PGS process technology with application as a binder quick lime, gypsum and some part of fly ash is studied [1]. The addition of the pozzolanic materials (silica fume, zeolite, metakaolin and granulated blast-furnace slag) can effectively improve the mechanical properties and decrease the water absorption of the autoclaved aerated concrete [6, 7].

The aerated concrete could be produced with using various aggregates, like sand, fly ash, flax straw or other industrial wastes [8]. Test results of thermal conductivity clearly shows that aerated concrete based on siliceous fly ash has better λ value than the sand aerated concrete of the same density. Papers [9, 10] show that replacement of sand for inactive silica to active silica sources such as silica fume and rice husk ash exhibits the highest mechanical properties aerated concrete compared to that containing silica sand. The possibility of autoclaved aerated concrete production with the addition of different kinds of glass cullet as a partial re-placement of sand was investigation [11]. The expanded perlite waste with the use of a quartz sand replacement in conventional autoclaved aerated concrete causes a unit weight decrease in the produced aerated concrete and change in it's properties [12].

The aerated concrete is one of the most sustainable building materials today. Aerated concrete is generally used to manufacture building envelopes (external and internal walls, floors, lintels), which together must meet the requirements for their bearing capacity and longevity, thermal resistance, soundproofing, fire resistance, water-, damp- and airproofing, as well as the requirements for the energy-saving, costs, and labor capacity [13]. Implementation of the effective building materials of the construction in terms of saving cost, energy, materials, and as a consequence, financial resources, with improved thermal insulation indicators is the strategic direction of the modern construction development.

2 Materials and Methods

The Portland cement CEM I 42,5 R produced at PJSC "Volyn-cement" with specific surface of 330 m^2/kg, the retaining on sieve №008 is 3,2%, the initial setting time is 1 h 50 min, the final setting time is 5 h 40 min, was used for the experiments. The fly ash (FA) from the Burshtyn TPP and light microsphere (cenosphere) are used for aerated concrete production as a finely dispersed filler. The fly ash from Burshtynska thermal power plant has the following properties: true density – 2.22 g/cm^3; bulk density – 910 kg/m^3; retaining on sieve № 008 – 10.2 mass%; chemical composition, mass%: $SiO_{2-5}3$; Al_2O_3 – 25.32; Fe_2O_3 + FeO – 13.91; MgO – 2.01; CaO – 4.98; SO_3 – 0.53; K_2O + Na_2O – 0.25. The light microsphere has the following properties: true density – 2.21 g/cm^3; bulk density – 450 kg/m^3; chemical composition, mass%: $SiO_{2-5}7$; Al_2O_3 – 26; Fe_2O_3 + FeO – 9; CaO – 5; K_2O + Na_2O – 3.

X-ray diffraction of light microsphere is presented in Fig. 1.

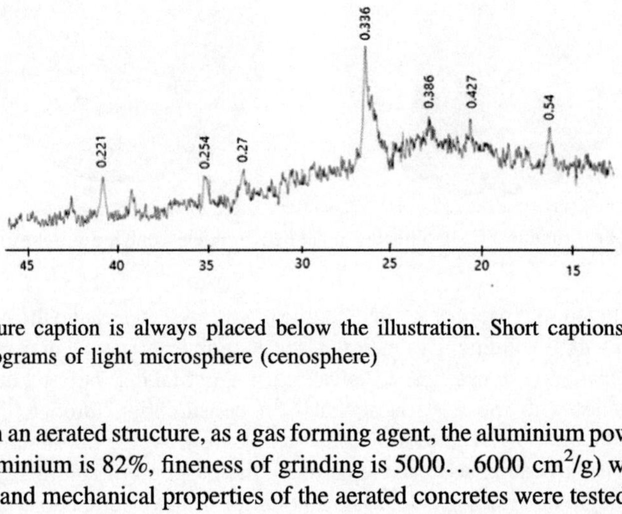

Fig. 1. A figure caption is always placed below the illustration. Short captions are centered, while Difractograms of light microsphere (cenosphere)

To obtain an aerated structure, as a gas forming agent, the aluminium powder (content of active aluminium is 82%, fineness of grinding is 5000...6000 cm^2/g) was used.

Physical and mechanical properties of the aerated concretes were tested by standard test methods.

3 Results and Discussion

Modern wall material should provide fast and economical construction, as well as ensure reliable, safe and again economical operation of the building. One of the best materials for cheapening and optimization, both construction and operation of houses, is lightweight and energy efficient aerated concrete.

The properties of aerated concrete obtained with the use of fly ash and light microspheres are studied. The performed investigations show that the replacement of fly ash in the composition of aerated concrete for the microspheres provides a decrease in the average density of aerated concrete from 969 to 440 kg/m^3. The flowability of the aerated concrete mixture and the consumption of cement do not change. It should be noted that after 28 days of hardening in normal conditions, the strength of aerated concrete obtained with using the microsphere is 1.2 MPa, while when using the fly ash – 3.5 MPa.

Studies of the aerated concrete pore structure show (Fig. 2) that the total porosity of aerated concrete obtained with using fly ash is 54.9%, and aerated concrete with a light microsphere – 82.3%. The aerated concrete with a light microsphere is characterized by an improved pore structure, the predominant pores are 1.2–2.1 mm in size. Thickness of partitions between pores is 0.15–0.22 mm. Increase in the porosity of aerated concrete promotes the reduction of thermal conductivity of the material, which for aerated concrete with a light microsphere is 0.11 W/(m K).

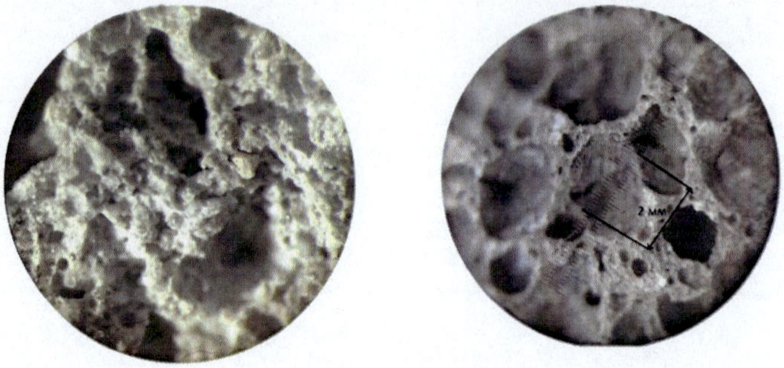

Fig. 2. Microstructure of non-autoclaved aerated concrete containing light microsphere

The developed microsphere aerated concrete was used to arrange the outer external walls of a residential building. Evaluation of the economic efficiency of aerated concrete used in the installation of external walls was carried out taking into account the cost of their construction and the cost of annual heat consumption through the enclosing structures of the building. The cost of construction work on the installation of external walls of various structures was calculated using the program of automated issuance of estimates for regional prices in 2019. The amount of heat consumption was determined in accordance with the normative indicators of thermal conductivity of structural elements of the wall and with account of the average temperature and duration of the heating season for the first climatic zone of Ukraine. Optimization of external walls structural elements was carried out by the method of mathematical planning of the experiment for the cost of a unit indicator of energy savings, according to the plan of a two-factor three-level experiment, with selected variables factors as the wall thickness (X1 = 0.4; 0.5; 0.6 m) and the thickness of the heat insulation (X2 = 0, 15, 30 cm).

The regression coefficients are calculated using a specially designed program in EXSEL, which uses a matrix approach to regression analysis and finding the coefficients of regression equations. Based on the obtained regression equations, isoparametric surfaces of the change in the value of the unit indicator of energy cost savings for aerated concrete and hollow brick walls (Y1, Y2 = const) are constructed (Fig. 3).

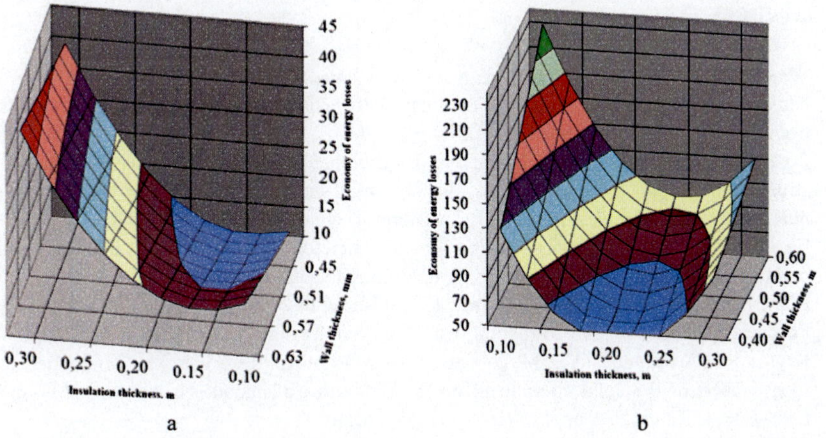

Fig. 3. Isoparametric surfaces of cost change of a unit indicator of economy of energy losses for walls from a hollow brick (a) and aerated concrete (b)

Determination of the cost of a unit indicator of energy - cost savings was carried out by dividing the cost of construction by the amount of heat loss savings for different wall designs. The lowest cost of a unit indicator testifies to the optimal selection of structures. According to this indicator, it is also possible to build a rating of priority energy saving measures for the building as a whole to implement an effective program of building renovation.

It is established that the most cost-effective brick wall thickness is 510 mm with 150 mm insulation, and the wall thickness of aerated concrete blocks is 400 mm with mineral wool insulation thickness of 150 mm. This method of mathematical planning of results can be used to evaluate any other variants of project decisions, which can estimate the cost of repairs, taking into account the durability, the cost of maintenance and operation of the building, the cost of liquidation, etc., on the basis of which it is possible to build a model of the cost of a building living cycle.

4 Conclusions

The non-autoclaved aerated concrete with microsphere, characterized by a strength of 1.2 MPa, average density of 440 kg/m^3 and thermal conductivity of 0.11 W/(m K) was obtained. The pore structure of aerated concrete samples was investigated. The economic efficiency of aerated concrete use in the installation of external walls was evaluated. Use of industrial wastes in aerated concrete demonstrated an important step towards the development of sustainable (environmentally friendly, energy-efficient and economic) infrastructure systems.

References

1. Walczak, P., Szymański, P., Różycka, A.: Autoclaved aerated concrete based on fly ash in density 350 kg/m^3 as an environmentally friendly material for energy - efficient constructions. Procedia Eng. **2**, 39–46 (2015)
2. Poznyak, O., Sanytsky, M., Zavadsky, I., Braichenko, S., Melnyk, A.: Research into structure formation and properties of the fiber-reinforced aerated concrete obtained by the non-autoclaved hardening. East.-Eur. J. Enterp. Technol. **3/6**(93), 39–46 (2018)
3. Fu, Y., Wang, X., Wang, L., Li, Y.: Foam concrete: a state-of-the-art and state-of-the-practice review. Adv. Mat. Sci. Eng. Article no. 6153602 (2020)
4. Marushchak, U., Sanytsky, M., Pozniak, O., Mazurak, O.: Peculiarities of nanomodified Portland systems structure formation. Chem. Chem. Technol. **13**(4), 510–517 (2019)
5. Krylov, E., Martynov, V., Mykolaiets, M., Martynova, O., Vietokh O.: Influence of modification of the solid component on the properties of nonautoclaved aerated concrete. East.-Eur. J. Enterp. Technol. **3/6**(99), 53–59 (2019)
6. Pachideh, G., Gholhaki, M.: Effect of pozzolanic materials on mechanical properties and water absorption of autoclaved aerated concrete. J. Build. Eng. **26**, 100856 (2019)
7. Dvorkin, L., Lushnikova, N., Bezusyak, O., Sonebi, M., Khatib, J.: Hydration characteristics and structure formation of cement pastes containing metakaolin. In: MATEC Web of Conferences, vol. 149, p. 01013 (2018)
8. Novosad, P., Pozniak, O., Melnyk, V., Braichenko, S.: Porous thermal insulation materials on organic and mineral fillers. In: Proceedings of CEE 2019. Lecture Notes in Civil Engineering. Advances in Resource-Saving Technologies and Materials in Civil and Environmental Engineering, vol. 47, pp. 354–360 (2020)
9. El-Didamony, H., Amer, A.A., Mohammed, M.S., AbdEl-Hakim, M.: Fabrication and properties of autoclaved aerated concrete containing agriculture and industrial solid wastes. J. Build. Eng. **22**, 528–538 (2019)
10. Kunchariyakun, K., Asavapisit, S., Sinyoung, S.: Influence of partial sand replacement by black rice husk ash and bagasse ash on properties of autoclaved aerated concrete under different temperatures and times. Constr. Build. Mater. **173**, 220–227 (2018)
11. Walczak, P., Małolepszy, J., Reben, M., Szymański, P., Rzepa, K.: Utilization of waste glass in autoclaved aerated concrete. Procedia Eng. **122**, 302–309 (2015)
12. Różycka, A., Pichór, W.: Effect of perlite waste addition on the properties of autoclaved aerated concrete. Constr. Build. Mater. **120**, 65–71 (2016)
13. Lesovik, V., Vorontsov, V., Glagolev, E., Pomochnicov, D., Pomochnicov, V., Volodchenko, A.: Increasing efficiency of composite thermal insulation foam concretes. Adv. Eng. Res. **133**, 414–419 (2017)

Analysis of the Current Methodology Disadvantage of the Consumed Thermal Energy Allocation Between Consumers for Heating of Multi-apartment Buildings and Ways of its Improvement

Serhii Protsenko(✉), Mykola Kizyeyev, Olha Novytska, and Nataliia Kravchenko

The National University of Water and Environmental Engineering, 11 Soborna St., Rivne city 33028, Ukraine
s.b.protsenko@nuwm.edu.ua

Abstract. The paper analyzes the current "Methodology of the consumed utilities allocation between consumers in the building" (hereinafter - Methodology) in the section of determining the consumed thermal energy for heating. The analysis shows the disadvantages of the current Methodology in the section of determining the consumed thermal energy allocation between consumers for heating of multi-apartment buildings. Such disadvantages are already confirmed by the initial experience of applying this Methodology into practice. The differences between the current Methodology and the Methodology developed by the National University of Water and Environmental Engineering (NUWEE) "Methodology of the consumed thermal energy calculating for common areas heating of multi-apartment buildings", namely the calculating principles of the thermal energy for heating of common areas and additional spaces of the building, minimum heat consumption in the heated space and etc. The analysis is given of the Methodology algorithm of heat energy allocation between consumers in those houses where the individual accounting of heat consumption for heating is organized by installing heat cost allocators on the heating units, which describes the disadvantages of the absence of the radiator coefficients' use. The current Methodology should be improved taking into account the above analysis using the approaches proposed by the authors, which are described in the developed methodology of the NUWEE.

Keywords: Thermal energy · Heat consumption · Heat energy allocation · Heat metering units · Common areas · Heat cost allocators

1 Introduction

Recently, there have been reports on social networks and the media in Ukraine that residents of multi-apartment buildings received receipts for district heating services of a very large sum. For example, on the official Facebook page of Kyivteploenerho it is reported that a tenant of the building (address: E. Sverstyuk Str., 52-B) has received

Z. Blikharskyy (Ed.): EcoComfort 2020, LNCE 100, pp. 353–361, 2021.
https://doi.org/10.1007/978-3-030-57340-9_43

receipts for heating a four-room apartment with a heated area of 127.3 m^2 in January 2020 in the amount of 52,932.18 UAH. This amount responds to a monthly heat consumption of 39.0178 Gcal (or 45.3777 MWh) and to an average monthly heating capacity of 61 kW per apartment (or 479 W/m^2). The district heating company is aware of the absurdity of such a bill, but sees the reason of such counting "in the imperfection of the regulatory framework, in particular the Resolution of the Cabinet of Ministers № 630, the requirements of which should be met by heating companies". It is very important to understand the reason of such large bills for district heating services, and what is wrong with the current method of thermal energy allocation consumed by the tenants of the multi-apartment buildings of individual apartments.

2 Methodologies of Thermal Energy Allocation in the Multi-apartment Buildings

The authors of this paper developed "Methodology of the consumed thermal energy calculating for common areas heating of multi-apartment buildings" in 2017–18 in the framework of economic contract research (agreement # 23-39/2017 between the Ministry of Regional Development, Construction, Housing and Communal Services of Ukraine and the National University of Water and Environmental Engineering). The main principles of this methodology where published in the research work report [1] and in papers [2, 3]. Unfortunately, the authors' proposals were not reflected in the "Methodology of the consumed utilities allocation between consumers in the building" [4], which was approved by the order of the Ministry of Regional Development of Ukraine from 22.11.2018 # 315. Consider the differences between the current Methodology and proposed by the authors.

2.1 The Allocation of the Total Amount of Thermal Energy Consumed in the Building Between Tenants

The total amount of thermal energy consumed in the building Q_{BLD} can be determined according to the readings of the commercial metering unit – thermal energy meter installed at the inlet of the building. The thermal energy is allocated between the total thermal energy Q_{BLD}^{HT} consumed for heating the building and the total thermal energy Q_{BLD}^{HW} for hot water supply (if there is such a system in the building).

The amount of thermal energy Q_{BLD}^{HT} consumed for heating the building can be expressed as the sum of the thermal energy amount for heating apartments (those equipped by heat metering units - $\sum Q_{iTU}$, and those not equipped with such units - respectively $\sum Q_{iWU}$), for common areas heating of Q_{CA} and for providing of the heating system functioning Q_{FS}.

$$Q_{BLD}^{HT} = \sum Q_{iTU} + \sum Q_{iWU} + Q_{CA} + Q_{FS} \quad (1)$$

Only value $\sum Q_{iTU}$ can be measured by instrumental methods with a certain error if the heat meters are installed on the heating pipelines' inlets of the apartments. This value cannot be measured directly in case of individual thermal energy accounting by installing the heat cost allocators on heating units. Unlike heat meters, which measure the consumed thermal energy in natural physical quantities (Gcal, GJ), allocators only integrally accumulate over a certain period of time the values of temperature pressure (temperature differences of heating unit surface and indoor air) in conventional (relative) units. The complex calculation algorithms are used in the world practice for the allocation of heat consumed in the building between individual consumers in proportion to the readings of the installed heat cost allocators [5].

Today there is no real possibility metrologically accurately to measure or calculate the other values in the right part of Eq. (1). It is possible only indirectly to calculate the sum of these values ($\sum Q_{iWU}$ + Q_{CA} + Q_{FS}) as the difference between the readings of the building heat meter Q_{BLD}^{HT} and all apartment heat meters $\sum Q_{iTU}$, with a certain error. The fact is that apartment heat meters have a significant disadvantage - all more or less affordable models give a significant measurement error (mostly to a lesser extent) [6, 8]. It is typical for low heat capacities, which take place in a modern insulated building and moderately cool weather. That is why according to the Law of Ukraine "On commercial accounting of heat energy and water supply" [7] heat meter installed at the inlet of the building or its part (entrance) is a commercial heat metering unit which readings are the base for bills' formation for district heating. Instead, apartment heat meters are heat metering units (HMU) that provide individual metering of heat consumption in buildings with two or more tenants. Based on their readings the heat energy allocation between tenants is carried out.

Therefore, the problem of correct solving of the Eq. (1) with three unknowns is very important - determining the amount of heat consumption $\sum Q_{iWU}$ for heating those apartments that are not equipped with HMU (it should be paid only by the tenants of such apartments), and the thermal energy amount (Q_{CA} + Q_{FS}) for common areas heating and the heating system's functioning in the building (it should be paid by all tenants of the building, including apartments which are equipped by individual heating systems according to the requirements of paragraphs 2, 3 and 6 of Article 10 of the Law [7]).

This complex problem is solved in the current Methodology [4] by the following method. The amount of thermal energy Q_{CA}, spent on heating common areas and additional spaces of the building, is taken as a share of the total building thermal energy consumption Q_{BLD}^{HT} at the level of: single-storey building - 20%; two-storey - 18%; three-storey - 16%; four-storey - 14%; five-storey - 12%; six-storey and above - 10% [4, sec. III, p. 2]. Similarly, the amount of thermal energy Q_{FS} delivered on providing the heating system functioning is taken as a percentage of Q_{BLD}^{HT}, depending on the type of source [4, sec. V, p. 2]. The thermal energy amount $\sum Q_{iWU}$ delivered for apartments' heating that are not equipped by heat metering units is calculated as the difference

$$\sum Q_{iWU} = Q_{BLD}^{HT} - \left(\sum Q_{iTU} + Q_{CA} + Q_{FS}\right) \tag{2}$$

This amount is allocated between apartments in proportion to their area. According to this method, the probability of error is very high. If the values of Q_{CA} and Q_{FS} are defined incorrectly - underestimated, then the calculated amounts of heat $\sum Q_{iWU}$ will be greater than the actual ones. In other words, the part of the expenses for Q_{CA} and Q_{FS} will be paid not by all tenants of the building, as required by Article 10 of the Law [7], but only by tenants with apartments without heat meters. The smaller a share of such apartments in the building, the more noticeable will be this financial burden for their residents. This is exactly the situation in the example given at the beginning of the paper: "only one apartment remained without registered heat metering unit from 177 ones in the building".

However, there is a rule in paragraph 2, section III of the current Methodology [4], that according to the tenants' decision, the shares of the consumed thermal energy on heating of common areas and additional spaces can be determined using a reduction factor from 0.2 to 0.9 depending on the thermal characteristics of enclosing structures, in particular the characteristics of filling window, balcony and door frames, the presence of vestibules, the quality of thermal insulation of engineering systems, etc.; or by decision of the tenants to determine the share, an increase factor from 1.1 to 2 can be used depending on the availability and technical condition of heating appliances in the common areas and additional spaces, the state of thermal characteristics of enclosing structures. The tenants of the building notify in writing the utility service or the allocation provider about the decision on factor's application or change during the non-heating period.

Thus, the value of Q_{CA} for a six-storey or higher building can be in the range from 2% to 20% of Q^{HT}_{BLD}, depending on specific conditions, but district heating providers are recommended to take it by default 10% of Q^{HT}_{BLD}. The tenants of the building should solve this difficult task of assessing the impact of all above factors on the value of Q_{CA} and make the decision of applying the reduction or increase factor. According to European experience the fixed share of heating costs varies in mostly countries from 30–50% [9, 10], and even for some countries this value is higher [10, 11].

Our proposed approach [1–3] is different from the described one. We assumed that the specific heat consumption for heating per unit area in apartments without individual meters should not differ significantly from its value q^{HT} in those apartments of the same building that are equipped with heat meters. Therefore, the thermal energy amount for heating apartments without individual meters can be approximately determined as the product of the value of q^{HT} per area of these apartments. In a somewhat simplified form (without correction factors that take into account the characteristics of the premises), our recommended method of calculation is as follows.

First, on the basis of the readings of the heat metering units, the average specific consumption of thermal energy q^{HT} for heating apartments equipped with such units is expressed as

$$q^{HT} = \sum Q_{iTU} / \sum F_{iTU} \tag{3}$$

where $\sum F_{iTU}$ – the total area of all apartments in the building, equipped with heat metering units.

Next, calculate the amount of heat consumption for heating apartments without heat metering units

$$Q_{iWU} = F_{iWU} \cdot q^{HT} \quad (4)$$

where F_{iWU} – the apartment area.

After that, calculate the amount of heat consumption for common areas heating and the heating system functioning of the building (Q_{CA} + Q_{FS}) as the difference between the readings of the commercial heat metering unit Q_{BLD}^{HT} and heat metering units $\sum Q_{iTU}$ and the calculated heat consumption for heating apartments without heat metering units $\sum Q_{iWU}$

$$(Q_{CA} + Q_{FS}) = Q_{BLD}^{HT} - \left(\sum Q_{iTU} + \sum Q_{iWU}\right) \quad (5)$$

The obtained difference (Q_{CA} + Q_{FS}) is allocated among all tenants of thermal energy of the building in proportion to the apartment area, as required by Article 10 of the Law of Ukraine "On commercial accounting of heat energy and water supply" [7].

The analysis and comparison of the computation results of the thermal energy allocation consumed by the multi-apartment building according to the current Methodology [4] and the methodology proposed by the authors [1] was carried out (see Tables 1 and 2). The 10-storey building has 140 apartments with total living area - 18358 m^2 and total area of common areas - 2456 m^2. The heating system provided from the centralized heat supply system. The consumed thermal energy for heating of the multi-apartment building per month is 260.204 Gcal.

The different possible variants of the ratio of the apartments' number with heat metering units HMU and without it are given in Table 1. The less the number of apartments without HMU, the greater the financial burden for tenants of such apartments according to the current Methodology [4]. The method proposed by the authors allows to balance this injustice, which in turn leads to the responsibility of tenants of apartments with HMU (for example, not to screw their thermostats, forcing neighbors to heat their apartments as well). From one side the developed methodology does not encourage the residents of the house to install HMU, and from the other side the increase of HMUs provides more accurate definition of the specific heat consumption and, consequently, more accurate allocation of thermal energy between tenants.

Table 1. The thermal energy allocation in the multi-apartment building for payment by the tenants without HMU according to the current Methodology [4] for different ratios of apartments

Number of apartments without HMU, pcs.	2	20	40	70	90	120	138
Area of apartments without HMU, m^2	351	2755	5510	9930	12038	16298	18007
Number of apartments with HMU, pcs.	138	120	100	70	50	20	2
Area of apartments with HMU, m^2	18007	15603	12848	8428	6500	2060	351
Total heat consumption by apartments without HMU, Gcal/month	22.18	5.05	3.80	3.24	3.10	2.97	2.93

Table 2. The amount of thermal energy consumed by the house on common areas and heating system functioning according to the current Methodology [4] and the proposed approach in the Methodology [1]

	Methodology	
	The current Methodology [4] (apartments with HTU)	The proposed by authors* [1] (with and without HTU)
The amount of thermal energy on common areas Q_{CA}, Gcal	26.020	
The amount of thermal energy on heating system functioning Q_{FS}, Gcal	20.816	
The thermal energy on common areas and heating system functioning, Gcal $Q_{FS} + Q_{CA}$		80.296
Total heat consumption, Gcal/month	2.554	2.925
Specific heat load including $Q_{FS} + Q_{CA}$, Gcal/m^2	0.0124	0.0142

According to the requirements of paragraph 4, section 2 of Article 10 of the Law [7] co-owners of the multi-apartment building or other building with two or more tenants may set correction factors Ki to allocate delivered thermal energy between individual consumers (in corner apartments (spaces), in apartments (spaces) located on the first and last floors of the building, above the passages, etc.). In this case, instead of Eq. (4) it is recommended to use the expression

$$Q_{iWU} = F_{iWU} \cdot q^{HT} \cdot K_i \tag{6}$$

We proposed the approach of determining the correction coefficients Ki, which can be found in [1–3].

2.2 The Minimum Consumption of Thermal Energy in Apartments

Another possible source of errors when charging for the district heating service may be incorrectly defined requirements for in the heated space. According to the Law [7], the amount of metered thermal energy in the apartments must not be less than the minimum share of average specific heat consumption among other tenants of the building. The current Methodology [4] also notes that "this share checks compliance with the thermal regime in these spaces during the heating period, which is not allowed to reduce the air temperature by more than 4 °C from the standard indoor air temperature".

The problem is that according to the current Methodology [4] "the minimum share of average specific heat consumption (q^{HT}_{min}) is 50% of the average specific heat consumption for heating (q^{HT}), so $q^{HT}_{min} = 0.5 \cdot q^{HT}$".

To verify the legitimacy of such a statement, let's make a simple calculation. Let's make a proportion

$$q^{HT}/q^{HT}_{min} = (t_i - \Delta t) - t_e/t_i - t_e \quad (7)$$

where t_n – the standard indoor temperature of the tenants' spaces; Δt – permissible decrease in indoor temperature between tenants' spaces during the period of their non-use; t_e – average external temperature of the heating period for the object area.

According to the state requirements [12] the indoor temperature of living rooms for healthy people should be 22 ± 2 °C, so not below t_i = 20 °C. According to state requirements [13] it is allowed to lower the indoor temperature in heated spaces during the heated period during of their non-use in residential buildings not more than Δt = 4 °C from the standard temperature, so not below 16 °C. The average external air temperature for the heating period (with an average daily air temperature not exceeding 8 °C) for most regions of Ukraine can be accepted (averaged) at the level of t_e = 0 °C [14]. Substituting the above data in Eq. (7), we obtain that the heat loss in residential areas with a minimum allowable air temperature of 16 °C on average during the heating period will be q^{HT}/q^{HT}_{min} = (20 – 4) – 0/20 – 0 = 0.8 from heat losses of the same spaces with standard indoor temperature 20 °C.

Therefore, the minimum share of the average specific consumption of thermal energy q^{HT}_{min} cannot be less than 80% of the average specific consumption of thermal energy for heating q^{HT}. If this value is taken equal to 50% of q^{HT}, as done in the current Methodology [4], it means the possibility of reducing the air temperature in heated rooms during their non-use on 10 °C that is contrary to the current standards [13], the reducing temperature can't be less than 4 °C. At such a low temperature, there will be significant heat flow into these rooms through interior partitions from adjacent rooms (apartments) with higher indoor air temperatures. In other words, the tenants of the nearest apartments (neighbors) will partially pay for heating apartments with non-standard air temperature instead of their owners.

According to our proposed Methodology the thermal energy value for HMU can't be less than

$$Q_{iTU} \geq K_{MIN} \cdot q^{HT} \cdot F_{iTU} \quad (8)$$

where K_{MIN} – the minimum permissible share of the average specific consumption of thermal energy among tenants in the building, expressed as in Eq. (7) and which cannot be less than 0.8 of the average specific consumption of thermal energy.

2.3 Heat Cost Allocators Installation

There are some disadvantages of the algorithm in paragraph 4, Section II of the current Methodology [4] of heat energy allocation between tenants in those buildings where individual accounting of heat consumption is organized by installing heat cost allocators on heating units. The Methodology [4] did not take into account the peculiarities of heating unites, in Eqs. (13), (14), (17) [4] there are no radiator coefficients which used in foreign methodologies (see, for example, [5, 15, 16]). The manufacturers provide the radiator coefficients for each type of heat cost allocator for all available heating units. This information is included in computer programs used for heat cost allocation between the tenants. Thus, performing calculations on the thermal energy allocation the use of radiator coefficients is important.

3 Conclusions

The above analysis shows significant disadvantages of the current Methodology [4] due to thermal energy allocation delivered for heating between tenants of multi-apartment buildings, which is also confirmed by the initial experience of its applying in practice. Therefore, this Methodology requires revision taking into account the above facts, which can be done, in particular, by using the proposed approaches of definition the values of thermal energy for common areas heating of Q_{CA} and for heating system functioning Q_{FS}, the minimum consumption of thermal energy on the heated space, and the importance of using the radiator coefficient for consumed thermal energy calculation using heat cost allocators.

References

1. Report on research work. Analysis of the domestic regulatory framework and world experience in calculating the amount of thermal energy consumed for heating of common areas and proposals development for changes to the Methodology for calculating the amount of thermal energy for heating common areas (final). Registration № 0117OU003267 RW. Rivne, NUWEE (2018)
2. Protsenko, S., Kizyeyev, M., Novytska, O.: Methodology of heat energy allocation for heating between tenants of the multi-apartment building. Energy, ecology, computer technology, life safety in construction: the collective monograph. Prydniprovska State Academy of Civil Engineering and Architecture, Dnipro, pp. 87–93 (2018)
3. Protsenko, S., Kizyeyev, M., Novytska, O.: Development of the methodology of heat energy allocation for heating between tenants of the multi-apartment building. In: Ventilation, Lighting, Heat and Gas Supply: Scientific and Technical Bulletin, Kyiv, KNUBA, vol. 8, pp. 48–52 (2019)
4. Methodology of the consumed utilities allocation between consumers in the building. Approved by order of the Ministry of Regional Development, Construction, Housing and Communal Services of Ukraine, 22 November, p. 315 (2018). https://zakon.rada.gov.ua/laws/show/z1502-18
5. MDK. Methods of the total thermal energy consumption allocation of the buildings on heating between the tenants based on readings of apartment heat meters. Viterra Energy Service LLC, Danfoss CJSC, Moscow, 4 April 2004 (2004)
6. Protsenko, S.: Estimation of methods of apartment-by-apartment accounting of heat consumption for heating in apartment buildings. Bull. NUWEE **1**(69), 216–226 (2015). http://ep3.nuwm.edu.ua/4654/
7. On commercial accounting of thermal energy and water supply: Law of Ukraine of 22.06.2017 № 2119-VIII. Information of the Supreme Council, № 34, Art. 370 (2017). http://zakon3.rada.gov.ua/laws/show/2119–19/print1504692425555062

8. An Investigation into Heat Meter Measurement Errors Final Report. AECOM House, pp. 63–77 Victoria Street, St Albans, Hertfordshire (2013). https://assets.publishing.service.gov.uk/government/uploads/system/uploads/attachment_data/file/375976/An_Investigaion_into_Heat_Meter_Measurement_Errors_Final_Report_AECOM.pdf
9. Robinson, S., Vogt, G. (empirica GmbH): Guidelines on good practice in cost - effective cost allocation and billing of individual consumption of heating, cooling and domestic hot water in multi-apartment and multi-purpose buildings, 42 p. (2016). https://ec.europa.eu/energy/sites/ener/files/documents/mbic_guidelines20170123_en.pdf
10. Castellazzi, L.: Analysis of Member States' rules for allocating heating, cooling and hot water costs in multi-apartment/purpose buildings supplied from collective systems. Implementation of EED Article, vol. 9, no. 3. EUR 28630 EN. Publications Office of the European Union, Luxembourg (2017). ISBN 978-92-79-69286-4. JRC106729. https://doi.org/10.2760/40665
11. Rose, J., Kragh, J.: Distribution of heating costs in multi-story apartment buildings. Energy Procedia **132**, 1012–1017 (2017). 11th Nordic Symposium on Building Physics, NSB2017, 11–14 June 2017, Trondheim, Norway
12. SBR B.2.2-15: 2019. Residential buildings. Main guidelines. Ministry of Regional Development of Ukraine, Kyiv, 44 p. (2019)
13. SBR B.2.5-67: 2013. Heating, ventilation and air conditioning. Ministry of Regional Development of Ukraine, Kyiv, 232 p. (2013)
14. SSTU-N B B.1.1-27: 2010. Construction climatology. Ministry of Regional Development of Ukraine, Kyiv, 123 p. (2011)
15. BR EN 834:2013. Heat cost allocators for the determination of the consumption of room heating radiators – Appliances with electrical energy supply, 36 p. http://home.aktor.qa/External%20Documents/Intenational%20Specifications/British%20Standards/BS%20EN/BS%20EN%2000834-2013.pdf
16. Csoknyai, I.: Methods of heat cost allocation. Periodica Polytechnica Ser. Mech. Eng. **44**(2), 227–236 (2000)

Impact of Undular Jump Characteristics on Erosion of Tailrace Channel

Oleksandr Riabenko(✉), Oksana Kliukha, Oksana Halych, and Dmytro Poplavskyi

National University of Water and Environmental Engineering, Rivne 33028, Ukraine
o.a.riabenko@nuwm.edu.ua

Abstract. The aim of article is to develop the method of calculating the free-surface profile of undular jump for correct evaluation of its characteristics influence on tailrace channel erosion. The research method is based on using the differential equations of wavelike near-critical flows free-surface profile with taking into account the possible non-hydrostatics in their initial cross-section, energy losses and wave attenuation along the length. The different methods of calculating the free-surface profile of undular jump were analyzed. It was paid special attention to consider this phenomenon as combination of solitary wave and cnoidal waves. The article emphasized the results of undular jump experimental researches which were made by authors on different laboratory setups. The comparison of theoretical and experimental data showed their good convergence and confirmed the conceptual correctness of developed method. The scientific novelty is taking into account total important factors such as possible deviation from hydrostatics in the initial cross-section, energy losses and wave attenuation along the length. The practice value is determining the length of wave attenuation that is needed to accept the length of tailrace channel paving.

Keywords: Undular jump · Solitary wave · Cnoidal waves · Channel erosion · Bank paving

1 Introduction

Undular jump is variety of hydraulic jump and characterized by wavelike form of its free-surface. This phenomenon occurs within different types of hydraulic constructions and channels under appropriate conditions. Many scientific works were devoted to the study this unique hydraulic phenomenon. Lots of publications show the increasing interest of scientists to this phenomenon during last decades.

While theoretical and experimental studying the undular jump it is usually determined the second conjugated h_2 and maximum h_{cr} depths, the length and conditions of existence, free-surface profile, kinetics and dynamics characteristics, the influence of slope or of curvature of elementary stream-lines on jump characteristics. During recent years, attention was paid to development of methods of calculating the free-surface profile of undular jump which are based on analytical and numerical solutions of the corresponding differential equations. Important results have been obtained in the study

Z. Blikharskyy (Ed.): EcoComfort 2020, LNCE 100, pp. 362–370, 2021.
https://doi.org/10.1007/978-3-030-57340-9_44

of the influence of the following factors on the considered phenomenon: lateral impact waves and the ratio of the width of a stream to its height [1], bottom friction and development of turbulent boundary layer [2, 3], asymptotic multiple scales analysis in a turbulent stream with using the k-ε model of turbulence [4], influence of Reynolds and Froude numbers [5] influence of slope of the bottom [6], detection of presence or absence of slope or of curvature of elementary stream-lines in initial cross-section and their influence upon the considered phenomenon [7, 8].

The very important aspects of theoretical and experimental studying of undular jump is necessity to take into account the whole complex of boundary conditions, i.e. the possible deviation from hydrostatic pressure distribution in depth in the initial cross-section of jump. The ignoration of aforesaid particularities can lead to wrong understanding the obtained results, paradoxes and mistakes [9]. Moreover, while calculating the theoretical free-surface profile of undular jump by various differential equations it is very often neglected the energy losses, therefore the free-surface profile has continuous mode.

The aim of article is to develop the method of calculating the free-surface profile of undular jump based on equations which take into account the possible non-hydrostatics in their initial cross-section, energy losses and wave attenuation for correct estimating the influence of mentioned factors on erosion of tailrace channel. The research methods use the differential equations of wavelike near-critical flows free-surface profile with taking into account these features influence.

2 Positive and Negative Aspects of Undular Jump

To positive aspects of undular jump belong next.

1. The undular jump occurrence ensures the operation of different types of hydropower and hydraulic constructions (hydropower plants, pumped storage power plants, pump stations, water spillways, etc.) in indicated range of level of headrace and tail water and allows electricity production, water supply, providing the flow pass in these conditions.
2. Knowing the value of maximal depth which is under one of the first waves crests allows reliably (without needless reserve) to determine the height of tunnels and pipes and assures the operating these structures in designing pressureless mode.
3. The undular jump occurrence with smooth wavelike surface ensures the unimpeded ice or wood pass in tail water that is all extremely relevant to the log-floating channels and tunnels.

To negative aspects of undular jump belong next.

1. The dynamic loads from wave action can lead to destruction of construction parts (Fig. 1).
2. Natural velocities under wave troughs can exceed the limit values and cause the bottom and bank erosion of tailrace channel (Fig. 1).
3. The exceedance of maximal depth h_{cr} of undular jump over the mean level of tail water requires the height increasing of protective dikes, free-flow tunnels and pipes.

Figure 1 shows real cases of destruction of constructions and erosion of tailrace channel. One of the reasons of it was undular jump formation in tail water of this channel.

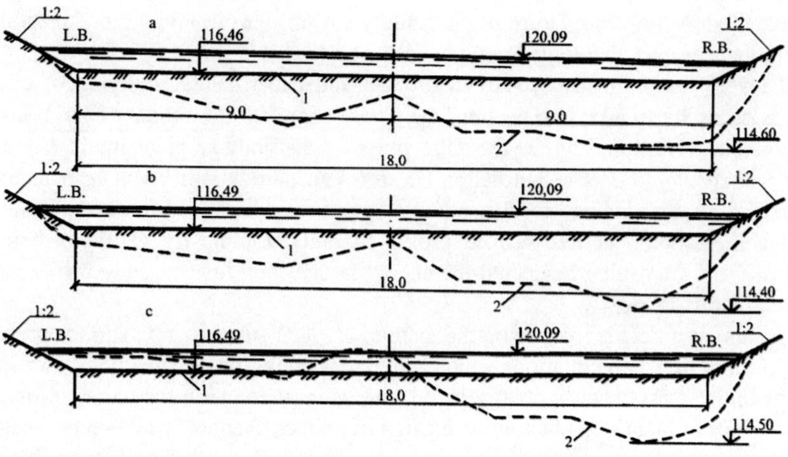

Fig. 1. Shape of plunge basin in tail water of sluice gate regulator № 3 of system Zdvyzh: a, b, c – representative cross-sections, which located about 3.0 m, 7.0 m, 12.0 m away from outlet wing walls of regulator; 1 – designed tail water profile; 2 – profile of plunge basin; L.B. – left bank, R.B. – right bank (the dimensions are in meters).

3 Theoretical Methods of Calculating the Free-Surface Profile of Undular Jump

One of common methods of calculating the theoretical free-surface profile of undular jump is representing this phenomenon as combination of solitary wave and cnoidal waves [10, 11]. Based on this model, it was developed the detailed methods of calculating for using in design practice [12].

Considered mathematical model of undular jump is based on "gluing together" the famous solutions of well-known different equations in the form of solitary wave

$$h = h_1 + (h_{cr} - h_1)sch^2\left[\sqrt{\frac{3g}{q^2}}(h_2 - h_1)\frac{x}{h_2}\right] \quad (1)$$

and cnoidal waves

$$h = h_1 + (h_{cr} - h_1)cn^2\left(\frac{x}{\Delta},\ k\right), \quad (2)$$

here h_{cr} and h_1 are maximal and minimal (initial) depths respectively, q is specific flow, g is acceleration of gravity, Δ and k are parameters of cnoidal waves.

Nevertheless, various authors use different variations of formulas (1) and (2).

"Gluing together" the indicated solutions is usually made by two variants – in cross-section that goes through the crest of first wave and in cross-section that goes through inflection point of free-surface curve placed after crest of first wave.

In the calculating the free-surface profile of undular jump using described methods it is often considered an ideal case when there are no energy losses along the length as a result this profile has continuous mode.

The significant shortcoming of described methods is neglect of possible presence of non-hydrostatic pressure distribution in the initial cross-section of undular jump, that is very important factor for small values of Froude number Fr_1 in the indicated cross-section. In work [8] it was derived the differential equation of free-surface profile of wavelike near-critical flows which takes into account explicitly the aforesaid deviation from hydrostatics:

$$\left(\frac{dh}{dx}\right)^2 = \frac{3}{Fr_1}\left[-\left(\frac{h}{h_3}\right)^3 + (2\beta_1 + Fr_1)\left(\frac{h}{h_1}\right)^2 - (2\beta_1 - 1 + 2Fr_1)\right], \tag{3}$$

here β_1 is coefficient of potential energy, $Fr_1 = q^2/(gh^3)$ is Froude number (index "1" means that indicated values refer to the initial cross-section of considered flows).

Integrating the differential "Eq. (3)" allowed to developed its general solution in form of equations system (4)–(7)

$$\eta = 1 + (\eta_{cr} - 1)cn^2\left(\frac{x}{\Delta}, k\right), \tag{4}$$

$$\Delta = 2h_1\sqrt{\frac{\eta_{cr}Fr_1}{3(\eta_{cr}^2 - Fr_1)}}, \tag{5}$$

$$k = \sqrt{\frac{\eta_{cr}(\eta_{cr} - 1)}{\eta_{cr}^2 - Fr_1}}, \tag{6}$$

$$\eta_{cr} = \frac{1}{2}\left[t_1 + Fr_1 + \sqrt{(t_1 - Fr_1)^2 - 4Fr_1}\right], \tag{7}$$

here $\eta = h/h_1$, $\eta_{cr} = h_{cr}/h_1$, t_1 is coefficient of hydrodynamic pressure.

To calculate the free-surface profile of undular jump by suggested method it is also used the equation of conjugated depths of near-critical flows [8]. It takes into account the possible deviation from hydrostatic pressure in their initial cross-section

$$\eta_2 = \frac{2}{\sqrt{3}}\sqrt{t_1 + 2Fr_1}\cos\left\{\frac{\pi}{3} - \frac{1}{3}\arccos\left[\frac{3\sqrt{3}\alpha_{02}Fr_1}{\sqrt{(t_1 + 2Fr_1)^3}}\right]\right\}, \tag{8}$$

here α_{02} is coefficient of kinetic moment in cross-section with second conjugated depth h_2.

Coefficients β_1 and t_1 are defined by using the coefficient of non-hydrostatics s_1

$$\beta_1 = \frac{1+2s_1}{3}, \quad t_1 = \frac{4s_1 - 1}{3}. \tag{9}$$

Suggested method is based on using the "Eq. (4)–(9)". Herewith it is assumed the maximal depth h_{cr} and second conjugated depth h_2 for undular jump, solitary waves and cnoidal waves are equal and constant along the length (Fig. 2).

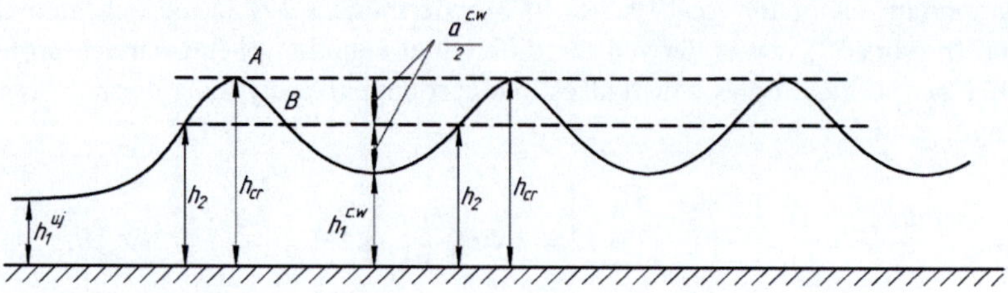

Fig. 2. Schema for calculating the free-surface profile of undular jump.

$$h_{cr}^{u.j} = h_{cr}^{s.w} = h_{cr}^{c.w} = h_{cr} = const, \tag{10}$$

$$h_2^{u.j} = h_2^{s.w} = h_2^{c.w} = h_2 = const. \tag{11}$$

The process of calculating free-surface profile of undular jump with taking into account the non-hydrostatics in its initial cross-section in the form of combination of solitary wave and cnoidal waves in case of absence the energy losses and wave attenuation restricts to is carried out by next points.

1. Firstly, the free-surface profile of undular jump on the section from initial depth h_1 to maximal depth h_{cr} is calculated. Herewith, the values of initial depth h_1, coefficient of non-hydrostatics s_1 and specific flow q are known. According to "Eq. (4)–(7)" the depth h_{cr} can be determined and free-surface profile of undular jump is calculated for the identified section. In case of non-hydrostatics presence in initial cross-section ($s_1 > 1$) the "Eq. (4)" describes the profile of soliton and in other case ($s_1 = 1$) it gives the profile of solitary wave.
2. Next step is determination of cnoidal waves characteristics. For given value of coefficient of kinetic moment α_{02} using "Eq. (8)" the second conjugated depth h_2 can be calculated, taking values of h_1, t_1, Fr_1 as for initial cross-section of undular jump.

The height of cnoidal waves equals

$$a^{c.w} = 2(h_{cr} - h_2). \tag{12}$$

Then initial depth of cnoidal waves can be found

$$h_1^{c.w} = h_{cr} - a^{c.w}. \tag{13}$$

Using the trial and error method the coefficient of non-hydrostatics $s_1^{c.w}$ in the initial cross-section of cnoidal waves can be determined from formula (7) when the value of maximal depth h_{cr} is known, the values of indicated characteristics $h_1^{c.w}$ and $Fr_1^{c.w}$ were calculated for considered cross-section.

3. The parameters of cnoidal waves Δ and k are determined by formulas (5) and (6) respectively and the free-surface profile of undular jump is calculated after point A taking the values of corresponding parameters as for initial cross-section of cnoidal waves (Fig. 3).

However, the undular jump occurs when Froude number is more than 1.8 as well. In this case practice shows that it is needed to take into account the energy losses and effect of wave attenuation along the flow length. To calculate the free-surface profile of undular jump with taking into account the energy losses and of wave attenuation along the flow length should be done next.

1. To calculate the maximal depth using formulas (14)–(16) which take into account energy losses

$$h'_{cr} = \frac{2}{3}h_1\left[\left(\frac{E_1}{h_1} - \frac{h_w}{h_1}\right) + \sqrt{4\left(\frac{E_1}{h_1} - \frac{h_w}{h_1}\right)^2 - 3\left(2\frac{E_1}{h_1} + Fr_1^2 - 1\right)}\right], \tag{14}$$

$$\frac{h_w}{h_1} = \left(\beta_1 - \eta_2 + \frac{Fr_1^2}{2}\left(1 - \frac{\alpha_2}{\eta_2^2}\right)\right), \tag{15}$$

$$\frac{E_1}{h_1} = \beta_1 + \frac{Fr_1^2}{2}. \tag{16}$$

In "Eq. (14), (15)" h_w is energy losses, α_2 is Coriolis coefficient, η_2 is the ratio of second conjugated depth h_2 with the first conjugated depth h_1 and can be obtained by formula (8).

2. To calculate the values of wave attenuation by equations [12]

$$\zeta_j = \zeta_0 \exp\left(2j \ln \sqrt{Fr_2}\right), \tag{17}$$

$$\zeta'_j = \zeta'_0 \exp\left(2j \ln \sqrt{Fr_2}\right), \tag{18}$$

here ζ_j and ζ'_j are values of wave attenuation from crest and trough of wave respectively (the wave attenuation is asymmetrical according line with second conjugated depth, and $\zeta_j \neq \zeta'_j$), j is wave number, Fr_2 is Froude number of cross-section with second conjugated depth (point B, Fig. 2).

3. Then next wave has height

$$a_j^{c.w} = a_{j-1}^{c.w} - \zeta_j - \zeta'_j. \tag{19}$$

The all necessary parameters of cnoidal waves can be calculated by "Eq. (4), (5), (6)" and the free-surface profile can be calculated by "Eq. 4".

4 Comparison of Obtained Results

To determine the accuracy of laboratory research, the level of reproducibility of experimental results and the adequacy of the described mathematical model of near-critical flows, a special series of experiments was carried out [8]. The reproducibility level was predicted using the Cochran's Q test. It was found that the calculated value of this test $Q_c = 0.2855$ is less than the table $Q_t = 0.4803$. This indicates a high level of reproducibility of experimental results on the applied experimental setup. The adequacy of the developed mathematical model was verified on the basis of Fisher's F test. The made calculations showed that the calculated value $F_c = 1.88$ is less than the table $F_t = 2.05$, which indicates the adequacy of the applied mathematical model. In this case, the accuracy of the reproducibility of the experiments is 0.61%, and the accuracy of depth measuring by water-gauge is estimated at 3–5%.

The comparison of theoretical free-surface profiles of undular jump with data of laboratory researches is shown in Fig. 3a and 3b.

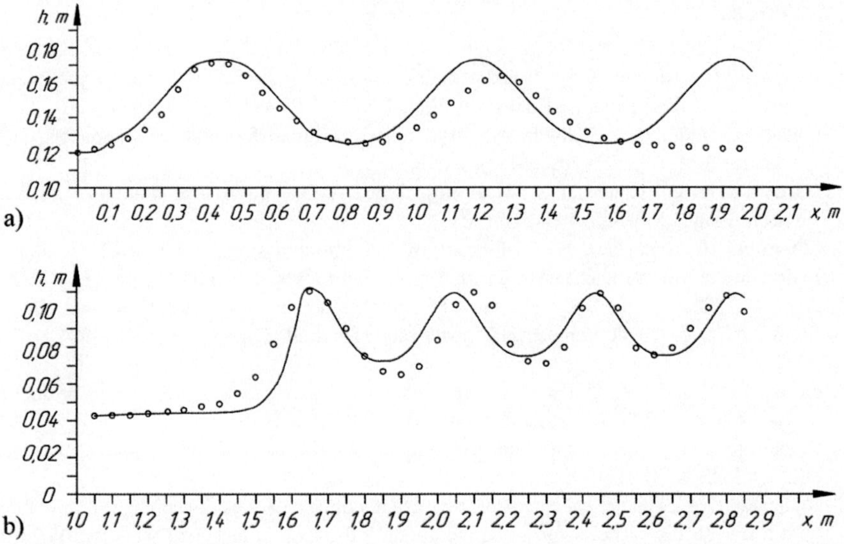

Fig. 3. Comparison of theoretical free-surface profiles of undular jump with data of laboratory researches: — –theoretical profile of free-surface of undular jump calculated by suggested method, O – data laboratory researches; a) Riabenko's experiment: q = 0,141 m^2/s, h_1 = 0,12 m, Fr_1 = 1,17; b) Halych's experiment: q = 0,05 m^2/s, h_1 = 0,042 m, Fr_1 = 3,4.

5 Conclusions

1. One of the reasons of tailrace channels erosion can be formatting different hydraulic regimes that have negative impact on the channel conditions.
2. Suggested method of calculating the free-surface profile of undular jump takes into account the possible non-hydrostatics in its initial cross-section, energy losses and effect of wave attenuation along the flow length.
3. In the calculation of the free-surface profile of undular jump in range of Froude number $1 < Fr_1 \leq 1.8$ it is possible not to take into the energy losses and wave attenuation along the flow length.
4. In case when Froude number $Fr_1 > 1.8$ it is necessary to take into energy losses and wave attenuation along the flow length.
5. The comparison of theoretical free-surface profiles of undular jump calculated by suggested method with data of laboratory researches showed their good convergence.

References

1. Chanson, H., Montes, J.S.: Characteristics of undular hydraulic jumps. Experimental apparatus and flow patterns. J. Hydraul. Eng. **121**(2), 129–144 (1995). ASCE
2. Ohtsu, I., Yasuda, Y., Gotoh, H.: Hydraulic condition for undular jump formation. J. Hydraul. Res. **39**, 203–209 (2001). IAHR
3. Castro-Orgaz, O., Chanson, H.: Near-critical free-surface flows: real fluid flow analysis. Environ. Fluid Mech. **11**, 499–516 (2011)
4. Steinrück, H., Schneider, W., Grillhofer, W.: A multiple scales analysis of the undular hydraulic jump in turbulent open channel flow. Fluid Dyn. Res. **33**(1–2), 41–55 (2003)
5. Auel, C., Albayrak, I., Boes, R.: Turbulence characteristics in supercritical open channel flows: effects of froude number and aspect ratio. Hydraul. Eng. **140**(4), 04014004 (2014). ASCE
6. Gotoh, H., Yasuda, Y., Ohtsu, I.: Effect of channel slope on flow characteristics of undular hydraulic jumps. WIT Trans. Ecol. Envir. **83**, 33–43 (2005)
7. Riabenko, O.A.: Current state and problems of undular jump theory. Appl. Hydromech. **18**(90(1)), 43–62 (2016)
8. Riabenko, O.A.: Theoretical elements and methods of calculations of near-critical fluid flows with a free surface. Thesis for academic degree of doctor of technical sciences: 05.23.16: Rivne, Ukraine (2003)
9. Riabenko, O.A.: Problems and paradoxes of near-critical fluid flow. Appl. Hydromech. **13**(85(4)), 37–51 (2011)
10. Iwasa, Y.: Undular jump and its limiting condition for existence. In: Proceedings 5th Japan National Congress for Applied Mechanics, Tokyo, pp. 315–319 (1956)
11. Tursunov, A.A.: Near-critical state of gravity water flows. Izvestiya VNIIG **90**, 20–1224 (1969)
12. Gunko, F.G., Leningrad, E. (eds.) Hydraulic calculations of constructions that manage the supercritical flow (1974)

Influence of Flexibility of Bolted Joints on Rigity of the Hingeless Frame

Volodymyr Romaniuk(✉) and Volodymyr Supruniuk

The National University of Water and Environmental Engineering, Rivne, Ukraine
v.v.romaniuk@nuwm.edu.ua

Abstract. The article investigates the influence of the flexibility of the rigid node of connection of the crossbar with the frame rack on the stress-strain state of the elements under the action of vertical evenly distributed load on the crossbar and horizontal evenly distributed load, applied to the frame racks. Since the bolts in the flange connection are stretched by the action of the bending moment, the nodes open, that is become partially hinged. Knowing some initial parameters of ideally rigid knots, it is possible to establish, what influence their change will have, both on the design scheme of a frame as a whole, and on its separate elements. The initial parameter method was used to determine the influence of the flange opening angles on the stress state of the elements. The proposed method allows us to determine the rigidity of any bolted flange connection, taking into account its actual operation. Taking into account the pliability of the nodes and, as a consequence, the structure as a whole, allows us to unload the elements and calculate the additional resource bearing capacity. In addition, the technique allows you to adjust the stiffness of the bolted joint by changing the diameter of the bolts, their number, distance between them or changing the thickness and height of the flanges, as well as use additional material resource by reducing the maximum stresses in the calculated cross sections.

Keywords: Bolt · Joint · Flange · Construction · Frame · Bearing capacity · Rigidity

1 Introduction

According to the current national building norms and practice of designing of building objects and constructions at first geometrical parameters of buildings or constructions are set, schemes of bearing constructions and their connections in knots are defined, and then on this basis the design scheme of a frame with application of all external force influences develops and set the boundary conditions for the connection of the node elements. Then the static calculation is performed and according to the efforts, defined in it, the constructive calculation of all elements and connections is performed. At this stage, in accordance with current domestic and European design standards [1, 2] boundary conditions of bolted and welded joints are accepted as idealized, that is rigid, hinged fixed, hinged movable etc., which does not fully meet the actual conditions of their work and of work of structures in whole. This is especially true of bolted

Z. Blikharskyy (Ed.): EcoComfort 2020, LNCE 100, pp. 371–377, 2021.
https://doi.org/10.1007/978-3-030-57340-9_45

connections and, in particular, flanged bolted connections, which are one of the most effective types of factory joints, and especially mounting joints of beams, frames, arches, trusses, etc., and which are very widely used in world design and construction practice of metal structures. This is confirmed by the practice of operation of existing structures, as well as numerous theoretical and experimental studies of domestic and foreign scientists [3–10]. In particular, tests of a steel perforated arch with a rigid flange node [3–5] revealed a significant difference in stresses, calculated theoretically and obtained experimentally, in cross sections near it (the difference was 56.6%–67.1%). This can be explained by the pliability of the flange bolted joint, which was opened in result of action in the node of bending moment and longitudinal force, upon in theoretical calculations, this connection was taken absolutely rigid. As a result, the rigid node worked not as rigid, but partly hinged, and the actual design scheme of the two-hinged arch was partly three-hinged, and therefore the value of the bending moment in the node was less, than the calculated, due to pliability. Experimental and theoretical studies [3, 5] found that the stiffness of the bolted connection was 0.412 compared to the stiffness of the ideal flange connection taken as a unit. It was also found that the stiffness of the node changes with the changes of parameters of its details. For example, with a change in bolt diameter from 10 to 24 mm the stiffness varies from 0.412 to 0.897 respectively.

2 Purpose

The purpose of these studies is to theoretically determine the actual stiffness of bolted flange joints of steel structures, as well as to develop an effective method for determining the influence of the opening angle of the flanges on the stiffness of the node as a whole.

3 Method

To achieve this purpose, the method of initial parameters is used, as well as the results of theoretical and experimental studies

4 Results

To study the influence of the pliability of the rigid node of connection of the crossbar with the racks on the stress-strain state of the elements, it is convenient to consider a hingeless transverse frame of the spatial framework of a single-span industrial building, which perceives vertical evenly distributed load q on the frame crossbar and horizontal evenly distributed wind load q_w applied to the frame racks (Fig. 1). The connection of the racks to the foundation is rigid with the use of anchor bolts, and with the crossbar is also rigid with the use of bolts and flanges. Since the anchor bolts and bolts in the flange connection are stretched under the action of the bending moment, these nodes are opened, that is become partially hinged. Knowing some initial parameters of ideally rigid knots, it is possible to establish, what influence their change will have, both on the design scheme as a whole, and on its separate elements. The method of initial parameters was used to determine the influence of the opening angles of the flanges on the stress-strain state of the elements.

$$\left.\begin{aligned} EI\theta_{xi} &= EI\theta_0 + \sum M_{xi}(x_i - a_1) + \sum F_{xi}\tfrac{(x_i-a_2)^2}{2} + \sum q_{xi}\tfrac{(x_i-a_3)^2}{6} \\ Ely_{xi} &= Ely_0 + EI\theta_0 x_i + \sum M_{xi}\tfrac{(x_i-a_1)^2}{2} + \sum F_{xi}\tfrac{(x_i-a_2)^3}{6} + \sum q_{xi}\tfrac{(x_i-a_3)^4}{24} \end{aligned}\right\}, \qquad (1)$$

where is θ_{xi}, y_{xi} – the angle of rotation and deflection in the cross section of the element under consideration; θ_0, y_0 – angle of rotation and deflection of the element at the beginning of the system (initial parameters); M_{xi}, F_{xi}, q_{xi} – respectively concentrated moments, concentrated forces and the value of evenly distributed load, which are applied to the structure; x_i – the distance from the beginning of the system to the cross section in which the deformation is calculated; a_1, a_2, a_3 – distances from the beginning of the system to the points of application, respectively, of concentrated moments, concentrated forces and to the beginning of areas of application of distributed load; I – moment of inertia of the section; E – modulus of elasticity of steel.

The design scheme of the left rack of a cross frame can be received, having replaced the discarded parts by reactions of internal efforts (Fig. 2).

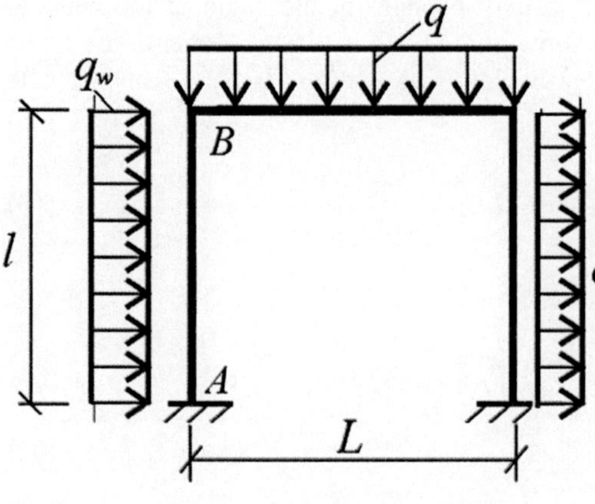

Fig. 1. Design scheme of a single-span transverse frame

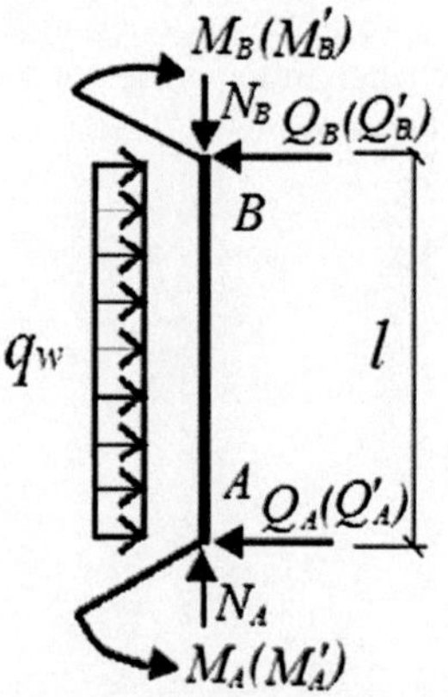

Fig. 2. Design scheme of the left rack of frame

Taking the initial parameters, namely that the angle of rotation in the idealized design scheme in the rigid flanged node B $\theta_B = 0$, using the first part of calculation Eq. (1) and taking the origin of coordinates in node B, the formula for the angle of rotation in node A will look like

$$EI\theta_A = EI\theta_B + M_B\, l - Q_B \frac{l^2}{2} + q_w \frac{l^3}{6}. \tag{2}$$

If the flange node B is opened, that is $\theta'_B \neq 0$, it will be partially hinged, due to which the design scheme of the whole frame will change and as a result there will be a redistribution of forces in its elements and design cross-sections. As a result, the values of the bending moment M_B on M'_B and the transverse force Q_B on Q'_B will change.

The angle of rotation in node A in this case will be determined from the formula

$$EI\theta'_A = EI\theta'_B + M'_B l - Q'_B \frac{l^2}{2} + q_w \frac{l^3}{6}. \tag{3}$$

Changing the angle of rotation in node B will slightly affect the value of the angle of rotation in node A, so for further calculations we can take $\theta_A = \theta'_A$. Equating Eqs. (2) and (3), we obtain a formula to determine the angle of rotation in node B

$$EI\theta'_B = (M_B - M'_B)l - (Q_B - Q'_B)\frac{l^2}{2} = \Delta M_B l - \Delta Q_B \frac{l^2}{2}. \tag{4}$$

That is, the value of the opening of the flanged node, namely the angle of rotation, is directly proportional to the value of the change of the bending moment and the transverse force in the support part.

Preliminarily accepting the condition that the ratio of the value of the bending moment to the value of the transverse force in the node is directly proportional to the change in the values of these forces after the deformation of the node, we can calculate the ratio

$$\frac{M_B}{Q_B} = \frac{\Delta M_B}{\Delta Q_B} = c \rightarrow \Delta Q_B \frac{\Delta M_B}{c}. \tag{5}$$

Considering Eq. (5), Eq. (4) will take the form

$$EI\theta'_B = \Delta M_B \left(l - \frac{l^2}{2c} \right). \tag{6}$$

Then

$$\Delta M_B = \frac{EI\theta'_B}{l - \frac{l^2}{2c}}. \tag{7}$$

A similar result can be obtained for other rigid nodes, that is the value of bending moment in a rigid node in case of its opening depends of the value of the angle of rotation θ'_B, stiffness of elements to be joined EI, length of these elements l and the ratio of the value of bending moment to the value of transverse force in the node c.

The value of the opening of the flange node B, that is the value of the angle of rotation θ'_B, can be determined by considering the design scheme of the bolted connection in this node under the efforts, determined by static calculation of the frame, namely bending moment M_B, longitudinal N_B and transverse forces Q_B (Fig. 3).

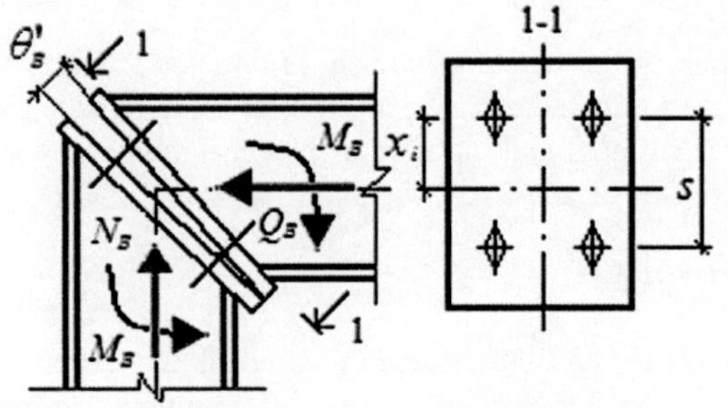

Fig. 3. Design scheme of node *B*

The force in the stretched bolt under the action of the bending moment

$$N_{bt} = \frac{M_B\, x_{max}}{2\sum x_i^2} \pm \frac{N_B}{n}. \tag{8}$$

where is x_{max} – the distance from the main axis to the axis of the ultimate bolt; x_i – the distances from the main axis to the i - th axis of each row of bolts; n – the number of bolts in the connection; N_B – longitudinal force in the element, that joins in the node.

Elongation of the most stretched bolt

$$\Delta l_6 = \frac{N_{bt} \sum t}{E\, A_{bn}}. \tag{9}$$

where is $\sum t$ – the sum of the thicknesses of the elements to be joined; A_{bn} – the cross-sectional area of the bolt netto.

The opening angle of the flange connection in the node, and accordingly, the angle of rotation of the left rack of the frame in this node, is

$$\theta'_B = arctg \frac{\Delta l_6}{S}. \tag{10}$$

where is S – the length of the plot of connection of two flanges between the most loaded and the least loaded bolts (see Fig. 3).

Thus, considering the design scheme of the flange connection in the node and determining the value of the opening angle of this connection, as well as having the selected cross section of the elements, their length, the ratio of bending moment and transverse force in the node, we can determine the change in bending moment ΔM due to of its disclosure.

The ratio $\frac{\Delta M}{M}100\%$ shows what percentage of stiffness the node has lost due to the opening of the flanges, and the new stiffness of the node will be

$$k = 1 - \frac{\Delta M}{M}. \tag{11}$$

The quantity of the dependence $\frac{EI\theta'_B}{l - \frac{l^2}{2c}}$ is constant value, and indicates, that the decrease in the value of the bending moment is directly proportional to the value of opening of the flange joint, and also depends on the length of the elements, connected in the node, of their stiffness and the ratio of bending moment to transverse force.

5 Scientific Novelty

The conducted researches and the received theoretical and experimental results [3–5] testify that the opening of flange nodes has essential influence on work of both nodes, and a frame design as a whole. As a result, the actual stiffness of the nodes in the design structures is reduced by 1.5–2.5% depending on the diameter of the bolts, their number and geometric parameters of the flanges. It is theoretically established and substantiated that if there are several such nodes in the frame structure, its total pliability will be determined as the sum of the reduction of stiffness of all nodes, that is the total stiffness of the structure decreases by the total stiffness reduction of all nodes. The performed calculations allow us to state, that for the hingeless frame as a whole (see Fig. 1), where there are four rigid nodes, the total pliability depending on the geometric parameters of all elements of the frame can be 6–10%.

6 Practical Significance

Since the cross sections of the elements, that connected in the node, near the rigid flange joints are usually the most tense, the accounting the real stiffness of bolted flange connections allows more rational design of such structural elements, namely to calculate their cross sections with accounting the redistribution of efforts along the length of the elements. As a result, the stresses in the calculated cross sections of the elements near the joints are reduced due to a certain pliability of both rigid nodes and the structure as a whole. The refined calculations, proposed by this method, make it possible to calculate the actual stresses in the cross sections, which will be less than those calculated by the traditional method, and as a result to obtain material saving on one frame structure as a whole about 3–5%.

7 Conclusions

The proposed method allows to determine the stiffness of any bolted flange connection, taking into account its actual operation on the basis of static calculation data. Accounting the pliability of the nodes, and, as a consequence, the structure as a whole, allows us to unload the elements and calculate the additional resource of bearing capacity during the design of the frame elements.

In addition, this technique allows us to adjust the stiffness of the bolted joint by changing the diameter of the bolts, their number, distance between them or changing the thickness and height of the flanges, as well as use additional material resource by reducing the maximum stresses in the calculated cross sections of elements.

References

1. Stalevi konstruktsiyi (Steel structures). DBN V.2.6 – 198: 2014 (2014)
2. Eurocode 3: Design of steel structures. EN 1993-1-8:2005 (2005)
3. Romaniuk, V., Supruniuk, V.: Mitsnist′ ta deformatyvnist′ perforovanykh elementiv stalevoyi arky (Strength and deformability of perforated elements of steel arch): monohrafiya, vol. 106. NUVHP, Rivne (2013)
4. Romaniuk, V., Supruniuk, V.: Osoblyvosti rozrakhunku prol′otnykh konstruktsiy z perforovanykh elementiv za skladnoho napruzheno-deformovanoho stanu (Features of the calculation of span structures from perforated elements in complex stress-strain state). Zbirnyk naukovykh prats′ Ukrayins′koho derzhavnoho universytetu zaliznychnoho transportu, vol. 175, pp. 98–108. Kharkiv, UkrDUZT, Vypusk (2018)
5. Romaniuk, V., Supruniuk, V.: Eksperymental′ni doslidzhennya prol′otnykh konstruktsiy z perforovanykh elementiv za skladnoho napruzheno-deformovanoho stanu (Experimental researches of flexible constructions from perforated elements at a complex stress-deformed state). Opir materialiv i teoriya sporud: nauk. – tekh zbirn, vol. 103, pp. 189–300. K: KNUBA. Vyp (2019)
6. Liu, X.-C., Cui, F.-Y., Jiang, Z.-Q., Wang, X.-Q.: Tension–bend–shear capacity of bolted-flange connection for square steel tube column. Eng. Struct. vol. 20115 (2019). Article 109798
7. Rezvani, F.H., Ronagh, H.: Span length effect on alternate load path capacity of welded unreinforced flange-bolted web connections. J. Constr. Steel Res. **138**, 714–728 (2017)
8. Wang, J., Uy, B., Thai, H.-T., Li, D.: Behaviour and design of demountable beam-to-column composite bolted joints with extended end-plates. J. Constr. Steel Res. **144**, 221–235 (2018)
9. Shardakov, I., Shestakov, A., Son, M., Zemlanuhin, A., Glot, I.: Beam to column flange connection: from elasticity to destruction (theory and experiment). Procedia Struct. Integr. **132018**, 1324–1329 (2018)
10. Shardakov, I., Shestakov, A., Glot, I.: Experimental and theoretical study of deformation processes in a flange connection of iron beams. Procedia Struct. Integr. **92018**, 207–214 (2019)

The Effect of Mechanical Activation on the Properties of Hardened Portland Cement

Myroslav Sanytsky[1(✉)], Alexandr Usherov-Marshak[2], Uliana Marushchak[1], and Alexey Kabus[2]

[1] Lviv Polytechnic National University, Lviv 79013, Ukraine
msanytsky@ukr.net
[2] Kharkiv National University of Civil Engineering and Architecture, Kharkiv, Ukraine

Abstract. The results of mechanical activation effect using vibro-milling on the properties of hardened Portland cement were presented. Comprehensive assessment of particle size distribution by volume and surface area of ordinary and activated Portland cement was carried out. The heat of hydration, initial and final setting times of the cement pastes, rheological properties and strength development of cement mortars were determined. The calorimetric data reveal that mechanical activation of ordinary Portland cement causes sharp increasing of heat hydration rate and the interaction at the boundaries of the "solid-liquid" intensifies at early hydration period as well as setting time reduction is observed. The research of workability and strength kinetics of mechanically activated Portland cement, modified with polycarboxylate superplasticizer are carried out. Modified mechanically activated Portland cement is characterized by early strength development (R_{c2}/R_{c28} = 0.74) as well as standard strength (R_{c28} = 59.6 MPa), which meet the requirements of rapid hardening high strength binders.

Keywords: Mechanical activation · Portland cement · Particle size distribution · Heat of hydration · Early hydration · Compressive strength

1 Introduction

Important criterions of innovative high performance and sustainable construction materials are providing by their high workability and intensive kinetics of strength development at the early age. Using of rapid hardening cement-based composites leads to possibility of loading structures earlier, increases of formwork turnover and reduces the time for the building construction [1]. With the aim to develop high performance construction materials, use of full cement potential would be the promising technique. Alkaline activation, hardening accelerator or nanoaddition such as nanosilica, C-S-H particles are used for early strength development of Portland cement [2–5]. Inclusion of the different nanoparticle in cement system exposes a higher surface area of the cementing material for the hydration reaction.

Z. Blikharskyy (Ed.): EcoComfort 2020, LNCE 100, pp. 378–384, 2021.
https://doi.org/10.1007/978-3-030-57340-9_46

The increasing of cement performance was achieved due to enhancement of the fineness with obtaining of ultrafine cement by physico-mechanical crushing of the conventional cement [6–8]. Technology of nanocement bases on use of fine cement with the nanoparticles contents of 24.0% and ultrafine cement with the 39.0% content of particles less than 1 μm, which are obtained in high-energy mills [9]. The ultrafine cement is attracting attention as the component for repairing deteriorated construction parts and cracks in concrete, as injection grouts, which provide cementation of fractured rocks and soft soils and prevent water ingress in underground construction structures [10]. Recently, blended and composite cements have been attracting attention due to their technical and environmental advantages. Such binders are alternative solutions aimed at decreasing resource use, saving energy as well as limiting greenhouse gas emissions. Fine grinding and mechanical activation have been suggested as effective methods to improve the reactivity of the blended and composite cement constituents [11, 12].

The disadvantage of ultrafine cements is increase of water demand, which can lead to a decrease of early strength. Therefore, a promising approach is the modification of such cement system with effective superplasticizers, which are characterized by high water reducing effect [13, 14].

The aim of this study is to investigate the effects of mechanical activation on the heat evolution and properties of Portland cement as well as effect of modification of activated Portland cement with polycarboxylate superplasticizer.

2 Experimental Program

2.1 Materials

Ordinary Portland cement CEM I 42.5R (OPC) provided by JSC "Ivano-Frankivskcement" (Ukraine) was used in investigation. The composition of the OPC presented by C_3S 64.44%, C_2S 12.88%, C_3A 5.65%, C_4AF 14.73%. The particles` size, density, and fineness of the used ordinary Portland cement are 10–30 μm, 3.15 g/cm^3 and 330 m^2/kg, respectively. CEM I 42.5R meets the requirements of EN 197. Polycarboxylate superplasticizer GLENIUM ACE 430 (PCE) was used for regulation of flowability of cement mortar based on mechanically activated Portland cement. The superplasticizer (1.0 mass%) was added together with mixing water.

2.2 Experimental Process and Methods

Mechanical activation of Portland cement was performed using laboratory vibro-mill for 60 and 90 min. The test program consisted of the following measurements: particle size distribution, isothermal calorimetry, setting time and strength measurements. The particle size distribution of Portland cement and activated cement was measured by a laser granulometer Mastersizer 3000. The effect of mechanical activation was estimated by degree of enhancement surface area of OPC and activated cement. It was determined as the ratio between surface area of particles and their volume (A/V). To assess the contribution of individual particles to total specific surface area, it was calculated a coefficient of incremental surface area (K_{isa}), which is determined by multiplying the A/V by the incremental volume of each fraction of the material.

Isothermal calorimetry was conducted to study the early hydration behavior in terms of the rate of heat release as well as the total heat released [15]. The heat of hydration was determined by isothermal microcalorimeter during the first 48 h of hardening at the temperature of 25 °C. The experiment was carried out for cement pastes with W/C = 0.4. Vicat needle tests were used to measure setting time. The highly flowable cement mortar (cement:sand = 1:3) based on natural sand (fineness modulus is 1.8) was used for research of strength development kinetics of the mechanically activated Portland cement. The consistency of fresh mortars was determined by flow table method.

3 Results and Discussion

The specific surface area of Portland cement CEM I 42.5R after 60 min of mechanical activation in the vibro-mill increases up to 510 m^2/kg, after 90 min – up to 640 m^2/kg. Particle size distribution of OPC with characteristic diameters D_{10} = 2.07 μm, D_{50} = 14.6 μm and D_{90} = 43.7 μm is shown in Fig. 1a. The content of particles of the ultrafine fraction (less than 1.0 μm) for CEM I 42.5R is 5.78%. The characteristic diameters D_{10} for mechanically activated cement with specific surface area 510 m^2/kg and 640 m^2/kg decrease to 1.46 and 1.26 μm, respectively. The maximum coefficient of incremental surface area of OPC (K_{isa} = 6.04 μm^{-1}·vol.%) corresponds to particle size 0.29 μm (Fig. 1b). The content of the ultrafine fraction of mechanically activated cements increases and the maximum of the differential coefficient of surface activity increases to K_{isa} = 8.21 μm^{-1}·vol.% for specific surface area of 510 m^2/kg and to K_{isa} = 9.86 μm^{-1}·vol.% for specific surface area of 640 m^2/kg.

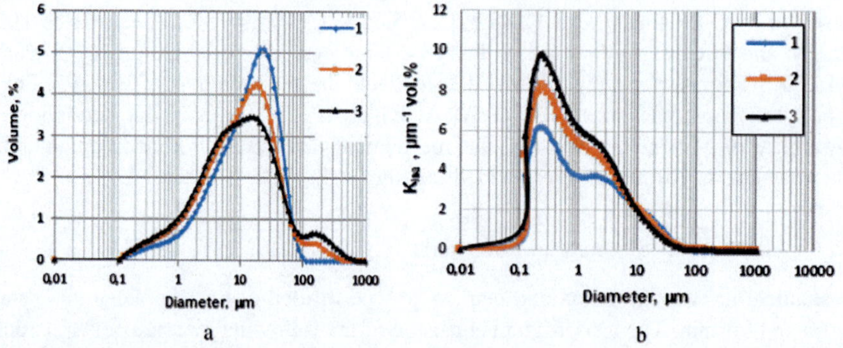

Fig. 1. Particle size distribution (a) and coefficient of incremental surface area (b) of Portland cements with a specific surface area, m^2/kg: 1–330; 2–510; 3–640

Figure 2 shows the variation of the rate of hydration of ordinary Portland cement and activated cement (specific surface area is 640 m^2/kg). The heat curves of Portland cement show two peaks. The appearance of the first exoeffect is due to the phenomena of wetting the binder particles with water, the second – due to hydrate formation and transitions of primary thermodynamically unstable hydrates. The first peak is observed after 6 min of ordinary Portland cement hydration and reaches 11.6 mW/g. The second exoeffect of 4.8 mW/g occurs after 9.2 h.

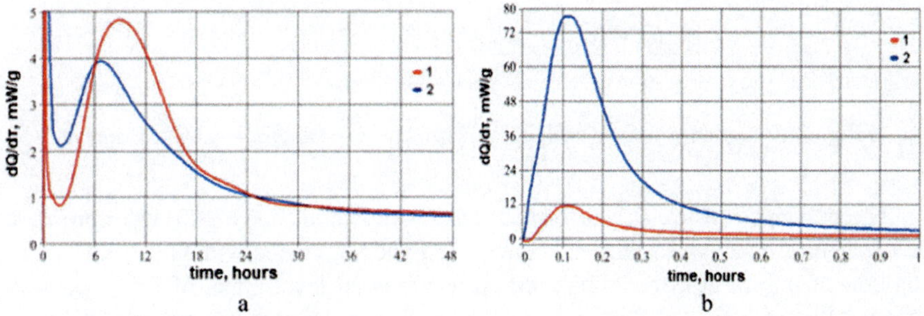

Fig. 2. Heat flow (a) and first exoeffect of heat flow (b) of cement: 1 – OPC, 2 – activated cement

The mechanical activation of Portland cement causes sharp increasing of first peak value due to the increase of binder fineness. As a result, the interaction at the boundaries of the "solid-liquid" intensifies. The rate of heat of hydration and cumulative heat of hydration release increase, while the duration of the induction period decreases during the hydration of mechanically activated Portland cement. Thus, the first exoeffect is observed after 6 min and reaches 77.7 mW/g. The peak rate of heat evolution of mechanically activated Portland cement increase by 7 time compared to OPC. The rate of heat evolution of activated cement after 48 h of hydration is recorded. The second peak – 3.9 mW/g is observed after 6.7 h.

It can be observed (Fig. 3) that the total heat of hydration evolved is higher for activated cement compared to the OPC. The total heat for OPC after 24 h is 229 J/g, and after 48 h – 294 J/g. If the total heat released at 24 and 48 h of hydration is compared, activated Portland cement resulted in 14.4% and 10.9% increase in the total heat. Increasing the values of heat of hydration indicates a more complete hydration process and their intensification with the rapid formation of complex hydrates such as ettringite and increasing the content of C-S-H phases.

The setting time results for ordinary and mechanically activated Portland cement demonstrate that the initial setting time for the Portland cement paste is 150 min whereas the final setting time is 240 min. When the specific surface area changes from 330 to 640 m^2/kg a reduction of setting time is revealed. The mechanically activated Portland cement (specific surface area of 510 m^2/kg) is characterized by initial and final setting time of 40 and 160 min respectively. When specific surface area increases to 640 m^2/kg initial and final setting time decrease to 30 and 130 min respectively. At the same time, increase of specific surface area causes increase of water demand from 30 to 33% to obtain constant test consistency.

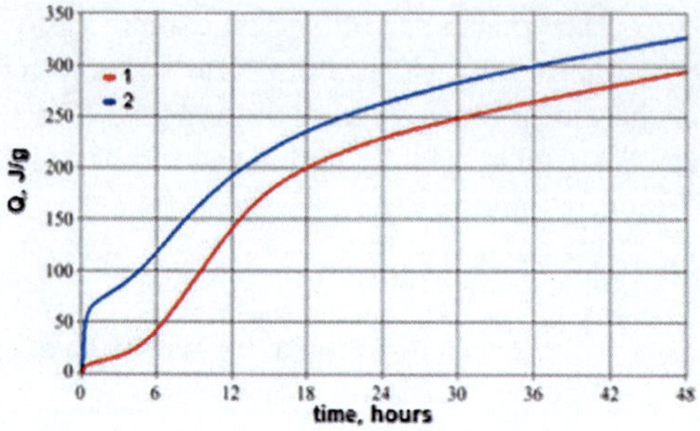

Fig. 3. Cumulative heat of hydration of cement: 1 – OPC, 2 – activated cement

The compressive strength of cement mortar (flowability F = 180–185 mm) based on mechanically activated Portland cement is higher compared with OPC (Fig. 4). The addition of polycarboxylate superplasticizer allows to level effect of increased water consumption of activated Portland cement to achieve the required workability.

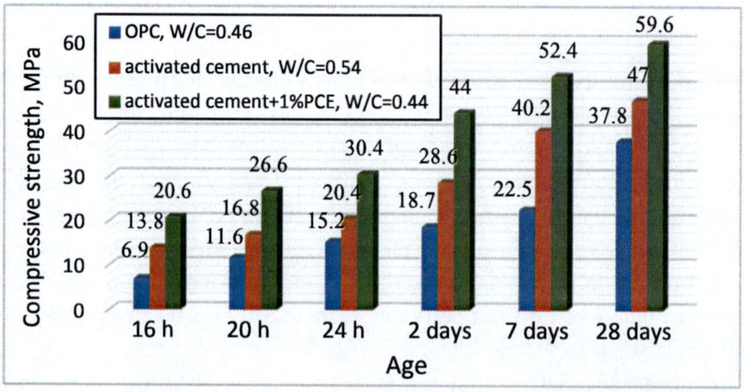

Fig. 4. Compressive strength of mortars based on the OPC and mechanically activated cement

The significant acceleration of the early strength of mechanically activated Portland cement is achieved due to water-reducing effect of the plasticizing admixture ($\Delta W/C = 18.5\%$) and the increased specific surface area of Portland cement ($640\ m^2/kg$). Modified mechanically activated Portland cement (F = 180 mm) is characterized by rapid strength development $R_{c2}/R_{c28} = 0.74$. The strength of modified mechanically activated Portland cement after 28 days ($R_{c28} = 59.6$ MPa) meets requirements to high-strength binders.

4 Conclusions

The mechanically activated Portland cement characterizes by increased activity caused by higher surface area. The coefficient of incremental surface area of mechanically activated Portland cement with specific surface area of 510 and 640 m^2/kg increases to 8.21 and 9.86 μm^{-1}·vol.%, respectively, compared to 6.04 μm^{-1}·vol.% for OPC. The mechanically activated cement exhibited high cumulative heat release in the initial stages. Mortar based on mechanically activated Portland cement modified with polycarboxylate superplastisizer characterizes by rapid strength development (R_{c2}/R_{c28} = 0.74). Its strength after 28 days increases by 57.7% compared to mortar based on OPC. Developed modified mechanically activated Portland cement meets the requirements to rapid hardening high strength binder.

References

1. Dvorkin, L., Babych, Y., Zhytkovsky, V., Bordyuzhenko, O., Filipchuk, S., Kochkarov, D., Kovalyk, I., Kovalchuk, T., Skrypnyk, M.: High-strength rapid hardening concretes and fiber reinforced concretes. NUVGP, Rivne (2017). (in Ukrainian)
2. Krivenko, P., Sanytsky, M., Kropyvnytska, T.: The effect of nanosilica on the early strength of alkali-activated Portland composite cements. Solid Stat. Phenom. **296**, 21–26 (2019)
3. Sanytsky, M., Marushchak, U., Olevych, Y., Novytskyi, Y.: Nano-modified ultra-rapid hardening Portland cement compositions for high strength concretes. In: Lecture Notes in Civil Engineering. Proceedings of CEE 2019, vol. 47, pp. 392–399 (2020)
4. Sikora, P., Lootens, D., Liard, M., Stephan, D.: The effects of seawater and nanosilica on the performance of blended cements and composites. Appl. Nanosci. (2020)
5. Savchuk, Y., Plugin, A., Lyuty, V., Pluhin, O., Borziak, O.: Study of influence of the alkaline component on the physico-mechanical properties of the low clinker and clinkerless waterproof compositions. In: MATEC Web of Conferences (2018)
6. Jo, B.-W., Chakraborty, S., Kim, K.H., Lee, Y.S.: The top-down nanotechnology in the production of ultrafine cement ($\sim$220 nm). J. Nanomater. 1–9 (2014)
7. Ivanova, I., Nefedov, S., Pustovgar, A., Adamtsevich, A., Eremin, A.: Comparison of laboratory methods for the design of injection grouts based on microfine cements. Procedia Engineering **165**, 1536–1541 (2016)
8. Chen, X., Zhou, J., Yuanyuan, Y.: Hydration of ultrafine and ordinary Portland cement at early ages. KSCE J. Civ. Eng. **18**, 1720–1725 (2014)
9. Chatterjee, A.: Chemistry and engineering of the clinkerization process – incremental advances and lack of breakthroughs. Cement Concr. Res. **41**(7), 624–641 (2011)
10. Bentz, D.P., Garboczi, E.J., Haecker, C.J., Jensen, O.M.: Effects of cement particle size distribution on performance properties of Portland cement-based materials. Cement Concr. Res. **29**(10), 1663–1671 (1999)
11. Sobolev, K., Lin, Z., Cao, Y., Sun, H., Flores, V.I., Rushing, T., Cummins, T., Weiss, W.: The influence of mechanical activation vibro-milling on the early age of hydration and strength development of cement. Cement Concr. Compos. **71**, 53–62 (2016)
12. Kropyvnytska, T., Rucinska, T., Ivashchyshyn, H., Kotiv, R.: Development of eco-efficient composite cements with high early strength. In: Lecture Notes in Civil Engineering. In: Proceedings of CEE 2019, vol. 47, pp. 211–218 (2019)

13. Marushchak, U., Sanytsky, M., Pozniak, O., Mazurak, O.: Peculiarities of nanomodified Portland systems structure formation. Chem. Chem. Tech. **13**(4), 510–517 (2019)
14. Marushchak, U., Sanytsky, M., Korolko, S., Shabatura, Y., Sydor, N.: Development of nanomodified rapid hardening fiber reinforced concretes for structure of special purpose. East. Euro. J. Enterp. Tech. **2/6**(92), 34–41 (2018)
15. Usherov-Marshak, A., Kabus, A.: Calorimetric monitoring of early hardening of cement in the presence of admixtures. Inorg. Mater. **49**(4), 449–452 (2013)

Thermal Renewal of Industrial Buildings Gas Supply System

Olena Savchenko(✉), Orest Voznyak, Khrystyna Myroniuk, and Oleksandr Dovbush

Lviv Polytechnic National University, Lviv 79013, Ukraine
o.savchenko@i.ua

Abstract. An important priority of the economic policy of Ukraine is the careful use of energy The article is devoted to decision of actual task of energy saving. Therefore, now a large-scale energy saving policy is being implemented in Ukraine, and energy efficiency objectives are comprehensive and cover both the legislative framework and technical innovations. The goal of the paper is to establish economic indicators of thermal renovation measures during the reconstruction of the gas supply system of the industrial buildings at condition of discount rate different magnitudes. It is consumed in the industrial buildings too much amount of gas. One of the effective ways to reduce energy consumption for boiler gas supply is to carry out the thermal modernization of the gas supply system. This article describes the economic indicators of thermal renovation measures during the reconstruction of the gas supply system of the industrial buildings. During the reconstruction of, the following thermal modernization measures were taken for comparison: installation of an energy separator, reconstruction of the gas supply system, and installation of automatics.

Keywords: Gas supply system · Energy saving · Energy audit · Thermal renewal · Heat utilization · Discount rate

1 Introduction

In our time, the issues of energy saving [5, 11], accounting for energy resources and managing their costs are extremely relevant. In the context of the acute economic crisis, the careful use of energy carriers is an important priority of Ukraine's economic policy. At present, as a priority task, a large-scale energy efficiency policy is being implemented in our country. The tasks of energy saving in Ukraine are complex and cover aspects of both external heat supply and internal engineering systems of buildings (heating, ventilation and air conditioning) [6–8], as well as the legislative framework [15] and technical innovation [4, 9]. Hence, a large amount of energy is spent on gas consumption in industrial buildings.

There is no doubt that the energy needs for gas supply system should be reduced as a result of thermal modernization. In order to achieve the maximum effect, it is necessary to determine the economically feasible level of thermal protection of gas supply systems, which should be optimal both in heat engineering and in economic terms.

Z. Blikharskyy (Ed.): EcoComfort 2020, LNCE 100, pp. 385–392, 2021.
https://doi.org/10.1007/978-3-030-57340-9_47

2 Analysis of Current Research and Problem Statement

Careful use of energy is an important priority for European economic policy. This problem is complex and covers both the legislative framework and technical innovation [14, 17, 18].

A large amount of thermal energy is consumed by the indoor climate systems [10, 20]. There is an objective need for energy savings [12, 16] and their reduction through energy audits. However, the issue of energy audit of gas supply systems was not considered. It is in this aspect that energy efficiency of the system as a whole can be achieved by reducing gas demand. The mathematic models [1–3] have been analyzed.

Therefore, thermal modernization of the gas supply system is an effective way of reducing energy costs. At the same time an important role is played by the technical and economic evaluation of the effectiveness of the measures. The use of modern methods of assessing the cost-effectiveness of thermo-modernization has been taken into account in the latest concept of economic calculations, in particular the recommendation of UNIDO (United Nations Industrial Development Organization) [19]. In this regard, it is advisable to use terminology, symbols and basic economic characteristics according to [19]. It is of interest to compare the results of the audit with different discount rates.

3 The Purpose and Objectives of the Study

The goal of this paper is to establish economic indicators of thermal renovation measures during the reconstruction of the gas supply system in the industrial buildings at condition of discount rate different magnitudes.

4 Methods, Materials and Research Results

In gas supply systems reconstruction such thermal efficient (energy saving) measures deserve our attention: installation of an energy separator, reconstruction of the gas supply system and installation of automatics.

The energy separator allows, without the presence of any additional primary energy sources, to heat natural gas at the gas distribution station before the reduction process to prevent the formation of crystalline hydrates in the gas pressure regulator [13]. The vortex tube was first investigated by J. Ranque. The design of the vortex tube is shown on Fig. 1.

The energy separator operates in the following way. The compressed gas is fed through a tangential nozzle duct into a pipe where intense circular motion is established. In this case, there is an uneven field of temperature. The gas layers near the axis are colder than the inlet gas, and the peripheral layers of the swirled flow are heated. Part of the gas in the form of a cold stream is drawn through the diaphragm, and the other part in the form of heated gas through the throttle valve is drawn from the other side of the pipe. This phenomenon is called the Ranque effect. When the throttle is closed gradually, the overall pressure level in the energy separator increases and the

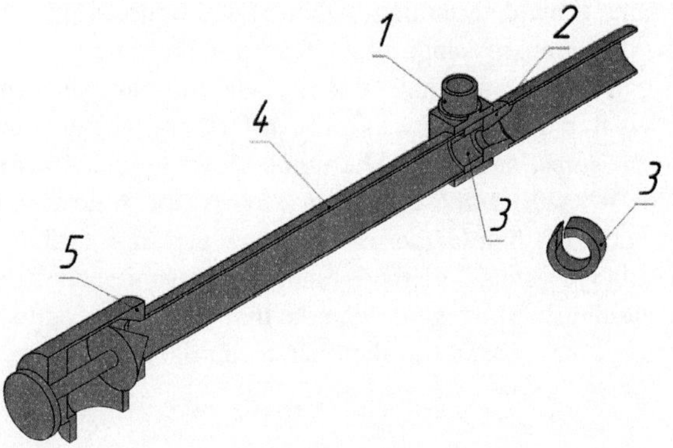

Fig. 1. Design of the vortex tube: 1 - case, 2 - diaphragm, 3 - nozzle channel, 4 - pipe, 5 - throttle valve.

flow of cold flowing through the aperture of the diaphragm increases with a corresponding reduction in the flow of hot flush. In this case, the temperature of the cold and heated streams also changes.

The gas distribution station with a vortex tube operates as follows. After the cleaning unit the gas flows from the main gas pipeline to the vortex tube entrance. In the vortex tube the gas is separated into two flows, i.e. heated and cold. The heated flow enters to the gas pressure regulator. The cold flow through the safety shut-off valve enters the distribution gas pipeline downstream of the gas pressure regulator. Pressure reduction of the heated natural gas avoids the formation of crystalline hydrates in the middle of the gas pressure regulator and accordingly ensures the reliability of gas supply. The throttle valve of vortex tube in this scheme of the gas distribution station is set to the specified quantity of the heated stream. Natural gas heating prevents the formation of crystalline hydrates inside the pressure regulator. Thus, the reliability of gas supply to consumers is ensured. On GDS the vortex tube throttle valve is fixed in the fixed position and does not allow regulate the amount of heated flow. The fraction of the heated stream is in the range 0.8–0.9 from the quantity inlet stream. The safety shut-off valve does not allow the gas pressure to increase downstream of the gas pressure regulator in the distribution pipeline.

The economic assessment implies the use of a modern methodology for assessing the economic efficiency of thermo-modernization systems, which takes into account the latest concepts of economic calculations, in particular UNIDO (United Nations Industrial Development Organization) recommendations. According to the concept of UNIDO, by introducing some symbols, for each simple thermal renewal measure the following basic economic characteristics are defined [19].

For the most qualified energy audit, it is expedient to consider the maximum possible number of the thermal renewal measures that can be operated by an energy auditor, that is, the so-called list of the thermal renewal measures should be as complete as possible. Therefore, there is a need to create such a method of carrying out the energy auditing, which would allow avoiding bulkiness when considering all possible thermal renewal variant, giving the opportunity to reduce their quantity in a reasonable way, and at the same time, determining without fail the most optimal final result - the re-commendation of the energy auditor to the customer.

Consequently, in order to optimize, it is necessary to compile a square matrix with the number of rows n and columns m, which, in fact, equals the number of all simple thermal renewal measures, namely n. The number of aggregate thermal renewal variants marked with "+" will increase by 1 in each of the following columns until it reaches the last total of simple thermal renewal measure (Table 1). In it Arabic numerals are numbered simple thermal renewal measure, and Roman - aggregate thermal renewal variant. In this regard, we note that the rows need to be filled with the appropriate thermosetting measures as their parameter S_i increases, i.e. from $SPBT_{\min}$ to $SPBT_{\max}$.

Table 1. Characteristics of energy saving measures.

No.	Measures	Energy costs for the "basic" option Qo, MJ/year	After the change Qi, MJ/year	Energy saving ΔQi ΔQi = Qo – Qi, MJ/year	Costs savings Ki Ki = ΔQi. Px, th. EUR/year
1	Use of the energy separator	23385	15170	8215	96.97
2	Reconstruction of the gas supply system	23385	18285	5100	56.10
3	Automatics installation	23385	20355	3030	33.33

Output data for energy audit of the GSS are: geographical position, technical drawings, annual energy consumption for the gas consumption of the GSS Q_x, MJ/year, cost of energy consumption P_x, EUR/MJ, data for counting of estimated cost of thermal modernization works Ii, EUR, the degree of discount r (economic analysis is carried out under conditions of constant prices and timing of investment consideration t = 15 years).

As a result, an optimal renovation option and its economic parameters are determined. Solution is obtained by the following algorithm:

1. Calculation of annual energy consumption for the needs of GSS Q_x, MJ/year and this option is considered as "basic".
2. Choosing a "list" of thermal renovation measures for this system.
   2.1. Use of the energy separator.
   2.2. Reconstruction of the gas supply system.
   2.3. Automatics installation.
3. Calculating the energy efficiency ΔQ_i for each thermal renewal measure as $\Delta Q_i = Q_o - Q_i$, and hence the annual savings K_i, EUR/year as $K_i = \Delta Q_i \cdot P_x$

The results of calculations are listed in Table 1.

4. Determination of each thermal renewal measure indexes (Table 2) at condition of different discount rates in Ukraine: r = 0,18 (2019) and r = 0,11 (2020).

Table 2. Economic indicators of the thermal renovation measures.

No	Measures	Ii	K'i	SPBTi	NPVRi r = 0,18	NPVRi r = 0,11
		th. EUR	th. EUR/year	year	EUR	EUR
1	Use of the energy separator	145.43	96.97	1.5	+72.55	+225.52
2	Reconstruction of the gas supply system	123.40	56.10	2.2	+15.43	+44.42
3	Automatics installation	96.67	33.33	2.9	+85.2	+8,0

5. Table 3 and Fig. 2 illustrates the dynamics of funds by years for the period of 15 years of operation of the event"use of the energy separator" taking into account the discount rate r = 0,11.

Table 3. Dynamics of funds by years.

t	0	1	2	3	4	5	6	7
NPVR	−145.4	−58.1	+11.0	+67.2	+109.7	+142.3	+167.4	+177.5
t	8	9	10	11	12	13	14	15
NPVR	+191.3	+196.0	+196.1	+193.0	+188.9	+177.4	+167.4	+157.6

6. Optimization to get the maximum economic effect due to thermal renovation.

Since the consideration of the thermal renewal measures in amount of 2^n is an extremely cumbersome process, it is advisable to simplify it, using a scientifically based methodology aimed at reducing the required amount of the thermal renewal measures, that is, to carry out appropriate optimization (Table 4).

We will optimize the thermal renewal variant taking into account the data of clause 6 and compile Table 3 ordered from the first thermal renewal "Use of the energy separator", in which the parameter $SPBT_1$ is minimal, until the last (third) "Reconstruction of the gas supply system" with a maximum $SPBT_3$.

The optimum, as noted, is that thermal renewal measure, in which value $NPVR_j$ is the maximum, namely, TRV_{III}. This means that the maximum economic effect will be in the case of the simultaneous application of three thermal renewal measures (Table 3).

The specific profit from the introduction of energy-saving technologies during of their operation period is approximately 220 th.EUR.

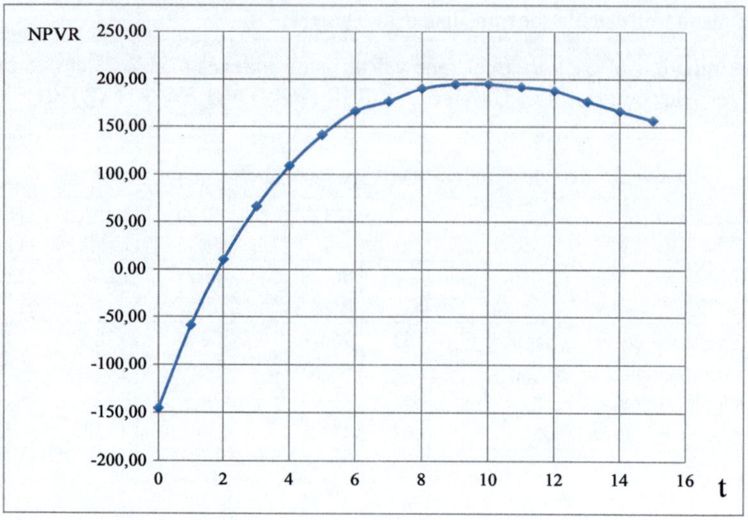

Fig. 2. Dynamics of funds by years.

Table 4. Optimization of options according to clause 6.

No.	Measures	Options		
		I	II	III
1	Use of the energy separator	+	+	+
2	Reconstruction of the gas supply system		+	+
3	Automatics installation			+
Indexes				
1	Investment expenses *I* (th.EUR)	145.43	268.87	365.53
2	Annual savings *K* (th.EUR/year)	96.97	153.07	186.40
3	Simple Payback Time - *SPBT* (year)	1.5	1.8	2.0
4	Net Present Value Ratio - *NPVR* (th.EUR) $r = 0{,}18$	+53.41	+64.76	+77.38
5	Net Present Value Ratio - *NPVR* (th.EUR) $r = 0{,}11$	+157.60	+194.50	+219.76

5 Conclusions

1. Use of gas separator will enable the design of energy-saving gas supply systems in the industrial buildings. Its simple payback time is 1,5 year.

2. The most profitable thermal renewal variant is aggregate action of three thermal renewal measures: using of the energy separator, reconstruction of the gas supply system, automatics installation. The specific profit from the introduction of energy-saving technologies during of their operation period is approximately 220 th.EUR.

References

1. Saleh, A.M.: Modeling of flat–plate solar collector operation in transient states. Purdue University, Fort Wayne, p. 73 (2012)
2. Basok, B., Davydenko, B., Farenuyk, G., Goncharuk, S.: Computational modeling of the temperature regime in a room with a two-panel radiator. J. Eng. Phys. Thermophys. **87**(6), 1433–1437 (2014)
3. Basok, B., Davydenko, B., Isaev, S., Goncharuk, S., Kuzhel, L.: Numerical modeling of heat transfer through a triple-pane window. J. Eng. Phys. Thermophys. **89**(5), 1277–1283 (2016)
4. Beshentseva, S.: Analysis of methods for preventing hydrate formation in pipelines. Bull. Cybern. **11**, 40–44 (2012)
5. Bilous, I., Deshko, V., Sukhodub, I.: Parametric analysis of external and internal factors influence on building energy performance using non-linear multivariate regression models. J. Build. Eng. **20**, 327–336 (2018)
6. Bilous, I., Deshko, V., Sukhodub, I.: Building inside air temperature parametric study. Mag. Civ. Eng. **68**(8), 65–75 (2016)
7. Buyak, N., Deshko, V., Sukhodub, I.: Buildings energy use and human thermal comfort according to energy and exergy approach. Energy Build. **146**(1), 172–181 (2017)
8. Deshko, V., Buyak, N.: A model of human thermal comfort for analysing the energy performance of buildings. East. Euro. J. Enterp. Tech. **4**(8–82), 42–48 (2016)
9. Fedoseyev, S.: Natural gas hydrates - prospects for study and use. Sci. Tech. Yakutia **1**(18), 14–18 (2010)
10. Kapalo, P., Sedláková, A., Košicanová, D.,Voznyak, O., Lojkovics, J., Siroczki, P.: Effect of ventilation on indoor environmental quality in buildings. In: The 9th International Conference "Environmental Engineering", Vilnius, Lithuania Selected Papers (2014). eISSN 2029-7092/eISBN 978-609-457-640-9
11. Klymchuk, O., Denysova, A., Shramenko, A., Borysenko, K., Ivanova, L.: Theoretical and experimental investigation of the efficiency of the use of heat-accumulating material for heat supply systems. EUREKA Phys. Eng. **2019**(3), 32–40 (2019)
12. Klymchuk, O., Denysova, A., Mazurenko, A., Balasanian, G., Tsurkan, A.: Construction of methods to improve operational efficiency of an intermittent heat supply system by determining conditions to employ a standby heating mode. East. Euro. J. Enterp. Tech. **6**(8–96), 25–31 (2018)
13. Koval, R., Banakhevych, Yu., Balinsky, I., Kashina, O.: Patent of Ukraine for invention, No. 43673, Gas distribution station, Bulletin, no. 11, 3p. 17 December 2001
14. Labay, V.Y., Savchenko, O.O., Zhelykh, V.M., Kozak, K.: Mathematical modeling of the heating process in a vortex tube at the gas distribution stations. Math. Model. Comput. **6**(2), 311–319 (2019)
15. National Commission for State Regulation of Energy and Utilities. http://www.nerc.gov.ua
16. Redko, A., Dzhyoiev, R., Davidenko, A., Pavlovskaya, A., Pavlovskiy, S., Redko, I., Kulikova, N., Redko, O.: Aerodynamic processes and heat exchange in the furnace of a steam boiler with a secondary emitter. Alexandria Eng. J. **58**(1), 89–101 (2019)
17. Savchenko, O., Zhelykh, V., Yurkevych, Yu., Shapoval, S., Kozak, K.: Using vortex tube for decreasing lossesof natural gas in engineering systems of gas supply. Pollack Periodica **13**(3), 241–250 (2018)

18. Smith, E., Pongjet, P.: Review of Ranque-Hilsch effects in vortex tubes. Renew. Sustain. Energy Rev. **12**, 1822–1842 (2008)
19. United Nations Industrial Development Organization. https://www.unido.org
20. Voznyak O., Korbut V., Davydenko B., Sukholova I.: Air Distribution Efficiency in a Room by a Two-Flow Device. In: Blikharskyy Z., Koszelnik P., Mesaros P. (eds.) Proceedings of CEE 2019. CEE 2019. Lecture Notes in Civil Engineering, vol. 47. Springer, Cham (2020)

The Sustainable Design of the Greenhouse by Criteria of Heat Losses and Solar Heat Gains

Mykola Savytskiy[1], Maryna Bordun[1(✉)], and Vitalii Spyrydonenkov[2]

[1] State Higher Education Establishment «Prydniprovska State Academy of Civil Engineering and Architecture», Dnipro, Ukraine
klmari@ukr.net
[2] Private Construction and Installation Enterprise "STROITEL-P", Dnipro, Ukraine

Abstract. The efficiency of greenhouse design is its ability to receive and accumulate heat from solar radiation. Lots of factors affect the efficiency of greenhouse. One of the main factors is the shape of the greenhouse. The greenhouse shape affects the amount of solar heat gain that received the greenhouse and the amount of heat losses during cold season. The purpose of this research is to determine the optimal shape of greenhouse by the criteria of solar heat gain and heat losses through the covering to reduce total energy consumption and improve energy efficiency of the greenhouse. The five most commonly used greenhouse shapes were considered. In order to determine the optimal shape of energy efficient greenhouse the calculations of the total solar heat gain during the year and heat losses during the cold period (October-April) for the five most common shapes of greenhouses were made. The comparison of solar heat gains and heat losses through enclosing constructions of greenhouses was made.

Keywords: Energy efficient greenhouse · Solar heat gain · Heat losses · Optimal greenhouse shape

1 Introduction

The main purpose of greenhouses is to continue the vegetation period and to change the environment in accordance with the needs of growing crops to ensure their high yield. Greenhouses are equipped the engineering systems, such as the heat supply, water supply, ventilation, lighting, etc. These systems allow creating and controlling an alternative environment for growing crops irrespective of external climatic conditions. Such systems usually require a lot of energy for their operation, especially if the greenhouse is operated during the year.

Therefore, more and more attention has recently been paid to search of the design solutions for sustainable and energy-efficient greenhouses that are operated with minimal energy consumption, but have high performance.

Z. Blikharskyy (Ed.): EcoComfort 2020, LNCE 100, pp. 393–401, 2021.
https://doi.org/10.1007/978-3-030-57340-9_48

2 Literature Review

The integrated approach must be used even at the design stage of the greenhouse to reduce energy consumption in the operation period and to improve energy efficiency of greenhouse. Namely, it is necessary to consider the factors, which affect the energy consumption and increase productivity of the greenhouse. The efficiency of a greenhouse design is its ability to receive and accumulate heat from solar radiation.

The main factors that have to be considered at the design and building stage of an energy efficient greenhouse are the location and orientation of the greenhouse according to the sides of the world; the shape of the greenhouse; the materials of the greenhouse enclosing constructions; engineering systems of the heat energy generation and accumulation.

The optimal location and orientation of the greenhouse are among the first factors which have to be considered during the design of the greenhouse. The amounts of solar heat and light which will get into greenhouse depend on its orientation. The greenhouse should have the most advantageous orientation to ensure maximum penetration of sunlight and heat during the day. The solar energy that the greenhouse obtains can be used not only in the lighting, but also as an additional alternative energy source in the heating system of the greenhouse [1, 2].

The next factor that affects the energy efficiency of the greenhouse is the design shape.

The most common greenhouse shapes are the rectangular shape with gable even-span and uneven-span transparent roofing, rectangular shape with gable transparent roofing and sloping walls, the shape with single-gable transparent roofing and opaque wall on one side, and arched greenhouses (Fig. 1).

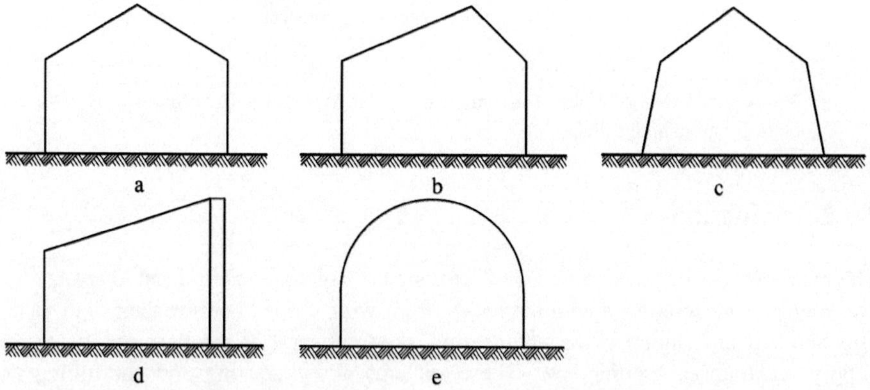

Fig. 1. The most common greenhouse shapes: a - the rectangular shape with gable even transparent roofing; b - the rectangular shape with gable uneven transparent roofing; c - rectangular shape with gable roofing and sloping walls; d - the shape with single-gable roofing and opaque wall on one side; e - the arched shape.

The rectangular greenhouse with gable even-span transparent roofing is traditional shape of the detached greenhouses. The most optimal inclination of the roofing surface depends on lighting and climatic conditions of the region and is from 25 to 40° (Fig. 1a).

The one side of the transparent roofing, which is oriented to the South, is usually increased in the greenhouse with gable uneven transparent roofing (Fig. 1b). It is considered that more solar energy enters the greenhouse through the enlarged south side of the transparent roofing. Sometimes, the other side is made of opaque construction materials to reduce heat losses [3].

The walls have low inclination in the greenhouse with gable transparent roofing and sloping walls. This increases the amount of solar energy that gets to greenhouse in winter, when the solar is low on the horizon (Fig. 1c) [3].

The shape with single-gable transparent roofing, which is usually oriented to the South has one opaque capital wall, which is oriented to the North (Fig. 1d). This shape helps to reduce the heat losses and to increase solar energy gain in such greenhouse [2, 3].

The arched shape of the greenhouse has a smaller reflective surface so more solar energy can enter the greenhouse. Moreover, such shape allows making the lower height of the greenhouse (Fig. 1e).

There are other forms of greenhouses, such as drop-shaped, polygonal, dome, prismatic, pyramidal, but they are less common. Construction cost of these greenhouses is higher than traditional and the construction is more difficult.

The construction materials also affect the overall energy efficiency of the greenhouse. The values of heat losses through the enveloping constructions depend on the type of material of the transparent surfaces (glass, polyethylene film or polycarbonate) [4]. When we choose materials for opaque greenhouse structures, such as the foundation, or the wall of the single-gable greenhouses, we'll have to consider that these elements can be used as heat accumulators [1].

The purpose of this research is to determine the optimal shape of a greenhouse by the criteria of solar heat gain and heat losses through the enclosing constructions to reduce total energy consumption and improve energy efficiency of the greenhouse.

3 Materials and Methods

To determine the optimal shape of an energy efficient greenhouse the calculations of the total solar heat gain during the year and heat losses during the cold period (October-April) for the five most common shapes of greenhouses were made: (Shape 1) - the rectangular shape with gable even transparent roofing; (Shape 2) - the rectangular shape with gable uneven transparent roofing; (Shape 3) - rectangular shape with gable roofing and sloping walls; (Shape 4) - the shape with single-gable roofing and opaque wall on one side; (Shape 5) - the arched shape.

All greenhouse models that are investigated have the same dimensions in plan 8.0 × 5.0 m, height - 4 m and the biggest transparent surface is orientated to the South. The transparent greenhouse roofing is made of cellular polycarbonate of 10 mm thickness. All greenhouses are located in Dnipro city (Ukraine). Despite the fact that the greenhouses have the same overall dimensions, the area of their transparent surface are different due to the different shape (Table 1). The total amount of sunlight will be different for each shape because each greenhouse has the different ratio between the transparent surface area and the floor area.

Table 1. Geometric characteristics of the greenhouse shapes.

Type of the greenhouse shape	Floor space greenhouse, m^2	Transparent surface area, m^2				Total transparent surface area, $\frac{m^2}{\%}$ *
		S	N	W	E	
Shape 1	40	42,1	42,1	14,8	14,8	$\frac{113,8}{92,9}$
Shape 2	40	51,1	34,7	14,8	14,8	$\frac{115,47}{94,3}$
Shape 3	40	39,5	39,5	12,9	12,9	$\frac{104,8}{85,6}$
Shape 4	40	60,8	–	14,8	14,8	$\frac{90,5}{73,9}$
Shape 5	40	44,1	44,1	17,3	17,3	$\frac{122,48}{100}$

* Note: The absolute values are given in numerator, the relative values are given in denominator (in percentage).

The following source data were adopted to calculate of the heat losses: outside air temperature in according to [5] for each month; the internal air temperature +16 °C [6]; the heat transfer coefficient on the surfaces that borders with the inside air is 8 W/(m·K) for the translucent surfaces and 8.7 W/(m·K) for non-transparent; the heat transfer coefficient on the surfaces that borders with the outside air is 23 W/(m·K) [7]. The soil temperature was set by layers for each month according to the study [8], where its constant value is at a depth of 3.2 m. The coefficient of heat conductivity soil (for loam) was assumed as 1.02 W/(m·K) according to [8]. The coefficient of the heat conductivity cellular polycarbonate of 10 mm thickness is 0,03 W/(m·K) [4].

The solar heat gains were defined according to [9], and were based on equivalent insolation areas which correspond to transparent surfaces of the greenhouses. The corrections from the sun shadings by external obstacles were also considered. The solar heat gains were calculated for each shape monthly, $Q_{sol}, kW \cdot h$ per year:

$$Q_{sol} = \left(\sum\nolimits_k \Phi_{sol,mn,k}\right)t \tag{1}$$

Where $\Phi_{sol,mn,k}$- is the time-average heat flow rate from solar heat source k, W;
t – is the length of the considered month or season, hours.

Heat flow by solar gains through the transparent greenhouse element is determined by formula [9]:

$$\Phi_{sol,k} = F_{sh,ob,k} \cdot A_{sol,k} \cdot I_{sol,k} - F_{r,k} \cdot \Phi_{r,k} \tag{2}$$

Where $F_{sh,ob,k}$ - is the shading reduction factor for external obstacles for the solar effective collecting area of surface k;

$A_{sol,k}$ - is the effective collecting area of surface k with given orientation and tilt angle in the considered zone, m^2;

$I_{sol,k}$ - is the total energy of the solar irradiation during the calculation period per m^2 of collecting area of surface k, with given orientation and tilt angle, is determined according to Annex A [9], W/m^2;

$F_{r,k}$ - is the form factor between the element and the sky;

$\Phi_{r,k}$ - is the extra heat flow due to thermal radiation to the sky from building element k.

The effective collecting area of surface k with given orientation and tilt angle is calculated by formula:

$$A_{sol,k} = F_{sh,gl} \cdot g_{gl}(1 - F_F) \cdot A_{w,p} \tag{3}$$

Where $F_{sh,gl}$ - is the shading reduction factor for movable shading, if there isn't movable shading, the shading reduction factor is $F_{sh,gl} = 1$;

g_{gl} - is the total solar energy transmittance of the transparent part of the element, determined by formula (4);

F_F - is the frame area fraction, ratio of the projected frame area to the overall projected area of the transparent element according to [9] for the transparent elements of greenhouse are $F_F = 0,2$;

$A_{w,p}$ - is the overall projected area of the transparent element, m^2.

$$g_{gl} = F_w \cdot g_n \tag{4}$$

Where F_w - is the correction factor for transparent glazing $F_w = 0,90$;

g_n - is the total solar transmittance factor at a normal angle of incidence.

4 Results

The total solar heat gains were calculated for each from five shapes monthly. The calculation results are presented in Table 2 and Fig. 2.

Table 2. The total annual solar heat gains to greenhouses.

Type of the greenhouse shape	The solar heat gains to greenhouses monthly, Q_{sol}, kWh												Total annual solar heat gains, $\frac{kW \cdot h}{\%}$ *
	I	II	III	IV	V	VI	VII	VIII	IX	X	XI	XII	
Sh. 1	1530	2297	3824	4902	6543	6942	6941	6221	4952	3180	1464	1084	$\frac{49880}{96,7}$
Sh. 2	1528	2295	3825	4906	6552	6954	6952	6229	4955	3179	1462	1082	$\frac{49919}{96,69}$
Sh. 3	1496	2248	3738	4789	6390	6777	6775	6072	4841	3114	1435	1062	$\frac{48737}{94,4}$
Sh. 4	1375	2051	3459	4449	5902	6186	6230	5709	4624	2967	1348	976	$\frac{45275}{87,7}$
Sh. 5	1584	2375	3959	5075	6775	7190	7192	6447	5125	3285	1510	1119	$\frac{51636}{100}$

* Note: The absolute values are given in the numerator, the relative values are given in the denominator in percentages.

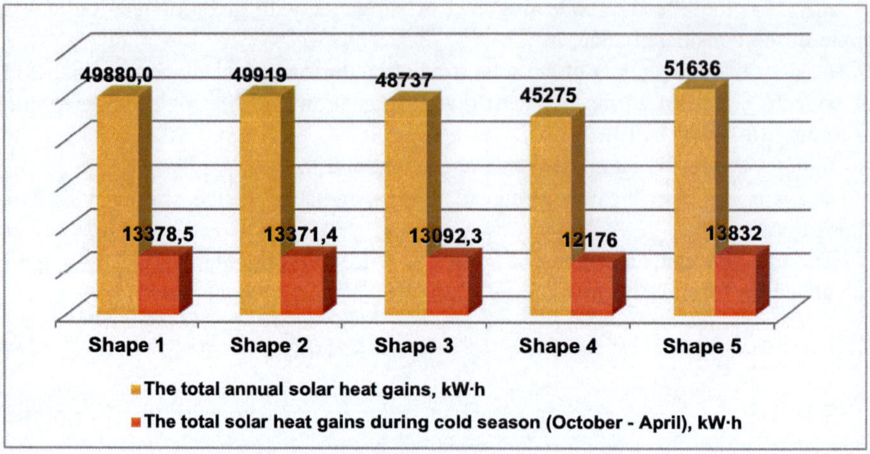

Fig. 2. The solar heat gains to the volume of greenhouses.

According to the calculation results, the Shape 5 receives the highest of solar radiation during the year - 51636 kW·h, or 1291 kW·h/m^2. The Shape 4 (greenhouse with single-gable roofing and opaque wall on one side) receives the least of solar radiation - 45275 kW·h, or 1132 kW·h/m^2, which is 12.3% less than the Shape 5. However, the area of the transparent surface of the Shape 4 is less on the 26.1% than the area of the Shape 5.

The amount of solar radiation that is received in the shape 1 and shape 2 is approximately at the same level - 49880 kW·h and 49919 kW·h respectively.

Calculation of heat losses was carried out for five greenhouse shapes by the program complex Elcut 5.1 Professional according to the method [7].The numerical values of the calculations are presented in Table 3.

Table 3. The heat losses through greenhouse enclosing constructions.

Type of the greenhouse shape	The monthly heat losses, kW·h							The annual heat losses, $\frac{kW \cdot h}{\%}$ *
	I	II	III	IV	X	XI	XII	
Shape 1	3787	3305	2820	1301	1309	2402	3363	$\frac{18287}{93,5}$
Shape 2	3840	3350	2857	1317	1327	2435	3408	$\frac{18534}{94,7}$
Shape 3	3508	3063	2619	1215	1207	2221	3113	$\frac{16946}{86,6}$
Shape 4	3245	2835	2429	1132	1115	2051	2877	$\frac{15684}{80,1}$
Shape 5	4055	3536	3012	1383	1405	2573	3601	$\frac{19565}{100}$

The analysis of the calculation results shows that the largest heat losses are observed in the shape 5 – 19565 kWh during the cold period. The least heat losses are observed in the shape 4 – 15684 kWh. It is 20% less than in the shape 5.

The comparison of solar heat gains and heat losses through enclosing constructions was made to determine the optimal shape of a greenhouse (Fig. 3).

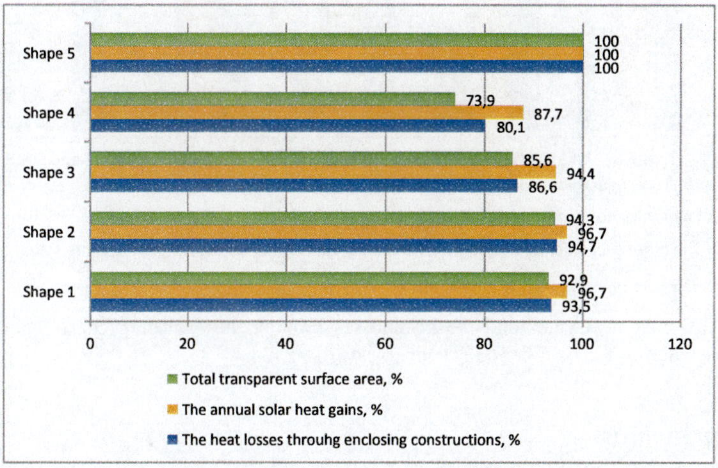

Fig. 3. The heat losses and solar heat gains for the each shape depending on the total transparent surface area, %.

The analysis of calculations shows that the heat losses and the solar heat gains nonlinearly decrease if the area of the transparent surface reduces. For example, the shape 4 has 26.1% less area of the transparent surface than the shape 5, but the solar heat gains reduce only 12.3%, and the heat losses reduce by 19.9%. Therefore, we can conclude that the greenhouse shape affects its efficiency and the shape 4 is the most profitable for the climatic conditions of Dnipro city (Ukraine).

Based on the fact that the main energy consumption at operation of the greenhouse is heating in the cold season so it is necessary to compare thermal losses and solar heat gains for each shapes during the cold season to define the optimal greenhouse shape (Fig. 4).

The difference between thermal losses and solar heat gains is offset by heating systems. Therefore, if it is decreased, the energy for heating of greenhouse and money are saved.

The data analysis shows that the most profitable form of the greenhouse is the shape 4. The difference between the heat losses and solar heat gain during the cold period is 3508 kW·h for this shape. The next is shape 3, which has the difference between heat losses and heat gains of 3854 kW·h. The next is shape 3, which has the difference between heat losses and heat gains of 3854 kW·h. The largest difference between heat gains and heat losses is observed in shape 5-5733 kW·h.

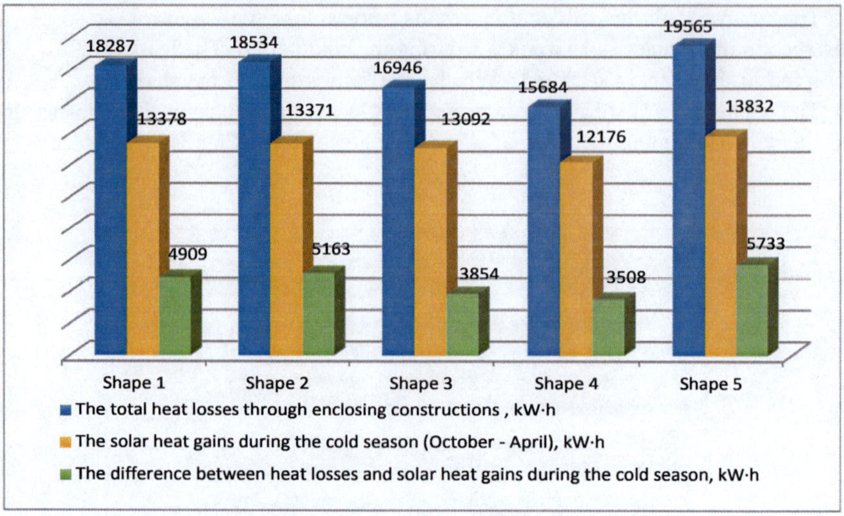

Fig. 4. The solar heat gains and the heat losses through the enclosing constructions of the greenhouses during the cold period (October - April), kW·h.

5 Conclusions

Based on the results we can make the following conclusions:

The greenhouse shape 5 receives the highest of the solar heat gains because it has the largest area of the transparent surface. But, at the same time, the greatest heat losses during the cold period are observed in the arched greenhouse.

The most profitable shape of a greenhouse by the criteria of solar heat gains and heat losses is the shape 4 -the greenhouse with the single-gable roofing and opaque wall on one side. The area of the transparent surface of the shape 4 is 26.1% less than of the shape 5. But, at the same time, the solar heat gains reduce only by 12.3% and the heat losses are reduced by 19.9%. The difference between the heat losses and solar heat gains during the cold season is the least and is 3508 kW·h.

References

1. Erat, B., Vulston, D.: Greenhouse in Your House. Handbook, 2nd edn. Stroyizdat, Moscow (1994)
2. Ivanko, A., Kalinichenko, A., Shmat, N.: The Solar Vegetariy. Popular science publication. MChP “Anfans”, Kuyv (1996)
3. Kubis, V.: The Design and Operation Experience of the Energy Efficient Greenhouses. Monograph. PGUAS, Penza (2014)
4. Savytskyi, M.: The rational design of greenhouse by criterion of the life cycle cost. In: 99th Construction, Materials Science, Mechanical Engineering, pp. 15–21. PSACEA, Dnipro (2017)

5. DSTU-N B V.1.1-27:2010: Building climatology. National Standard of Ukraine. Minregion, Kyiv (2010)
6. DBN V.2.2-2-95 1995: Buildings and Structures. Greenhouses and Hotbed. State Construction Rules of Ukraine. Ukrarkhbudinform, Kuyv (1995)
7. DBN V.2.6-31:2016: Thermal insulation of buildings. State building rules of Ukraine. Minregionbuild, Kyiv (2017)
8. Nikiforova T.: Scientific bases and methods of calculation of structures of earth sheltered buildings taking into account external influences. The thesis. PSACEA, Dnipro (2016)
9. DSTU B.A.2.2-12:2015: Energy performance of buildings. Method for calculation of energy use for space heating, cooling, ventilation, lighting and domestic hot water. National Standard of Ukraine. Minregion, Kyiv (2015)

Clarification of Thermal Characteristics of the Solar Collector Integrated into Transparent Facade

Stepan Shapoval, Iryna Venhryn(✉), Khrystyna Kozak, and Hanna Klymenko

Lviv Polytechnic National University, Lviv 79013, Ukraine
iryna.venhryn@gmail.com

Abstract. The article focuses on current issues and the state of development of energy-efficient construction. The paper notes the need to implement alternative technological solutions in energy-efficient buildings. The need to improve the design of solar collectors integrated into transparent facades of buildings is analyzed. Also, the article proposes and studies an experimental design of a solar collector integrated into the transparent facade of the building. Data on the thermal characteristics of the proposed solar collector integrated into the transparent facade of the building are presented. The article analyzes the effect of simulated solar radiation on the process of heating the heat carrier in the mode of natural circulation in the solar heat supply system. The results obtained allow us to establish the thermal efficiency of the transparent facade integrated into the structure. It is established that the efficiency of the solar heating supply system in the mode of natural circulation of the heat carrier under the intensity 300 W/m^2 average 26%, and under the intensity 900 W/m$^{2–64}$%.

Keywords: Solar collector · Mode of natural circulation of the heat carrier · Solar heat supply system · Temperature of the heat carrier · Efficiency coefficient · Energy-efficient house

1 Introduction

Scientists from The Fraunhofer Institute for Solar Energy Systems ISE (Germany) in their research emphasize the importance of integrating solar collector elements into the construction of fencing. This solution can be basis for future progress in the field of Solar energy. [1, 2]

A lot of thesis in this area claim that such systems are approximately 40% cheaper than traditional solar thermal installations. Scientists also described methods for better integration of such systems in the construction process.

Current research at the Passive House Institute (Germany) is the integration of heating and ventilation systems in the building facade during their reconstruction or the introduction at the state level of passive houses with similar structures for social housing [3, 4]. However, their articles and research are scattered in different directions and without specific methods.

Z. Blikharskyy (Ed.): EcoComfort 2020, LNCE 100, pp. 402–408, 2021.
https://doi.org/10.1007/978-3-030-57340-9_49

A priority new area of research in solar energy is the application of technology for changing the phase of materials for storage and conversion of solar energy. This technology was developed during the Second world war, but it became relevant when it was possible to use modern technologies [5]. Conversion of solar energy in construction is described in the work [6], where in addition to scientific hypotheses there are also practical recommendations.

Also, at present, more and more definitions are used ‘energy-efficient building’ [7]. It should be noted that the mandatory requirement for providing such buildings is the use of alternative sources and methods of energy supply such as: heat pumps, solar collectors, wind generators, solar power plants, the use of various storage systems of heat and electricity. It should be note that under using such alternative energy sources there is a need for additional useful space for their installation. However, providing the building with alternative energy sources ensures its energy independence and environmental friendliness.

The problem of installing and using flat solar collectors, which are one of the types of solar collector designs, is given in the paper [8]. In addition, most papers claim that it is necessary to improve thermal solar collectors in order to obtain a high-temperature heat carrier in the final result [9, 10, 11].

Combining the design of the solar collector with the aesthetic appearance of the building requires a high understanding of the operation of the solar heat supply system. In addition, it is necessary to understand that the functioning of the solar heat supply system can be necessary for the consumer at any time of the year [12, 13].

A number of scientific papers concern the research and determination of rational parameters for the use of solar heat supply systems [14, 15]. The paper [16] allows to understand the demand and supply for solar heating systems equipment in Ukraine, which is important for the deeper equipment understanding and its priority for improvement in research.

2 Objectives the Formulation of the Problem

As a result of examining the relevance of research in the world, the prospects for additional research require solar collectors integrated into the design of energy-efficient buildings.

In addition, there is a need for complex experimental studies to establish the thermal characteristics of solar heat supply system designs.

3 The Analysis of Recent Research and Publications

It is necessary to analyze current research and hypotheses concerning solar heat supply systems for the better understand the validity of experimental studies. In the case of a passive solar heating supply system (SHSS), the house performs the role of the heat storage battery [17, 18].

Solar energy is delivered directly to the premises in the case of an open SHSS, whereas a closed SHSS receives energy through the sun-sensing material (heat absorber). [19, 20, 21] In addition to the appropriate type of SHSS, research should pay attention to the mode of circulation of the heat carrier in the SHSS. Since there are different modes of the heat carrier circulation, causing the need for additional investment and/or different possibilities of generating heat energy for the consumer.

A new direction for improving solar collectors is their integration into the construction process and research of the mode of heat energy extraction to the consumer for using such a structure [22, 23].

For example, the window in combination with elements of the solar collector design is the promising area of research. However, the disadvantages of this design include significant heat loss through glass, which is in contact with the heated air and possible cooling of the building at night [24, 25]. Solar heating systems can also be used for heating greenhouses [26].

A number of works lead to an understanding of the circulation mode of the heat carrier in the SHSS. For example, the mode of natural circulation of the heat carrier (MNCHC) in SHSS it is easy to install the necessary components. Also, MNCHC it does not require a circulation pump, which allows the buyer of an energy-efficient house to be even more energy-independent. For this mode, the ability to self-regulate is inherent, which is the positive point to ensure a uniform temperature in the room. This mode is also characterized by thermal stability in the system. However, the MNCHC needs constant correct operation by the consumer. In addition, for the mode of natural circulation of the heat carrier, the radius of action (that is, the heat transfer of heat energy from the center of the system to the consumer) must be no more than 30 m in the horizontal plane.

4 The Main Material

For experimental studies of the SHSS, it was proposed to design the window solar collector as part of the building facade to install its thermal characteristics (Fig. 1).

Simplification of the experimental solar heat supply system was made due to a number of reasons typical for the MNCHC. In the SHSS, changes in the temperature and density of the heat carrier cause changes in the heat carrier flow rate through an increase/decrease in the circulation pressure in the system.

The design of the window solar collector as part of the facade has an inherent verticality, which causes the fairly stable circulation variable in the SHSS. However, with the increase in the total length of all pipes in the SHSS (the proposed design of the solar collector has additional circulation pipes directly in the window space), which is the reason for the pressure drop in the SHSS. This requires an increase in pipeline diameters and the smaller the use of shut-off and control valves and fittings.

The increase in pipeline diameters for the proposed experimental design of the solar collector was eliminated due to an increase in the volume of the circulating heat carrier in the system and, accordingly, an increase in the inertia of the such system.

Such decision would be the main reason for slowing down the start of the system and getting it into stabilization mode, and ultimately this would lead to an increase in the use of organic fuel and increase the cost of the SHSS in whole.

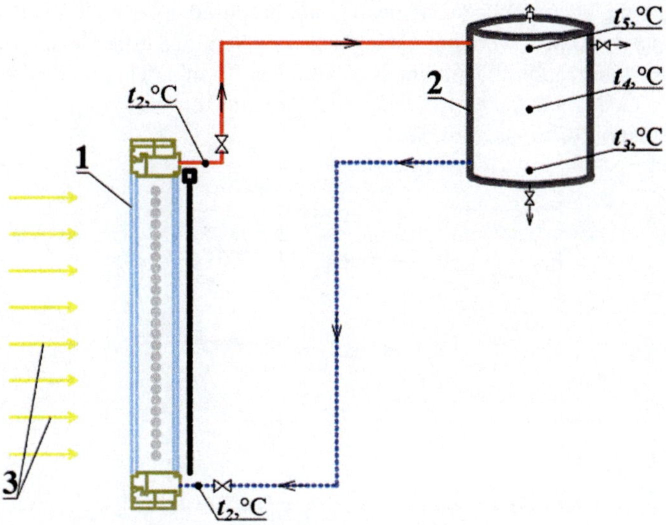

Fig. 1. The experimental solar heat supply system containing the solar collector structure integrated into the facade, where 1 is the solar collector structure, 2 is the tank for storing thermal energy, and 3 is simulated solar radiation

Therefore, the experimental design of the solar collector was marked by a fairly small number of fittings and the SHSS itself had shut-off and control valves only for adjusting the flow in the system.

According to research on the design of the solar collector was applied 300 W/m^2 та 900 W/m^2 the intensity of solar radiation.

Figure 2 shows the temperature characteristics of the SHSS, where t_1, °C – temperature of the heat carrier at the entrance to the solar collector, t_2, °C – temperature of the heat carrier at the outlet of the solar collector. The average temperature of the output under the intensity 900 W/m^2 was higher by 25%.

The average temperature change of the heat carrier in the heat storage tank for these intensities differed by 16.5% and was greater for greater intensity (Fig. 3). In addition, at the higher intensity in the storage tank of thermal energy, the accelerated increase in temperature was observed in comparison with the lower intensity of thermal energy.

The graphical dependence of the instantaneous power of the solar collector during the experiment was constructed based on experimental studies. It is established that for 900 W/m^2 the power was almost 5.5 times greater than under the intensity 300 W/m^2 (Fig. 4).

An important heat engineering parameter of the solar heat supply system is its efficiency. As a result of experimental studies of efficiency of the SHSS in the MNCHC under the intensity 300 W/m^2 it was the average of 26%, and under the intensity 900 W/m$^{2–64}$% (Fig. 5).

The heat engineering characteristics for the proposed solar collector design in the solar heat supply system were not distinguished by significant inertia and the speed of reaching the stabilization mode, which occurred in 20 min. This time of entering the stabilization mode is sufficient for effective operation of the system in the mode of natural circulation of the heat carrier.

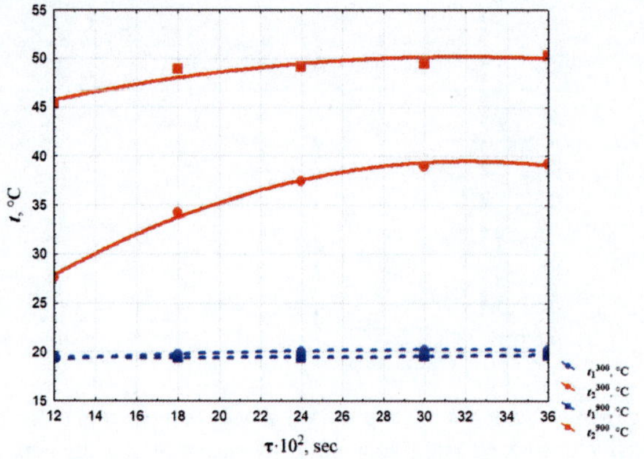

Fig. 2. Changes in the temperature of the heat carrier in the solar collector design according to the intensity of the simulated solar energy input volume 300 W/m^2 (t_1^{300}, °C; t_2^{300}, °C) and 900 W/m^2 during the experiment

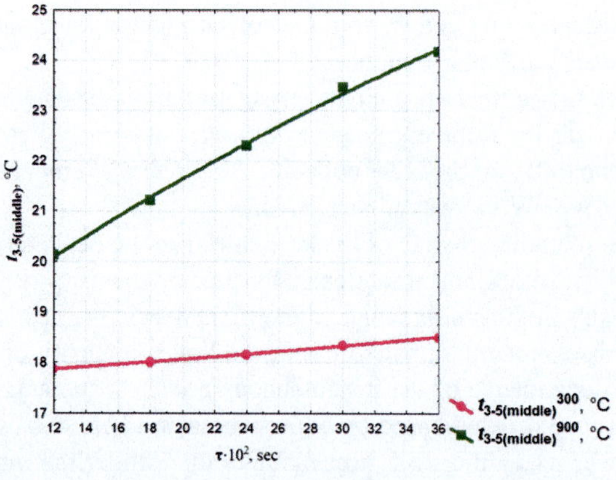

Fig. 3. The average temperature of the heat carrier in the heat storage tank during the experiment, where $t_{3\text{–}5(\text{middle})}^{300}$, °C – the average temperature of the heat carrier in the storage tank of thermal energy according to the intensity of the simulated solar energy input 300 W/m^2, $t_{3\text{–}5(\text{middle})}^{900}$, °C – the average temperature of the heat carrier in the storage tank of thermal energy according to the intensity of the simulated solar energy input 900 W/m^2

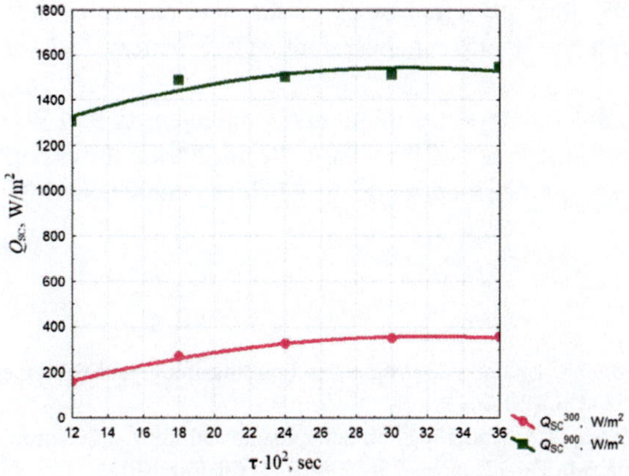

Fig. 4. Change in the instantaneous power of the solar collector during the experiment, where Q_{SC}^{300}, °C – the thermal power of the solar collector at the intensity of simulated solar energy input 300 W/m^2, Q_{SC}^{900}, °C – the thermal power of the solar collector at the intensity of simulated solar energy input 900 W/m^2

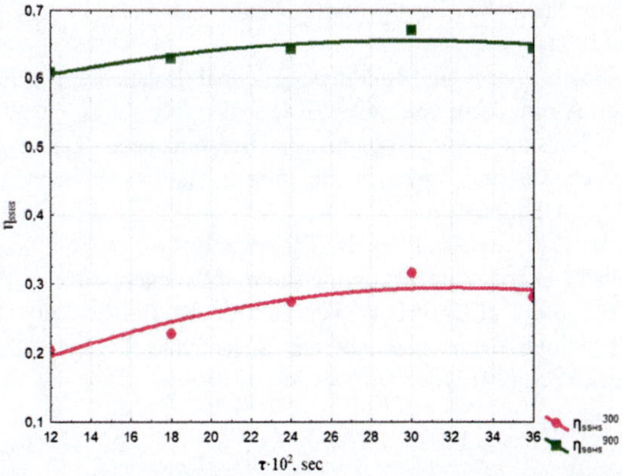

Fig. 5. The efficiency of the solar heat supply system in the mode of natural circulation of the heat carrier during the experiment, where η_{SHSS}^{300}, °C – the efficiency coefficient of the solar collector under the intensity of the simulated solar energy 300 W/m^2, η_{SHSS}^{900}, °C – the efficiency coefficient of the solar collector under the intensity of the simulated solar energy 900 W/m^2

5 Conclusions

The proposed design of the solar collector integrated into the window as part of the transparent building facade was studied in the mode of natural circulation of the heat carrier in the solar heat supply system.

It is established that the temperature of the heat carrier by intensity 900 W/m^2 reached 50 °C, and the average temperature of the battery tank of thermal energy reached 24 °C.

The average temperature of the output under the intensity 900 W/m^2 was higher by 25%. The instantaneous power of the solar collector during the experiment differed by 5.5 times for the tested intensities, while the efficiency coefficient differed by 2.5 times.

References

1. Maurer, C., Cappel, C., Kuhn, T.: Progress in building-integrated solar thermal systems. Sol. Energy **154**, 158–186 (2017)
2. Maurer, C., Kuhn, T.: Modelling of active solar building envelopes for cost-effective evaluation. J. Facade Des. Eng. **5**, 75–82 (2017). (in English)
3. Schnieders, J., Feis,t W., Rongen, L.: Passive houses for different climate zones. Energy Build. **105**, 71–87 (2015)
4. Dermentzis, G., Ochs, F., Siegele, D., Feist, W.: Renovation with an innovative compact heating and ventilation system integrated into the facade. Energy Build. **165**, 451–463 (2018)
5. Khudhair, A., Farid, M.: A review on energy conservation in building applications with thermal storage by latent heat using phase change materials. Energy Convers. Manage. **45**(2), 263–275 (2004)
6. Chwieduk, D.: Solar Energy in Buildings: Thermal Balance for Efficient Heating and Cooling. Elsevier, Amsterdam, vol. 362 (2014)
7. Energy saving in housing stock: problems, practice, prospects. Guide «NDI Proektrekonstruktsiia» . Deutsche Energie-Agentur GmbH(dena), Instituts Wohnen und Umwelt GmbH, vol. 144 (2006). (in Ukrainian)
8. Manyala, R.: Solar Collectors and Panels, Theory and Applications. Sciyo, Croatia (2010)
9. Denk, S.: Renewable energy sources: on the shore of the energy ocean. Publishing house of the Perm state technological University, vol. 288 (2008). (in Russian)
10. Denzer, A.: The Solar House: Pioneering Sustainable Design. Rizzoli International Publications, New York (2013)

Determination of the Charring Rate of Timber to Estimate the Fire Resistance of Structures at Real Temperature Modes of Fires

Taras Shnal[1(✉)], Serhii Pozdieiev[2], Stanislav Sidnei[1], and Andrii Shvydenko[1]

[1] Lviv Polytechnic National University, Lviv, Ukraine
taras.m.shnal@lpnu.ua
[2] Cherkassy Institute of Fire Safety Named After Heroes of Chornobyl of National University of Civil Defense of Ukraine, Cherkasy, Ukraine

Abstract. The article presents the results of researches of fire resistance of timber structures, taking into account the realistic temperature modes of fires. The proposed method of determining the charring rate is based on the main parameters of fire spread that are taken into account in the calculations, based on the analysis of the heating temperature of unprotected elements of timber structures, where an approach on solving an time-dependant equation of thermal conductivity is used. The research of the influence of the depth of the charring on the limit of fire resistance depends to a lesser extent on the coefficient of slots but to a greater extent it depends on the density of the fire load. Thus, the difference for the beam with a cross section of 150 × 300 was 15 min under varying density of the fire load, and for different slot coefficients it is not more than 5 min. Provided that the density of the fire load is less than 690 MJ/m2 for a beam with a cross section of 150 × 300, the limit of fire resistance does not occur at all. The proposed method allows more accurate determining of the limit of fire resistance of timber structures as it takes into account both the main parameters of fire spread and thermal-physical characteristics of the timber.

Keywords: Fire resistance · Charring rate · Timber construction · Parametric fires · Structural fire design

1 Introduction

1.1 Problem Statement

In recent decades, the development of temperature models of realistic modes gained momentum. Temperature models in the form of parametric curves, zone models, and models based on numerical fluid and gas dynamics allow to predict the temperature mode of the fire more accurately and adequately, which, in its turn, allows to assess the temperature impact on building structures more accurately and determine their fire resistance compared to nominal temperature curves, including standard temperature-time curve.

Z. Blikharskyy (Ed.): EcoComfort 2020, LNCE 100, pp. 409–418, 2021.
https://doi.org/10.1007/978-3-030-57340-9_50

At the same time, methods for estimation of the fire resistance of building structures, according to the current standards, are mainly focused on the use of simplified methods with standard temperature mode, strength and thermal-physical characteristics of materials that do not meet the temperature models of fire spreading in premises. The above data led to the development of the research direction of fire resistance of building structures using real temperature modes of fires. This is especially regarding the timber structures and flammable timber-based materials, so their practical use requires fire protection and in some cases it can be limited.

The research of fire resistance of timber structures and timber-based materials, taking into account the temperature of realistic fire modes will allow to determine their fire resistance more accurately, which will help to avoid the use of expensive fire protection systems and expand their application for civil and industrial construction.

1.2 Analysis of Recent Studies and Publications

There is a significant amount of works devoted to researches of fire resistance of timber structures.

The work [1] contains the researched results of samples of cross sections of timber beams both protected and unprotected for which the temperature and rate of charring were determined. On the basis of the research there was developed a method of interpolation of temperature profiles in cross sections and method of building the charred zone in the researched samples under conditions of standard temperature mode.

In the work [2] its authors presented the results of field tests on the fire resistance of laminated timber beams, determined the charring rate, and researched the influence of fire protection and temperature distribution in the cross section of a timber beam.

In the work [3], the behavior of protected and unprotected elements of the upright frame of timber structures in the parametric mode of fire, according to [4] there was conducted numerical analysis and comparison with experimental data.

Reinforced concrete is one of the most popular materials in construction [5–9]. Nowadays, there is the great number of works which analyze possibilities to improve the reinforced concrete structures' properties and increase their durability [10, 11]. However, a significant number of buildings and structures require renovation and strengthening [12, 17]. There could also often appear problems with fire resistance in steel-concrete and composite-reinforced structures, which is not typical for structures strengthened with a reinforced concrete casing [18–20].

2 Purpose and Objectives of the Research

The purpose of this work is to develop a method of fire resistance of timber structures in which the timber charring indirectly took into account the temperature mode of realistic fire.

Based on the experimental data, using the obtained regression dependence of the maximum charring rate and the maximum time of the charring process, to build corresponding surface dependencies on the selected most significant parameters of the premises where the fire occurs.

3 Methods for Determining the Heating of Cross Sections of Timber Structures

To conduct a calculated estimation of the fire resistance of timber structures, the heating temperature of the elements of these structures at any time of the fire should be determined. To determine the heating temperature of the unprotected elements of timber structures, the main approach is used, which is based on the solution of the time-dependant equation of thermal conductivity, which has the form [21–29]:

$$c_p(\theta)\rho_w(\theta)\frac{\partial\theta}{\partial t} = \frac{\partial}{\partial x}\left(\lambda_w(\theta)\frac{\partial\theta}{\partial x}\right) + \frac{\partial}{\partial y}\left(\lambda_w(\theta)\frac{\partial\theta}{\partial y}\right) \tag{1}$$

where $\rho_{w(\theta)}$ - timber density, depending on temperature, kg/m3; $c_p(\theta)$ - specific heat of timber, depending on temperature, J/(kg·°C); λw(θ) is thermal conduction coefficient of timber, W/(m·°C) ; t - hour, s.

The thermal problem in this formulation involves the application of boundary conditions of the third kind:

$$-\lambda(\theta)\frac{\partial\theta}{\partial r} = \alpha(\theta_{fi} - \theta_w) \tag{2}$$

where α is the heat transfer coefficient, W/(m2 · °C ; $_{fi}$, - temperature near the surface of the element of the timber structure, °C; - surface temperature of the timber structure element, °C; r - current spatial coordinate, m.

The heat transfer coefficient is determined taking into account convection and heat transfer by heat radiation and should be determined by the following expression [28, 29]:

$$\alpha = \alpha_B + \alpha_K \tag{3}$$

where: α_B - heat transfer coefficient by radiation; α_K - convective heat transfer coefficient.

The value of the components of the heat transfer coefficient is calculated according to the recommendations of the standards [28, 29] using the formulas:

α_K = 25 W/(m2 · °C);

$$\alpha_B = \varepsilon \cdot \sigma \cdot \frac{\theta_W^4 - \theta_P^4}{\theta_W - \theta_P} \tag{4}$$

where ε = 0.8 - the degree of blackness of the element surface of the timber structure; σ = 5.67·10 − 8 W/(m^2·°C ^{4}) – the Stefan-Boltzmann constant; α = 9 W/(m^2·°C) competitive heat transfer coefficient on the non-heating side, which takes into account convection and radiation simultaneously.

The initial temperature of the material and the environment θ_0 = 20 °C.

Thermal and physical characteristics, i.e. the coefficient of thermal conductivity, specific heat and density, depending on the temperature are calculated according to the recommendations of the standard [30].

The temperature mode of the fire is determined according to the approach proposed in [14]. The following formulas are used there:

$$\Theta_p = 20 + \frac{\Theta_{emax} - 20}{\Theta_{max}} 1325\left(1 - 0.324exp\left(-0.333 \cdot 10^{-3}t\right) - 0.204exp\left(-0.028t\right) - 0.472exp\left(-0.317t\right)\right) \quad (5)$$

where the maximum fire temperature with standard mode is determined by the following formulas:

$$\Theta_{\max} = 20 + 1325\left(1 - 0.324\exp\left(-0.333 \cdot 10^{-3}t_m\right) - 0.204\exp\left(-0.028t_m\right) - 0.472\exp\left(-0.317t_m\right)\right) \quad (6)$$

To describe the descending branch of the fire, it is proposed to use a formula that expresses the linear law of temperature decrease and has the following form:

$$\Theta_c = \Theta_{emax} - \frac{\Theta_{emax} - 20}{\tau_{max} - t_m}(t - t_m) \quad (7)$$

Here, the parameters $\Theta_{e\ max}$, t_m та τ_{max} are the maximum medium-volume temperature, the time to reach the maximum medium-volume temperature and the duration of the fire, respectively. It is defined by the following expressions:

$$\Theta_{\mathrm{emax}} = 120.791 + 4946.172 \cdot O - 0.640 \cdot q_{t,d} + 1.196 \cdot O \cdot q_{t,d}; \quad (8)$$

$$t_m = 128.081 + 3129.187 \cdot O - 0.148 \cdot q_{t,d} + 3.1898 \cdot O \cdot q_{t,d}; \quad (9)$$

$$\tau_m = 74.941 + 4612.440 \cdot O - 0.052 \cdot q_{t,d} + 2.552 \cdot O \cdot q_{t,d}; \quad (10)$$

The differential equation of thermal conductivity was solved by the finite difference method, as described in the works [30].

4 Results

To describe the dependence of the charring rate, there were used the results of solving the thermal conductivity equation with the line cutting the charring zone, corresponding to the isotherm with the critical value of temperature θ_{cr} = 250 °C at which the charring process begins. Under such conditions, the thickness of the charred zone in the cross section of the element of the timber structure is determined.

Varying the parameters according to the matrix of the plan of the first order of the complete factorial experiment and the ranges of variation of factors according to Table 1.

Table 1. Intervals of variation of factors in a numerical experiment.

Slot coefficient O, $m^{0.5}$			Fire load, $q_{t,d}$, MJ/m^2		
Least value	Average value	Largest value	Least value	Average value	Largest value
0.043	0.054	0.065	500	850	1200

According to the results of the experiments, the corresponding values of the charring rate were obtained using the mathematical apparatus (1)–(9) and using the value of the initiation of the charring process θ_{cr}. The obtained data are given in Table 2.

Table 2. Parameters of fires in model premises in the conditions of full factorial experiment according to the accepted planning matrix.

Experimental situation	1	2	3	4
Maximum charring rate, v_{max}, mm/min	2.3	0.4	2.4	0.8
Maximum time of charring process, t_{vmax}, min	58	6	64	21

Applying the data in Table 2, the coefficients of regression dependence of the type $y = b_0 + b_1x_1 + b_2x_2 + b_3x_1x_2$ [14] were determined, which are given in Table 3.

Table 3. Coefficients of regression for charring rate parameters.

Model	b_0	b_1	b_2	b_3
Maximum charring rate, v_{max}, mm/min	1.475	−0.125	0.875	0.075
Maximum time of the charring process, t_{vmax}, min	37.25	−5.25	23.75	2.25

Using the obtained regression dependence of the maximum charring rate and the maximum time of the charring process, there were built the corresponding surfaces of the dependence on the selected most significant parameters of the premises where the fire occurs, as shown in Fig. 1.

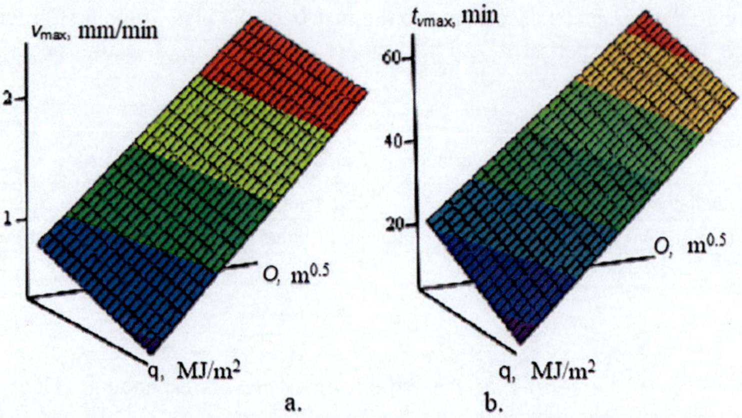

Fig. 1. Surface dependence of charring rate parameters on the most significant parameters of premises with fires: a - maximum charring rate; b - the maximum time of the charring process.

The obtained data allow determining the dependence of the charring rate of timber.

To describe the dependence of the charring rate on the time of fire, the following expression is used:

$$v(t) = v_{max}(1 - t/t_{vmax}) \tag{11}$$

The maximum charring rate and the maximum time of the charring process are determined by the expressions obtained from the determined regression dependencies with the converted coefficients from the coded coefficients of Table 2:

– for the maximum charring rate:

$$v_{max} = 0.648 + 30.742 \cdot O - 0.00178 \cdot q_{t,d} + 0.0239 \cdot O \cdot q_{t,d} \tag{12}$$

– for the maximum time of the charring process:

$$t_{vmax} = 26.915 + 1058.6 \cdot O - 0.0446 \cdot q_{t,d} + 0.718 \cdot O \cdot q_{t,d} \tag{13}$$

For determining the load-bearing capacity of timber beams, it is convenient to use a simplified calculation method, which is recommended in EN 1995-1-2: 2012 Eurocode 5 [31]. According to this method, the load-bearing capacity of the cross section of a timber beam should be determined applying the following expression:

$$f_{c,d\,fi} = k_{mod\,fi} f_{c,0,k} \tag{14}$$

wood for bending under normal conditions (parallel to its fibers) [15]; where $f_{c,0,k}$ = 65 MPa – strength

$k_{mod,fi} = \frac{W_{ef}}{W_r}$ – strength reduction factor [31]

here W_{ef} is the effective axial moment of resistance of the cross section of the beam, which is determined taking into account the charring zone; Wr is the reduced axial moment of resistance of the cross section of the beam under normal conditions.

The coefficient of reduction of the cross-sectional strength of the beam, taking into account the charring is determined by the following formula:

$$k_{fi} = \frac{(b - x_c(t))(h - x_c(t))^2}{bh^2} x_c(t) = v_{max}(t - 0.5 \cdot t^2 / t_{vmax}) \tag{15}$$

In this formula as a time parameter can be used the number of the standard class of fire resistance. Under such conditions, the condition of preserving the bearing capacity, which is expressed by the inequality [31], can be checked:

$$k_{mod,fi} \geq \mu_f \tag{16}$$

where μ_f is the load coefficient that expresses the ratio of stresses from the applied forces to the load-bearing capacity of the cross section.

After performing all necessary calculations, there were obtained the dependences of the coefficient of strength reduction of beams with different geometric parameters of cross-sections, depending on the time of fire for different values of the coefficient of slots and the density of the fire load. Figure 2 shows the obtained dependences.

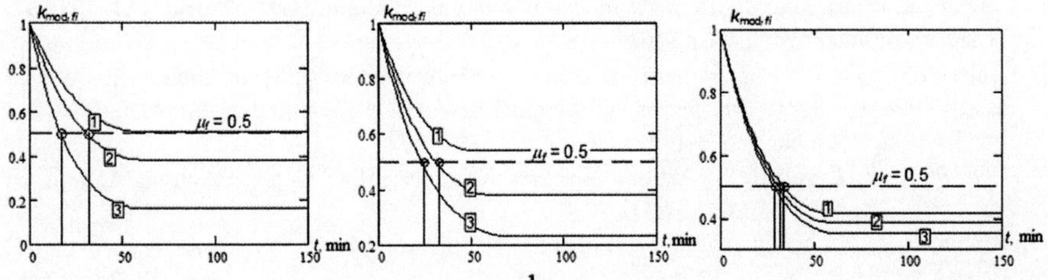

Fig. 2. Graphs of dependences of the coefficient of strength reduction of beams with different geometric parameters of cross-sections from the time of fire for different values of the coefficient of slots and density of fire load: a - O = 0.0045 м$^{0.5}$, $q_{t,m}$, = 1200 MJ/m^2 (1 – slot №1, 2 – slot №2, slot №3); b - slot №2, O = 0.0055 m0.5 (1 – $q_{t,m}$, = 600 MJ/m^2, 2 – qt,m, = 800 MJ/m^2, $q_{t,m}$, = 1000 MJ/m^2); c - slot №2, $q_{t,m}$, = 800 MJ/m^2 (1 – O = 0.0045 m $_{0.5}$, 2 – O = 0.0055 m$^{0.5}$, O = 0.0065 m$^{0.5}$).

5 Conclusions

Given the conducted researches, the following conclusions can be drawn:

1. The research of the influence of the depth of the charring on the limit of fire resistance depends to a lesser extent on the coefficient of slots but to a greater extent it depends on the density of the fire load. Thus, the difference for the beam with a cross section of 150 × 300 was 15 min under varying density of the fire load, and

for different slot coefficients it is not more than 5 min. Provided that the density of the fire load is less than 690 MJ/m2 for a beam with a cross section of 150 × 300, the limit of fire resistance does not occur at all. The proposed method allows more accurate determining of the limit of fire resistance of timber structures as it takes into account both the main parameters of fire spread and thermal-physical characteristics of the timber.

2. There were identified the regularities of influence of parameters of premises with fire on the parameters defining the given linear dependence of charring rate in the form of regression dependences for the maximum charring rate $v_{\max} = 0.648 + 30.742 \cdot O - 0.00178 \cdot q_{t,d} + 0.0239 \cdot O \cdot q_{t,d}$ and the maximum time of the charring process $t_{v\max} = 26.915 + 1058.6 \cdot O - 0.0446 \cdot q_{t,d} + 0.718 \cdot O \cdot q_{t,d}$.. Basing on the identified regularities there was developed a calculation method of the estimation of the limit of fire resistance of the bent elements of timber constructions.

References

1. Pozdieiev, S.V., Nekora, O.V., Horbachenko, Ya.V., Fedchenko, I.V.: Geometry of charring zone in cross sections of fire protected timber beams in conditions of fire. Collection of scholarly works, no. 37, pp. 168–177 (2015)
2. Pelekh, A.B., Demchyna, B.H., Shnal, T.M., Bula, S.S., Krochak, O.V.: Field tests of the timber frame structure on fire resistance in conditions of real fire, no. 627, pp. 167–171. NU "Lvivska politekhnika", Visn (2008)
3. Bula, S.S., Shnal, T.M.: Charring depth of fire protected studs in timber frame wall assemblies exposed to fire. Series: Theory and practice of construction, no. 912, pp. 3–11. Ed. Lvivska politekhnika, Lviv (2019)
4. EN 1991-1-2. Eurocode 1: Actions and Structures, Part 1-2: General Actions-Actions on Structures Exposed to Fire (2002)
5. Blikharskyy, Z., Vashkevych, R., Vegera, P., Blikharskyy, Y.: Crack Resistance of RC Beams on the Shear. In: Lecture Notes in Civil Engineering, vol. 47, pp. 17–24 (2020). https://doi.org/10.1007/978-3-030-27011-7_3
6. Bobalo, T., Blikharskyy, Y., Kopiika, N., Volynets, M.: Serviceability of RC Beams Reinforced with High Strength Rebar's and Steel Plate. In: Lecture Notes in Civil Engineering, vol. 47, pp. 25–33 (2020). https://doi.org/10.1007/978-3-030-27011-7_4
7. Khmil, R., Tytarenko, R., Blikharskyy, Y., Vashkevych, R.: Influence of load level during strengthening of reinforced concrete beams on their reliability. IOP Conf. Ser. Mater. Sci. Eng. **708**(1), 012054 (2019)
8. Krainskyi, P., Vegera, P., Khmil, R., Blikharskyy, Z.: Theoretical calculation method for crack resistance of jacketed RC columns. IOP Conf. Ser. Mater. Sci. Eng. **708**(1), 012059 (2019)
9. Vegera, P., Vashkevych, R., Blikharskyy, Z.: Fracture toughness of RC beams with different shear span. In: MATEC Web of Conferences **174**, p. 02021 (2018). https://doi.org/10.1051/matecconf/201817402021
10. Bobalo, T., Blikharskyy, Y., Kopiika, N., Volynets, M.: Theoretical analysis of RC beams reinforced with high strength rebar's and steel plate. IOP Conf. Ser. Mater. Sci. Eng. **708**(1), 012045 (2019)

11. Krainskyi, P., Blikharskyy, Y., Khmil, R., Vegera, P.: Crack resistance of RC columns strengthened by jacketing. In: Lecture Notes in Civil Engineering, vol. 47, pp. 195–201 (2020). https://doi.org/10.1007/978-3-030-27011-7_25
12. Blikhars'kyi, Y.Z.: Anisotropy of the Mechanical Properties of Thermally Hardened A500s Reinforcement. Mater. Sci. **55**(2), 175–180 (2019). https://doi.org/10.1007/s11003-019-00285-0
13. Blikharskyy, Z., Brózda, K., Selejdak, J.: Effectivenes of strengthening loaded RC beams with FRCM system. Arch. Civ. Eng. **64**(3), 3–13 (2018)
14. Kharchenko, Y.V., Blikharskyy, Z.Y., Vira, V.V., Vasyliv, B.D.: Study of structural changes in a nickel oxide containing anode material during reduction and oxidation at 600 °C. In: Fesenko, O., Yatsenko, L. (eds.) NANO 2018. SPP, vol. 221, pp. 595–604. Springer, Cham (2019). https://doi.org/10.1007/978-3-030-17759-1_42
15. Kharchenko, Y., Blikharskyy, Z., Vira, V. Vasyliv, B., Vasyliv, B: Study of nanostructural changes in a Ni-containing cermet material during reduction and oxidation at 600 °C. Appl. Nanosci. (2020). https://doi.org/10.1007/s13204-020-01391-1
16. Lobodanov, M., Vegera, P., Blikharskyy, Z.: Planning experiment for researching reinforced concrete beams with damages. In: Lecture Notes in Civil Engineering, vol. 47, pp. 243–250 (2020). https://doi.org/10.1007/978-3-030-27011-7_31
17. Zhang, Q., Mol'kov, Y.V., Sobko, Y.M., Blikhars'kyi, Y.Z., Khmil', R.E.: Determination of the mechanical characteristics and specific fracture energy of thermally hardened reinforcement. Mater. Sci. **50**(6), 824–829 (2015). https://doi.org/10.1007/s11003-015-9789-9
18. Krainskyi, P., Blikharskyy, Y., Khmil, R., Blikharskyy, Z.: Experimental study of the strengthening effect of reinforced concrete columns jacketed under service load level. In: MATEC Web of Conferences, vol. 183, p. 02008 (2018). https://doi.org/10.1051/matecconf/201818302008
19. Selejdak, J., Blikharskyy, Y., Khmil, R., Blikharskyy, Z.: Calculation of reinforced concrete columns strengthened by CFRP. In: Lecture Notes in Civil Engineering, vol. 47, pp. 400–410 (2020). https://doi.org/10.1007/978-3-030-27011-7_51
20. Vegera, P., Khmil, R., Vegera, P.: Shear strength of reinforced concrete beams strengthened by PBO fiber mesh under loading. In: MATEC Web of Conferences, vol. 116, p. 02006 (2017). https://doi.org/10.1051/matecconf/201711602006
21. Bushev, V.P., Pchelintsev, V.A., Fedorenko, V.S., Yakovlev, A.I.: Fire resistance of buildings/Under the total. In: Pchelintsev, V.A. (ed.), 2nd edn., Revised and add. Stroyizdat, M (1970)
22. Pozdieiev, S., Sidnei, S., Nekora, O., Fedchenko, S.: Research of wooden bearing structures behavior under fire condition with use advanced methods of fire resistance calculation considering eurocode 5 recommendation. In: International Scientific Conference on Woods & Fire Safety, WFS: Wood & Fire Safety, pp. 326–332 (2020)
23. Demeshok, V., Zalevs'ka, A., Tychenko, E., Zmaga, Y.: The study of carrying capacity of timber slabs with use the finite elements method. In: MATEC Web of Conferences, vol. 116, p. 02010. https://doi.org/10.1051/matecconf/201711602010 (2017)
24. Nyzhnyk, V.V., Tarasenko, O.A., Kyrychenko, O.V., Kosiarum, S.O., Pozdieiev, S.V.: The criteria of estimating risks of spreading fire to adjacent building facilities. In: IOP Conference Series: Materials Science and Engineering, Reliability and Durability of Railway Transport Engineering Structures and Buildings, vol. 708, 20–22 November, Kharkiv, Ukraine (2019)
25. Nekora O., Slovynsky V., Pozdieiev S. The research of bearing capacity of reinforced concrete beam with use combined experimental-computational method. In: MATEC Web of Conferences, vol. 116, art. no. 02024. https://doi.org/10.1051/matecconf/201711602024 (2017)

26. Vlasova, E.A.: Approximate methods of mathematical physics: [Textbook. for universities / ed. V.S. Zarubina, A.P. Krishchenko]/Vlasova EA, Zarubin V.S., Kuvyrkin G.N. - M.: Bauman MSTU (2001)
27. EN 1995-1-2:2004. Eurocode 5: Design of timber structures: General-Structural fire design (2004)
28. Shnal, T.M.: Fire resistance and fire protection of timber structures. Ed. Lviv Polytechnic National University, Lviv (2006)
29. Lie, T.T.: A procedure to calculate fire resistance of structural members. In: International Seminar on Three Decades of Structural Fire Safety, 22/23 February, pp. 139–153 (1983)
30. Shnal, T.M.: Development of scientific bases for the estimated fire resistance of building structures under conditions of influence of parametric temperature modes of fires: thesis by Dr. of technical science: 21.06.02 (2019)
31. Konig, J., Walleig, L.: Timber frame assemblies exposed to standard and parametric fires. Stocholm (2000)

Development of a Mathematical Model of Fire Spreading in a Three-Storey Building Under Full-Scale Fire-Response Tests

Taras Shnal[1], Serhii Pozdieiev[2], Roman Yakovchuk[3(✉)], and Olga Nekora[2]

[1] Lviv Polytechnic National University, Lviv, Ukraine
[2] Cherkassy Institute of Fire Safety named after Heroes of Chornobyl, National University of Civil Defense of Ukraine, Cherkasy, Ukraine
[3] Lviv State University of Life Safety, Lviv, Ukraine
yakovchukrs@ukr.net

Abstract. The purpose of the work was to develop a mathematical model of fire spreading in a three-story building during full-scale fire-response tests; research of accuracy and reliability of parameters of temperature modes of fire in separate premises of the building. The Pyrosim computer system, a user shell for the Fire Dynamics Simulator program, was used to calculate the temperature in the models of premises under fire. A numerical experiment was conducted to model full-scale tests of premises with fire in a three-story building using computer gas-hydrodynamics methods. The nature of the fire development and the time dependences of its main parameters were revealed, which in turn allowed analyzing the adequacy of the modelling results and investigating their adequacy and accuracy. The obtained results of the research on the accuracy of modeling the full-scale tests of premises with fire in a three-story building revealed that the error, determined when comparing experimental and calculated data, was not significant. A relative error did not exceed 28%, and root-mean-square deviation did not exceed 51 °C. This means that the modeling results are adequate, which allows to use this approach for predicting the behavior of building structures in a fire under realistic conditions.

Keywords: Computer fire modeling · Fire resistance · Fire dynamics simulator · PyroSim · Field fire model

1 Introduction

1.1 Problem Statement

Fire safety of buildings and structures depends on an adequate estimation of fire resistance of their structures. A proper estimation of fire resistance of structures is possible only when using for calculations such temperature models of fires, which most accurately correspond to fire under realistic conditions.

Mathematical modeling of fire in premises involves the availability of data on the size of the premises, its openings, the density of fire load, the value of the maximum

Z. Blikharskyy (Ed.): EcoComfort 2020, LNCE 100, pp. 419–428, 2021.
https://doi.org/10.1007/978-3-030-57340-9_51

temperature of the fire and the time of its occurrence, and so on. Some research data are to some extent contradictory, due to differences in the objectives of particular researches. When carrying out the calculated estimation of fire resistance of building structures, taking into account the real thermal impact of fire, the analysis of the fire resistance problem is reduced to solving three separate tasks. These are the tasks of determining the temperature in a premise with fire near the structural elements; the task of calculating the temperature in the inner layers of structural elements and the task of thermopower response of structural elements to a given thermal impact.

The situation is complicated here by the need to use powerful commercial software, which requires appropriate qualified specialists and significant material costs for the license acquisition required for using this software. Another problematic issue for application of this approach is the need to consistently perform these steps in the absence of the possibility of using general mathematical relations to obtain the final result.

This will increase the productivity of calculations without indicating the advantages of this approach to the temperature modes of fires, namely greater accuracy and reliability with reduced rigidity of the requirements for fire resistance of the researched structures.

Elimination of the specified shortcomings of the approach of calculated estimation of fire resistance of the elements of building structures on the basis of the temperature modes of fires under realistic conditions, allows solving an actual scientific and technical problem.

1.2 Analysis of the Recent Studies and Publications

The research results of the temperature mode during a fire in a residential premise are described in [1–3]. Recently, quite popular became the use of special software – the Fire Dynamics Simulator (FDS) - for computer modeling of the results of tests of building structures for fire resistance.

This software allows to perform numerical simulations of the temperature distribution in structures during fire-response tests, as well as other parameters that are important for predicting the probability of progressive collapse in case of fire-caused damage [4–6].

The work [7] presents some results of several researches obtained using the FDS (version 4.0 Fire Dynamics Simulator), which are compared with experimental data. The aim of the work was to test the capabilities of the FDS program to model the flame spreading, as well as to determine the optimal values of the parameters of the combustible load of the material for the engineering use of the FDS. The authors of the works [8, 9] have researched experimentally and numerically the fire resistance of facade structures. The experimental setup modeled a three-story residential building. The numerical model was built in the CFD program by the FDS with similar geometry and instrumentation. The authors were able to properly reproduce the real test conditions in the model, but the temperatures close to the fire source could not be properly taken into account in the model.

The works [10, 11] compare the results of full-scale fire-response tests according to the Swedish (SP Fire 105) and British (BS 8414-1) methods. The presented results of experimental researches and computer modeling take into account different variations of fire exposure, fire load and the type of fire load. The modeling in the FDS made it possible to reproduce the experimentally determined temperature values both qualitatively and quantitatively.

In the work [12], fire dynamics models were performed using the FDS for two full-scale experiments conducted by the Efectis France laboratory. The main purpose of this research was to evaluate the ability of the numerical model to reproduce the quantitative results of temperatures and heat flux for further estimation of the characteristics of the impact of fire on the facade insulation. When comparing the experimental data with numerical calculations, there were obtained satisfactory results of temperature and heat flux.

The issue of fire resistance calculation is highly relevant, especially for the most widely used in the world reinforced concrete structures [13–18]. The problem of defects and damages in such structures significantly reduces fire resistance [19–22]. In particular, one of the most influential factors is the corrosion of reinforcement, which leads to structural defects and reduced performance [23–26]. Today there is the great variety of strengthening methods. Among the most popular are metal and reinforced concrete casing, as well as the external carbon tape, FRCM systems and other methods [27]. Similarly, this problem is typical for structures strengthened with external reinforcement. Additionally, interesting is the issue of fire resistance calculation for structures strengthened with reinforced concrete casing [28].

The authors of the works [29–31] investigated the influence of horizontal obstacles at different heights between unprotected openings on the facade of the building on the spread of external fire experimentally and compared the data using the numerical FDS instrument. It was concluded that the FDS version 6.2.0 can reproduce experimental results with a high level of detail. Authors of the work [32] attempted to reproduce the scenario of an external fire in facade insulation systems and proposed a method of quantitative assessment of the fire risk using the FDS software.

2 Purpose and Objectives of the Research

The purpose of the work is to develop a mathematical model of fire spread in a three-storey building during full-scale fire-response tests; a research of accuracy and reliability of the parameters of temperature modes of fire spread in separate premises of the building. To achieve this purpose, it is advisable to use the methods of computational gas-hydrodynamics, which allow determining the limits of application of this approach for predicting the behavior of building structures exposed to fire.

To achieve this purpose, the following objectives were set:

- to conduct a set of numerical experiments on the development and spread of fire in the premises of the full-scale model of a three-story building with the help of mathematical modeling by the methods of computer gas-hydrodynamics;
- to estimate the influence of factors of statistical error in the experiment on the basis of model results;

- to estimate the limits of application of this approach for forecasting the behavior of building structures in fire under realistic conditions, taking into account the obtained data of statistical error of the model results.

3 Materials and Methods of the Research of the Temperature Modes Parameters for Fire Development by Means of Mathematical Modeling

We used the Pyrosim computer system, a user shell for the Fire Dynamics Simulator (FDS) program, to calculate the temperature in the premises models in fire. This FDS system uses numerical algorithms to solve the complete system of the Navier-Stokes differential equations to determine the temperature and other dangerous factors during a fire [33]. The model results of these premises should be used to research the influence of the scale of premises' models in fire, the influence of automatic fire extinguishing systems and artificial ventilation systems.

To visualize the calculation results, the software module of the PyroSim Smoke-view system is used. It allows building appropriate graphical representations for the temperature distributions. This system also allows monitoring the dynamics of temperature fields and reproducing the heating process with animation. This system also allows obtaining pictures of smoke condition, and the distribution of the concentration of combustion products [33, 34].

The FDS algorithms are based on the numerical solution of the Navier-Stokes differential equations, assuming that fire flows have a low velocity and dependence on temperature, with the corresponding patterns of formation. In this case, for the numerical approximation of differential equations of heat and mass transfer, the method of finite differences on regular grids is used according to the computational explicit scheme "predictor-corrector" of the second order of accuracy in coordinates and time [33, 34]. When modeling the premises with fires, it should be taken into account that the calculation grid is rectangular and this can affect the accuracy of calculations in curvilinear calculation areas [33, 34].

4 Results

4.1 Geometric Configuration of the Calculation Area

The Fig. 1 shows a geometric diagram of the calculation area in a three-story building.

The Fig. 2 shows a 3-D model with the location of temperature control points in the premises on the first (Fig. 2a) and second floor (Fig. 2b). The thermocouples were installed at a distance of 0.1 m from the inner surface of the wall, which were modeled in the software system. It is important to estimate the influence of the temperature of the fire mode on building structures. There were identified areas with thermocouples P1… P4, which are located at a distance of 0.1 m from the inner surface of the fencing

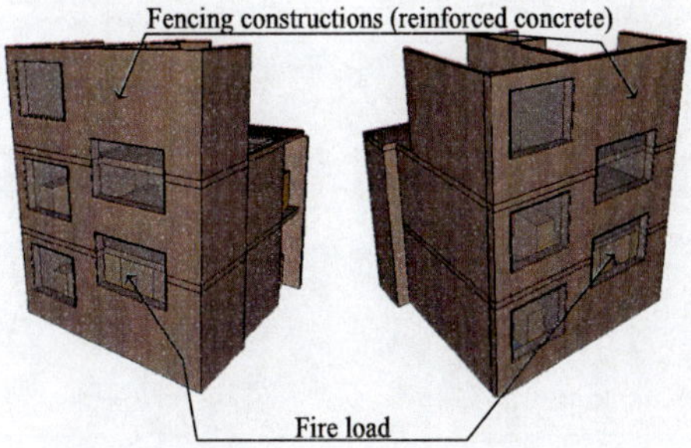

Fig. 1. Complete geometric diagram of the calculation area in a three-story building

constructions, according to which it was convenient to estimate the influence of the temperature field on the fencing constructions.

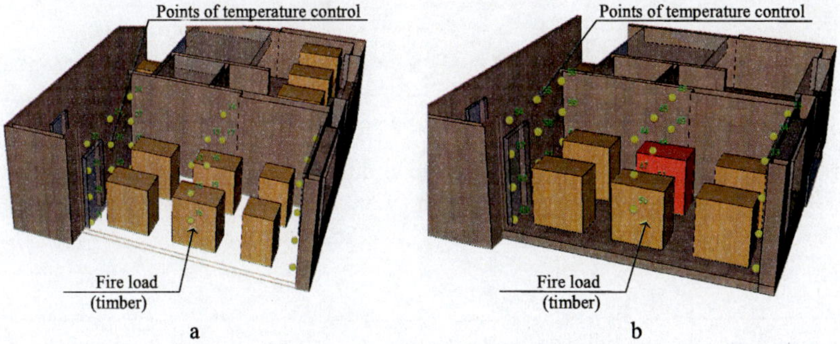

Fig. 2. Location of the point for recording the temperature indicators: a - on the first floor; b - on the second floor of the experimental building

4.2 Results of Temperature Calculation During a Fire in a Three-Storey Building

As a result of the calculation, the temperature distributions on the areas on the corresponding areas were obtained. The obtained results are shown in Fig. 4.

According to the temperature sensors (thermocouples), which were installed according to the schemes of location of thermocouples (Fig. 2), areas with thermocouples P1…P6 were identified and graphs of temperature dependence as a function of time were obtained. The results are shown in Fig. 5.

To research the reliability of the model results of temperature modes for full-scale tests in the premises of a three-story building, the main statistical parameters of the obtained data were researched. The data on the mean absolute deviations, mean relative deviations, mean square deviations of thermocouple indicators for the researched relevant tests are given in the Table 1. The data given in Table 1 showed that the error due to the difference between the calculated and experimental data reveals an

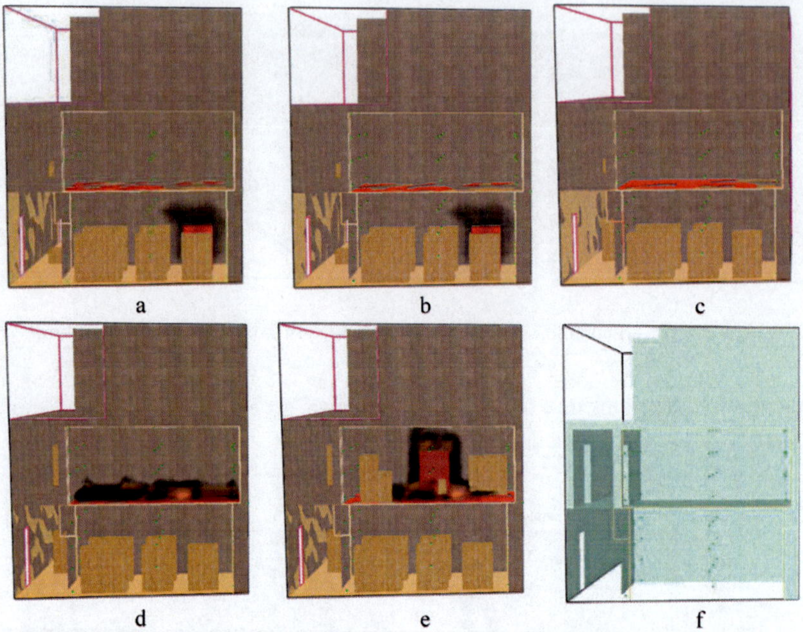

Fig. 4. The position of the flame jet of the fire at different moments of its development and spread in the section of the building: a - 600 s; b - 1200 s; c - 1800 s; d - 2400 s; e - 3000 s; f - 3600 s

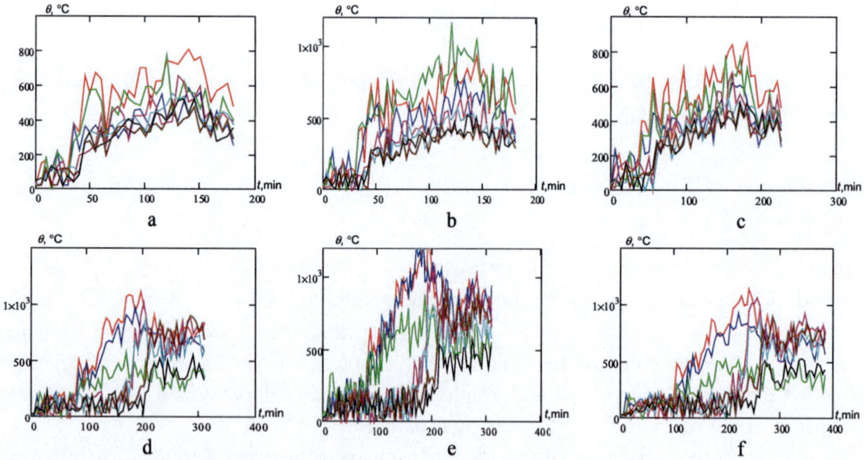

Fig. 5. Temperature indicators in the measured areas: a – 1; b – 2; c – 3; d – 4; e – 5; f – 6

acceptable accuracy of the determined mean volume temperature by mathematical modeling, since the relative error does not exceed 28% and the square deviation does not exceed 51 °C. This means that the temperature modeling results for full-scale tests in a three-story building are sufficiently accurate.

Table 1. Absolute deviations, relative deviations, square deviations of temperature modes compared with the experimental data

№ n/o	Position of the premise	Absolute deviations, °C	Relative deviations, %	Mean square deviations, °C
1.	The first floor	130.1	24.8	56.4
2.	The second floor	145.5	27.4	44.8
Mean values		137.8	26.1	50.6

The data on statistical criteria of temperature indicators for the corresponding series of tests are given in the Table 2.

Table 2. Statistical criteria of temperature indicators of model results of full-scale tests

№ n/o	Position of the premise	Cochrane's criterion	Student's criterion	Fisher's criterion
1.	The first floor	0.96	0.69	1.002
2.	The second floor	0.98	0.84	1.008

The results given in the Table 2 showed that the values of statistical criteria, due to the difference between the calculated and experimental data, do not exceed the tabular values. This means that the modeling results are adequate.

To verify the adequacy of the obtained results, the trend lines of the dependence of the medium-volume temperature on time and the corresponding variances of the deviations were built.

The built graphic dependences are given below in Fig. 6.

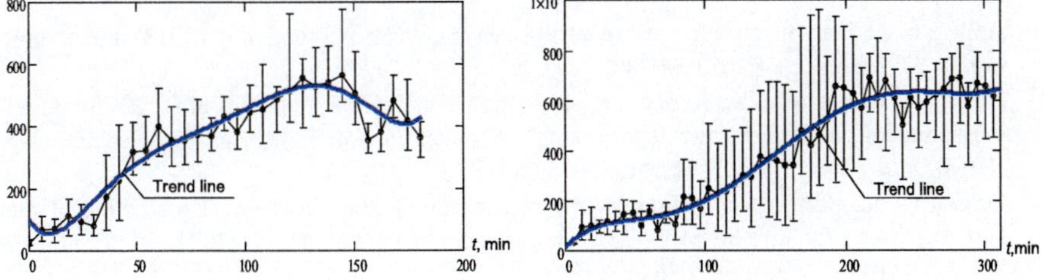

a b

Fig. 6. Graphs of dependences of medium-volume temperature on time and corresponding variances of deviations for the premises: a - on the first floor; b - on the second floor.

5 Conclusions

There was conducted a numerical experiment on modeling full-scale tests of premises with fire in a three-storey building with application of methods of computer gas-hydrodynamics. The identified spread of fire and the time dependences of its main parameters were revealed, which in turn allowed analyzing the adequacy of the modeling results and researching their adequacy and accuracy.

References

1. Shnal, T., Synenko, I., Yasinsky, D.: Analysis of methods of conducting field tests for fire resistance of buildings and structures. Theory Build. Pract. **664**, 357–362 (2010). Bulletin of the National University of Lviv Polytechnic
2. Dankevich, I., Schnal, T., Demchina, B.: Investigation of the temperature regime of fire in a model room. Theory Build. Pract. **742**, 46–51 (2012). Bulletin of the National University of Lviv Polytechnic
3. Dankevich, I., Prokhorenko, S., Shnal, T., Yuzkiv, T., Koval, O.: Research of the heating process of fencing constructions at the time of fire within a living space. Sci. Bull. **27**, 167–172 (2013). Ukrainian Research Institute of Civil Protection
4. Nekora, O., Slovynsky, V., Pozdieiev, S.: The research of bearing capacity of reinforced concrete beam with use combined experimental-computational method. In: Matec Web of Conferences, vol. 116, (2017). https://doi.org/10.1051/matecconf/201711602024. art no. 02024
5. Semerak, M., Pozdeev, S., Yakovchuk, R., Nekora, O., Sviatkevych, O. Mathematical modeling of thermal fire effect on tanks with oil products. In: Matec Web of Conferences, Fire and Environmental Safety Engineering, vol. 247, p. 00040 (2018). https://doi.org/10.1051/matecconf/201824700040
6. Pozdieiev, S., Nekora, O., Kryshtal, T., Zazhoma, V., Sidnei, S.: Method of the calculated estimation of the possibility of progressive destruction of buildings in result of fire. In: MATEC Web of Conferences, vol. 230 (2018). https://doi.org/10.1051/matecconf/201823002026. art no. 02026
7. Hietaniemi, J., Hostikka, S., Vaari. J.: FDS simulation of fire spread – comparison of model results with experimental data (2004). http://www.vtt.fi/inf/pdf/workingpapers/2004/W4.pdf
8. Jansson, R., Anderson, J.: Experimental and numerical investigation of fire dynamics in a facade test Rig. In: Proceedings of Fire Computer Modeling, p. 247 (2012)
9. Anderson, J., Jansson, R.: Fire dynamics in façade fire tests: measurement and modeling. In: Proceedings of Interflam 2013, p. 93, Royal Holloway College, University of London, UK (2013)
10. Anderson, J., Jansson, R.: Facade fire tests – measurements and modeling (2013). https://doi.org/10.1051/matecconf/20130902003
11. Anderson, J., Boström, L., Jansson, R., Milovanovi´c, B.: Fire dynamics in facade fire tests: measurement, modeling and repeatability. Applications of Structural Fire Engineering (2015). https://doi.org/10.14311/asfe.2015.059
12. Dréan, V., Schillinger, R., Auguin, G.: Fire exposed facades: Numerical modelling of the LEPIR2 testing facility. In: MATEC Web of Conferences, vol. 46, p. 03001 (2016). https://doi.org/10.1051/matecconf/20164603001
13. Blikharskyy, Z., Brózda, K., Selejdak, J.: Effectivenes of strengthening loaded RC beams with FRCM system. Arch. Civil Eng. **64**(3), 3–13 (2018)

14. Bobalo, T., Blikharskyy, Y., Kopiika, N., Volynets, M.: Serviceability of RC beams reinforced with high strength rebar's and steel plate. In: Lecture Notes in Civil Engineering, vol. 47, pp. 25–33 (2020). https://doi.org/10.1007/978-3-030-27011-7_4
15. Khmil, R., Tytarenko, R., Blikharskyy, Y., Vashkevych, R.: Influence of load level during strengthening of reinforced concrete beams on their reliability. IOP Conf. Ser. Mater. Sci. Eng. **708**(1), 012054 (2019)
16. Krainskyi, P., Blikharskyy, Y., Khmil, R., Blikharskyy, Z.: Experimental study of the strengthening effect of reinforced concrete columns jacketed under service load level. In: Matec Web of Conferences, vol. 183, p. 02008 (2018). https://doi.org/10.1051/matecconf/201818302008
17. Selejdak, J., Blikharskyy Y., Khmil R., Blikharskyy, Z.: Calculation of reinforced concrete columns strengthened by CFRP. In: Lecture Notes in Civil Engineering, vol. 47, pp. 400–410 (2020). https://doi.org/10.1007/978-3-030-27011-7_51
18. Vegera, P., Vashkevych, R., Blikharskyy, Z.: Fracture toughness of RC beams with different shear span. In: MATEC Web of conferences, vol. 174, p. 02021 (2018). https://doi.org/10.1051/matecconf/201817402021
19. Blikharskyy, Z., Vashkevych, R., Vegera, P., Blikharskyy, Y.: Crack resistance of RC beams on the shear. In: Lecture Notes in Civil Engineering, vol. 47, pp. 17–24 (2020). https://doi.org/10.1007/978-3-030-27011-7_3
20. Bobalo, T., Blikharskyy, Y., Kopiika, N., Volynets, M.: Theoretical analysis of RC beams reinforced with high strength rebar's and steel plate. IOP Conf. Ser. Mater. Sci. Eng. **708**(1), 012045 (2019)
21. Krainskyi, P., Blikharskyy, Y., Khmil, R., Vegera, P.: Crack resistance of RC columns strengthened by jacketing. In: Lecture Notes in Civil Engineering, vol. 47, pp. 195–201 (2020). https://doi.org/10.1007/978-3-030-27011-7_25
22. Lobodanov, M., Vegera, P., Blikharskyy, Z.: Planning experiment for researching reinforced concrete beams with damages. In: Lecture Notes in Civil Engineering, vol. 47, pp. 243–250 (2020). https://doi.org/10.1007/978-3-030-27011-7_31
23. Blikhars'kyi, Y.Z.: Anisotropy of the mechanical properties of thermally hardened A500s reinforcement. Mater. Sci. **55**, 175–180 (2019). https://doi.org/10.1007/s11003-019-00285-0
24. Kharchenko, Y.V., Blikharskyy, Z.Y., Vira, V.V., Vasyliv, B.D.: Study of structural changes in a nickel oxide containing anode material during reduction and oxidation at 600 °C. In: Fesenko, O., Yatsenko, L. (eds.) Nanocomposites, Nanostructures, and their Applications NANO 2018. Springer Proceedings in Physics, vol. 221. Springer, Cham (2019). https://doi.org/10.1007/978-3-030-17759-1_42
25. Kharchenko, Y., Blikharskyy, Z., Vira, V., Vasyliv, B.: Vasyliv, B: Study of nanostructural changes in a Ni-containing cermet material during reduction and oxidation at 600 °C. Appl. Nanosci. (2020). https://doi.org/10.1007/s13204-020-01391-1
26. Zhang, Q., Mol'kov, Y.V., Sobko, Y.M. et al.: Determination of the mechanical characteristics and specific fracture energy of thermally hardened reinforcement. Mater Sci. **50**, 824–829 (2015). https://doi.org/10.1007/s11003-015-9789-9
27. Vegera, P., Khmil, R., Vegera, P.: Shear strength of reinforced concrete beams strengthened by PBO fiber mesh under loading. In: MATEC Web of Conferences, vol. 116, p. 02006 (2017). https://doi.org/10.1051/matecconf/201711602006
28. Krainskyi, P., Vegera, P., Khmil, R., Blikharskyy, Z.: Theoretical calculation method for crack resistance of jacketed RC columns. IOP Conf. Ser. Mater. Sci. Eng. **708**(1), 012059 (2019)

29. Nilsson, M., Mossberg, A., Husted, B., Anderson, J.: Protection against external fire spread - horizontal projections or spandrels. In: 14th International Fire Science & Engineering Conference, Royal Holloway College, University of London, UK, Vol. 2, pp. 1163–1174 (2016). https://www.researchgate.net/publication/306078631_Protection_against_external_fire_spread_-_Horizontal_projections_or_spandrels
30. Nilsson, M., Nilsen, J., Mossberg, A.: Validating FDS against a large-scale fire test for facade systems. In: 3rd Fire and Evacuation Modelling Technical Conference (FEMTC), Melia Costa del Sol in Torremolinos, Spain (2016). https://www.researchgate.net/publication/311264945_Validating_FDS_against_a_large-scale_fire_test_for_facade_systems
31. Nilsson, M.: The impact of horizontal projections on external fire spread - a numerical comparative study, Report nr. 5510, Lund University, Division of Fire Safety Engineering, Lund (2016)
32. Zhang, G., Zhu, G., Zhao, G.: Analysis of the influence of construction insulation systems on public safety in China. Int. J. Environ. Res. Public Health **13**, 861 (2016)
33. McGrattan, K., et al.: Fire Dynamics Simulator User's Guide, FDS Version 6.2.0, SVN Repository Revision: 22352, NIST Special Publication 1019, National Institute of Standards and Technology, Gaithersburg, MD USA (2015)
34. McGrattan, K.B., Baum, H.R., Rehm, R.G., Hamins, A., Forney, G.P., Floyd, J.E., Hostikka, S., Prasad, K.: Fire Dynamics Simulator Technical Reference Guide, 6 edn. Volume 1: Mathematical Model. NIST Special Publication 1018 (2016)

Cohesion of Slurry Surfacing Mix on Bitumens of Different Acid Numbers at Different Curing Temperatures

Iurii Sidun(✉), Sergii Solodkyy, Oleksiy Vollis, and Volodymyr Gunka

Lviv Polytecnic National University, Lviv 79013, Ukraine
siduniurii@gmail.com

Abstract. Cohesion of Slurry Surfacing mix is an important technological indicator that allows you to start traffic on time. In this article Slurry Surfacing mix curing is studied in three modes: 10 °C (high humidity), 20 °C (normal humidity) and 30 °C (high humidity). As binders for Slurry Surfacing mix there were used oxidized bitumens produced from light crude-oil and distillation bitumen produced from heavy crude-oil. There was confirmed that the first bitumens are characterized by low acid numbers, while the latter ones – by high ones. On the base of the bitumens there were produced cationic slow-setting bitumen emulsion and Slurry Surfacing mix. It was found that high-acid number bitumen is definitely optimum version in comparison with low acid number for the usage in Slurry Surfacing mix at different temperature modes – by criterion of cohesion strength build-up rate for the mix. This criterion consists of three periods (times) on mix curing: Set («Set» Torque), Traffic (Early Rolling Traffic) and Cure Time (Cured Slip Torque).

Keywords: Cohesion · Slurry-surfacing mix · Curing temperature · Bitumen emulsion · High and low acid number bitumens

1 Introduction

Slurry-surfacing mix (SSM) is high-efficient material for motor-roads paving, but the process of SSM production and paving strongly depends upon such weather-climate factors as air temperature and ambient humidity. It is known that SSM shall not be applied if either the pavement or air temperature is below 50 °F (10 °C) and falling, but may be applied when both pavement and air temperatures are above 45 °F (7 °C) and rising. No SSM shall be applied when there is the possibility of freezing temperatures at the project location within 24 h after application. The mixture shall not be applied when weather conditions prolong opening to traffic beyond a reasonable time [1, 2]. A lot of articles are dedicated to study of SSM [3–8]. Some of them [9–11] touch the issue of influence upon the SSM of such natural factors as change of temperature, air humidity and wind. These external environmental factors change both during the road season and time of the day, and they will influence directly on SSM production and paving, as well as on its curing. Besides of that, as the research has found the surface

Z. Blikharskyy (Ed.): EcoComfort 2020, LNCE 100, pp. 429–435, 2021.
https://doi.org/10.1007/978-3-030-57340-9_52

temperature affects the rate at which an emulsion breaks [9] As depicted in в Caltrans Standard Specifications Section 37 [12] the basic requirement for success SSM is that the emulsion must be able to break and form continuous films, as it is the only way a slurry mixture can become cohesive. As a result, humidity, wind conditions, air and pavement surface temperature are important and need to be considered. Modifications to additives should be made according to the changing environment during application. In any case, application of a SSM is generally not suitable for night work. This is due to the lower evaporation rate at night, which results in longer breaking and curing times. As stated in [13] SSM shall only be placed when the ambient temperature is 8 °C (for a conventional SSM project, air temperature should be a minimum of 10 °C and rising. Humidity should be 60% or less and a slight breeze is advantageous. Work should not be started if rain is imminent. SSMs will typically resist rain induced damage after as little as one hour but typically require at least three hours to cure to a fully waterproof state. Additionally, breaking time for a slurry is affected by ambient temperature. Work should not be started if freezing temperatures are anticipated within 24 h of construction. Ideal microsurfacing weather conditions are those with low humidity, a slight breeze, and with sustained high temperatures into the forthcoming days [14] High humidity is a detriment to any microsurfacing owing to its acting to retard the breaking of the emulsion [15].

Timely curing of the mix determined the traffic time for the slurry pavement. Therefore, the studies done were targeted on determination of how does the various-binders-based-SSM's cohesion strength build-up rate depend upon the temperature mode of preparation, paving and curing of the mix. Under the notion "various binders" we mean high- and low acid bitumens. The internal factor of binder is decisive in the rate of SSM formation, as far as we are aware of the fact that the mix cures fast when high acid bitumen is used and slowly – when low acid bitumen is used [16, 17]. One more important internal factor, having influence upon the curing time, is reactivity of the aggregate used, which is determined by methylene blue method. When high-reactive aggregates are used, in the mix there shall be contained the substantial quantity of mix extending additive (admix), which extends mixing time (breaking time) and allows production of the mix in good time. But the most popular for usage as mix extending additive are those emulsifier solutions, which (besides of extending the mix time) also extend the curing time. Such dependence is not fatal for SSM on high acid bitumen, but is important for low acid bitumen. Thus, the objective of the study was to test the change of influence of external factors (temperature and humidity) upon the SSM curing – with application of high and low acid bitumen.

2 Methods and Materials

Interestingly, in [18] they note that currently there are no practical methods to determine SSM curing time right on the road. The engineers, in their turn, use two empirical tests (Stick Test, Shoe Test). In laboratory conditions cohesion of SSM was determined by standard method according to ISSA, TB139 [19], while mix time was determined according to ISSA, TB113 [20]. According to ISSA classification, there were distinguished three periods (times) on mix curing: Set («Set» Torque), Traffic

(Early Rolling Traffic) and Cure Time (Cured Slip Torque). For the Slurry Surfacing mix design there were used Ukrainian aggregates from granite quarries: JSC «Polonskiy Gorniy Combinat» (Khmelnitskiy Region). The Slurry Surfacing mix grading was designed for type 2 with maximum particle size for the coarse filler 10 mm [2]. In order to avoid influence of grading peculiarities upon the indices of cohesion strength build-up rate the Slurry Surfacing mix design was performed based on the exactly the same narrow fractions. The Slurry Surfacing mix grading selected refers to type 2 SSM (Table 1) and corresponds to the limit values according to [2].

For the production of bitumen emulsion for Slurry Surfacing there was used oxidized bitumen 70/100 obtained from light crude-oil from PJSC "UkrTatNafta" (Kremenchuk Refinery, Ukraine), 70/100 from JSC Mozyr Refinery (Belarus, Gomel Region, town of Mozyr) and distillation bitumen 70/100 (Nynas, Sweden) obtained from heavy crude-oil. All the bitumens used correspond to [21].

Table 1. The estimated grading 0–10 for the type 1 mix ISSA.

Orifices Ø for round-shaped sieves, mm	Average grading 0–10 for the type 2 mix ISSA, % by mass	
	Partial sieve residues	Complete passing through sieves
<0.5	30	0
0.5	40	30
2	20	70
5	10	90
10	0	100

Table 2. Main physical mechanical parameters of bitumens

Index and values	70/100 UkrTatNafta	70/100 JSC Mozyr	70/100 Nynas
Penetration at 25 °C, (0,1 mm)	72	75	80
Softening point, °C	48	48	47
Ductility at 25 °C, (10 mm)	>100	>100	>100
Fraaß breaking point, °C	−16	−15	−10
Flash point, °C	290	295	230
Solubility, %	99,3	99,4	99,5
Total acid no. mg KOH/g	0.5	0.6	3.0

Penetration at 25 °C determined according to [22], softening point [23], ductility at 25 °C [24], fraaß breaking point [25], flash point [26], solubility [27], total acid no. [28].

While analyzing Table 2 one can see that the bitumens chosen are similar by the main quality indices, but substantially differ by total acid number. Therefore, the statement was asserted that bitumens obtained from light crude-oil by oxidation method are characterized by low acid numbers, whereas bitumens obtained from heavy crude-oil by distillation method are characterized by high acid numbers [29]. Based on the bitumens, emulsifier Redicote 404 (Nouryon. Sweden), hydrochloric acid and water, by means of laboratory bitumen emulsion plant SEP-0.3R of Danish company DenimoTech – there were produced cationic slow-setting bitumen emulsions for further SSM preparation (Table 3). Besides, as a component for SSM there was included Portland cement, produced by JSC «NicolayevCement» (Ukraine, town of Nikolayev), tap water and extending additive – 10% solution of emulsifier Redicote E-11 (Nouryon. Sweden).

Table 3. Bitumen emulsion formulations for SSM

Components	Emulsion formulation (CSS), % mass		
	70/100 UkrTatNafta	70/100 JSC Mozyr	70/100 Nynas
Bitumen	62		
Emulsifier	Redicote 404 – 1.1		
Water phase pH (acid)	pH = 2.5 (HCI)		
Water	Till 100		

3 Results and Discussions

SSM curing was studied in three modes: 10 °C (high humidity), 20 °C (normal humidity) and 30 °C (high humidity). In Table 4 there are presented the designed SSM compositions – by criterion of mix time on the basis of the bitumens used at different temperatures.

Table 4. Mix time of SSM

Mix No, bitumen		Components, g				Mix time, s	T °C
		Cement	Water	Admix	Emulsion		
1.70/100UkrTatNafta	1.1	0.5	10,0	0,5	14.0	125	10
	1.2	1.0	10.0	1.0	14.0	123	20
	1.3	1.25	11.0	1.75	14.0	124	30
2.70/100 JSC Mozyr	2.1	0,75	10,0	0,5	14.0	126	10
	2.2	1.25	10,0	1.0	14.0	129	20
	2.3	1.25	11,0	2.0	14.0	127	30
3.70/100 Nynas	3.1	0.75	10.0	0.75	14.0	125	10
	3.2	1.0	10.0	1.25	14.0	121	20
	3.3	1.5	12.0	2,25	14.0	122	30

Data analysis in Table 4: distillation and oxidized bitumens shows that the increase of testing temperature requires the increase of Portland cement and extending additive content in the compositions. The difference in Portland cement content at T = 10 °C and T = 30 °C constitutes 0.75 part, while extending additive 1.25–1.5 part. Besides, at T = 30 °C SSM requires somewhat more water.

The investigation results – on determination of slurry-surfacing mix cohesion strength build-up rate – are presented in Table 5 and in Fig. 1.

Table 5. Slurry-surfacing mix cohesion strength build-up rate

Mix No, bitumen		Periods (times) on mix curing, h			
		Set	Traffic	Cure	T °C
1.70/100UkrTatNafta	1.1	0.50	5.0	7.0	10
	1.2	0.50	4.0	6.0	20
	1.3	0.50	3,5	5.5	30
2.70/100 JSC Mozyr	2.1	0.50	5,0	7,0	10
	2.2	0.50	4,0	6,0	20
	2.3	0.50	3,5	5,5	30
3.70/100 Nynas	3.1	0.42	0.75	1.0	10
	3.2	0.17	0.25	0.75	20
	3.3	0.08	0.17	0.5	30

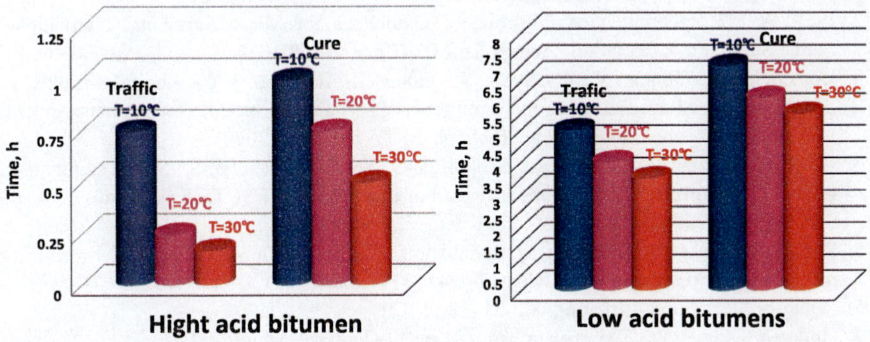

Fig. 1. Periods (times) of Slurry-surfacing mix curing depending on the temperature

On Fig. 1 can see that the mix on high acid bitumen at the temperature of 20 °C and normal relative air humidity is characterized by: Set Time (0.17 h), Traffic Time (0.25 h) and Cure Time (0.75 h), while on low acid bitumen: Set Time (0.5 h), Traffic Time (4.0 h) and Cure Time (6.0 h). Application of high acid bitumen allows production of the mix even at the temperature of 10 °C and increased relative air humidity without substantial time consumption for its curing (Traffic Time comes in 0.75 h and Cure Time in 1.0 h). The fastest hardening of the mixture is observed at temperature −20 °C.

4 Conclusions

It was found that the decrease of ambient temperature brings down the rate of mix curing. That is due to the slower evaporation and water separation from bitumen emulsion and the mix in general. At high temperatures SSM requires larger quantity of extending additive, while it worsens the mix curing indices. Still, this peculiarity does not play so important role as temperature and air humidity in SSM curing pro-cess and acidity of the bitumen. There is brilliantly witnessed the slow curing of the mix at limit-low temperature for SSM paving (T = 10 °C). Still, the usage of high acid in SSM allows production of the mix even at low temperatures and increased air humidity without substantial time consumption for its curing. SSM on the base of oxidized bitumen even at high temperature (T = 30 °C) is not efficient as by the development of the mix cohesion.

It was found that high-acid distillation bitumen is definitely optimum version in comparison with oxidized (low-acid) one – for the use in SSM at different temperature modes.

References

1. ISSA A105 Recommended Performance Guidelines for Emulsified Asphalt SSM, International Slurry Surfacing Association, Annapolis, MD (Revised) (2010)
2. ISSA A 143 Recommended Performance Guidelines for Micro Surfacing, International Slurry Surfacing Association, Annapolis, MD (Revised) (2010)
3. Demchuk, Y., Gunka, V., Pyshyev, S., Sidun, I., Hrynchuk, Y., Kucinska-Lipka, J., Bratychak, M.: Slurry Surfacing mixes on the basis of bitumen modified with phenol-cresol-formaldehyde resin. Chem. Chem. Technol. **14**(2), 251–256 (2020)
4. Grilli, A., Graziani, A., Carter, A., Sangiorgi, C., Specht, L., Callai, S.: Slurry surfacing: a review of definitions, descriptions and current practices. RILEM Tech. Lett. **40**, 103–109 (2019)
5. Robati, M., Carter, A., Perraton, D.: Evaluation of a modification of current microsurfacing mix design procedures. Can. J. Civ. Eng. **42**, 319–328 (2015)
6. Robinson, H.: Slurry surfacing. **82**, 31–32 (2013)
7. Holleran, G.: ABC's of slurry surfacing. Asphalt Contract. Mag. (2001)
8. Parish, J., Mackenzie, J.: The use of thin film slurry surfacing on military airfield pavements in the UK. In: Proceedings of the Institution of Civil Engineers-Transport (1994)
9. Moulthrop, J.: Slurry/micro-surface mix design procedure. Northeast Pavement Preservation Partnership, Warwick, R.I. (2007)
10. Senadheera, S., Gransberg, D., Kologlu, T.: Seal coat field manual, Research Report TX-1787-P, Texas Department of Transportation, Austin, p. 244 (2001)
11. Wood, T.: Minnesota experiences with innovative microsurfacing. In: Proceedings, Midwest Pavement Preservation Partnership, Missoula, Montana (2007)
12. California Department of Transportation/ Standard Specifications, Sacramento, California, May 2006
13. California Department of Transportation. Maintenance Technical Advisory Guide (TAG): Volume I – Flexible Pavement Preservation, Second Edition, 7 March 2008
14. National Highway Institute. Pavement Preservation Treatment Construction Guide, Chapter 8—"Microsurfacing". Federal Highway Administration, Washing-ton, D.C. (2007)

15. Asphalt Institute, Asphalt Surface Treatments—Construction Techniques, Educational Series No. 12 (ES-12), Lexington, Kentucky, 28 p. (1988)
16. Pyshyev, S., Grytsenko, Y., Solodkyy, S., Sidun, I., Vollis, O.: Using bitumen emul-sions based on oxidized, distillation and modified oxidized bitumens for SSM production. Chem. Chem. Technol. **9**(3), 359–366 (2015)
17. Sidun, I., Vollis, O., Solodkyy, S., Gunka, V.: Cohesion of slurry surfacing mix with slow setting bitumen emulsions. In: Blikharskyy, Z., Koszelnik, P., Mesaros, P. (eds.) Proceedings of CEE 2019. CEE 2019. Lecture Notes in Civil Engineering, vol. 47. Springer, Cham (2020)
18. Gransberg, D.: Microsurfacing. Technical Report NCHRP SYNTHESIS 411Affiliation: National Academies (2010)
19. TB139 Test Method to Determine Set and Cure Development of Slurry Surfacing Systems by Cohesion Tester. International Slurry Surfacing Association (2017)
20. TB 113 Test Method for Determining Mix Time for Slurry Surfacing Systems. International Slurry Surfacing Association (2017)
21. EN 12591: 2009 Bitumen And Bituminous Binders - Specifications For Paving Grade Bitumens
22. EN 1426:2000 European Standard. Bitumen and bituminous binders. Methods of tests for petroleum and its products. Determination of needle penetration
23. EN 1427:2007 European Standard. Bitumen and bituminous binders. Determination of the softening point. Ring and Ball method
24. DSTU 8825:2019 Ukrainian standard. Bitumen and bituminous binders. Method of determining elongation
25. EN 12593 European Standard. Bitumen and bituminous binders. Determination of the Fraas breaking point
26. EN 22592:1994 European Standard. Methods of test for petroleum and its products. Petroleum products. Determination of flash and fire points. Cleveland open cup method
27. EN 12592:2014 European Standard. Bitumen and bituminous binders. Determination of solubility
28. ASTM D664 - 18e2 Standard Test Method for Acid Number of Petroleum Products by Potentiometric Titration
29. Sobol, K., Blikharskyy, Z., Petrovska, N., Terlyha, V.: Analysis of structure formation peculiarities during hydration of oil-well cement with zeolitic tuff and metakaolin additives. Chem. Chem. Technol. **8**, 461–465 (2014). https://doi.org/10.23939/chcht08.04.461

Influence of Climatic Factors on Runoff Formation and Surface Water Quality of the Stryi River Basin

Volodymyr Snitynskyi[1], Petro Khirivskyi[1], Ihor Hnativ[1], and Roman Hnativ[2](✉)

[1] Lviv National Agrarian University, Dubliany, Lviv 79000, Ukraine
[2] Lviv Polytechnic National University, Lviv 79013, Ukraine
gnativ.roman.m@gmail.com

Abstract. Global changes in climatic characteristics affect the hydrochemistry of surface waters, apart from direct economic intervention. Such changes in the hydrological regime will affect the change in the qualitative characteristics of surface waters and the concentration and content of major components in river waters. Natural processes of formation of the chemical composition of river waters also violates anthropogenic influence. The purpose of the conducted research is to analyze the factors influencing the runoff formation and surface water quality of the foothills of the Ukrainian Carpathians and the choice of optimal flood protection of inhabited territories from destruction. The abnormally warm weather caused the lack of snow reserves as a major factor in spring runoff. There was no permanent snow cover until the end of the second decade of February except the Carpathian highlands. The persistent lack of rainfall in the autumn and winter has determined the low water content of most rivers in the country. Studies have shown that the overall water-management and ecological situation may be complicated, especially in water bodies with significant anthropogenic load, with a deterioration in water quality due to the limitation of the dilution of polluted waste water and the inability of natural washing of river beds to dry up.

Keywords: Runoff formation · Water quality · Climatic factors

1 Introduction

Global warming, which leads to increased intensity and duration of droughts, complicates environmental conditions for the development of society in all countries. This causes a decrease in crop yields and fisheries productivity. Climate change trends are driving humanity to use renewable and environmentally friendly energy sources. A better understanding of the variability of the climate system can help agriculture and forestry increase their productivity and efficiency [1].

The state of the biosphere depends on the heat and moisture formed by the transformation of solar radiation during the interaction of the components of the climate system. The plant world largely determines the reflectivity of the planet and is the main source of oxygen, participates in the processes of moisture circulation, regulation of

Z. Blikharskyy (Ed.): EcoComfort 2020, LNCE 100, pp. 436–442, 2021.
https://doi.org/10.1007/978-3-030-57340-9_53

carbon dioxide content in the atmosphere and the formation of its temperature regime. The anthropogenic activity has a special influence as a result of which the properties of both individual components and the climate system as a whole change.

The supply of fresh groundwater in the Carpathian region is uneven and depends on the location of the terrain in a certain hydrogeological basin and aquifers. According to the results of monitoring observations of the groundwater regime in the Lviv region, direct dependencies of groundwater levels fluctuations on climatic and geomorphological factors were revealed [2].

2 Main Part

Global changes in climatic characteristics affect the hydrochemistry of surface waters, apart from direct economic intervention. Scientists note that the quantitative parameters of modern warming of the regional climate of Ukraine at the rate of increase of average annual ground temperatures correspond to the global warming. This leads to an increase in low-water runoff, especially in winter, as well as average annual costs due to a decrease in evaporation and ice cover thickness and a reduction in the period of freezing [2]. Such changes in the hydrological regime will affect the change in the qualitative characteristics of surface waters and the concentration and content of major components in river waters. The natural processes of the formation of the chemical composition of river waters are also disturbed by anthropogenic impact. More and more chemical components formed as a result of economic activity are washed off the surface of the catchment, increasing the salt concentration by 3–5 times in comparison with the natural value [3].

Most fresh groundwater is the main source of water supply for urban and rural settlements in the region. The worst conditions for fresh water supply are in the Turka, Skole, Staro Sambir and Sambir districts. However, even in the most affluent Sokal district, not all water is suitable for drinking water supply because of its pollution in large areas [4]. Recently, there has been a gradual increase in freshwater extraction volumes in most water intakes in the region [5].

Atmospheric precipitation absorbs the chemical composition of natural sources and changes as a result of purification of atmospheric air from components of natural and technogenic origin [6]. To analyze the patterns of formation of the chemical composition of natural waters, one of the most important factors is the amount and chemical composition of precipitation, which are closely related to the climatic conditions of the Stryi River basin. According to the vertical climatic zonality of the Carpathian Mountains, a regular decrease in temperature indicators and atmospheric pressure and an increase in the amount of precipitation are observed (Fig. 1). Their mean long-term values change according to the geomorphological vertical zoning of the territory. The gradients of changes in annual precipitation vary with altitudes of 80–90 mm for every 100 m of absolute height above sea level. At the mouth, precipitation amounts to 650 mm at an absolute height of 239 m and more than 1200 mm in the mountainous part at altitudes exceeding 1000 m [7].

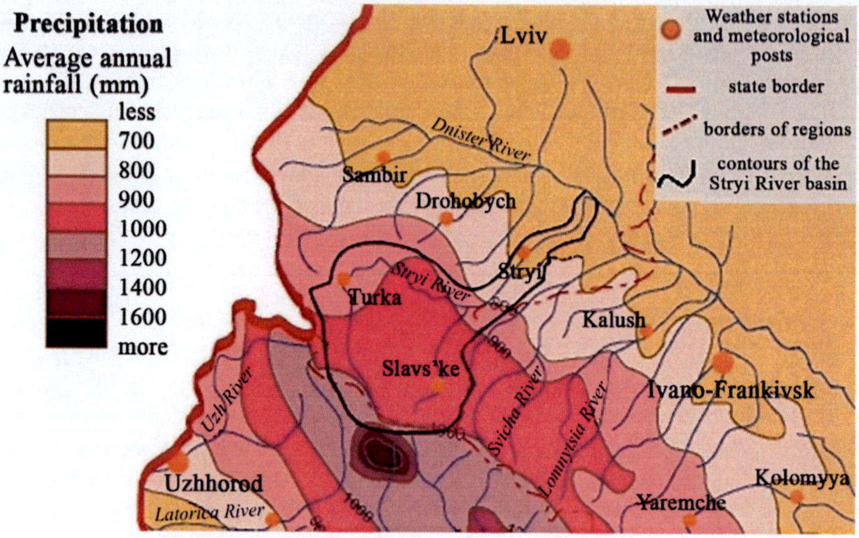

Fig. 1. Average annual rainfall in the Stryi River basin and adjacent territories

Atmospheric precipitation is the most important factor in the purification of atmospheric air from various gas and dust pollution, but they are an extremely important source of water-soluble components and the formation of the chemical composition of natural waters. Within the Carpathian region, the chemical composition of atmospheric precipitation has been studied by many scientists [8, 9]. The results of the research showed the influence of the chemical composition of atmospheric precipitation in the formation of hydro-chemical content of natural waters and the calculation of the flow of chemical substances with precipitation in the Stryi River basin [10].

3 Water Supply of the Rivers of the Stryi River Basin, Their Water Regime of Levels and Runoff

Taking into account the long-term observations of the water supplying of the rivers of the Stryi River basin, we observe that up to 50% of the annual total water balance of the runoff is caused by snow supplying, due to the receipt of melt water during the snowmelt period (Fig. 2). Up to 44% occupy rain water due to direct runoff from the catchment area. Groundwater inflows in the Stryi River basins account for only 6%, but it is the longest in time and prevailing in periods of no precipitation. The river Stryi is characterized by considerable instability in the level regime [11].

The spring flood on the mountain rivers and the river Stryi, although it is clearly expressed on a long-term annual weighted average hydrograph, but it very rarely has catastrophic consequences. This is due to the afforestation, the high zoning of the main catchment area and the exposure of the slopes. The peak of the spring waterfall falls in April (Fig. 2).

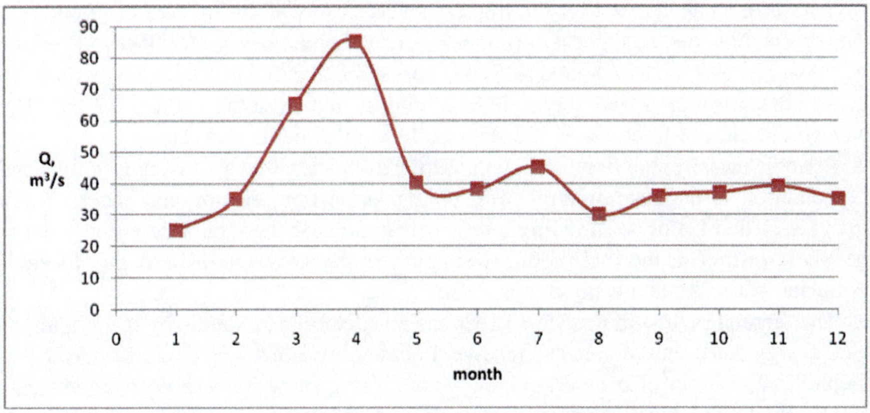

Fig. 2. Long-term weighted average hydrograph of the river Stryi at the meteorological station Verkhnye Sinyvidnoye according to observations from 1991 to 2010 [12]

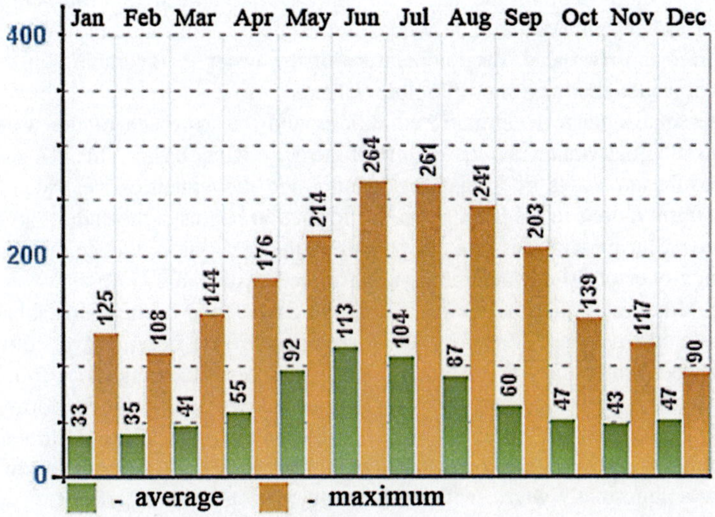

Fig. 3. Average monthly and maximum rainfall (mm) with wetting correction (climatic data for Stryi city for the period from 1899 to 2019 [1])

In some years, the spring flood is poorly expressed, but in summer and autumn high rainfall is observed (Fig. 3). Discharge and volumes of runoff in the Stryi River basin in some years are much greater than the corresponding characteristics of the spring flood [12].

In general, the summer-autumn period accounts for 40–50% of annual runoff. However, there are years with continuous flooding. The hydrological regime of the

Stryi River is characterized by the highest levels observed during summer floods. In some years, they can be highest during autumn and sometimes winter floods. Then the water rises from 8 to 170 cm, and in some years more than 3.5 to 5.5 meters per day. In the summer-autumn period there are 3–5 floods, and in some years - 12–15. The average duration of floods is 10–25 days and the maximum is 55 days.

Periodic catastrophic floods of 1955, 1969, 2008 and 2010 significantly influenced the condition of engineering structures, coastal protection, erosion and accumulation processes. The idea of constructing a reservoir flood near the village of Pidhorodtsi in the Skole district at the end of the twentieth century was not realized due to environmental warnings and economic problems.

The thermal regime of the Stryi River basin is determined mainly by the amount of solar energy that a unit of area can receive. In case of windless and cloudless weather it is up to 160 kcal/cm^2. However, in cloudy conditions, snow cover, it decreases by 3–7 times, due to the high value of scattering and reflection. According to long-term meteorological observations, the light part of the day with such conditions averages 70–80%, resulting in the average effective radiation not reaching 40 kcal/cm^2.

The aforementioned climatic features of the Stryi River basin determine its thermal regime, which, by virtue of its vertical zonation, is territorially substantially different. In the mountainous part of the basin, in places of riverhead of brook-creeks and springs, the water temperature in July does not exceed 4–9 °C and only in the case of distance from the riverhead, the temperature of the water in favorable climatic conditions approaches the maximum possible values.

The seasonal change in climatic conditions and the approach of the water temperature to 0 °C determines the ice regime of the Stryi River basin. This is a change in the time of the processes of formation, opening and movement of ice. According to long-term observations in the Stryi basin, ice formation begins at the end of November. Freezing-over during winter is often disturbed by the breaking of the ice during thaws. The freezing-over of 20 days or more is considered sustainable [1, 13].

It should be noted that the ice regime of the Stryi River basin is unstable. In the upper and middle reaches of the river, there are often several freezing-overs during the winter, between which there is drifting of ice and temporary clearing rivers from ice. In the lower reaches, freezing-over is incomplete, in many places polynyas form [13].

During the autumn-winter period of 2019–2020 in the river basins of Ukraine there is an extremely unfavorable hydrometeorological situation for the formation of spring flood. It was abnormally warm, with frequent alternation of short periods of decrease and increase in air temperature. There was no steady transition of the air temperature through zero degrees towards negative temperatures. The average winter temperature was 3.5–4.9 °C above the climatic norm. January 2020 was especially warm, when the average air temperature exceeded the monthly norm by 5–7 °C. Rainfall was generally less than normal [1, 14].

The abnormally warm weather caused the lack of snow reserves as a major factor in spring runoff. There was no permanent snow cover until the end of the second decade of February except the Carpathian highlands. The persistent lack of rainfall in the autumn and winter has determined the low water content of most rivers in the country. The water flow of the Carpathian region rivers, as a result of the formation of melt-rain floods, increased during the two decades of February and amounted to 20–70% of the monthly norm on the Dniester and its flat tributaries.

Our studies have allowed us to establish the dependencies shown in Fig. 4.

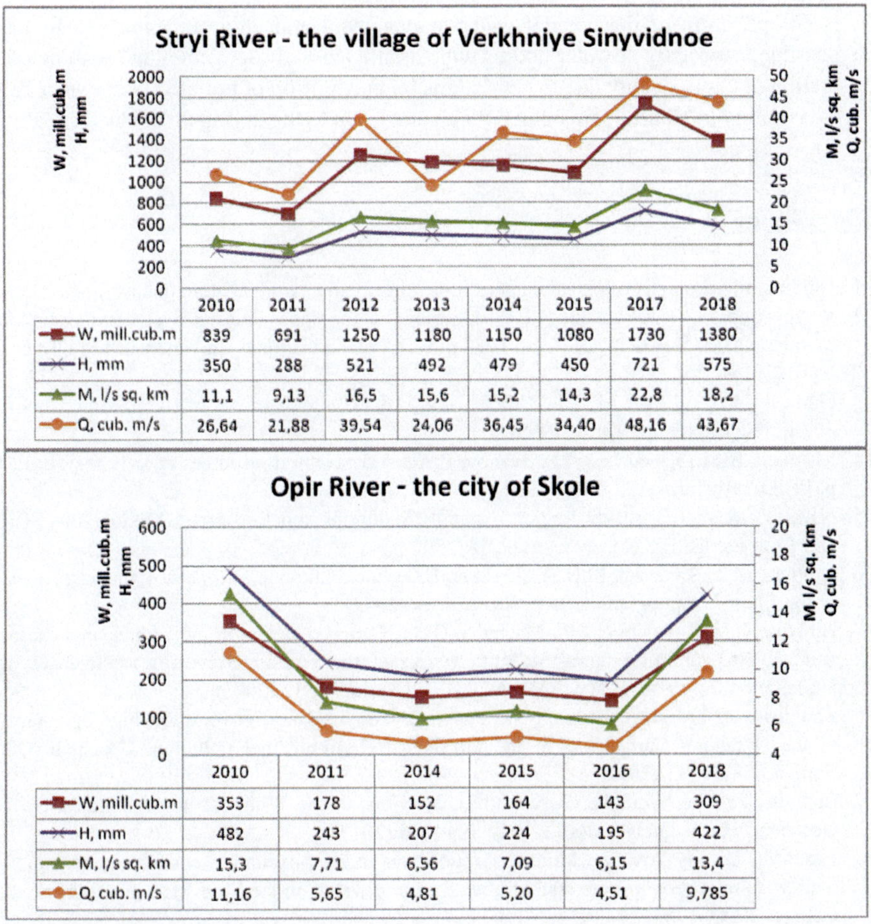

Fig. 4. Change of runoff parameters for 2010–2018 from natural factors in the Stryi River basin: W - runoff volume; H - drainage layer; M - drain module; Q - average flow of surface water (The catchment area at the village of Verhnye Sinyvidne F = 2400 km^2, at the point of Skole F = 733 km^2)

4 Conclusions

1. In the absence of snow reserves, as a major factor in spring runoff, the indices and nature of the hydrological regime on the rivers of the Carpathian region in 2020 will

largely be determined by changes in air temperature and precipitation. Spring flood on most small and medium-sized rivers in the country may not be pronounced.

2. In the Carpathian rivers, which are an area of potential flood danger, in case of heavy rainfall during the snowmelt period, dangerous melt-rainfall floods of different altitudes can be rapidly formed. This can adversely affect the water-protective dykes, suburban areas and the objects on them.
3. A complication of the general water management and environmental situation is possible, especially at water bodies with significant anthropogenic load with deterioration of water quality due to restrictions on the dilution of polluted wastewater and the inability to naturally flushing river channels, up to the drying up of small streams.

References

1. Ukrainian hydrometeorological center. https://meteo.gov.ua/ua/33345/climate/climate/
2. Pylypovych, O.V., Kovalchuk, I.P.: Geoecology of the upper dniester river basin system: a monograph. In: Kovalchuk, I.P. (ed.) 284 p. Ivan Franko National University of Lviv, Lviv – Kyiv (2017)
3. Goreev, L.M., Peleshenko, V.I., Khilchevsky, V.K.: Hydrochemistry of Ukraine: a textbook, 307 p. Higher School, Kyiv (1995)
4. Matolich, B.M., Kovalchuk, I.P., Ivanov, E.A., et al.: Natural resources of Lviv region, 120 p. PE Lukashchuk V.S., Lviv (2009)
5. Nazaruk, M.M. (ed.): Lviv Region: Natural Conditions and Resources: Monography. The Old Lion Publishing House, Lviv (2018). 592 p
6. Kulbida, M.I., Barabash, M.B., Elistratova, L.O., et al.: Ukraine's climate: in the past … and in the future?: monograph, 234 p. Steel, Kyiv (2009)
7. Biblyuk, N.I., Kovalchuk, I.P., Machuha, O.S.: Natural disasters in the carpathians: causes and ways to minimize them. Scientific works of the Forestry Academy of Sciences of Ukraine, no. 6, pp. 105–119. RVV NLTU of Ukraine, Lviv (2008)
8. Zhovinsky, E.Ya., Kryuchenko, N.O., Papariga, P.S.: The snow cover of the high mountains of the Ukrainian Carpathians is an indicator of environmental pollution. Geochem. Ore Format. **29**, 89–93 (2011)
9. Carabin, V.: The hydrochemistry of the major ions of the White Cheremosh River. Geol. Geochem. Fossil Fuels **1–2**(162–163), 101–106 (2013)
10. Borutska, Y., Dyakov, V.: Mineral-sorption and oxygen-cavitation complex geochemical barriers to migration of potentially polluted natural waters in the Stryi River basin. Mineral. Collect. **2**(64), 195–214 (2014)
11. Strutinskiy, V., Yakhno, O., Machuga, O., Hnativ, I., Hnativ, R.: Analysis of interaction between a configurable stone and a water flow. Eastern-Eur. J. Enterp. Technol. **6**(10(96): Ecology), 14–20 (2018). https://doi.org/10.15587/1729-4061.2018.148077. ISSN 1729-3774. Scopus
12. Report on the research work on the spatial analysis of changes in the water regime of surface water bodies in Ukraine due to climate change [Electronic resource] (2013). 228 c. http://uhmi.org.ua/project/rvndr/avr.pdf
13. Gerenchuk, K.I. (ed.) Nature of Lviv Region, 152 p. View of Lviv. un-ty, Lviv (1972)
14. Hauer, F.R., Hill, W.R.: Temperature, Light, and Oxygen. Methods in Stream Ecology, 2nd edn., pp. 103–117. Elsevier, Amsterdam (2007). https://doi.org/10.1016/B978-012332908-0.50007-3

The Methodology of Experimental Bending Moments Determination in Bridge Span Structures

Yuriy Sobko(✉)

Lviv Polytechnic National University, Lviv 79013, Ukraine
sobko.yuriy@ua.sika.com

Abstract. The article presents results of the study of the experimental bending moments in the nonlinear area of section deformation determined by the proposed algorithm and showed practically accurate using for close to the real size constructions of the full scale T-beam length 13,5 m with rigid/frame reinforcement. This beam was separated from adjacent beams of the bridge deck superstructure by cutting of welded metal plates of T-beam shelves and diaphragms. Then a testing metal structure system of individual design and execution, which consisted of a truss length of 15,0 m and a height of 2,5 m was installed. By means of ties and traverses system the external testing load was transferred to the existing bridge supports. The technique and algorithm of experimental bending moments determining by the values of beams cross sections vertical displacements obtained in field tests of bridge deck structures, which allow to take into account the physical nonlinearity of RC elements deformation, as well as the presence of existing defects and initial deformed and strained state of spatial structure.

Keywords: Real size constructions · Rigid/Frame reinforcement · Bridge span structures · Reinforce concrete · T-girder bridge deck structures

1 Introduction

One of the most common types of bridges constructions are T-girder bridge deck structures with welded rigid so called "frame" reinforcement cages [1–5]. These are road bridges with and without of transverse beams and main beams, simply supported, regular structure, small and medium span, which were designed for loads of class H-18, NK-80. The specificity of such girder structures is that with a rather thin shell of T-beams geometry, i.e. with low own weight, the cross sections are supersaturated with reinforcement steel (percentage of reinforcement up to 5%), which is not a conventional feature of classical reinforced concrete elements. Along the girder welded reinforcement cages are made with breaks in the rods [6–10]. It means that the amount of reinforcement and, accordingly, the stiffness in the cross sections where the rods was broken is variable, i.e. the spatial system of the girder structure along the girder is irregular. These girder structures have been in operation for about 30–40 years. They were designed by the method of allowable stresses, which indicates the presence of

Z. Blikharskyy (Ed.): EcoComfort 2020, LNCE 100, pp. 443–450, 2021.
https://doi.org/10.1007/978-3-030-57340-9_54

certain strength reserves that need to be used rationally, for example in determining how to reconstruct, widening and increase of structure load bearing capacity. To do this, it is important to have a real values of the girder structures cross-sections stress-strained state of the elements.

This task has already been solved earlier in KADI, LPI, BildorNDI and in the other institutions [11–13]. There were studies the bridge girder structures on Plexiglas models and reinforced concrete as well. As a result of tests, experimental data of sections vertical displacements were obtained, which determined the experimental particles of external force/bending moments, while mistakenly believing that the stiffness of the sections remains constant and equal to the design value Ecired according to Eq. (1):

$$f_{ij} = \frac{Fw_{ij}}{\sum_{i,j=1}^{n,m} w_{ij}}, \tag{1}$$

where F is the external test load; fij - the share of external test load on the i-th beam in the j-th section; wij- vertical component of displacement of the i-th beam in the j-th section; n is the number of beams in cross section; m is the number of calculated cross-sections along the span.

The proportion of external force (experimental bending moments) was also determined by another known method from the measured support reactions of beams with regular and irregular stiffness in the cross section of girder bridge deck structures, assuming that the share of external load transmitted to the i-th beam of the span system is equal to the sum of the support reactions of this beam according to Eq. (2):

$$f_i = R_{Ai} + R_{Bi};\ i = 1, 2, 3\ldots n, \tag{2}$$

where R_{Ai} and R_{Bi} – support reactions of i-th beam.

It should be noted that the Eq. (1), (2) are valid for elastic spatial beam systems with regular in cross section stiffness, assuming that the fraction of force F_i acts only in the transverse cross section of bridge deck where external force F in the form of concentrated force is applied.

As can be seen, the previous investigations did not determine the experimental bending moments and stiffness, taking into account the nonlinear behavior of reinforced concrete sections deformation with cracks. This issue was first solved in the laboratory R&D-88 of National University "Lviv Polytechnic" [14, 15] by developing a new method for determining the experimental bending moments, taking into account cracking and nonlinearity of RC sections deformation [16–18]. The essence of the method was that the bending moments in the cross sections of the reinforced concrete beam under a certain scheme of application of external load and certain conditions of supporting are a function of the average relative deformations of concrete and reinforcement steel of these sections in Eqs. (3), (4):

$$M_i = F(\varepsilon_{cmi}), \quad (3)$$

$$M_i = F(\varepsilon_{smi}), \quad (4)$$

where ε_{cmi}, ε_{smi} experimental values of average relative deformations of concrete and steel; i = 1, 2, 3,…, n - the number of cross sections of the beam that was investigated; M – beam's bending moment in the i-th section.

Therefore, after experimental testing the bridge span structure according to a similar loading scheme and obtaining experimental values of average relative deformations of concrete and steel in beams cross sections in the girder structure found the appropriate experimental bending moments over the entire load range – in the elastic and elastic-plastic stages performance.

Further development the method of experimental moments determining in girder span structures sections taking into account the nonlinearity of RC elements based on deformation diagrams of bridge reference beams should be continued in the direction of using the differential equation of bent axis with curvature determining by measured vertical displacements of sections and experimental average strains of concrete and steel as well. These issues were addressed in this paper.

2 Experimental Research Methodology and Main Results

Tests of the full scale T-beam length 13,5 m with rigid/frame reinforcement were made (Fig. 1). This beam was separated from adjacent beams of the bridge deck superstructure by cutting of welded metal plates of T-beam shelves and diaphragms. Then a testing metal structure system of individual design and execution, which consisted of a truss length of 15,0 m and a height of 2,5 m was installed. By means of ties and traverses system the external testing load was transferred to the existing bridge supports (Fig. 1a, b). Measurements of deformations were performed with mechanical microindicators based on 250 mm distance, which allowed to obtain the values of the average relative deformations of the RC sections. To measure the deformations of the reinforcement steel, special holders were welded at the level of the lower reinforcing rod center of gravity (Fig. 1, c). Vertical displacements were measured by deflectors 6PAO in the same sections. Crack formation was observed to determine the width of crack opening with a microscope. If possible, all measuring instruments were duplicated. As a result, according to the experimental data in the investigated sections, reference diagrams of concrete and steel deformations were received, as well as graphical dependences the change in the initial stiffness of the cross sections of the beams. Figure 2 presents these diagrams for the most characteristic section 2-2 in the middle part of the beam.

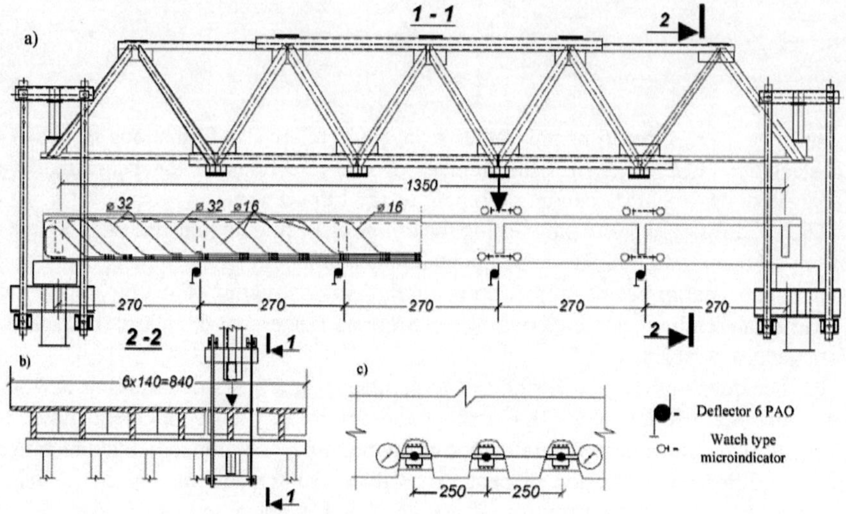

Fig. 1. The bridge beam with rigid/frame reinforcement and testing steel truss system general view: a - facade; b - cross section; c - a general view of mechanical watch type micro indicators installation.

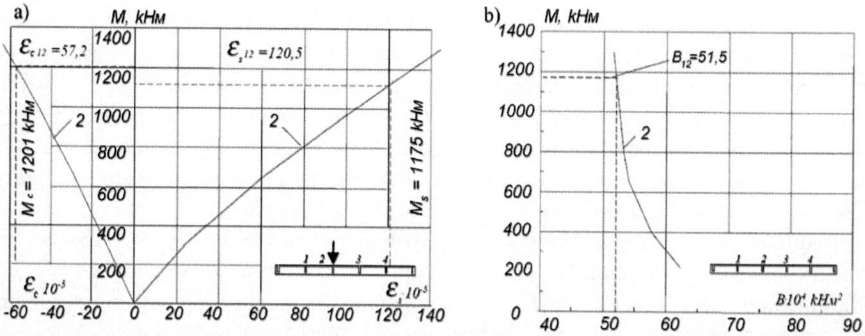

Fig. 2. The main experimental results of reference beam: a - reference graphs of concrete and steel relative deformations; b - reference beam stiffness diagram.

The bridge deck superstructure was tested using a similar method. The experimental data matrix of the relative deformations ε_c, ε_s and vertical displacements w_z in the cross sections of the beams was obtained. Consider as an example the most characteristic section 2–2 in the middle part of the span under loading with a concentration force F = 800 kN. In this section, the following values of relative deformations were obtained: ε_c = 0,0005721, ε_s = 0,001205. Next, according to the deformation diagrams of reference beam (Fig. 2,a) determine the experimental bending moments in the elastic-plastic stage of operation, acting in the first beam from the force F = 800 kN: $M_{s2\text{-}2}$ = 1175 kNm (for ε_s), $M_{c2\text{-}2}$ = 1201 kNm (for ε_c,). The average experimental bending moment was $M_{exp2\text{-}2} = 1186$ kNm. Similarly it was determined for all sections thought the deck superstructure [1].

Consider the differential Eq. (5) of the beam bended axis:

$$y'' = \frac{M}{B}, \tag{5}$$

where y'' - cross section curvature under bending; B - stiffness of the section in the area with cracks.

The real experimental cross sections curvature can be determined from experimental data of average relative deformations Eq. (6):

$$y'' = \frac{1}{r} = \frac{\varepsilon_{cm} + \varepsilon_{sm}}{h}, \tag{6}$$

where ε_{cm} and ε_{sm} - experimental values of average relative deformations of concrete and steel; h - the distance from the concrete upper fiber to the center of gravity of the reinforcing welded cage lower rod.

Having found the value of the experimental curvature and the corresponding stiffness of the section in the area with cracks, from Eq. (5) it is easy to determine the value of M_{exp} in this section.

The experimental curvature can also be determined by the measured vertical displacements by drawing their longitudinal graphs in the beams of the deck girder structure [2]. According to the experimental graph of vertical displacements (Fig. 3,a), it is possible to write the equation of curvature, which is based on regression analysis (Fig. 3,b). For the first beam of the deck girder superstructure when loaded with a force F = 800 kN in section 2–2, the Eq. (7) will have the form:

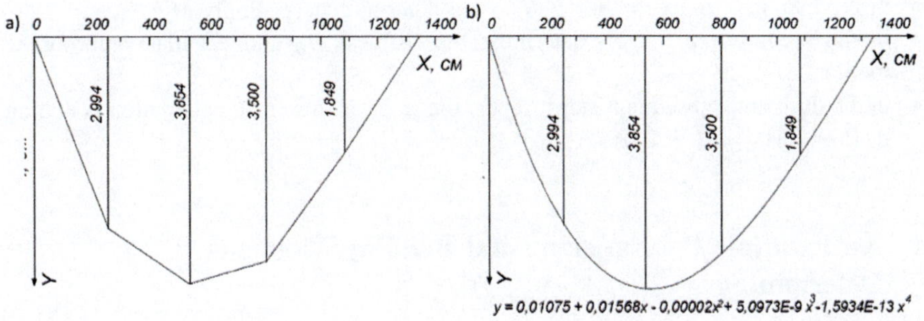

Fig. 3. Vertical displacement of first beam of deck superstructure: a – experimental; b – by regression analysis.

$$y = 0.010750 + 0.01568x - 0.00002x^2 + 5.0973e - 9x^3 - 1.5934e - 13x^4, \quad (7)$$

where y - vertical displacement, cm; x - cross-sectional distance, cm.

By differentiating the Eq. (7) twice by x, we obtain the following Eq. (8) of curvature:

$$\frac{d^2y}{dx^2} = y'' = -0.00004 + 30.582e - 9x - 19.121e - 13x^2. \quad (8)$$

Substituting the section coordinate x = 540 cm, we obtain y'' = 2,355·10^{-5} cm^{-1}. The experimental curvature value Eq. (9) in the same cross section by the Eq. (6):

$$y'' = \frac{1}{r} = \frac{\varepsilon_{cm} + \varepsilon_{sm}}{h} = 2.235 \cdot 10^{-5}\text{cm}^{-1}. \quad (9)$$

As could see, these values obtained by two independent methods are almost the same, which indicates the possibility of their use independently of each other.

3 Algorithm of Determining the Experimental Bending Moments by Experimental Curvature

From tests data of a reference bridge beam similar on a design, conditions of supporting and loading to beams of a girder deck superstructure, to draw the diagram (M–B_{exp}).

By the tests of the girder deck superstructure get experimental data of the average relative deformations and displacements; determine the experimental curvature of the bended axis of the sections by Eqs. (6), (7). By iterative method from the diagram (M–B_{exp}) determine the appropriate experimental cross-sectional stiffness in the area with cracks, while:

- in the first step to accept $B = E_c\, I_{red}$ and determine accordingly M_I;
- in the second step for M_I determine the stiffness B_{II} and calculate appropriate moment M_{II};
- in the third and subsequent steps, repeat the steps as in step II; calculation is ending if: $(M_i - M_{i\text{-}1})/M_i < 0{,}05$.

4 An Example the Experimental Bending Moments Determining

As an example, we define the experimental bending moments in section 2-2 according to the diagram (M-B_{exp}) (Fig. 2, b) and the above algorithm:

- when determined the y$''$ by the formula (6)
  - I-st step M_I = 2,235960 = 2145 kNm;
  - II-nd step M_{II} = 2,235505 = 1128,6 kNm;
  - III-rd step M_{III} = 2,235510 = 1139,8 kNm.

As can be seen, already in the third step of iterations the stiffness practically does not change (because the diagram has a gentle character, which is a characteristic feature of the reinforced beams with frame cage reinforcement) and the necessary convergence of results is reached.

– when determined the y″ by the Eq. (7)
  - I-st step $M_I = 2{,}355960 = 2310$ kNm;
  - II-nd step $M_{II} = 2{,}355505 = 1215$ kNm;
  - III-rd step $M_{III} = 2{,}355510 = 1201$ kNm

Comparing the obtained values of the experimental moments in the cross section 2–2 of the first beam of the bridge deck girder structure, determined by the methodology and the above mentioned method found that the discrepancy is: for Eq. (6) – 3,5%, for Eq. (7) - + 1,6%, which indicates a practical coincidence of results according to the proposed algorithm and method.

5 Conclusions

The experimental bending moments in the nonlinear area of section deformation determined by the proposed algorithm showed practically accurate and close to the real object results. In addition, this technique has its advantages: there is no need to determine the experimental average relative deformations ε_{cm}, ε_{cm} when testing girder deck structures and, as a consequence, reducing the complexity of work in preparation for testing, as well as reducing the time of testing and in-house work important in engineering practice.

References

1. Blikharskyy, Z., Brózda, K., Selejdak, J.: Effectivenes of strengthening loaded RC beams with FRCM system. Arch. Civ. Eng. **64**(3), 3–13 (2018)
2. Blikharskyy, Z., Vashkevych, R., Vegera, P., Blikharskyy, Y.: Crack resistance of RC beams on the shear. In: Lecture Notes in Civil Engineering, vol. 47, pp. 17–24 (2020). https://doi.org/10.1007/978-3-030-27011-7_3
3. Krainskyi, P., Blikharskyy, Y., Khmil, R., Vegera, P.: Crack resistance of RC columns strengthened by jacketing. In: Lecture Notes in Civil Engineering, vol. 47, pp. 195–201 (2020). https://doi.org/10.1007/978-3-030-27011-7_25
4. Selejdak, J., Blikharskyy Y., Khmil R., Blikharskyy Z.: Calculation of reinforced concrete columns strengthened by CFRP. In: Lecture Notes in Civil Engineering, vol. 47, pp. 400–410 (2020). https://doi.org/10.1007/978-3-030-27011-7_51
5. Vegera, P., Vashkevych, R., Blikharskyy, Z.: Fracture toughness of RC beams with different shear span. In: MATEC Web of Conferences, vol. 174, p. 02021 (2018). https://doi.org/10.1051/matecconf/201817402021
6. Blikhars'kyi, Y.Z.: Anisotropy of the mechanical properties of thermally hardened a500s reinforcement. Mater. Sci. **55**, 175–180 (2019). https://doi.org/10.1007/s11003-019-00285-0

7. Kharchenko, Y.V., Blikharskyy, Z.Y., Vira, V.V., Vasyliv, B.D.: Study of Structural changes in a nickel oxide containing anode material during reduction and oxidation at 600 °C. In: Fesenko, O., Yatsenko, L. (eds.) Nanocomposites, Nanostructures, and Their Applications. NANO 2018. Springer Proceedings in Physics, vol. 221. Springer, Cham (2019). https://doi.org/10.1007/978-3-030-17759-1_42
8. Kharchenko, Y., Blikharskyy, Z., Vira, V. Vasyliv, B., Vasyliv, B: Study of nanostructural changes in a Ni-containing cermet material during reduction and oxidation at 600 °C. Appl. Nanosci. (2020). https://doi.org/10.1007/s13204-020-01391-1
9. Lobodanov, M., Vegera, P., Blikharskyy, Z.: Planning Experiment for researching reinforced concrete beams with damages. In: Lecture Notes in Civil Engineering, vol. 47, pp. 243–250 (2020). https://doi.org/10.1007/978-3-030-27011-7_31
10. Zhang, Q., Mol'kov, Y.V., Sobko, Y.M., et al.: Determination of the mechanical characteristics and specific fracture energy of thermally hardened reinforcement. Mater. Sci. **50**, 824–829 (2015). https://doi.org/10.1007/s11003-015-9789-9
11. Blikharskyy Z., Khmil, R., Vegera, P.: Shear strength of reinforced concrete beams strengthened by PBO fiber mesh under loading. In: MATEC Web of Conferences, vol. 116, p. 02006 (2017). https://doi.org/10.1051/matecconf/201711602006
12. Bobalo, T., Blikharskyy, Y., Kopiika, N., Volynets, M.: Serviceability of RC beams reinforced with high strength rebar's and steel plate. In: Lecture Notes in Civil Engineering, vol. 47, pp. 25–33 (2020). https://doi.org/10.1007/978-3-030-27011-7_4
13. Khmil, R., Tytarenko, R., Blikharskyy, Y., Vashkevych, R.: Influence of load level during strengthening of reinforced concrete beams on their reliability. In: IOP Conference Series: Materials Science and Engineering, vol. 708, no. 1, p. 012054 (2019)
14. Bobalo, T., Blikharskyy, Y., Kopiika, N., Volynets, M.: Theoretical analysis of RC beams reinforced with high strength rebar's and steel plate. In: IOP Conference Series: Materials Science and Engineering, vol. 708, no. 1, p. 012045 (2019)
15. Krainskyi, P., Vegera, P., Khmil, R., Blikharskyy, Z.: Theoretical calculation method for crack resistance of jacketed RC columns. In: IOP Conference Series: Materials Science and Engineering, vol. 708, no. 1, p. 012059 (2019)
16. Blikharskyy, Y., Kopiika, N., Selejdak, J.: Non-uniform corrosion of steel rebar and its influence on reinforced concrete elements' reliability. Prod. Eng. Arch. **26**(2), 67–72 (2020). https://doi.org/10.30657/pea.2020.26.14
17. Krainskyi, P., Blikharskyy, Y., Khmil, R., Blikharskyy, Z.: Experimental study of the strengthening effect of reinforced concrete columns jacketed under service load level. In: MATEC Web of Conferences, vol. 183, p. 02008 (2018). https://doi.org/10.1051/matecconf/201818302008
18. Mohsem El.Sh.: Feasibility study of transversely prestressed double tee bridges. PCI J. **35**(5), 56–69 (1990)

Research of Temperature Regime in the Module for Poultry Growing

Nadiia Spodyniuk[1(✉)] and Anna Lis[2]

[1] National University of Life and Environmental Sciences of Ukraine, Kiev, Ukraine
n_spoduniuk@meta.ua
[2] Częstochowa University of Technology, Częstochowa, Poland

Abstract. The task of modern systems of heat supply in the poultry houses are the maintenance of the necessary temperature parameters in the habitats of the poultry. On the basis of researches of infrared heating systems in poultry house an engineering technique for calculating the temperature regime in module for growing poultry has been developed. The module was equipped with an infrared heater and an exhaust outlet and allows designing efficient infrared heating systems in agricultural premises to ensure standard temperature regime. In this case, the technique is only resolved when the conditions for providing the required temperature and air speed parameters are met. If the temperature of the module t_{in} changes from 16 °C to 35 °C, and inflow air speed varies from 0.2 m/s to 0.3 m/s, the infrared heating system and the ventilation system completely provide the normalized air temperature in the irradiation area. Based on an algorithm using a developed technique a computer program was developed. This program allows, through the use of a simple method of substituting input factors, to determine the parameters of the temperature regime of the module for poultry breeding.

Keywords: Module four poultry breeding · Infrared heater · Exhaust outlet · Thermal power · Engineering calculation

1 Introduction

Air heating systems are widespread among existing poultry farms. To provide the necessary air exchange and maintaining a stable air temperature in the enclosure indoor poultry house when using the air heating system, the heating of external supply air is accompanied by the loss of a large share of energy resources. It is known that the use of primary energy in Ukraine has been steadily increasing over the years. Therefore, if not take action to save them, Ukraine will continue to be in an energy-dependent state and this concern in particular to the agricultural sector (Fig. 1) [1].

Therefore, the task of modern systems of heat supply of poultry premises is to maintain the required temperature parameters in the poultry location area with quality control of air temperature in the process of poultry growth [2] and [3]. This will significantly reduce energy consumption during an economically justified period of normal operation [4–6].

Z. Blikharskyy (Ed.): EcoComfort 2020, LNCE 100, pp. 451–458, 2021.
https://doi.org/10.1007/978-3-030-57340-9_55

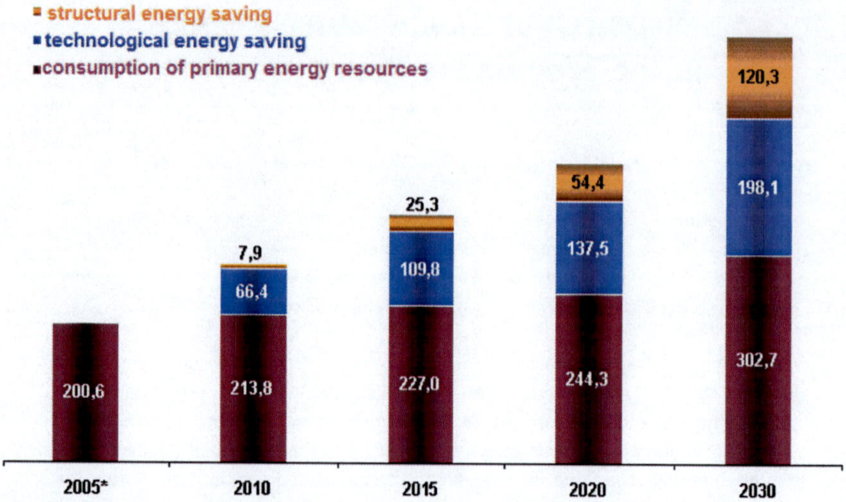

Fig. 1. Forecast of consumption of primary resources dynamics levels of structural and technological energy savings by 2030, million tonnes of conventional fuel (baseline scenario)

Thus, it is advisable to use infrared heating systems for such premises. When using these systems, the local heating of the technological zone for poultry breeding is carried out. However, it is important to ensure a constant supply of fresh air through ventilation system. Therefore, a new method of poultry breeding with mutual influence on the temperature of the heating and ventilation system is suggested. Such a solution is possible with the modular breeding of poultry as variety of breeding poultry in cells.

2 Aim of a Paper

Since there are some difficulties in calculating infrared heating systems in modules for poultry breeding, the purpose of the article is to develop an engineering calculation method of microclimate parameters of the technological zone of module for poultry breeding. This calculation method will help to determine the temperature of the air in technological zone with infrared heating.

3 Research Methods

The proposed method of calculating of the temperature regime is based on a scientific approach to the tasks, namely physical modeling of thermal processes in a module [7] and [8].

On Fig. 2 the design diagram of the module for poultry breeding is shown. The infrared heater 1 is intended for local heating of the process zone, exhaust outlet 2 removes contaminated air from the module, through the air distributor 4 and the static pressure chamber 5 is ensured a constant flow of fresh supply air, lighting fixture 3 is intended to provide normalized module lighting.

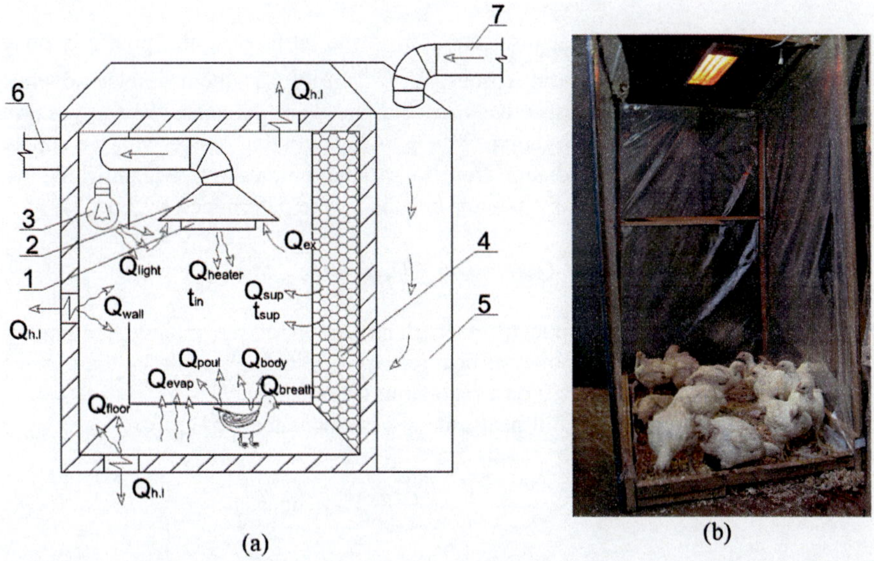

Fig. 2. Calculation scheme (a) and a photo of the experimental setup (b) of module for poultry breeding. 1 – infrared heater; 2 – exhaust outlet; 3 – lighting fixture; 4 – supply air distributor; 5 – static pressure chamber; 6 - exhaust air duct; 7 – influx air duct

To ensure a normalized temperature regime in the module the system of balance equations with determination of temperature of all surfaces and air in the module was calculated. In order to assimilate the heat and humidity surpluses, the air exchange was calculated, taking into account heat input and heat losses in the module. The calculation of air exchanges was performed according to the principles described in the papers [9] and [10].

The initial data for the engineering calculation are:

- heat output of the infrared heater, its design features;
- installation height of the infrared heater;
- the blackness degree of the surface of the heated floor;
- air speed in the module for poultry breeding.

The calculation algorithm is as follows:

1. Determination of heat input in the module for poultry breeding in winter and summer periods, W are shown in Eqs. (1) and (2):

$$Q_{in}^{Cold} = Q_{body}^{Cold} + Q_{breath}^{Cold} + Q_{light} + Q_{h.s} + Q_{sup}, \tag{1}$$

$$Q_{in}^{Warm} = Q_{body}^{Warm} + Q_{breath}^{Warm} + Q_{light} + Q_{solar} + Q_{sup}, \tag{2}$$

where Q_{body}^{Cold}, Q_{body}^{Warm} - heat dissipation from the surface of the poultry's body respectively in cold and warm seasons, W; Q_{breath}^{Cold}, Q_{breath}^{Warm} - the heat released while the poultry is breathing respectively in cold and warm seasons, W; Q_{light} – heat dissipation from artificial lighting, W; Q_{sup} - heat coming inside with the supply air, W; Q_{solar} – thermal radiation from solar radiation due to overlapping, W.

2. Heat losses in the module for poultry breeding were determined by Eq. (3):

$$\sum Q_{loss} = Q_{h.l} + Q_{evap} + Q_{ex}, \ \mathrm{W}, \tag{3}$$

where $Q_{h.l}$ - heat losses due to external protection of the module for poultry breeding, W; Q_{ex} – the amount of heat removed from the module by the exhaust outlet, W; Q_{evap} - heat losses on evaporation of moisture, W.

3. The total air exchange by full heat excess was calculated by Eq. (4):

$$G_{Q_{excess}^{full}} = \frac{3,6 \cdot Q_{excess}^{full}}{I_{in} - I_{out}}, \ \mathrm{kg/hour}, \tag{4}$$

where $Q_{excess}^{full\ Cold} = Q_{poul}^{full\ Cold} + Q_{light} + Q_{h.s} - Q_{h.l}$ - the amount of air required to assimilate the heat excess in the cold season, W; $Q_{excess}^{full\ Warm} = Q_{poul}^{full\ Warm} + Q_{light} + Q_{solar} - Q_{h.l}$ - the amount of air required to assimilate the heat excess in the warm season, W; I_{in}, I_{out} – specific enthalpy of air, respectively in the technological zone of the module for poultry breeding and the environment, kJ/kg.

4. The total air exchange by the assimilation of moisture residues was calculated by Eq. (5):

$$G_{moist} = \frac{1000 \cdot W_{tot}}{d_{in} - d_{out}}, \ \mathrm{kg/hour}, \tag{5}$$

where d_{in}, d_{out} – the moisture content, respectively, of the air in the module for poultry breeding and ambient air, g/kg.

5. The total moisture content in the module for poultry breeding (Eq. (6)):

$$W_{tot} = W_{drop} + W_{poul} + W_{evap}, \ \mathrm{kg/hour}. \tag{6}$$

W_{drop} – the amount of moisture that evaporates from the droppings and the deep litter, kg/hour, W_{poul} – moisture emissions from the poultry, kg/hour, W_{evap} – the amount of moisture that evaporates from the enclosure, was assumed equal to 10% of the total moisture content, released by poultry, kg/hour.

6. The total air exchange for carbon dioxide in the module for poultry breeding was calculated by Eq. (7):

$$G_{\text{carbon}} = \frac{\text{k}_{\text{carbon}}}{\text{B}_{\text{in}} - \text{B}_{\text{out}}} \cdot \rho_{\text{in}}, \text{ kg/hour} \tag{7}$$

where B_{in}, B_{out} - accordingly the permissible concentration of gas in indoor air and ambient air, l/m^3; ρ_{in} - density of internal air at the calculated temperature, kg/m^3. k_{carbon} - the amount of carbon dioxide released when poultry was breathing, l/hour.

7. The maximum air exchange from the calculated total air exchanges was selected for the calculation: $G_{calc} = G_{max}$.
8. The calculated air exchange determines the amount of heat, which is removed from the module for poultry breeding by an exhaust outlet Q_{ex}, W.
9. The temperature on the surface of the heater was determined on the basis of experimental studies by Eq. (8):

$$t_{\text{heater}} = (30x - 50x^2) + t_a, \text{ °C} \tag{8}$$

where x – running coordinate along the length of the heater, m; t_a – temperature of the lateral surfaces of the heater, ºC.

10. The speed of air flow through the air distributor to the technological zone was determined by Eq. (9):

$$v_{\text{sup}} = \frac{G_{\text{calc}}}{F_{\text{sup}} \cdot \rho_{\text{in}} \cdot 3600}, \text{ m/s} \tag{9}$$

F_{sup} –area of the inlet opening of the air distributor, m^2.

11. The air flow temperature in the module for poultry breeding is taken to be equal to the normalized air temperature in the poultry module (Eq. (10)):

$$t_{\text{sup}} = t_{\text{in}}^{\text{norm}}, \text{ °C} \tag{10}$$

12. Whereas the physical model does not fully cover all possible factors of influence on the temperature regime, the results of the experimental field studies were also taken into account for the development of the calculation methodology. Based on the obtained results the relative air temperature in the module for poultry breeding was determined by Eq. (11):

$$\begin{aligned} \bar{t}_{\text{in}} = 1.41 + 0.07\frac{Q_{\text{heat}} - 1000}{500} - 0.045\frac{H - 1.25}{0.25} - 0.075\frac{v_{\text{sup}} - 0.2}{0.1} \\ - 0.02\frac{Q_{\text{heat}} - 1000}{500} \cdot \frac{v_{\text{sup}} - 0.2}{0.1} + 0.0125\frac{H - 1.25}{0.25} \cdot \frac{v_{\text{sup}} - 0.2}{0.1} \end{aligned} \tag{11}$$

Q_{heat} – the thermal power of the infrared heater, W; H – installation height of the heater, m; v_{sup} – air speed of inflow air, m/s.

13. The air temperature in the module was equal to Eq. (12):

$$t_{\text{in}} = \bar{t}_{\text{in}} \cdot t_{\text{sup}},\ ^{\circ}\text{C} \tag{12}$$

On the basis of the discrepancies between the results of the experimental and theoretical studies, the correction factor θ was determined. This allows adjusting the air temperature in the module, taking into account at least significant factors influencing the temperature regime, which were previously simplified (Eq. (13)).

$$\theta = \frac{t_{\text{in}}^{\text{exp}}}{t_{\text{in}}^{calc}}, \tag{13}$$

where $t_{\text{in}}^{\text{exp}}$ - air temperature, determined experimentally by Eq. (11), °C; t_{in}^{calc} - air temperature determined on the basis of analytical studies, °C.

14. Whereas the area of module for poultry breeding is small, then the formation of the temperature regime in it is affected by the surface temperatures of all the protections τ_{i}, °C. The air and surface temperatures in the module form the so-called temperature level, or temperature of the module.
The radiation temperature of the surfaces in the module was determined by Eq. (14):

$$t_R = \frac{\sum \tau_i \cdot F_i}{\sum F_i},\ ^{\circ}\text{C}, \tag{14}$$

where τ_i - temperature of the *i*-th surface in the module, °C; F_{i} – the area of the *i*-th surface in the module, m^2.

15. Whereas the area of the heated surfaces of the module is small enough, then the module temperature is assumed to be approximately equal to the arithmetic value t_{in} and t_{R}, °C (Eq. (15)).

$$t_{\text{mod}} \approx \frac{t_{\text{in}} + t_R}{2},\ ^{\circ}\text{C} \tag{15}$$

16. As a result of comparison of temperature indicators values, obtained on the basis of analytical and experimental studies under the same conditions, the correction factor was determined, taking into account the influence of variable parameters on its value. On Fig. 3 shows graphical dependency of determination of the correction factor θ from the height of the heater installation *H*, m and its thermal power Q_{heat}, W.

4 Research Results

To use the proposed dependencies to further creation a computer program the algorithm of engineering calculation of temperature mode parameters of the module was offered. Based on an algorithm using an existing technique a computer program was developed, which allows determining the temperature conditions of the poultry breeding module.

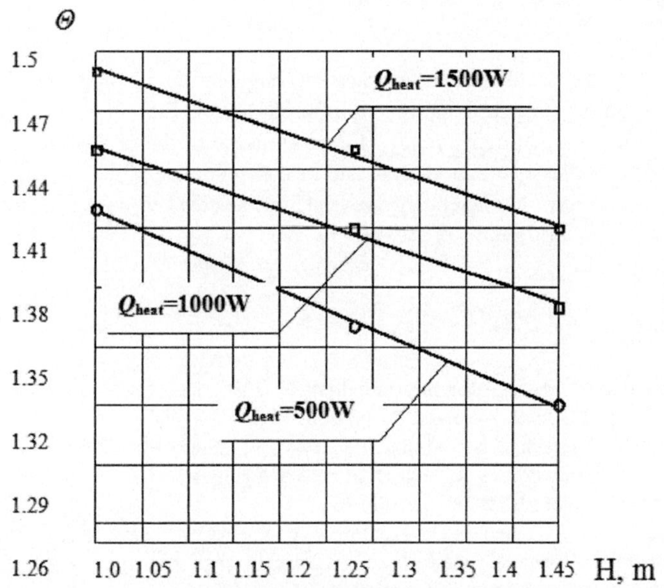

Fig. 3. Dependence of the correction factor θ from the thermal power of the heater Q_{heat}, W and the height of its installation *H*, m

The technique is only resolved when the conditions for providing the required temperature and air speed parameters are met:

- if the air temperature in the module t_{in} belongs to the range 16 °C $\leq t_{in} \leq$ 35 °C, and supply air speed is within range $0.2\,\mathrm{m/s} \leq v_{sup} \leq 0.3\,\mathrm{m/s}$ then the infrared heating system and the general ventilation system fully provide the normalized air temperature in the irradiation area and the calculation is complete;
- if the air temperature $t_{in} <$ 16 °C, or $t_{in} >$ 35 °C, and the air speed $v_{sup} <$ 0.2 m/s, or $v_{sup} >$ 0.3 m/s, then it is necessary to change the thermal power of the infrared heater, the height of its installation or underlying floor surface and repeat the calculation, until equalities 16 °C $\leq t_{in} \leq$ 35 °C and $0.2\,\mathrm{m/s} \leq v_{sup} \leq 0.3\,\mathrm{m/s}$ are fulfilled.

This method of engineering calculation can be used in the field of heating, in particular for the heating of premises of agro-industrial complexes, such as poultry houses. Using it, it is easy to determine the temperature and air speed in the irradiative zone of infrared heater. In particular, analytical dependencies can be used for the design of thermal systems of module for poultry breeding with indoor air temperature 16–35 °C and air speed 0.2–0.3 m/s.

5 Conclusions

On the basis of the conducted researches the engineering method of calculation of module for poultry breeding using an infrared heater and an exhaust outlet was developed. The graphical dependency of determination of the correction factor θ from the height of the heater installation H, m and its thermal power Q_{heat}, W was developed. This allows to design of efficient infrared heating systems to ensure a standard temperature regime in agricultural premises.

References

1. Ukraine's energy strategy for the period up to 2030. https://de.com.ua/uploads/0/1703-EnergyStratagy2030.pdf. Accessed 24 July 2013
2. Kocaman, B., Esenbuga, N., Yildiz, A., Laçin, E., Macit, M.: Effect of environmental conditions in poultry houses on the performance of laying hens. Int. J. Poult. Sci. **5**(1), 26–30 (2006). https://doi.org/10.3923/ijps.2006.26.30
3. Candido, M.G.L., Tinoco, I.D.F.F., Pinto, F.D.A.D.C., Santos, N.T., Roberti, R.P.: Determination of thermal comfort zone for early-stage broilers. Engenharia Agricola **36**(5), 760–767 (2016). https://doi.org/10.1590/1809-4430-Eng.Agric.v36n5p760-767/2016
4. Baxevanou, C., Fidaros, D., Bartzanas, T., Kittas, C.: Energy consumption and energy saving measures in poultry. Energy Environ. Eng. **5**(2), 29–36 (2017). https://doi.org/10.13189/eee.2017.050201
5. Alaw Qotbi, A.A., Najafi, S., Ahmadauli, O., Rahmatnejad, E., Abbasinezhad, M.: Investigation of poultry housing capacity on energy efficiency of broiler chickens production in tropical areas. Afr. J. Biotech. **10**(69), 15662–15666 (2011). https://doi.org/10.5897/AJB10.2662
6. Davoud Heidari, M., Omid, M., Akram, A.: Optimization of energy consumption of broiler production farms using data envelopment analysis approach. Mod. Appl. Sci. **5**(3), 69–78 (2011). https://doi.org/10.5539/mas.v5n3p69
7. Lorencena, M.C., Puttow Southier, L.F., Casanova, D., Ribeiro, R., Teixeira M.: A framework for modeling, control and supervision of poultry farming. Int. J. Prod. Res. 1–19 (2019). https://doi.org/10.1080/00207543.2019.1630768
8. Trokhaniak, V., Rutylo, M., Rogovskii, I., Titova, L., Luzan, O., Bannyi, O.: Experimental studies and numerical simulation of speed modes of air environment in a poultry house. INMATEH – Agric. Eng. **59**(3), 9–18 (2019). https://doi.org/10.35633/INMATEH-59-01
9. Spodyniuk, N., Zhelykh, V., Dzeryn, O.: Combined heating systems of premises for breeding of young pigs and poultry. FME Trans. **46**, 651–657 (2018). https://doi.org/10.5937/fmet1804651S
10. Shcherbovskykh, S., Spodyniuk, N., Stefanovych, T., Zhelykh, V., Shepitchak, V.: Development of a reliability model to analyze the causes of a poultry module failure. Eastern-Eur. J. Enterp. Technol. **4**(3(82)), 4–9 (2016). https://doi.org/10.15587/1729-4061.2016.73354

Development of Component Composition of Engineered Cementitious Composites

Nazar Sydor(✉), Uliana Marushchak, Serhii Braichenko, and Bohdan Rusyn

Lviv Polytechnic National University, Lviv 79013, Ukraine
nazar.sydor@ukr.net

Abstract. The use of engineering cementitious composites – a specially developed cement-based material reinforced with fibers – allows to enhance the bearing capacity, stability under static and dynamic influences, as well as durability of building structures. The effect of component composition of engineering cementitious composites on workability, flexural and compressive strength was investigated from micromechanics principles point of view. It is shown that the optimal cement to sand ratio, partial replacement cement by fly ash, incorporation of polycarboxylate superplasticizer, as well as reinforcement of the engineered composites structure with dispersed fibers contribute to their mechanical properties both at early and later hardening period. The modified engineered cementitious composites are characterized 28-days compressive strength 61 MPa, flexural strength – 14.5 MPa and crack resistance coefficient 0.24. Partial replacement of Portland cement by fly ash causes formation of needle and fibrous hydration products in unclinker part that reinforce the matrix on the micro- and nanolevel and the phenomena of "self-reinforcement" is realized.

Keywords: Engineered Cementitious Composite · Fiber reinforcement · Fly ash · Polycarboxylate superplasticizer · Flexural strength

1 Introduction

The progress in the construction industry gives a new impetus to the development of hybrid, layered, thin-walled profile and other types of building structures of the new generation. Strict requirements to the safety and reliability of buildings and structures lead to the need of performance and durability increase of concrete used in the construction, reconstruction and repair of construction [1, 2]. However, high strength concrete is a brittle material in which strains are localized in place of first crack appearance after the limit loads. Cracking, which evolves in concrete structures subjected to in-service loading, reduces their bearing capacity as well as corrosion resistance, increases the possibility of water and other chemicals penetrating, which can lead to decrease durability of composites.

The destruction of conventional concrete due to its brittle behaviour was important factor for the development of Engineered Cementitious Composites (ECC). The design of ECC of Engineered Cementitious Composites is based on principles of

Z. Blikharskyy (Ed.): EcoComfort 2020, LNCE 100, pp. 459–465, 2021.
https://doi.org/10.1007/978-3-030-57340-9_56

micromechanics and fracture mechanics to provide high tensile strength and ductility [3–6]. The advantage of ECC compared with fiber-reinforced concrete is the ability to use them in thin-walled structures and for infrastructural applications where they are majorly fit comfortably for the repairs and retrofitting of existing structures [3].

The theory of micromechanics involves the optimization of the component composition and microstructure of material, taking into account the interaction of cement matrix and fibers, which provides cross linking of the structure [4]. However, after the appearance of the first crack, the load-bearing capacity of the ECCs does not change, which leads to deformation strengthening and accompanied by multiple cracking. To control this process, ECC do not contain coarse aggregate, as well as a limited content of fine aggregate, as ones lead to increase in the opening width of crack [5]. However, the increased cement consumption in the ECC compared to the concrete can lead to increase of crack formation, which is caused by increased heat release and shrinkage as well as cause negative effect on material cost [3].

Reduction of heat release and shrinkage is provided by partially replacement of cement by supplementary cementitious materials, in particular fly ash, crushed glass, chalk, zeolite [7–12]. The effect of multiple cracks is provided by dispersed reinforcement. The fiber in the composites works to stretch or prevent displacement, ensuring the integrity of the system. The fibers perceive the load and absorb energy of external loading. Polypropylene fibers are characterized by a low modulus of elasticity, therefore, they prevent particle displacement, dampens secondary strains and increase impact resistance [5, 13, 14]. To control the rheological properties of ECC mixtures, high effective polycarboxylate superplasticizers are used, which allows to reduce the porosity, increase the strength of composites [15, 16]. The aim of present study is to development ECC compositions focused on flexural ductility of cementitious composites.

2 Experimental Program

2.1 Materials

The Portland cement CEM I 42.5 R JSC "Ivano-Frankivskcement" (Ukraine), fly ash from Burshtyn thermal power plant, fine aggregate – natural quartz sand ($M_f = 1.24$) were used to prepare Engineered Cementitious Composites. Polypropylene fiber (f) 5 mm in length was used for disperse reinforcement. The Glenium ACE430 polycarboxylate (PCE) with dosage by 0.7 wt% was used as a modifier of rheological properties of composites.

At the first investigation stage the mix proportions (Portland cement and sand) are variated from 1: 0 to 1: 1.5 with step 0.5. The binary binder system contained Portland cement and fly ash in a ratio 1:1 and polypropylene fiber were used at second investigation stage. Then effect of polycarboxylate superplasticizer on consistency, flexural and compressive strength was investigated.

2.2 Experimental Process

The ECC components were mixed according to EN 196-1 procedure. Fibers are added into the mortar matrix and mixed until all fibers evenly distributed. The consistency of the ECC mixture was determined using the flow table test according to EN 1015-3. The flowing of compositions variated from 150 to 158 mm. The samples 20 mm × 20 mm × 80 mm were unformed after 24 h and cured in normal condition (90–100% RH at 20 ± 2 °C). After 2, 7 and 28 days the samples were tested on flexural and compressive strength. The scanning electron microscopy was carried out to investigate the fibermatrix bonding and the hydration characteristics of the mix.

3 Results and Discussion

The consistency test result of cement-sand compositions is shown that increase sand content cause decrease workability fresh mix. In this case the water consumption increase by 13.3; 30.0 and 43.3% respectively for mixes with cement:sand ratio 1:0.5; 1:1 and 1:1.5 for obtaining alike mix consistency (150–158 mm) (Fig. 1a). Using of binary binder system contained Portland cement and fly ash in a ratio 1:1 provides increase mix workability compared to mix based on CEM I 42.5 due to the roller bearing effect of spherical shape of fly ash particles (Fig. 1b). The water consumption decrease by 17.6% in this case. Therefore, when fiber was added, the viscosity of mixes increases and water consumption enhances. It is necessary to use superplasticizer. Modifying of fiber reinforced composite based on binary binder system (binder to sand ratio is 1:1) with polycarboxylate superplasticizer provide necessary workability at W/B ratio 0.24.

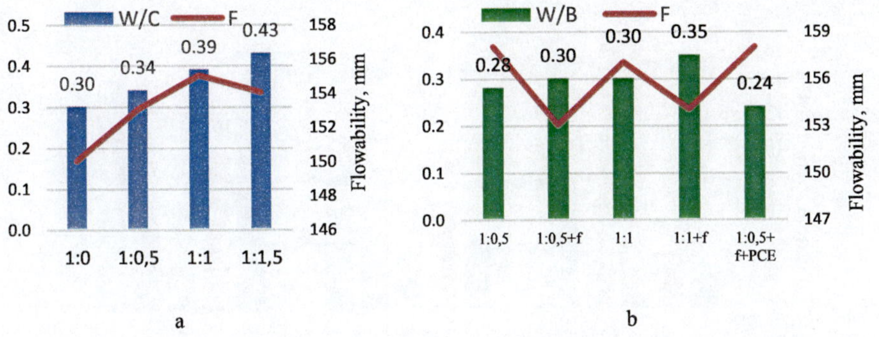

Fig. 1. Flowability and water-cement W/C (a), water-binder W/B (b) ratio of composites

Result of flexural test shows that increasing of sand content causes decreasing of flexural strength (Fig. 2). Thus, mix, in which cement to sand ratio is 1:1, characterizes by flexural strength reducing by 9.3% and 19.7% respectively after 2 and 28 days compared to cement paste. At same time, mix with cement to sand ratio is 1:0.5 shows insignificant strength reducing by 5–7%. For further research compositions with a cement to sand ratio of 1:0.5 and 1:1 were used.

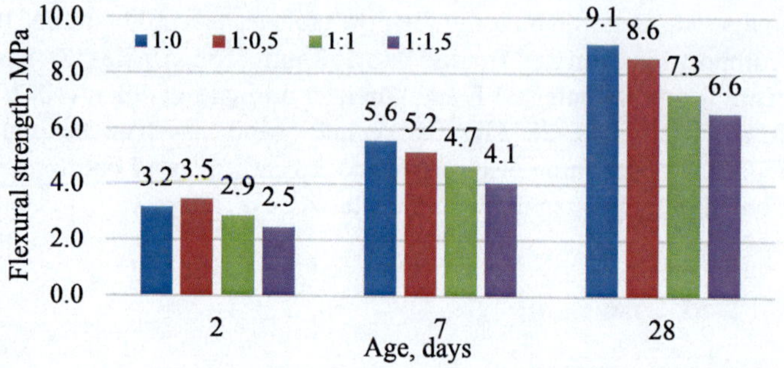

Fig. 2. Flexural strength of cement composites

From the point of view of micromechanical modelling the fly ash content effect on material microstructure and mechanical properties altering process was investigated. Cement in the amount of 50% was replaced by fly ash and binary binder system (the cement to fly ash ratio is 1:1) was used. The flexural strength of this composite after 2 days decrease by 17.1% (Fig. 3). Experimental results shown that mix with addition high volumes of fly ash characterize by decrease long-term flexural strength by 9–11%. The significant enhancement of flexural strength is observed for ECC fiber reinforced by polypropylene fiber and modified by polycarboxylate superplasticizer. In this case flexural strength after 2 days is 6.2 MPa and after 28 days – 14.5 MPa. The strength ratio f_{fl2}/f_{fl28} which indicates the flexural strength development is 0.43.

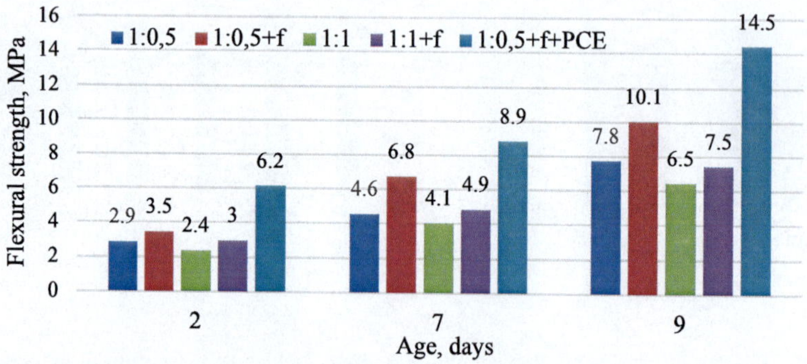

Fig. 3. Flexural strength of cementitious composites based on binary binder system

Compressive strength of disperse reinforced ECC increases by 6–14% compared to nonreinforced composite. Compressive strength of ECC modified by polycarboxylate superplasticizer increases by 1.5 time after 2 days and by 1.1 time after 28 days compared to ECC without modifier due to more amount of particle contact in hydrated system. The 28-days strength of modified ECC is 61 MPa and crack resistance coefficient 0.24. The enhancement of compressive strength of modified ECC could be related with efficient dispersion effect of polycarboxylate ether based superplastisizer and important microstructural aspects: the packing effect, pozzolanic reaction of fly ash (Fig. 4).

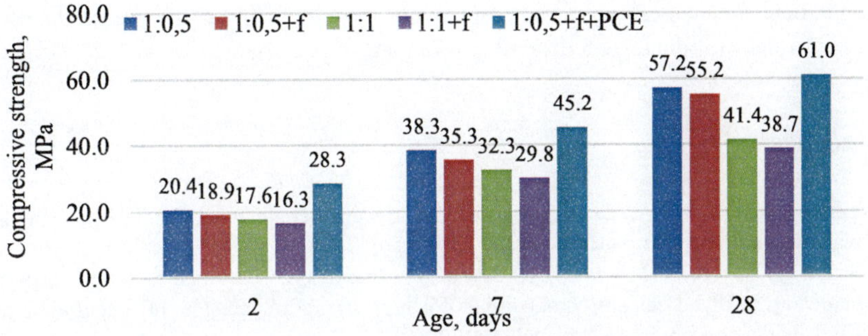

Fig. 4. Compressive strength of cementitious composites based on binary binder system

The microstructure of the ECC modified with polycarboxylates superplastisizer after 7 days is homogeneous. Polypropylene fiber provide three-dimensional reinforcement of cementitious matrix (Fig. 5a).

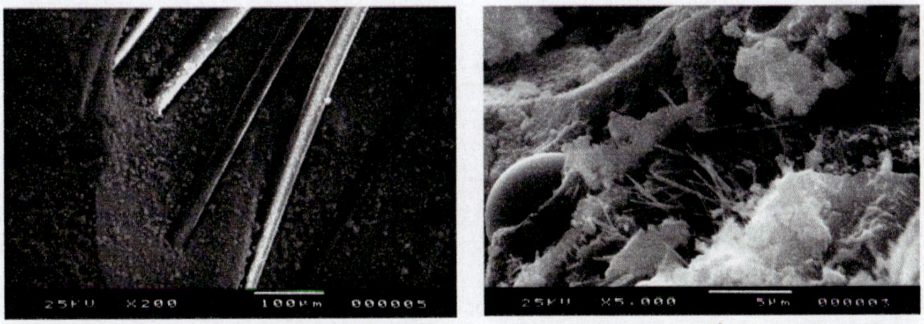

a b

Fig. 5. SEM images of the ECC

Partial replacement of Portland cement by fly ash causes binding of $Ca(OH)_2$ with formation of needle and fibrous habitus hydration products in unclinker part that reinforce the matrix on the micro- and nanolevel (Fig. 5b). In this case the phenomena of "self-reinforcement" is appeared [17].

4 Conclusions

Micromechanical modelling analysis was used for Engineered Cementitious Composites composition development. The compressive strength of ECC becomes higher by 1.1–1.5 time and flexural strength – by 1.4–1.8 time when superplasticizer is incorporated into cement matrix. It is revealed that a high volume faction of fly ash tends to reduce the fiber and matrix interface due to multilevel reinforcement. Enhance of matrix toughness is in the favour of attaining high flexural strength. ECCs show an improvement mechanical properties and while the use of industrial waste instead of cement results in reducing environmental impact.

References

1. Torres, A., Burkhart, A.: Developing sustainable high strength concrete mixtures using local materials and recycled concrete. Mater. Sci. Appl. **7**, 128–137 (2016)
2. Sanytsky, M., Marushchak, U., Olevych, Y., Novytskyi, Y.: Nano-modified ultra-rapid hardening Portland cement compositions for high strength concretes. In: Lecture Notes in Civil Engineering, vol. 47, pp. 392–399 (2020)
3. Chethan, V.R., Ramegowda, M., Manohara, H.E.: Engineered cementitious composites – a review. Int. Res. J. Eng. Technol. **2**(5), 144–149 (2015)
4. Li, V.C.: On engineered cementitious composites (ECC). A review of the material and its applications. J. Adv. Concr. Technol. **1**(3), 215–230 (2003)
5. Kewalramani, M.A., Mohameda, O.A., Syed, Z.I.: Engineered cementitious composites for modern civil engineering structures in hot arid coastal climatic conditions. Proc. Eng. **180**, 767–774 (2017)
6. Halvaei, M., Jamshidi, M., Latifi, M., Behdouj, Z.: Performance of low modulus fibers in engineered cementitious composites (ECCs): flexural strength and pull out resistance. Adv. Mater. Res. **687**, 495–501 (2013)
7. Ondova, M., Stevulova, N., Estokova, A.: The study of the properties of fly ash based concrete composites with various chemical admixtures. Proc. Eng. **42**, 1863–1872 (2012)
8. Kotsay, G.: Pozzolanic activity diagnostics of fly ash for Portland cement. Chem. Chem. Technol. **10**(3), 355–360 (2016)
9. Chung, S.-Y., Elrahman, M.A., Sikora, P., Rucinska, T., Horszczaruk, E., Stephan, D.: Evaluation of the effects of crushed and expanded waste glass aggregates on the material properties of lightweight concrete using image-based approaches. Materials **10**(12), 1354 (2017)
10. Chepurna, S., Borziak, O., Zubenko, S.: Concretes, modified by the addition of high-diffused chalk, for small architectural forms. Mater. Sci. Forum **968**, 82–88 (2019)
11. Borziak, O.S., Plugin, A.A., Chepurna, S.M., Zavalniy, O.V., Dudin, O.A.: The effect of added finely dispersed calcite on the corrosion resistance of cement compositions. In: IOP Conference Series: Materials Science and Engineering, vol. 708, p. 012080 (2019)
12. Markiv, T., Sobol, K., Petrovska, N., Hunyak, O.: The effect of porous pozzolanic polydisperse mineral components on properties of concrete. In: Lecture Notes in Civil Engineering, vol. 47, pp. 275–282 (2020)
13. Marushchak, U., Sydor, N., Braichenko, S., Margal, I., Soltysik, R.: Modified fiber reinforced concrete for industrial floors. In: IOP Conference Series: Materials Science and Engineering vol. 708, no. 1, p. 012094 (2019)

14. Plugin, A., Kostiuk, T., Plugin, O., Bondarenko, D., Sukhanova, Y., Partala, N.: Interaction of mineral and polymer fibers with cement stone and their effect on the physical-mechanical properties of cement composites. Int. J. Eng. Res. Afr. **31**, 59–68 (2017)
15. Tolmachov, S., Belichenko, O., Zakharov, D.: Influence of additives on flexural strength of concrete. In: MATEC Web of Conferences, vol. 116, p. 01019 (2017)
16. Marushchak, U., Sanytsky, M., Pozniak, O., Mazurak, O.: Peculiarities of nanomodified portland systems structure formation. Chem. Chem. Technol. **13**(4), 510–517 (2019)
17. Marushchak, U., Sanytsky, M., Sydor, N., Braichenko, S.: Research of nanomodified engineered cementitious composites. In: Proceedings of the 2018 IEEE 8th International Conference on Nanomaterials: Applications and Properties, p. 8914835 (2018)

Physico-Chemical Investigations of Water Suspensions Microfillers

Serhij Tolmachov(✉) and Olena Belichenko

Kharkov National Automobile and Highway University, Kharkiv 61002, Ukraine
tolmachov.serg@gmail.com

Abstract. At present, microfillers are widely used in heavy concretes. Mineral composition and dispersity of microfillers differ. There is no single method for assessing the effectiveness of microfillers. The views of different researchers on the need to use these materials are different. There are many contradictions that need clarification. In the technology of transport concretes, carboxylate based superplasticizers are used, as well as melamine formaldehyde and naphthalene formaldehyde compounds. Firms that produce such superplasticizers indicate in the annotations only the basis of the additive component, and not the whole of its composition. The use of superplasticizers in different cements differs greatly in the effect of the action. A traditional method for determining the effectiveness of microfillers and superplasticizers in concretes involves the preparation of samples and their subsequent testing at various times of hardening. It is advisable to use simpler physicochemical methods. These are methods for determining the pH and electromotive force (EMF) of aqueous suspensions of cements and microfillers. Instruments for measuring these indicators are simple, and the results can help predict the effectiveness of the use of additives in concrete. This article presents the results of studies of the pH and EMF of aqueous suspensions of cements and microfillers with and without superplasticizers. We used ground quartz and limestone microfillers, as well as carboxylate and naphthalene formaldehyde superplasticizers. Dependences of changes in these indicators over time are obtained. Based on the data obtained, it is possible to determine how the electrochemical properties of suspensions vary with time.

Keywords: Aqueous suspensions · Electromotive force · Microfiller · Organomineral complex · Superplasticizer

1 Introduction

In the practice of world construction is widely used high-performance concrete (HPC). The composition of such concrete, in addition to traditional aggregates and chemical additives, includes finely ground mineral particles - microfillers (MCF). Most researchers of the HPC believe that the main tasks of microfillers are to change the rheological characteristics of concrete mixtures and compact the structure of concrete. In concrete, fillers of different mineral and chemical composition are used. However, their use in concretes is based on the experimentally obtained laws. Experimental studies take a lot of time and do not allow to formulate theoretical ideas about the mechanism of action of microfillers, or the microfiller + superplasticizer (SP) complex.

Z. Blikharskyy (Ed.): EcoComfort 2020, LNCE 100, pp. 466–473, 2021.
https://doi.org/10.1007/978-3-030-57340-9_57

This can lead to large errors in predicting the properties of HPC and assessing their durability. Experimental and practical results from different authors show that the number of MCF and SP, which is used in concrete, is different. Therefore, it is relevant to assess the effectiveness of the use of MCF and the MCF + SP complex using physicochemical research methods.

2 State of the Question

Microfillers usually have a high specific surface and surface activity. Therefore, the interaction of active surface centers and microfillers, can determine the properties of hardening concrete. [1]. Despite the rather large and long-term experience of using MCF in concrete technology, there are contradictions. On the one hand, it is undoubted that the introduction of MCF leads to a change in the parameters of porosity, increases the density, strength and other characteristics of concrete. On the other hand, the use of MCF increases the water demand of concrete mixes and leads to a decrease in the strength of concrete. To overcome the shortcomings of MCF, it was proposed to use a complex of additives: MCF + SP. In addition, other ways to improve the efficiency of MCF have been proposed. For example, the disadvantage of the use of silica fume in the composition of the organic-mineral additive is its negative effect on the workability of mixtures. Therefore, along with a superplasticizer and silica fume, it is proposed to introduce ash microspheres into the organic-mineral complex [2]. It helps to improve the rheological characteristics of concrete mixes. In addition, the pozzolanic activity of ash microspheres makes it possible to bind $Ca(OH)_2$ into hydrosilicates. The density of the contact zone increases, which leads to an improvement the strength and durability of concrete. In the work of E. Sohoshko and N. Zajchenko [3] shows that the use of pozzolanic additives, along with superplasticizers of a new generation, makes it possible to obtain high-strength and durable concrete.

In the work of R. Feldman [4] it was shown that with the introduction of silica fume in the concrete, the pore radius decreases, which leads to compaction of the structure of the cement stone. It is known that the porosity of concrete varies over time. In the initial period of hardening, the porosity may be high, but later, due to the reaction of microfillers with new growths of cement, the pores will become overgrown with reaction products, and the concrete will be compacted. This mechanism is described in many articles, for example, in the article by I. Markovic, where the compaction model is given [5]. According to some authors, the compaction of hardened cement stone depends of the distance between the opposing surface areas of the interacting particles, which are the centers of crystallization [6]. In the system, which contains an organic-mineral complex consisting of SP and microfiller, the concentration of microparticles of the filler plays an important role. SP will reduce the amount of water in the concrete mix and thus the distance between the particles. The microfiller will saturate the system with microparticles. This leads to compaction of cement stone, creating cramped conditions in which directional formation and growth new formations in set cement takes place. There is an optimum content of microfiller particles at which the density is maximum. It is known that an increase in the content of fine fractions in the composition of sand, as well as dust particles in the composition of aggregates, reduces the effectiveness of the use of superplasticizers. To achieve the desired effect, the amount of superplasticizer is usually increased. It is logical to assume that SP is adsorbed on the surface of small particles. Other researchers argue that the adsorption of anion-active SP on the surface of silica is impossible because the adsorbent and adsorbate have the same sign of charge. Most researchers believe that the mechanism of action of SP of naphthalene and melamine series is adsorption and has an ion-electrostatic nature [1, 7–10].

The ion-electrostatic mechanism of action of SP is confirmed by the following phenomena: various selective adsorption of superplasticizers with respect to mineral powders, which depends on the potential-determining ions and the difference in the polarities of water and the solid phase; selectivity of superplasticizers during adsorption on minerals.

This suggests the possibility of selective adsorption of different SP on particles with different polarities of the surface.

We assumed that the mechanism of influence of the organomineral complex is not limited to the fact that the structure of the cement stone is compacted. With the simultaneous presence of a SP and a microfiller in the solution, interaction between them is possible. The presence of interaction can be assessed by changing the colloidal-chemical and chemical properties of solutions of microparticles.

Given this, it can be assumed that the naphthalene formaldehyde superplasticizers adsorption is linear, and that of carboxylates is bulk [11]. At the same time, the nature of adsorption is different: in the first case, chemical adsorption may predominate, in the second - physical adsorption.

In the work [12] it was established that polycarboxylates have high surface activity and their inherent electrostatic and steric mechanism of action. Due to this, there is an increase in the number of contacts in the early period of hydration with the formation of nanoscale phases C - S - H.

According to E.D. Shchukin two cases of adsorption are possible [13].

1 case - the concentration of surface-active reagent (SAR), which are these SP is low. Then, during the adsorption of SAR on the solid polar surface, the SAR molecules are located by the polar group on the surface of the solid, and the hydrocarbon chain is

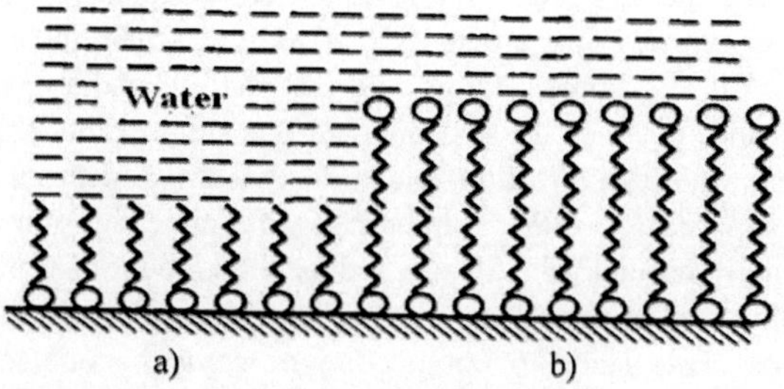

Fig. 1. The orientation of SAR molecules at the "polar solid - aqueous solution" phase interface: a) at low SAR concentrations; b) at high SAR concentrations

turned into an aqueous medium (Fig. 1, a). This adsorption can be chemical or physico-chemical.

2 case - high SAR concentration. When the entire surface is covered with a layer of SAR molecules, re-adsorption occurs. This adsorption is physical. A second layer of SAR is formed, in which the hydrocarbon chain faces the hydrocarbon chain of the first layer, and the polar group is in water (Fig. 1, b). With this arrangement of SAR molecules, water easily penetrates to the surface and the wettability of the solid body improves.

It was found that the presence on the surface of microfillers of strong Lewis centers increases the content of OH- ions in the zone of contact between the additive and cement, locally lowering the pH and facilitating the transfer of calcium ions into the solution to compensate the negative charge [1, 14, 15].

At the initial stage of hydration of the cement suspension, the surface of the microfiller, interacting mainly with water, changes two main indicators of the activity of such a suspension - pH and Electromotive force (EMF). Activators of the hydration process of clinker minerals are microfillers, which can shift the pH of the system to lower values. A shift in EMF towards more positive values should speed up the hydration. Therefore, the aim of the work is to establish the nature of changes in the physicochemical properties of aqueous solutions of cements and microfillers with and without superplasticizers.

3 Experimental

3.1 Materials

As microfillers were used microsilica fume and crushed chalk. Microfillers were obtained by grinding in a ball mill. The specific surface of microfillers is: silica fume - $S_{ss} = 281\ m^2/kg$, $S_{ss} = 980\ m^2/kg$; chalk - $S_{ss} = 328\ m^2/kg$, $S_{ss} = 1040\ m^2/kg$. Superplasticizers: naphthalene formaldehyde Sika 20 Gold and polycarboxylate Sika Plast 2508.

3.2 Methods and Procedure

From microfillers were prepared 10% aqueous suspensions and the pH and EMF of suspensions were determined using a pH meter device. The viscosity of aqueous solutions of SP was determined using a viscometer with a capillary diameter of 0.54 mm. When conducting experiments, suspensions of microfillers and solutions of SP were kept at a constant temperature of +20 °C.

4 Results and Discussion

Studies have shown that with an increase in the concentration of superplasticizers, a decrease in pH occurs in the region of acidic solutions. Intensive decrease of pH occurs to a concentration of C = 0.6% (Table 1). This decrease is 0.56 for Sika Plast 2508 (carboxylate) superplasticizer and 1.91 for Sika Gold superplasticizer (naphthalene formaldehyde). Further, the decrease of pH in solutions with Sika Gold occurs to a lesser extent, and in solutions with Sika Plast 2508 a slight increase of pH is observed.

Table 1. Change of pH and EMF of aqueous solutions of additives

Type of additive	Sika Gold		Sika Plast 2508	
The concentration of the aqueous solution of the additive,%	pH	EMF, mV	pH	EMF, mV
Water	6.68	+29	6.29	+52
0.05	5.54	+89	6.18	+54
0.1	5.57	+93	6.16	+59
0.15	5.59	+96	6.12	+64
0.2	5.12	+120	5.93	+73
0.3	5.10	+121	5.90	+74
0.4	4.97	+129	5.81	+79
0.5	4.82	+137	5.85	+77
0.6	4.77	+140	5.73	+84
0.7	4.80	+138	5.71	+85
0.8	4.69	+145	5.70	+86
0.9	4.67	+146	5.69	+86
1.0	4.67	+146	5.73	+84
1.2	4.58	+149	5.78	+88
1.5	4.52	+152	5.83	+92
1.8	4.48	+156	5.89	+94

When the concentration of the additive Sika Gold is C = 0.05%, a minimum pH value is observed. The decrease of pH is 1.14. To a much lesser extent, this decrease is noticeable for the solution of Sika Plast 2508. Additive to concentration C = 0.15% on the graphs, the pH changes slightly. Further, when the concentration range of additives is C = 0.15…0.2%, a second sharp decrease of pH is observed. It is possible that a greater decrease of pH in solutions with the addition of Sika Gold can be explained by the greater charge of associates of the molecules of this additive in comparison with the molecules of the additive Sika Plast 2508. Studies of changes of the EMF (electromotive force) value showed an intensive increase of the EMF in solutions with the addition of Sika Gold at first with an increase in concentration from 0 to 0.05%, and then from 0.15 to 0.2% (Table 1). On the graph of the change of EMF in solutions with the addition of Sika Plast 2508, a significant increase in EMF is evident only in the region from C = 0 to C = 0.2%. The numerical values are given in Table 1.

The kinematic viscosity studies (Fig. 2) of aqueous solutions of additives were conducted, which showed that the viscosity decreases sharply at concentrations C = 0.05% (Sika Gold additive) and C = 0.1% (Sika Plast 2508 additive). It is known that such a decrease in the viscosity of aqueous solutions corresponds to a critical concentration of micelle formation (CCMF) of surface-active reagents (SAR).

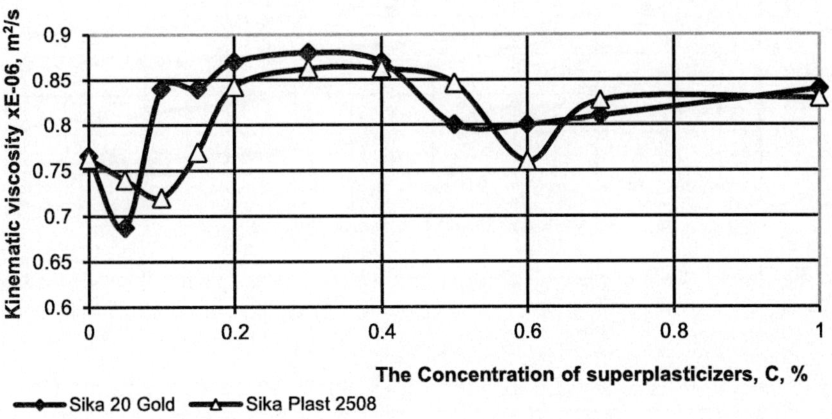

Fig. 2. The effect of the concentration of superplasticizers on the kinematic viscosity of aqueous solutions

The obtained results indicate that using analysis of changes of pH and EMF of aqueous solutions of SAR, it is possible to determine not only CCMF, but also, probably, other colloid-chemical characteristics of SAR. Studies have been conducted to change the pH and EMF of organic-mineral complexes, which consist of a superplasticizer and a microfiller. Microfillers were chalk and silica fume. The specific surface of the microfillers was the same $S_{ss} = 300\ m^2/kg$.

Studies have shown that in suspensions containing silica fume and chalk, with an increase the concentration of superplasticizers, a decrease in pH occurs (Fig. 3). At the same time, the initial pH value is almost two times higher than the pH of aqueous solutions of superplasticizers. On curves with Sika Gold superplasticizer, two extremes can be seen: a minimum at the concentration of superplasticizers C = 0.2% and a maximum at the concentration of superplasticizers C = 0.3%.

The minimum pH of suspensions with additives corresponds to the second minimum pH of aqueous solutions of superplasticizers (Table 1). Extremes for suspensions with the addition of Sika Plast 2508 are missing. Studies of the EMF of suspensions with superplasticizers showed that the sign of the EMF of pure suspensions changes to negative (Fig. 4). The EMF reaches values of - 190 mV, which in absolute value is significantly higher than the values of EMF of aqueous solutions of superplasticizers (Table 1). An increase the concentration of superplasticizers leads to an adequate reduction in the negative electromotive force. This dependence has an extreme character in the case of a suspension of microsilica in an aqueous solution of the Sika Gold superplasticator. There is a pronounced maximum that corresponds to the concentration of Sika Gold C = 0.2% and a minimum that corresponds to the concentration of Sika Gold C = 0.3%. The maximum coincides with the extremum in Fig. 3.

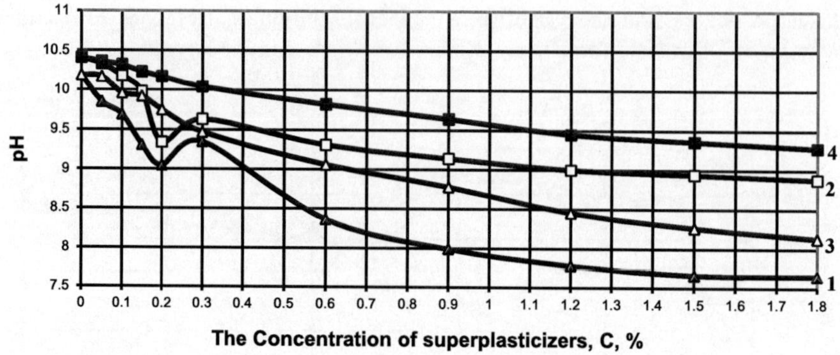

Fig. 3. Change of pH of microfiller suspensions with additives: 1) silica fume + Sika Gold; 2) chalk + Sika Gold; 3) silica fume + Sika Plast 2508; 4) chalk + Sika Plast 2508

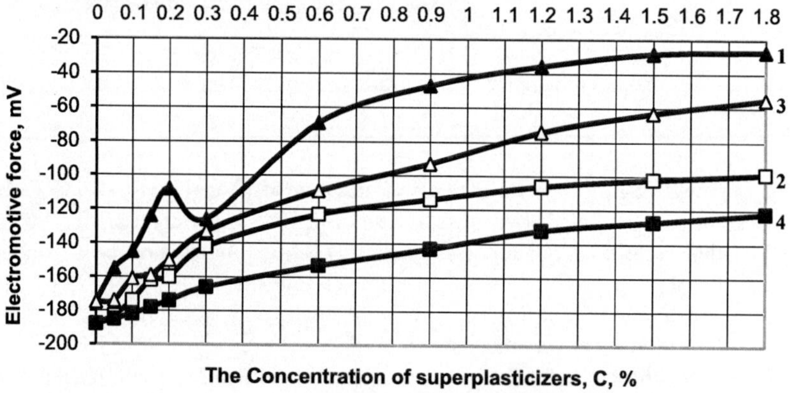

Fig. 4. Change of EMF of microfiller suspensions with additives: 1) silica fume + Sika Gold; 2) chalk + Sika Gold; 3) silica fume + Sika Plast 2508; 4) chalk + Sika Plast 2508

5 Conclusions

1. It is shown that with increasing concentration of superplasticizers, the pH of aqueous solutions decreases. In this case, the EMF of aqueous solutions increases.
2. With the introduction of superplasticizers in suspension microfillers, a decrease in the pH of the solutions can be seen. The introduction of superplasticizers in suspensions of microfillers reduces the electronegativity of the EMF. The value of EMF increases in the direction of positive values. The increase in the content in the aqueous suspension of superplasticizers leads to a gradual neutralization of the negative sign of the EMF charge.
3. Based on the above results, it is possible to consider the adsorption of superplasticizers on mineral fillers possible regardless of their mineral composition and the sign of the surface charge.

4. The points of extreme changes in pH and EMF in the plots are the points of critical concentration of micelle formation of superplasticizers in the presence of microfillers.

References

1. Juchnevskyj, P.I.: Vlyjanye chymycheskoj pryrodi dobavok na svojstva betonov. BNTU, Mynsk, (2013)
2. Sagradyan, A.A., Simakova, G.A.: Izuchenie svojstv tyazhelogo betona modificirovannogo organomineralnoj dobavkoj, vklyuchayushej zolnye mikrosfery. Izvestiya Vuzov. Stroitelstvo Arkhit. **4**, 26–31 (2012)
3. Sokhoshko, E.V., Zajchenko, N.M.: Samouplotnyayushijsya beton v sovremennom monolitnom domostroenii. Visnyk Donba'koji derzhavnoji akademiji budivnictva i arxitektury: Suchasni budivel'ni materially **1**(75), 112–116 (2009)
4. Feldman, R.F.: Pore structure, permeability and diffusivity as related to durability. In: 8th International Congress on the Chemistry of Cement, Rio de Janeiro, Brazil, pp. 1–21 (1986)
5. Markovic, I.: High-performance hybrid–fiber concrete – development and utilisation. DUP Science, The Netherlands (2006). ISBN 90-407-2621-3
6. Sheynfeld, A.V.: Osobennosti formirovaniya ierarhicheskoj mikro- i nanostruktury cementnyh sistem s kompleksnymi organomineralnymi modifikatorami. Beton zhelezobeton **2**, 16–21 (2016)
7. Neville, A.M.: Svoystva betona. Stroiizdat, Moscow (1972)
8. Rauen, A.Y.: Wirkungsmechanism von Betonßverflussigern auf der Basis von Wasserloslichen Melaminharzen. Cem. Concr. Res. **6**(1), 57–61 (1976)
9. Ratinov, V.B., Rosenberg, T.I.: Dobavki v beton. Stroiizdat, Moscow (1989)
10. Batrakov, V.G.: Modifitsirovannyye betony. Teoriya i praktika. Stroiizdat, Moscow (1998)
11. Krasovsky, P.S.: Fiziko-khimicheskiye svoystva formirovaniya struktury tsementnykh betonov. Publishing House DVGUPS, Khabarovsk (2013)
12. Marushchak, U., Sanytsky, M., Pozniak, O., Mazurak, O.: Peculiarities of nanomodified Portland systems structure formation. Chem. Chem. Technol. **13**(4), 510–517 (2019). https://doi.org/10.23939/chcht13.04.510
13. Schukin, E.D., Pertsova, A.V., Amelina, E.A.: Kolloidnaya khimiya. The higher school, Moscow (2007)
14. Shangina, N.N.: Prognozirovaniye fiziko-mekhanicheskikh kharakteristik betonov s uchetom donorno-aktseptornykh svoystv poverkhnosti napolniteley i zapolniteley. Diss. Doctors of technical sciences (Engineering), St. Petersburg (1998)
15. Komokhov, P.G.: Upravlenie svojstvami cementnyh smesej prirodoj napolnitelej. Izvestiya Vuzov. Seriya «Stroitelstvo» **9**, 51–54 (1997)

Crack Resistance of Concretes Reinforced with Polypropylene Fiber

Yurii Turba(✉) and Sergii Solodkyy

Lviv Polytechnic National University, Lviv 79013, Ukraine
yurii.v.turba@lpnu.ua

Abstract. The article presents the results of the study of the influence of dispersed reinforcement of cement concrete with polypropylene fiber with different variations of fiber consumption, fine and coarse aggregates, as well as coarse aggregates' maximum size on the strength and crack resistance. The analysis of complete state diagrams of the studied concrete series shows a slight difference in the subcritical stage of concrete failure (until the beginning of the main crack movement) between reinforced and non-reinforced concrete. The main advantage of the introduced polypropylene fiber could be observed in the supercritical fracture stage - the fiber inhibits the fracture process of the sample after the moment of the main crack development (when the maximum destructive load was already applied). The fracture toughness increases with grain size of coarse aggregate increasing from 15 to 20 mm, with subsequent stabilization of its value. Increasing of the cement-sand mortar amount in concrete leads to increase in the fracture toughness, while the specific effective energy consumption for static fracture does not change significantly and reaches the maximum value at grain spacing coefficient of 1.4. As the amount of introduced fiber increases from 4 to 7 kg per 1 m3 of concrete, the fracture toughness increases and remains at the same level with the maximum fiber content of 10 kg.

Keywords: Dispersed reinforced concrete · Fiber concrete · Fiber · Crack resistance · Fracture mechanics · Complete state diagrams

1 Introduction

The global tendency towards the usage of concrete as the main material due to fast and large-scale modern construction pace is constantly increasing, which is associated with the successful solution of economic, environmental and technical problems [1–8]. The use of concretes with improved properties is becoming especially relevant [9–12].

One of the perspective construction materials, which enables to eliminate such concrete disadvantages as low tensile strength and high destruction fragility are dispersed reinforced concretes - fiber-reinforced concretes (FRC) [13–17]. Fiber-reinforced concrete in contrast to steel bars has the number of advantages, in particular corrosion resistance. Therefore, due to this property strengthening and renovation of structures is not required [18–21].

The presence of reinforcing fibers in concrete, if their optimal content is ensured, increases the density, homogeneity and reduces the risk of cracking, which allows to

Z. Blikharskyy (Ed.): EcoComfort 2020, LNCE 100, pp. 474–481, 2021.
https://doi.org/10.1007/978-3-030-57340-9_58

predict higher durability, dynamic resistance, abrasion, frost resistance and durability of cement concrete in general [22–24].

In addition to the optimal fiber content, the process of preparation of the fiber concrete mixture with its uniform distribution throughout the matrix volume is important, which is achieved by the use of additives and modern methods of mixtures' preparation. Due to the concrete strength increase by introduction of fiber, the reduction of size and weight of concrete products could be achieved. Fiber concrete can be used in reinforced concrete structures. Nowadays there are various ways to improve the properties of reinforced concrete structures, using high-strength concrete, high-strength reinforcement and concrete admixtures. The combined use of fiber concrete in reinforced concrete reduces the metal consumption of these elements.

2 Purpose of the Research and Problem Formulation

Since the scientific and technical literature emphasizes that fiber concrete has increased crack resistance, the purpose of this paper is to test the effectiveness of dispersed reinforcement, using methods and criteria of fracture mechanics, which are based on critical values of stress-strain state parameters in the cross sections of structural elements - fracture toughness and fracture energy.

3 Materials and Methods of Research

Preparation and testing of concrete samples was carried out in accordance with the requirements [23]. For concrete mixtures following materials were used:

- Portland cement PC II/A-Sh-500 of general construction purpose;
- aggregates: fine – quartz sand with size module of 1,29, crushed stone sieving of 1.25–2.5 mm fraction; coarse - granite crushed stone of 5–30.0 mm fraction.

Polypropylene fiber Enduro HPP45 is used in order to ensure its higher performance in the concrete according to results of previous research of this article authors' [7, 8], actual properties are given in Table 1.

Table 1. Chemical and physical properties of polypropylene fiber Enduro HPP45.

Property	Value
Fiber length	45 mm
Type/form	Macro/monofibers
Specific weight	0.91 g/cm^3
Melting temperature	164 °C
Ignition temperature	>550 °C

The aggregate mixture composition was designed by the method of absolute volumes with continuous particle size distribution.

As technological factors which have influence on crack resistance of concrete was considered the maximum size of coarse aggregate, spacing coefficient of coarse aggregates' grains and fiber consumption.

Composition of concrete mixtures of studied series' is given in Table 2.

Table 2. Concrete mixtures' composition.

Series' marking	Cement consumption, kg/m^3	W/C	Aggregates, kg/m^3		Fiber consumption, kg/m^3	Spacing coefficient of coarse aggregates' grains	Maximum size of coarse aggregate, mm
			Fine	Coarse			
B	350.0	0.44	690	1247	–	1.4	20.0
A_{d-}			745	1181	7.0		15.0
A_{d+}			693	1249			30.0
A_0			690	1247			20.0
$A_{\alpha-}$			528	1403		1.1	
$A_{\alpha+}$			820	1123		1.7	
A_{f-}			690	1247	4.0	1.4	
A_{f+}					10.0		

4 Research Results

Crack resistance parameters of concretes were identified by equilibrium mechanical tests of prisms with previously created crack of normal separation according to the three-point bending scheme of with simultaneous recording of a complete load-deflection diagram (*F-V*) on the special installation [9] in the age of 28 days. General view of installation is given on Fig. 1.

According to experiment results were formulated fully equilibrium deformation diagrams (FEDD) and complex of force and energy characteristics of concrete crack resistance was calculated.

Strength of concretes on tension with bending was calculated on the basis of maximum load value according to diagram (Fig. 2).

Results of study of strength and deformability parameters are given in Table 3.

The highest strength was indicated for concretes of basic series B (non-reinforced) and series A_0, with average technological factors values. Strength of other series' concretes were on average 10% lower, except $A_{\alpha+}$ series with maximum cement-sand mortar (CSM) content.

Fig. 1. General view of testing installation

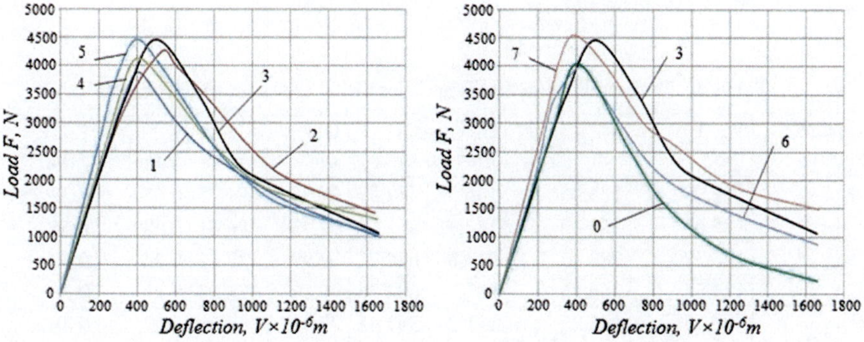

Fig. 2. Fully equilibrium deformation diagrams of concrete series: 1 - A_{d-}; 2 - A_{d+}; 3 - A_0; 4 - $A_{\alpha-}$; 5 - $A_{\alpha+}$; 6 - A_{f-}; 7 - A_{f+}; 0 - B.

Table 3. Strength parameters of studied concretes.

Series' marking	Compression strength, R_b, MPa	Strength on tension with bending, R_{tb}, MPa	Fragility criteria, $X_{tb} = (G_f E_{tb}/R_{tb}^2) \cdot 10^{-3}$ m
B	45.2	8.54	173
A_{d-}	40.8	8.20	251
A_{d+}	40.6	8.96	210
A_0	44.2	9.43	194
$A_{\alpha-}$	43.8	8.73	249
$A_{\alpha+}$	32.2	9.41	263
A_{f-}	40.0	8.44	228
A_{f+}	39.6	9.60	267

The highest strength on tension with bending is provided by concretes with maximum fiber content (A_{f+} series) with average all parameters' values (A_0) and maximum content of CSM ($A_{\alpha+}$). This indicates the positive effect of fiberglass and macrostructure homogeneity of concrete on strength on tension with bending. The lack of fiber in concrete increases its fragile destruction nature.

Analysis of fully equilibrium deformation diagrams (FEDD) of the studied series (Fig. 2) shows insignificant difference in the subcritical stage of concrete destruction (before the beginning of the main crack movement) between reinforced and non-reinforced concrete. The main advantage of the introduced polypropylene fiber could be observed in the supercritical destruction stage - the fiber inhibits the destruction process of the sample from the moment of the main crack development (when the maximum destructive load has been already applied).

Among the main force and energy characteristics of concretes' crack resistance are critical stress intensity factor K_i (fracture toughness) and specific energy consumption for static destruction G_F (Table 4). Character of their changes depending on technological factors is given on Fig. 3.

Table 4. Force and energy characteristics of concretes' crack resistance.

Concrete series	W_i, 10^{-2} N · m	W_l, 10^{-2} N · m	G_i, J/m^2	G_F, J/m^2	Gce, J/m^2	J_i, J/m^2	K_i, MPa·$^{1/2}$	K_c, MPa·$^{1/2}$
B	87.92	184.06	146.54	437.59	34.75	81.75	0.65	0.32
F_{d15}	85.63	263.30	142.72	552.52	99.45	86.41	0.66	0.55
F_{d30}	128.32	282.46	213.86	624.71	37.99	137.62	0.76	0.32
F_0	124.64	266.16	207.74	604.56	49.87	128.03	0.77	0.38
$F_{\alpha 1,1}$	90.75	285.21	151.24	599.41	72.98	89.80	0.69	0.48
$F_{\alpha 1,7}$	98.48	285.35	164.14	591.69	66.66	106.64	0.80	0.51
F_4	95.90	238.32	159.84	513.18	55.96	102.40	0.71	0.42
F_{10}	98.86	328.17	164.76	678.29	87.61	99.71	0.77	0.56

With the increase of coarse aggregates' size takes place increase in value of G_F, simultaneously for fracture toughness the increase occurs with the increase of coarse aggregates size from 15 to 20 mm, with further stabilization of its value. The increase in amount of CSM in concrete results in increase of fracture toughness; on the other hand energy consumption G_F does not change significantly and reaches the highest value at spacing coefficient 1.4.

If the fiber consumption is increased from 4 to 7 kg for 1 m^3 of concrete the value K_i also increases and is kept on the same value at maximum fiber content of 10 kg. Almost proportional dependency of factor G_F could be obtained by the amount of used fiber.

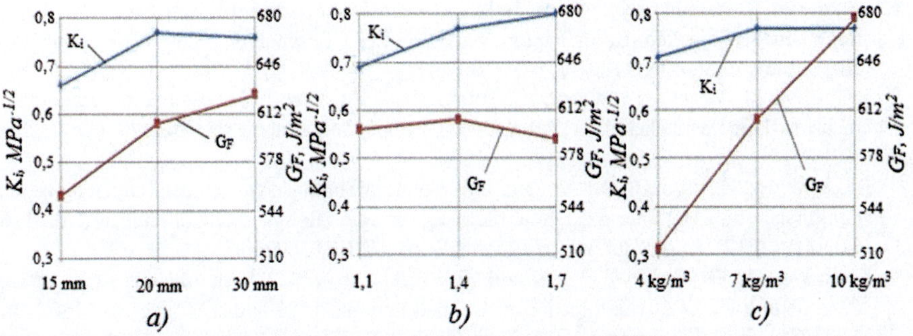

Fig. 3. Values of static critical stress intensity factor Ki and specific energy consumption for static destruction GF for concretes with following parameters: a – maximum size of coarse aggregates; b – spacing coefficient; c – fiber content

5 Conclusions

Reinforcement of concrete with polypropylene fiber increases strength on tension with bending, as well as force and energy characteristics of concretes' crack resistance. Significant qualitative influence of fiber in concrete is observed at the formed main crack in a supercritical stage of concrete destruction. With the increase of the coarse aggregate size and the amount of fiber, the specific energy consumption for static destruction increases significantly. Higher content of cement-sand mortar is characterized by increased fracture toughness. The optimal values of technological factors are shown by results of crack resistance of concretes reinforced with polypropylene fiber at the average and maximum level.

References

1. Blikharskyy, Z., Brózda, K., Selejdak, J.: Effectivenes of strengthening loaded RC beams with FRCM system. Arch. Civ. Eng. **64**(3), 3–13 (2018)
2. Blikharskyy, Z., Vashkevych, R., Vegera, P., Blikharskyy, Y.: Crack resistance of RC beams on the shear. In: Lecture Notes in Civil Engineering, vol. 47, pp. 17–24 (2020). https://doi.org/10.1007/978-3-030-27011-7_3
3. Bobalo, T., Blikharskyy, Y., Kopiika, N., Volynets, M.: Serviceability of RC beams reinforced with high strength rebar's and steel plate. In: Lecture Notes in Civil Engineering, vol. 47, pp. 25–33 (2020). https://doi.org/10.1007/978-3-030-27011-7_4
4. Bobalo, T., Blikharskyy, Y., Kopiika, N., Volynets, M.: Theoretical analysis of RC beams reinforced with high strength rebar's and steel plate. IOP Conf. Ser. Mater. Sci. Eng. **708**(1), 012045 (2019)
5. Khmil, R., Tytarenko, R., Blikharskyy, Y., Vashkevych, R.: Influence of load level during strengthening of reinforced concrete beams on their reliability. IOP Conf. Ser. Mater. Sci. Eng. **708**(1), 012054 (2019)

6. Krainskyi, P., Blikharskyy, Y., Khmil, R., Vegera, P.: Crack resistance of rc columns strengthened by jacketing. In: Lecture Notes in Civil Engineering, vol. 47, pp. 195–201 (2020). https://doi.org/10.1007/978-3-030-27011-7_25
7. Krainskyi, P., Vegera, P., Khmil, R., Blikharskyy, Z.: Theoretical calculation method for crack resistance of jacketed RC columns. IOP Conf. Ser. Mater. Sci. Eng. **708**(1), 012059 (2019)
8. Blikharskyy, Z., Khmil, R., Vegera, P.: Shear strength of reinforced concrete beams strengthened by PBO fiber mesh under loading. In: MATEC Web of Conferences, vol. 116, p. 02006 (2017). https://doi.org/10.1051/matecconf/201711602006
9. Krainskyi, P., Blikharskyy, Y., Khmil, R., Blikharskyy, Z.: Experimental study of the strengthening effect of reinforced concrete columns jacketed under service load level. In: MATEC Web of Conferences, vol. 183, p. 02008 (2018). https://doi.org/10.1051/matecconf/20181830200
10. Lobodanov, M., Vegera, P., Blikharskyy, Z.: Planning experiment for researching reinforced concrete beams with damages. In: Lecture Notes in Civil Engineering, vol. 47, pp. 243–250 (2020). https://doi.org/10.1007/978-3-030-27011-7_31
11. Selejdak, J., Blikharskyy, Y., Khmil, R., Blikharskyy, Z.: Calculation of reinforced concrete columns strengthened by CFRP. In: Lecture Notes in Civil Engineering, vol. 47, pp. 400–410 (2020). https://doi.org/10.1007/978-3-030-27011-7_51
12. Vegera, P., Vashkevych, R., Blikharskyy, Z.: Fracture toughness of RC beams with different shear span. In: MATEC Web of Conferences, vol. 174, p. 02021 (2018). https://doi.org/10.1051/matecconf/201817402021
13. Brandt, A.M.: Fibre reinforced cement-based (FRC) composites after over 40 years of development in building and civil engineering. Compos. Struct. **86**(1–3), 3–9 (2008)
14. Kakooei, S., Akil, H.M., Jamshidi, M., Rouhi, J.: The effects of polypropylene fibers on the properties of reinforced concrete structures. Constr. Build. Mater. **27**(1), 73–77 (2012)
15. Karahan, O., Atis, C.D.: The durability properties of polypropylene fiber reinforced fly ash concrete. Mater. Des. **32**(2), 1044–1049 (2011)
16. Mazaheripour, H., Ghanbarpour, S., Mirmoradi, S.H., Hosseinpour, I.: The effect of polypropylene fibers on the properties of fresh and hardened lightweight self-compacting concrete. Constr. Build. Mater. **25**(1), 351–358 (2011)
17. Vairagade, V.S., Kene, K.S., Deshpande, N.V.: Investigation on compressive and tensile behaviour of fibrillated polypropylene fibers reinforced concrete. Int. J. Eng. Res. Appl. **2**(3), 1111–1115 (2012)
18. Blikhars'kyi, Y.Z.: Anisotropy of the mechanical properties of thermally hardened A500s reinforcement. Mater. Sci. **55**, 175–180 (2019). https://doi.org/10.1007/s11003-019-00285-0
19. Kharchenko, Y.V., Blikharskyy, Z.Y., Vira, V.V., Vasyliv, B.D.: Study of structural changes in a nickel oxide containing anode material during reduction and oxidation at 600 °C. In: Fesenko, O., Yatsenko, L. (eds.) Nanocomposites, Nanostructures, and Their Applications, NANO 2018. Springer Proceedings in Physics, vol. 221. Springer, Cham (2019). https://doi.org/10.1007/978-3-030-17759-1_42
20. Kharchenko, Y., Blikharskyy, Z., Vira, V. Vasyliv, B., Viktoriya, P.: Study of nanostructural changes in a Ni-containing cermet material during reduction and oxidation at 600 °C. Appl. Nanosci. (2020). https://doi.org/10.1007/s13204-020-01391-1
21. Zhang, Q., Mol'kov, Y.V., Sobko, Y.M. et al. Determination of the Mechanical Characteristics and Specific Fracture Energy of Thermally Hardened Reinforcement. Mater Sci 50, 824–829 (2015). https://doi.org/10.1007/s11003-015-9789-9

22. Solodkyy, S., Kahanov, V., Hornikovska, I., Turba, Y.: A study of fracture toughness of heavy-weight concrete and foam concrete reinforced by polypropylene fibre for road construction. Eastern Eur. J. Enterp. Technol. **4**(5(76)), 40–46 (2015)
23. Turba, Y., Solodkyy, S., Markiv, T.: Strength and fracture toughness of cement concrete, dispersedly reinforced by combination of polypropylene fibers of two types. In: Lecture Notes in Civil Engineering, vol. 47, pp. 488–494. Springer, Cham (2020). https://doi.org/10.1007/978-3-030-27011-7_62
24. Xu, Z., Hao, H., Li, H.N.: Experimental study of dynamic compressive properties of fibre reinforced concrete materials with different fibres. Mater. Des. **33**, 42–45 (2012)

Check for updates

Assessment of the Economic Feasibility of Using Alternative Energy Sources in Ukraine

Malgorzata Ulewicz[1], Vasyl Zhelykh[1], Yurii Furdas[2], and Khrystyna Kozak[2](✉)

[1] Czestochowa University of Technology, Czestochowa 42-201, Poland
[2] Lviv Polytechnic National University, Lviv 79013, Ukraine
cr__i@ukr.net

Abstract. The article deals with an analysis of the popular types of alternative energy sources and prospects for their application in countries with temperate climates, particularly in Ukraine. The need to compare the potential of alternative energy in Ukraine with the cost of equipment required for the use of alternative energy sources is substantiated. According to the specifics of each region, the concept of choosing efficient energy sources is defined, in particular, attention should be paid to comparing the cost of materials, capital operating and costs for each type of energy, in particular, Sun energy, biogas, biomass, air potential and geothermal energy. In the form of tables and nomograms presented data regarding to technically achievable energy potential for different regions of Ukraine. Analyzing the obtained data, it can be stated that geothermal energy, air potential and solar energy deserve special attention.

Keywords: Alternative energy sources · Sun energy · Energy potential · Energy technologies

1 Introduction

The problem of using energy sources that would become a full-fledged replacement for organic energy is becoming increasingly acute. Mankind is constantly looking for an alternative. There is a clearly defined category of "alternatives" to energy sources, which are: permanent, renewable and recurring [1]. Analyzing the current trends in the world energy supply system, the beginning of a qualitative stage of energy market development is observed. There is a process of reorientation of the energy strategy, not only of the European Union, but also of the Middle East. The strategy of energy development is being revised in the conditions of formed partnerships with the largest importers of natural resources, in particular Norway, Russia and Algeria, as well as in the conditions of a well-established industry for processing traditional energy sources. Alternative energy remains a priority for the European Union's economy. Continued government stimulus since 2004 has led to a rapid increase in the share of renewables in gross final energy consumption, as well as to significant investment by powerful Asian companies [2]. Although recently, at the global level, investment in renewable energy in 2017, investment decreased by 23% compared to 2015.

Z. Blikharskyy (Ed.): EcoComfort 2020, LNCE 100, pp. 482–489, 2021.
https://doi.org/10.1007/978-3-030-57340-9_59

According to the United Nations Environment Program (UNEP), the sector received less than $ 242 billion. The organization's report drew attention to the global share of electricity from renewable energy sources (RES), which increased from 10.3% in 2015 to 11.3% in 2016. Regarding renewable energy sources in terms of generation types, investments in solar energy in 2014 amounted to $ 114 billion, which is 34% less than in 2015. As for wind energy, it was financed in the amount of $ 113 billion, which is 9% less than in 2015. The policy of reducing state sup-port for RES has a negative impact on investment. The main reason lies in the "green" tariffs. In many EU countries, companies have received state guarantees for the sale of "green" electricity at fixed prices, while over time, prices have changed. Regardless of the development of events, energy consumption will continue to grow [3–5]. Of course, this will depend on state regulation of alternative energy development [6, 7].

In accordance with the classification of the International Energy Agency, renewable energy sources include the following categories [7]:

- renewable energy sources (RES), which are burned, and biomass waste:
- solid biomass and animal products: biological mass, including any material of vegetable origin used directly as fuel or converted into other forms before incineration (wood, vegetable and animal waste; charcoal obtained from solid biomass);
- biomass gas/liquid: biogas obtained in the process of anaerobic fermentation of biomass and solid waste, which is burned to produce electricity and heat;
- municipal waste: materials incinerated for the production of heat and electricity (waste from the residential, commercial and public sectors). Disposed of by the municipal authorities for the purpose of centralized destruction;
- industrial waste: solid and liquid materials, which are incinerated directly, usually at specialized enterprises, for the production of heat and electricity;
- hydropower: potential, or kinetic, energy of water converted into electrical energy by hydropower plants, both large and small;
- geothermal energy: thermal energy coming from the earth's interior, usually in the form of hot water or steam;
- solar energy: solar radiation used to produce hot water and electricity;
- wind energy: kinetic wind energy used to generate electricity in wind turbines;
- tidal energy, sea waves and ocean energy: the mechanical energy of tidal currents or waves used to produce electricity.

Appropriate conditions must be created for the production, supply, transportation, storage, supply and consumption of energy produced from alternative sources. The main measures to be outlined by the state remain organizational and legal, financial, economic and technical and technological [8, 9]. This is a forced policy, because alternative energy sources are environmentally friendly and directly affect the economic security of states [10, 11]. The Road Map of Energy until 2050 is proposed, which reflects the key role of RES in the energy supply of transport, industry and other consumers [12].

The potential of renewable energy sources is available throughout Ukraine. The main components of renewable energy include: solar energy, geothermal energy, solid biomass energy, biogas energy and energy of the upper layer of soil and air.

2 Types of Alternative Energy Sources Used in Ukraine

The leading environmentally friendly source of energy is the Sun. The potential of solar energy in Ukraine is high enough for the use of both solar collectors and photovoltaic panels in almost all areas. The use of flat solar collectors, which use both direct and scattered solar radiation, is effective for solar heat supply. Conversion of solar energy into electricity in the conditions of Ukraine it is expedient to use photovoltaic panels [13, 14].

Geothermal energy is the heat of the Earth, which is mainly formed due to the decay of radioactive substances in the earth's crust and mantle. The temperature of the earth's crust rises by 2.5–3 °C every 100 m (the so-called geothermal gradient). Thus, at a depth of 20 km it is about 500 °C, at a depth of 50 km – about 700 … 800 °C, and at the core of the Earth – about 5000 °C. In certain places, especially along the edges of the tectonic plates of the continents, as well as in the so-called "hot spots", the temperature gradient is almost 10 times higher, and then at a depth of 500–1000 m the temperature of the rocks reaches 300 °C. However, even where the temperature of the earth's rocks is not so high, geothermal energy resources are sufficient.

Geothermal energy is used for heating, water supply and air conditioning in residential and public buildings and structures in cities and rural areas.

The most appropriate at present is the use of energy from geothermal waters.

One of the directions of geothermal energy development is the use of combined systems for heat and electricity contained in geothermal heat carriers.

Currently, about 70% of wood waste in the form of sawdust, wood chips, pellets and briquettes is used as biofuel. Energy crops are some species of trees and plants that are specially grown for the production of solid biofuels. They are divided into three groups:

- fast-growing trees;
- perennial grasses;
- annual grasses.

Energy crops also include traditional agricultural plants grown for the production of biodiesel (rapeseed, sunflower), bioethanol (corn, wheat) and biogas (corn).

An effective way to supplement and replace traditional fuels is the production and use of biogas, which is formed as a result of anaerobic fermentation of organic biomass.

Biogas is a mixture of gases: methane, carbon dioxide, hydrogen sulfide, ammonia and other gases. Biogas can be obtained regardless of climatic and weather conditions. This type of fuel enables the production of both thermal and electrical energy, which makes it a competitive type of fuel.

Natural energy sources of the environment include atmospheric air, water of rivers, seas, topsoil and groundwater.

Thermal energy in the warm period of the year accumulates in the top layer of soil. This layer between the heating depth and the isothermal surface can be considered as a natural seasonal accumulator of thermal energy, and the energy that was used in the winter will accumulate in the warm period of the year. This also applies to groundwater located in the upper soil layers [15].

Geographical location, climatic conditions and the specifics of economic development of the region dictate the conditions for the prospects of using a particular type of alternative energy. Therefore, the research is based on scientific justification for the use of alternative energy sources for different geographical areas of Ukraine.

3 The Main Material

The purpose of the study is to create a map of rational alternative energy sources, taking into account the cost of equipment for the implementation of these measures and the geographical location of the region.

Ukraine occupies a large area and has a sufficient length from West to East, the climate in different regions can vary significantly. Therefore, there is a need to explore the prospects of using different alternative energy sources not for the country as a whole, but separately by region. The map is presented on the basis of the database of energy indicators of renewable energy sources and the distribution of their energy potential on the territory of Ukraine: 1. solar energy; 2. geothermal energy; 3. biomass energy; 4. biogas energy; 5. Environmental energy (topsoil and air potential) (Fig. 1).

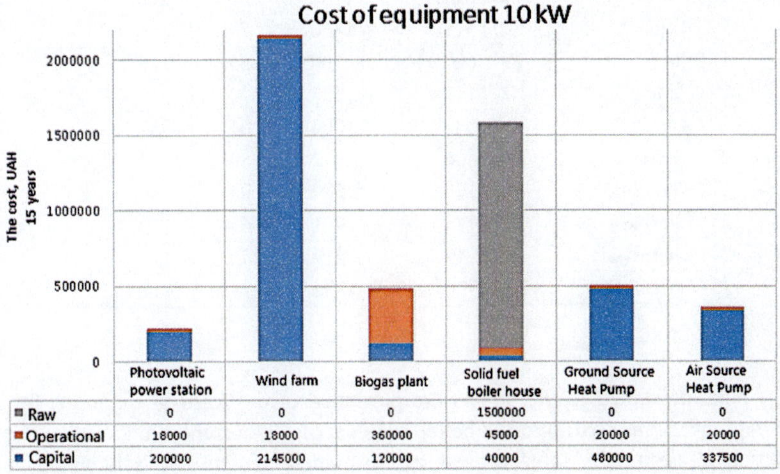

Fig. 1. Expenditures for the implementation of measures for the use of alternative energy sources, taking into account capital and operating costs

The calculation took into account the service life of 15 years.

Taking into account the capital costs for the implementation of measures for the production of alternative energy sources, the technically achievable energy potential for: solar energy, geothermal energy, biomass, biogas and air potential is presented.

The most efficient systems for obtaining energy are: 1. Solar power plant; 2. Biogas plant; 3. Ground heat pump; 4. Air heat pump (see Table 1).

Table 1. Technically achievable energy potential

	Regions	Sun	Geothermal	Biomass	Biogas	Air potential
		K, t_{oe}/year				
1	Crimea	270	775	827	111	280
2	Vinnytsia	170	217	1042	114	70
3	Volyn	130	168	348	73	84
4	Dnipro	220	266	1113	167	840
5	Donetsk	190	224	717	168	1029
6	Zhytomyr	180	252	779	73	84
7	Transcarpathian	100	596	144	56	56
8	Zaporizhzhia	200	252	932	78	350
9	Ivano-Frankivsk	90	123	190	76	84
10	Kyiv	180	245	873	187	861
11	Kropyvnytskyi	160	203	1019	61	140
12	Luhansk	190	224	670	86	406
13	Lviv	150	554	472	109	175
14	Mykolaiv	190	203	752	61	126
15	Odessa	260	284	946	99	231
16	Poltava	180	614	1158	90	210
17	Rivne	120	518	360	77	56
18	Sumy	150	600	700	55	126
19	Ternopil	100	119	465	70	42
20	Kharkiv	200	632	1000	101	567
21	Kherson	220	606	872	50	70
22	Khmelnytsky	140	175	689	94	84
23	Cherkasy	150	175	901	165	112
24	Chernivtsi	60	49	217	53	112
25	Chernihiv	200	326	836	73	112

The data, presented in the table were presented in the form of diagrams for individual regions, which will allow a better analysis of the current situation (see Fig. 2). The article presents the most typical cases, namely technically achievable energy potential for Lviv, Kyiv and Mykolayiv regions.

According to the obtained data, it is possible to identify the most appropriate energy sources for each region. Taking into account the cost of capital and operating costs, we obtain that: efficient energy sources for Ukraine are geothermal energy, air potential and solar energy, whereas, for example, wind power plants require significant investment, which leads to a longer payback period. However, the undisputed leader is solar energy.

The study proves that the use of certain alternative energy sources is a promising area of energy in Ukraine.

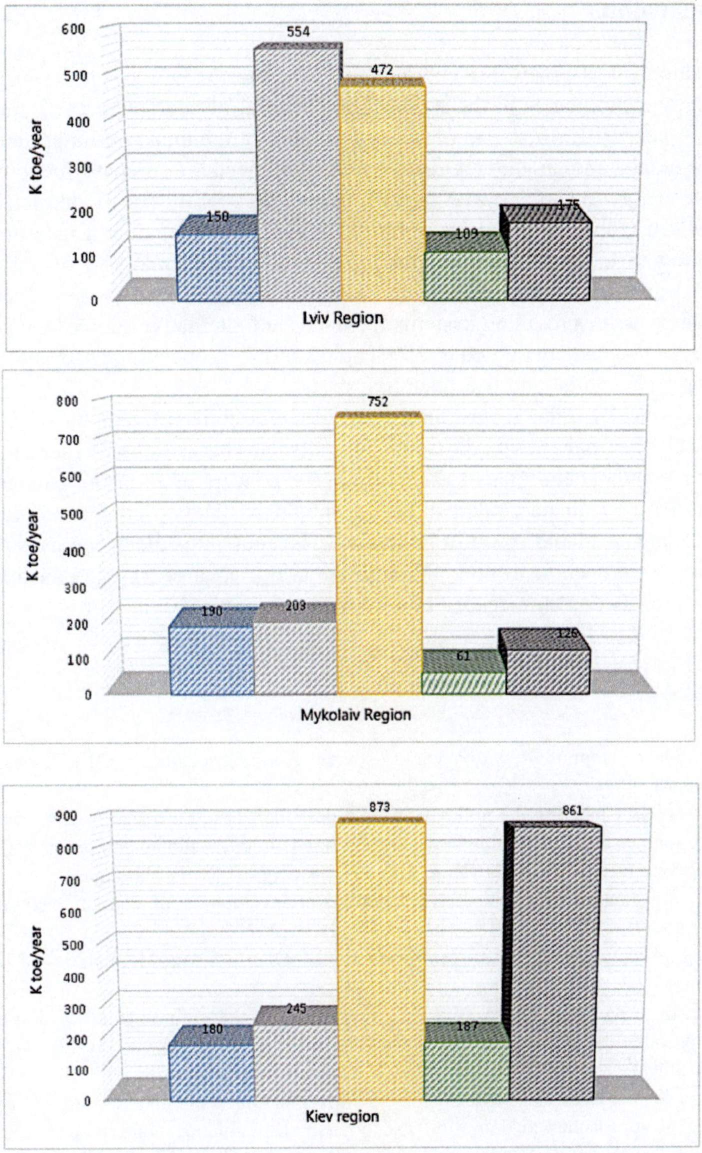

Fig. 2. Technically achievable energy potential for different regions of Ukraine ▧ – Sun energy; ▧ – Geothermal energy; ▧ – Biomass; ▧ – Biogas; ▧ – Air potential

4 Conclusions

The development of renewable energy shows the most dynamic development among other energy technologies in the world. The potential of alternative energy sources in Ukraine is presented, which can provide up to 50% of the total consumption of energy resources today, and in the future – 100%. The need to compare the potential of alternative energy in Ukraine with the cost of measures for the use of alternative energy sources is substantiated. The costs of funds for the implementation of measures for the use of alternative energy sources, taking into account capital and operating costs, determined that geothermal energy, air potential and solar energy are effective. The percentage costs for providing systems for the use of alternative sources are presented, so for the biogas plant amounted to 25% capital and 75% operating costs and for the air heat pump 94% capital and 6% operating costs.

The technically achievable energy potential for different regions of Ukraine is analyzed. Geothermal energy, air potential and solar energy deserve special attention. According to the peculiarities of each region, the concept of choosing efficient energy sources is defined, in particular, attention should be paid to comparing the cost of equipment and available resource potential. The most expedient energy sources for each region have been identified, although the use of solar energy and air heat pumps remains a priority for the territory of Ukraine.

References

1. Ye, S.: The concept of alternative energy sources A young scientist, no. 4(07), 44 (2014). (in Ukrainian)
2. Danilova, N.: Changing the policy of state regulation of the European market of alternative energy sources under the influence of current trends in international competition. Bull. Taras Shevchenko Nat. Univ. Kyiv Econ. **10**(162), 54–58 (2014). (in Ukrainian)
3. Ye, K.: Substantiation of perspective directions of development of alternative sources. Econ. Innov. no. 60, 217–225 (2015). (in Ukrainian). book 1
4. Holweg, M.: The genealogy of lean production. J. Oper. Manag. **25**(2), 420–437 (2007). (in Ukrainian)
5. Sokhatska, O.M., Strelbytska, N.Y.: Current trends in the global market of non-traditional and renewable energy sources. Energy saving. Energy. Energy Audit **11**(93), 38–52 (2011). (in Ukrainian)
6. Maistro, S., Voloshin, O.: Mechanisms of state regulation of alternative energy development: theoretical approaches to definition and content. In: Efficiency of Public Administration, Collection of Scientific Works, vol. 43, pp. 36–43 (2015). (in Ukrainian)
7. Matviychuk, L.Yu., Gerasymchuk, B.P.: Economic feasibility of using alternative energy sources. Econ. Forum (4), 12–16 (2013). (in Ukrainian). Lutsk National Technical University
8. Kaletnik, G.M., Pindyk, M.V.: The concept of alternative energy sources and their place in the implementation of energy efficiency policy of Ukraine. Econ. Financ. Manag. Curr. Issues Sci. Pract. **8**, 7–18 (2016). (in Ukrainian)
9. Onoshko, O.S.: Alternative energy in the system of economic security of Ukraine. Economics **1**, 32–39 (2013). (in Russian)

10. Stepanova, A.: Diversification of energy dependence of Ukraine. J. Taras Shevchenko Nat. Univ. Kyiv, Econ. **7**(172), 69–73 (2015). (in Ukrainian)
11. Geletukha, G.G., Zhelezna, T.A., Drozdova, O.I.: Analysis of the main provisions of the EU energy roadmap to 2050. Ind. Heat Eng. **34**(6) (2012). (in Ukrainian)
12. Voitko, S.V.: System analysis of energy security of countries: the use of renewable energy sources. Econ. Forum **4**, 29–35 (2013). (in Ukrainian)
13. Mysak, Y., Pona, O., Shapoval, S.: Evaluation of energy efficiency of solar roofing using mathematical and experimental research. Eastern Eur. J. Enterp. Technol. **3**(8(87)), 26–32 (2017)
14. Shapoval, S., Shapoval, P., Zhelykh, V.: Ecological and energy aspects of using the combined solar collectors for low-energy houses. Chem. Chem. Technol. **11**(4), 503–508 (2017)
15. Adamski, M.: Lean manufacturing: MathModelica in modeling of countercurrent heat exchangers. In: Proceedings - 8th EUROSIM Congress on Modelling and Simulation, EUROSIM 2013, vol. 21, no. 2, pp. 439–442 (2013)

Estimation of the Ecological Flow of Mountain River in Ukrainian Carpathians for Small Hydropower Projects

Svitlana Velychko(✉) and Olena Dupliak

Kyiv National University of Construction and Architecture, Kyiv 03680, Ukraine
velychko.sv@knuba.edu.ua

Abstract. In recent years the construction of small hydropower plants has increased in the mountainous part of the Carpathian region. The modern requirements for the ecological flow of rivers are significantly increased. There are no regulations for determining the minimum ecological flow for small mountain rivers in Ukraine. The magnitude of the minimum environmental flow significantly affects the assessment of the hydropower potential of the river. Ecological flow calculations are complicated by the lack of hydrological stations in the upper part of mountain rivers. The purpose of the work is development of a simplified engineering method for determining the minimum ecological flow under limited hydrological data for small mountain river of Ukraine. It was carried out the minimum ecological flow calculation for two sections of the Irshava River in the Tisza basin to estimate the proposed method. For comparison, it was used method of the minimum ecological acceptable flow, wetted perimeter method, BFM, IHA, Tennant. Based on the analysis of European and Ukrainian methods, the simplified engineering method for determining the minimum ecological flow is proposed: for the mountain river it is the mean annual flow of 90% probability.

Keywords: Ecological flow · Ecological acceptable flow · Minimum flow rate · Hydropower · Water reservoir · Hydrological regime

1 Introduction

The last decades in Ukraine the use of hydropower resources of small hydropower plants has increased by 22% and will continue to grow following the global trends in developing of renewable energy. One of the most important goals of the implementation of the Water Directive is to achieve a "good" ecological status of surface water bodies. Hydropower projects (even small ones) producing environmentally friendly energy, have a significant impact on the water bodies. Intensive construction of diversion-type hydropower plants by private investors in the Carpathian region and the lack of a unified regulatory for determining of the ecological flow for small rivers leads to violations in the field of environmental legislation by investors, on the one hand, speculations with intimidation of village communities about environmental damage from the introduction of any kind of planned activity in the water body.

Z. Blikharskyy (Ed.): EcoComfort 2020, LNCE 100, pp. 490–498, 2021.
https://doi.org/10.1007/978-3-030-57340-9_60

In the European Union each country has own methodology for minimum ecological flow calculation [1], some of them presented in Table 1. In general, the methods of assessing the environmental flow can be divided into: hydrological, hydraulic, assessment of habitats and holistic methods.

Table 1. Ecological flow according EU and neighbouring countries regulation.

Country	Ecological regulation
Belarus	75% of annual flow of 95% probability
France, North Macedonia, Spain	Not less than 0.1 of mean annual flow
Germany	From 1/3 to 1/6 mean minimum flow
Portugal	2.5–5% of mean annual flow
United Kingdom	Mean annual flow of 95% probability
Tennant method (USA)	30%/60% of mean annual flow

The hydrological method is determining of seasonal fluctuations. This method is easy to use, because it requires only a statistical analysis of the hydrological data. The oldest one is the Tennant method [2]. According to this method, the ecological flow, which is 10% of the mean annual flow is necessary to survive most forms of life, 30% of the mean annual flow is considered as a flow that creates favorable conditions for the development of living organisms and 60% - provides excellent conditions for the development of living beings and satisfying all needs. The Tennant method is improved by calculating different proportions of intra-annual flow [3].

Indicators of Hydrologic Alteration (IHA) is based on the hypothesis that not only certain quantities of water are needed for the development of the ecosystem, but also droughts, floods, periods of water shortage [4, 5]. It is necessary to analyze 32 flow parameters for each year of observation and determines an acceptable interval of variation between values that are within the range of 25%–75% probability.

Basic Flow method (BFM) is widely used in Spain [6]. BFM is based on studying the change of hydrological data and obtaining the base flow for each month applying the simple moving average model.

Hydraulic methods determine the relationship between the parameters of the flow and geometry of river channel. The simplest one is wetted perimeter method which described in article [7]. According this method the relationship between the flow and the wetted perimeter forms the break points, that characterize the necessary flood level for the river ecosystem. Obodovsky O. [8] determined term "ecologically acceptable flow" (EAF) as a flow that does not violate the hydromorphology of the channel: provide the sediment transit during floods and prevent siltation during low flow season. The hydraulic method for assessing ecological flow is undoubtedly one of the components of the environmental flow assessment, but these methods do not take into account water inhabitants demands.

Habitat simulation methods determine the relationship between water flow in the river and the habitat requirements for pre-selected species. Recommendations for implementation are developed in articles [9, 10].

Holistic methods [11–13] consist in combining different methods of ecological flow estimating into a single system: sediment transport, river channel morphology, bank stabilization, aquatic biotopes, tributaries bases, floods, runoff regime, conditions required for fish, habitat distribution, and socio-economic aspects of water use. Yatsyk A. proposed methodology for determining the ecological flow for small plain rivers with sandy sediments in Ukraine, depending on the limiting factors: water velocity, flood covered floodplain, pollution of rivers with sewage, oxygen content [14]. The application of this method to mountain rivers that are not contaminated with wastewater, leads to low flow and low levels of water in the river, and therefore not suitable for calculating ecological flow in mountain rivers.

The evolution of assessing ecological flow is directed towards the complication of calculations by the introduction of an increasing number of components, which may be justified in the implementation of large projects.

Our research was aimed to identify the main factors and propose the simplified engineering method of determining the ecological flow for the implementation of small hydropower projects on the example of the Irshava River.

2 Method

2.1 Study Area

The Irshava River is the right tributary of the Borzhava River in the Transcarpathian oblast of Ukraine. The length of the river is 48 km, the catchment area is 346 km^2, the average slope is 18 m/km. The valley of the Irshava River is a little twisty, V-shaped, the slopes are steep and convex. There are rapids in the upper reaches of the Irshava River, the bottom is uneven pebble-rocky, the depth of water is 0.1–0.6 m, the average velocity is 0.6–1.2 m/s. In this work the site near village Zagattia with catchment area of 51,1 km^2 were studied. The runoff is not regulated in the upper reaches of the river.

The materials of long-term observations (1955–1987) of the hydrometric station on the Irshava River near the Irshava town were used to calculate the main characteristics of the Irshava River runoff on the site, the methodology is described in our work [15]. Observations of the sediment runoff were not carried out on the Irshava River basin. The monitoring of the sediment runoff is carried out on the hydrometric station Shalanky below the confluence of the Irshava River in the Borzhava River which is geomorphologically similar to the Irshava River. Topographic material of scale 1: 2000 was used to plot cross-sections and to determine the river channel parameters. Settlements, industries are absent and environment has not undergone anthropogenic influence, water quality is defined as "excellent" above researched sites.

The mountainous part of the Irshava River has a fishery value, it is a natural habitat, a spawning place and a feeding place for the Ukrainian brook lamprey, brown trout, alburnoides, phoxinus, the barbus. Potential fish productivity of the Irshava River, taking into account all components of the feeding base (algae, invertebrates and various types of substrates), is in averages 12.1 kg/(ha·year) [16].

2.2 Estimation of Ecological Flow Using the Basis Flow Method (BFM)

The basic flows for each month of the year (from 1955 to 1987) were calculated using daily average flows and simple moving average model for each year. The ecological flow for each month was determined according the Eq. (1). If the flow in the river is less than the base flow (Q_b), then the base flow is assumed to be natural [6].

$$Q_{BMF,i} = Q_b \left(\frac{Q_{mean,i}}{Q_{min,i}} \right)^{0,5}, \tag{1}$$

where $Q_{BMF,i}$ – maintenance basic flows for each month of the year; Q_b – basic flows is average of the moving average values with the largest increment; $Q_{mean,\ i}$ – mean flow for the each month; $Q_{min,i}$ – minimum monthly flow.

2.3 Estimations of Ecological Flow Using the Indicators of Hydrologic Alteration (IHA)

The 32 components of the environmental flow for dry, medium and wet years were identified with IHA Software [5] based on the simulated daily average flows using 32 years of every day observation.

2.4 Calculation of Ecological Flow Using Hydraulic Methods

Breakpoint which characterized the critical value of the flow for the development of aquatic organisms, was defined using method of wetted perimeter. The section of the river between sites 1 and 2 was divided into four sections, and cross sections were plotted. The relationship between flow and wetted perimeter is plotted according to the equation of uniform flow in open channels. The value of the critical flow is determined at the first breakpoint by the method of maximum curvature.

Ecologically acceptable flow (EAF) which provides the movement of suspended sediments during the dry season and transport of suspended and bottom sediments in the period of flood, was determined for months with low flow (July-October and January) and for months with flood flow (April-June and November-December, February) according Eq. (2).

$$V_{ps} < V_{EAF}, \tag{2}$$

where V_{ps} – velocity preventing siltation, m/s; V_{EAF} – velocity corresponding to the passage of ecological acceptable flow, m/s.

3 Results

Ecological flow was calculated using different methods from the simplest as wetted perimeter to more complicated hydrological methods, the results are shown in Fig. 1. To compare the obtained ecological flow with natural river regime, mean annual flow (50%) is added to the Fig. 1.

The minimum ecological flow that correspond to critical state of productive biomass calculated by the wetted perimeter method is 0.6 m^3/s. Similarly result is obtained using Tennant method: for low flow is 0.4 m^3/s and for flood flow is 0.8 m^3/s. These two methods are the simplest with minimal amount of output data, but they define a constant flow during dry and flood seasons, as shown by the analysis of the hydrological regime of the river, it does not correspond to reality.

The analysis of the Fig. 1 shows that results are vary and repeat seasonal variation in different proportions. Calculated ecological flow by IHA method in mean year is higher by 50% than results of another methods, dry year variation more comparable with another methods.

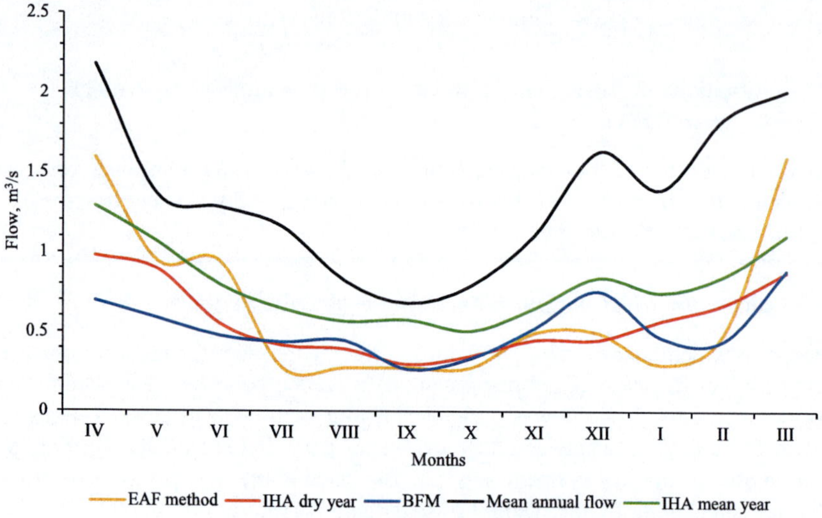

Fig. 1. Minimum ecological flow.

The EAF method, that based on the transport capacity of the river channel, well describes the necessary flushing by flood during the flood period (March-June), but gives low flow in the dry season. The weakness of this method is the lack of reliable measured data of sediment runoff over a long period, the monitoring data exist for stations with large catchment areas in the flat part of the rivers. Consequently, calculations using site-analogue are difficult and inaccurate.

During the low flow season ecological flow calculated by IHA method dry year, BFM and EAF method give similar results. But to ensure the bottom sediment transport during the flood period higher flow rate is required than calculated by hydrological methods, taking into account the method of the EAF.

The mountainous rivers of the Ukrainian Carpathians are mostly not regulated, the main task is to preserve their ecological status and maintain biodiversity. There are Brown trout, Danube lamprey, Common barbel in the upper reaches of the Irshava River, which impose additional restrictions on the preservation of ecological flow in the river. Unfortunately, studies of the impact of the flow and the parameters of the channel on the ichthyofauna and feed base in the Irshava River was not carried out. The studies performed on other Danube tributaries [17, 18] show that according to plotted Habitat Suitability Curves (HSCs), optimum depths for the Brown trout is depth from 0.2 to 0.6 m, and velocity from 0.2 to 0.7 m/s.

The analysis of the velocity characteristics of the ecological flow shows that all used methods satisfy the optimal requirements. Requirements of optimal depth is satisfied using IHA method for mean year (Fig. 2), other methods provide a depth of 15 cm in summer-autumn period.

The greatest difficulty in using IHA and BFM method is getting reliable mean daily discharge during long period (at least 20 years). Modelling daily flow takes a lot of time and the results are not reliable. In addition, the use of separate curves for dry, mean and wet years in practice cannot be realized because the absence of a hydrometric station above the hydropower station.

Therefore, in order to assess the hydropower potential of small mountain rivers, it is necessary to use a simplified method. For the considered sites on the Irshava River the mean annual flow of 90% probability may be used as ecological flow. This flow reflects the natural change of water level, provide the required depth (>0.2 m) and velocity for Brown trout in low flow period and the necessary flushing of the river channel during the winter and spring floods (Fig. 2).

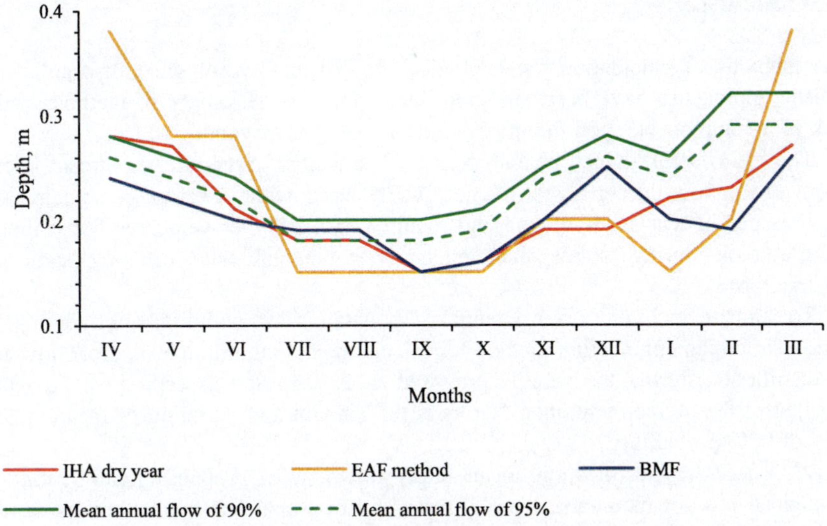

Fig. 2. The depths variation of the ecological flow.

Mean annual flow of 95% probability will be enough to transport of sediment in the low flow and flood periods, but the water depth in summer-autumn time does not exceed 18 cm. Although in work [19] for some mountain rivers with a small catchment area Habitat Suitability Curves were plotted with optimal depth of 15 cm. Therefore, the possibility of using mean annual flow exceeded 95% should be proved by ichthyological study with clarification of optimal fish habitat area.

If taken into account the economic component electricity output under different methods of ecological flow, the implementation of mean annual flow of 95% probability as ecological flow allows to increase production by 21% compared with the flow of 90%. So, the cost of ichthyological study will be fully justified by increasing of electricity output. Application of IHA method and mean annual flow of 90% probability allows producing virtually the same amount of energy, and therefore is economically equivalent. The largest production in the mean water year provides BFM due to using of runoff during the flood period.

4 Scientific Novelty and Practical Significance

Scientific novelty is to show authors' view on comprehensive approach to determine the main factors that should be taken into account when calculating ecological flow for the Transcarpathian mountain rivers in small hydropower projects. Practical significance of paper is to proposed simplified method to accesses ecological flow for small hydropower project with limits of hydrological data, that let more accurate hydropower capacity calculations.

5 Conclusions

The methods of calculation the ecological flow using hydrological, hydraulic and holistic approaches have been analyzed, indicating a wide variety of methods and a lack of a common view on this problem in the EU and beyond.

The main factors which characterize small mountain rivers are significant fluctuations of the flow during a year, higher quality water without sewerage contaminant, the presence of rare ichthyofauna and requirement of higher velocities for sediment transportation during floods than plain rivers and absence of long series of observations.

To estimate the hydropower potential of mountain rivers with limits of hydrological data, a simplified engineering method for calculating the minimum ecological flow and its distribution during the year is proposed. It is proposed as ecological flow that satisfied of the above-mentioned factors is the amount and variation of mean annual flow of 90% probability.

With necessary justification, mean annual flow of 95% probability can be used as ecological flow for mountain rivers, if research of optimal fish habitat area is carried out, that can significantly increase electricity output.

References

1. Blinkov, I.: New approach on environmental flow assessment of mountain streams in the Republic of Macedonia. Section of Natural. Math. Biotech. Sci. **35**(1), 57–66 (2014). https://doi.org/10.20903/csnmbs.masa.2014.35.1.55
2. Karimi, S, Yasi, M., Eslamian, S.: Use of hydrological methods for assessment of environmental flow in a river reach. Int. J. Environ. Sci. Technol. **9**, 549–558 (2012). https://doi.org/10.1007/s13762-012-0062-6
3. Li, C., Kang, L.: A new modified tennant method with spatial-temporal variability. Water Resour. Manag. **28**(14), 4911–4926 (2014). https://doi.org/10.1007/s11269-014-0746-4
4. Richter, B.D., Baumgartner, J.V., Powell, J., Braun, D.P.: A method for assessing hydrologic alteration within ecosystems. Conserv. Biol. **10**, 1163–1174 (1996). https://doi.org/10.1046/j.1523-1739.1996.10041163.x
5. The Nature Conservancy, Indicators of Hydrologic Alteration Version 7.1 User's Manual (2009). https://www.conservationgateway.org/Documents/IHAV7.pdf. Accessed 03 May 2020
6. Palau, A., Alcázar, J.: The Basic Flow method for incorporating flow variability in environmental flows. River Res. Appl. **28**, 93–102 (2012). https://doi.org/10.1002/rra.1439
7. Gippel, C., Stewardson, M.: Use of wetted perimeter in defining minimum environmental flows. Regul. Rivers Res. Manag. **14**, 53–65 (1998). https://doi.org/10.1002/(SICI)1099-1646(199801/02)14:1<53::AID-RRR476>3.0.CO;2-Z
8. Obodovskyi, O.H.: Hydrological and ecological assessment of river processes (on the example of rivers in Ukraine), vol. 274, Kyiv (2001). (in Ukrainian)
9. Bovee, K.D.: Development and evaluation of habitat suitability criteria for use in the Instream Flow Incremental Methodology, vol. 235, Washington, D.C (1986)
10. Dunbar, M.J., Alfredsen, K., Harby, A.: Hydraulic-habitat modelling for setting environmental river flow needs for salmonids. Fish. Manag. Ecol. **19**, 500–517 (2012). https://doi.org/10.1111/j.1365-2400.2011.00825.x
11. Zeiringer, B., Seliger, C., Greimel, F., Schmutz, S.: River hydrology, flow alteration, and environmental flow. In: Riverine Ecosystem Management. Aquatic Ecology Series, vol. 8, pp. 67–89. Springer, Cham (2018). https://doi.org/10.1007/978-3-319-73250-3_4
12. King, J., Brown, C., Sabet, H.: A scenario-based holistic approach to environmental flow assessments for rivers. River Res. Appl. **19**, 619–639 (2003). https://doi.org/10.1002/rra.709
13. Poff, N., Richter, B., Arthington, A., Bunn, S., Naiman, R., Kendy, E., Warner, A.: The ecological limits of hydrologic alteration (ELOHA): a new framework for developing regional environmental flow standards. Freshw. Biol. **55**, 147–170 (2009). https://doi.org/10.1111/j.1365-2427.2009.02204.x
14. Yatsik, A.V.: Ecologic Foundations for Rational Water Management, vol. 640. Henesa, Kiev (1997). (in Russian)
15. Velychko, S., Dupliak, O.: Assessment of flow parameters the Irshava River in the absence of observations on the site. In: STC Problems of Water supply, Sewerage and Hydraulic, vol. 31, pp. 15–24 (2019)
16. Ustych, V.I.: The ichthyofauna of the Irshava River and its recovery strategy. Doctoral dissertation. Kyiv, vol. 19 (2011). (in Ukrainian)
17. Vismara, R., Azzellino, A., Bosib, R., Crosa, G., Gentili, G.: Habitat suitability curves for brown trout (salmo trutta fario l.) in the river Adda, northern Italy: comparing univariate and multivariate approaches, regulated rivers. Res. Manag. **17**, 37–50 (2001). https://doi.org/http://doi.org/10.1002/1099-1646(200101/02)17:1<37::AID-RRR606>3.0.CO;2-Q

18. Stefunková, Z., Belcakova, I., Majorosova, M., Skrinar, A., Vasekova, B., Neruda, M., Macura, V.: The impact of the morphology of mountain watercourses on the habitat preferences indicated by ichtyofauna using the IFIM methodology. Appl. Ecol. Environ. Res. **16**(5), 5893–5907 (2016). https://doi.org/10.15666/aeer/1605_58935907
19. Macura, V., Stefunkova, Z., Majorosova, M., Halaj, P., Skrinar, A.: Influence of discharge on fish habitat suitability curves in mountain watercourses in IFIM methodology. J. Hydrol. Hydromech. **66**(1), 12–22 (2018). https://doi.org/10.1515/johh-2017-0044

Influence of Orientation of Buildings Facades on the Level of Solar Energy Supply to Them

Vasyl Zhelykh, Pavlo Shapoval, Stepan Shapoval(✉), and Mariana Kasynets

Lviv Polytechnic National University, Lviv 79013, Ukraine
shapovalstepan@gmail.com

Abstract. Energy-efficient building forms take into account the fundamentals given for passive houses and zero energy houses. One of them is to account for the amount of solar energy received on the facade surface. The paper describes the study of the influence of the orientation of buildings facades on the level of solar energy supply to them. The paper presents theoretical dependencies for determining the amount of solar energy input to the facade surface due to its orientation relative to the sides of the horizon. In addition to the appropriate orientation of houses for the use of solar energy, the paper considers the need to use this energy by equipment on walls or windows. It was calculated that the difference in solar energy received by the surface of the Southern, South-East and South-West is no more than 5%. The paper suggests approximation equations that allows to calculate the amount of solar energy received on the different orientation facade surface.

Keywords: Solar energy · Solar collector · Solar heat supply system · Building facade · Heating period

1 Introduction

The modern level of science and technology development makes it possible to effectively manage all components of modern buildings, in particular energy supply systems. The use of energy saving technologies to improve the energy characteristics of a building is a promising development direction [1]. For buildings, taking shape into account takes an important place in the design. Therefore, scientists and engineers are looking for new ways to implement effective solutions in construction. Such a solution, for example, is Smart Energy Networks (SEN). SEN this is a combination of renewable energy sources with cogeneration systems that are highly efficient. This solution allows you to integrate unconventional technologies into large-scale applications during construction [2–4].

It should be mentioned that one of the first energy-efficient buildings in Ukraine was built in Kiev taking into account the fundamental principles of energy-efficient building forms according to orientation on the sides of the horizon. In Ukraine, a number of projects for the construction of environmental, energy-efficient buildings have been implemented [5].

Z. Blikharskyy (Ed.): EcoComfort 2020, LNCE 100, pp. 499–504, 2021.
https://doi.org/10.1007/978-3-030-57340-9_61

Solar energy is a powerful source of life on Earth. Therefore, a number of works draw attention to the amount of energy received by this type of energy on the earth's surface. In works [6] is described the method developed for determining the potential of solar energy in large cities based on the typology construction. As a result of this method, the solar heat supply system on the roof of the building can be calculated taking into account shading from obstacles and with mounting distances.

Solar energy reaches the atmosphere in a directed flow, but the Earth's surface receives both direct flow and scattered atmospheric radiation. Solar radiation falling normally on the Earth's surface changes due to: changes in the distance between the Earth and the Sun; atmospheric dispersion of air, water vapor and dust by molecules; atmospheric absorption by oxygen, ozone, water and carbon dioxide [7]. The receipt of solar radiation on the surface of an energy-efficient form should be taking into account such moments.

2 Objectives the Formulation of the Problem

Improving the energy performance of the building is possible by implementing solutions at the design stage to account for the amount of solar energy received on the facade surface. In this regard, the actual direction of research is to determine the amount of solar energy input to the facade surface.

3 The Analysis of Recent Research and Publications

One of the publications that optimizes the geometric parameters of energy-efficient buildings is the thesis of V. Martynov. The paper provides recommendations on the use of polar diagrams in solving problems of designing the thermal insulation shell of buildings [8].

In practice, installations that are part of the construction of an external fence become popular [9]. The model shown in the work [10] is the structure is designed for a part of the building and is recommended for installation on the northern and southern parts of the coating. The similar design has the wall for heating the house described in the work [11]. Patented is the design of the solar panel of the house, which contains the partition, the circulation channel, thermal insulation made in the form of the conical frame with the base that is beveled towards the lower opening of the partition [12]. The solar wall structure described by Charlene Riegger has the perforated metal plate. The wall solar collector is effective on cloudy days, although on a smaller scale [13]. The design of the house is presented in the patent [14], which reduces heat loss due to horizontal channels. These horizontal channels are placed in the ground under the floor and others with North and South orientation in the house structure. The disadvantage of this design is the high cost.

It is worth noting that all the above-mentioned installations for converting solar energy require the preliminary calculation of the receipt of solar energy on the surface of the facade. As well as choosing among them the most optimal design for the particular region of construction.

4 The Main Material

Due to the movement of the Earth around the axis and around the Sun, the supply of solar energy is uneven and depends on many factors. Therefore, there is a need to optimally orient the solar collectors, that is, to find such optimal angles at which the maximum possible amount of solar energy will be obtained.

The maximum annual amount of direct solar energy is achieved when the surface is tilted $\beta = 0{,}9\ \varphi$. Moreover, the surface should be oriented to the equator with the slope for summer $\varphi + 10^{o}$ and for winter $\varphi - 10^{o}$. It is known about the results of studies of the influence of the azimuthal angle γ of irradiation on the surface of solar energy. Furthermore, when $\gamma = 0^{o}$ and $\gamma = 22.5^{o}$ for latitude to $\varphi = 45^{o}$ the difference in relative annual exposure differs only by 2%. It is also indicated that every 15° the azimuth angle causes the shift in the daily distribution of solar energy by about 1 h in the direction of the morning hours, if γ there is a positive, and in the afternoon, if γ is negative [7].

The average monthly value of the total solar energy received on the horizontal surface was calculated taking into account the normative data on the receipt of solar energy on the Ukraine territory. Within Ukraine this value is approximately equal 345 MJ/m^2 (Fig. 1).

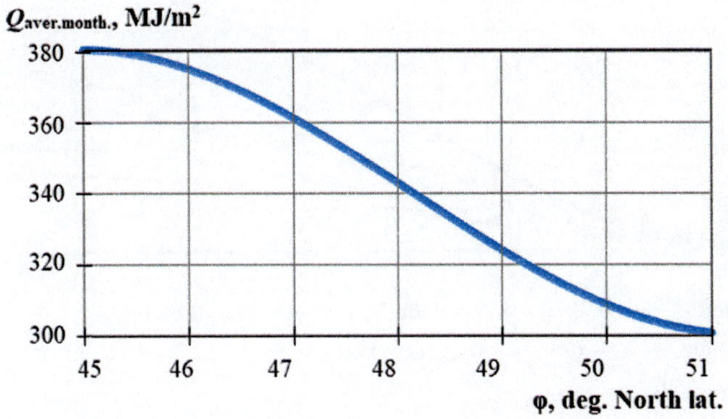

Fig. 1. The average monthly total solar energy $Q_{\text{aver.moth.}}$, MJ/m^2, received on a horizontal surface on the Ukraine territory

Based on the data in Fig. 1, was obtained the Eq. (1):

$$Q_{\text{aver. moth.}} = 0{,}6389 \cdot \varphi^3 - 92{,}2857 \cdot \varphi^2 + 4{,}4243 \cdot 10^3 \cdot \varphi - 7{,}0054 \cdot 10^4, \mathrm{J/m^2} \tag{1}$$

where, $Q_{\text{aver. moth.}}$ – annual average monthly total solar energy, MJ/m^2; φ – the value of latitude, deg. North. lat.

According to the integral estimation of the amount of solar energy, it is advisable to separate its receipt on the horizontal surface seasonally and in the heating period. After all, installing solar collectors on the facades of buildings requires an additional assessment of the solar energy amount to be able to use them during the heating period. As a result of analysis of seasonal fluctuations of solar energy on the territory of Ukraine for a year, the total solar energy increases most in the spring and summer season. In the winter season, the minimum total solar energy is 87 MJ/m^2.

The total solar energy is sufficient to partially cover the load on fuel and energy costs during the heating period. For example, at latitude 51, the average monthly total solar energy during the heating period is 164 MJ/m^2, which is 26% less than the 45th geographical latitude of Ukraine.

In autumn, the significant decrease in direct solar radiation from September to November leads to the decrease in total solar energy by up to 45%. The summer season is characterized by the increase in the receipt of direct solar radiation by about 10 times in relation to the winter period. Therefore, the efficiency of flat solar collectors increases during the summer period.

The amount of total solar radiation (direct and scattered) received on the surface during the year for Ukrainian cities was obtained (Fig. 2).

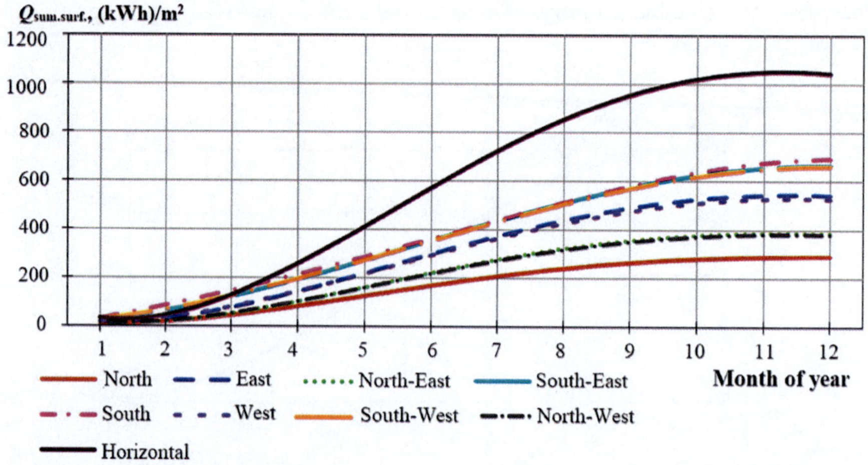

Fig. 2. Amount of total solar radiation $Q_{sum.surf.}$, kWh/m^2, which comes on the surface of different orientation for Kiev

It is analyzed that the horizontal surface receives more than 1000 kWh/m^2. The surface of the Southern orientation receives the same amount of total solar energy as the surfaces of the Southwest and Southeast, and the difference between them is no more than 5%.

On the basis of Fig. 2 approximation equations of solar radiation input on the surface of different orientation are obtained. Also, on the example of the city of Kiev, the formula for the total amount of solar radiation received on vertical and horizontal surfaces is obtained.

$$Q_{\mathrm{North}} = -1{,}26 \cdot x^2 + 45{,}77 \cdot x - 59{,}38 \quad (2)$$

$$Q_{\mathrm{North-East}} = -1{,}63 \cdot x^2 + 61{,}22 \cdot x - 87{,}09 \quad (3)$$

$$Q_{\mathrm{East}} = -1{,}86 \cdot x^2 + 80{,}01 \cdot x - 110{,}02 \quad (4)$$

$$Q_{\mathrm{South-East}} = -1, 61 \cdot x^2 + 85, 61 \cdot x - 94, 35 \quad (5)$$

$$Q_{\mathrm{South}} = -1{,}21 \cdot x^2 + 80{,}68 \cdot x - 68{,}26 \quad (6)$$

$$Q_{\mathrm{South-West}} = -1{,}58 \cdot x^2 + 84{,}33 \cdot x - 88{,}84 \quad (7)$$

$$Q_{\mathrm{West}} = -1{,}87 \cdot x^2 + 77{,}9 \cdot x - 102{,}57 \quad (8)$$

$$Q_{\mathrm{North-West}} = -1{,}59 \cdot x^2 + 59{,}93 \cdot x - 84{,}2 \quad (9)$$

$$Q_{\mathrm{Horizontal}} = -3{,}7 \cdot x^2 + 157{,}65 \cdot x - 233{,}05 \quad (10)$$

$$Q_{\mathrm{Sum.+Horizontal}} = -16{,}3 \cdot x^2 + 733{,}11 \cdot x - 927{,}75 \quad (11)$$

where, x – serial number of the month (from 1 to 11).

Approximation Eqs. (2)–(11) allows to calculate the amount of solar energy received on the facade surface. In addition, equations could calculate the amount of energy in the month during which the installation will account for the greatest generation of energy, depending on the orientation of the facade relative to the sides of the horizon.

5 Conclusions

The maximum annual amount of direct solar energy is achieved when the surface is tilted $\beta = 0{,}9\ \varphi$. The surface should be oriented to the equator with the slope for summer $\varphi + 10^{\circ}$ and for winter $\varphi - 10^{\circ}$.

It is advisable to calculate the amount of solar energy on the surface during the season and during the heating period. It was found that direct and scattered solar radiation is most important in the spring-summer season as the result of the analysis of post-season fluctuations in solar energy on the Ukraine territory. In the heating period, the total solar energy is 60% more than in winter and sufficient to partially reduce the use of traditional energy sources.

The approximation equations described in this paper allows to calculate the amount of energy received on the facade surface. If the building's facade contains the solar collector, this will allow to take into account the seasonal operation of the solar collector.

References

1. Andreas, K.: NSERC smart net-zero energy buildings strategic research network. SNEBRN Newslett. **1** (2012)
2. Lund, H., Andersen, A., Ostergaard, P., Mathiesen, B., Connolly, D.: From electricity smart grids to smart energy systems – a market operation based approach and understanding. Energy **42**, 96–102 (2012)
3. Mathiesen, B., Lund, H., Connolly, D., Wenzel, H.: Smart energy systems for coherent 100% renewable energy and transport solutions. Energy **145**, 139–154 (2015)
4. Sig Chai, D., Wena, J., Nathwani, J.: Simulation of cogeneration within the concept of smart energy networks. Energy Convers. Manag. **75**, 453–465 (2013)
5. Synytsia, S.: Energy efficiency in Germany-opportunities for Ukraine. Friedrich Ebert Stiftung, 21 (2010). (in Ukrainian)
6. Horvath, M., Kassai-Szoó, D., Csoknyai, T.: Solar energy potential of roofs on urban level based on building typology. Energy Build. **111**, 278–289 (2016)
7. Daffi, D.zh., Bekman, U.: Thermal processes using solar energy. Translated from English by the editor Malevskyi Yu. Myr, 420 (1977). (in Russian)
8. Martynov, V.: Dissertation of doctor of science «Optimization of geometric parameters of energy-efficient buildings» (2013). (in Ukrainian)
9. Shapoval, S.: The potential of solar energy in Ukraine. In: Litteris et Artibus: Proceedings of the 5th International Youth Science Forum, pp. 122–123. Lviv Polytechnic National University (2015)
10. Patent 1288459. Solar installation for heating of the building (1987). (in Russian)
11. Patent 1333995. Solar system for building air heating (1987). (in Russian)
12. Patent 1444594. Solar panel of the building (1988). (in Russian)
13. Riegger, C.: Transpired solar collector walls: use solar, save green. Interface, 5–8 (2008)
14. Patent 1451480. Building with solar heating system (1989). (in Russian)

Methodology for Calculating the Composition of Fine-Grained Concrete with High Resolution

Vadim Zhitkovsky(✉), Leonid Dvorkin, Vitaliy Marchuk, and Mykhailo Fursovych

National University of Water and Environmental Engineering, Rivne, Ukraine
v.v.zhitkovsky@nuwm.edu.ua

Abstract. The article presents the methodology of designing the composition of fine-grained concrete, taking into account the method of compaction, the parameters of the aggregate and the binder obtained on the basis of a complex of experimental studies. The results of studies of the effect of the type of concrete mixtures and their method of compaction on the strength of fine-grained concrete and the relationship of strength with water-cement ratio are presented. An original method of determining the thickness of a cement paste film on aggregate grains is proposed, which allows one to calculate the ratio between the components of a concrete mix based on the physical model of the concrete structure. The developed calculation methodology can be used both on computer systems and manually thanks to the built nomograms. This methodology is suitable for practical application in determining the composition of concretes for fine-walled and thin-walled products, as well as for thick-reinforced, reactive-powder concretes and fiber concrete.

Keywords: Fine-grained concrete · Design · Composition · Strength · Compacting

1 Introduction

A significant increase in the static and dynamic properties of concrete is observed in the case of replacing concrete with a large aggregate with fine-grained [1–3]. In such concretes, the size of the aggregate does not exceed 10 mm and provides much higher homogeneity [4], deformability and crack resistance [5].

For fine-grained concrete is characterized by an increased ratio of tensile strength in bending to compressive strength [6]. With the same compressive strength, bending strength for fine-grained concrete is 10…15% higher than conventional ones. Accordingly, the indicators of dynamic and impact strength and endurance of concrete increase. This is due to the greater homogeneity of the structure of fine-grained concrete [7].

The high specific surface of the aggregate in concrete leads to an increased (by 20…40%) consumption of cement, which is necessary to fill inter-granular pores and create a sufficient coating of cement paste [8]. Reducing cement consumption is achieved by choosing the optimal particle size distribution of the aggregate, the introduction of active mineral additives and micro-fillers, the use of superplasticizers

Z. Blikharskyy (Ed.): EcoComfort 2020, LNCE 100, pp. 505–513, 2021.
https://doi.org/10.1007/978-3-030-57340-9_62

and effective methods of compaction. Yu.M. Bazhenov [9] proposed the dependence of the strength of sand concrete in the form of an empirical Eq. (1):

$$f_{cm} = AR_c\left(\frac{C}{W + V_{e.a.}} - 0.8\right), \quad (1)$$

where C is the cement consumption, W is the water consumption, $A = 0.8$ for high quality materials, 0.75 and 0.65 for medium and low quality, respectively; $V_{e.a}$ is the entrained air volume.

It has been shown that for each FGC component there is an optimal value of W/C, providing the highest strength values. At low values of W/C (W/C $\leq$ 0.4) the concrete strength varies linearly depending on the cement - sand ratio (C/S).With a decrease of C/S below the optimal values, the FGC mixtures workability is reduced, with increasing C/S the amount of excess water in concrete is higher, which also leads to a decrease in strength.

Numerous experimental data show that the compressive strength of fine-grained concrete, in addition to C/W, cement strength and aggregate quality, is influenced by many other factors, such as workability of the mixture, conditions of concrete hardening, the presence and amount of active mineral additives, etc. In addition, the method of compaction of the mixture also has a significant effect on the properties of fine-grained concrete [10].

For mixtures whose stiffness cannot be determined by conventional methods (ultra-rigid or semi-dry (bulk) mixtures), as well as for concrete mixtures that are compacted by special force methods, the volume of trapped air depends on the parameters and features of a particular method of compaction. In the last years fine-grained concrete (FGC) is more and more used for high-slump, pressed and vibro-pressed elements.

Numerous experimental data show that fine-grained concrete compressive strength depends on cement-water ratio C/W, cement strength and aggregate quality as well as many other factors, like mixture workability, hardening conditions, type and quantity of mineral admixtures, etc. Additionally, the mixture compaction type has also a significant influence on fine-grained concrete properties [10, 11].

2 Aim and Scope of the Research

The aim of current research is to develop a method of designing the composition of fine-grained concrete, which takes into account the structural features of this type of concrete, the concrete mixture workability and the influence of the method of concrete mix compaction on the strength.

3 Materials and Methods

The development of a method for designing the composition of fine-grained concrete was performed by mathematical processing of experimental data obtained by the authors, as well as data from other authors in the literature [10, 11]. On the basis of the received regression dependences the technique of calculation of structure is offered and the algorithm of its application is developed.

4 Results and Discussion

Numerous experimental data show that the strength of fine-grained concrete under compression, besides C/W, cement strength and aggregate quality, is influenced by many other factors, such as workability of the mixture, concrete hardening conditions, the presence and amount of active mineral additives and etc. [7, 9, 11]. Along with this, the method of compaction of the mixture also has a significant effect on the properties of fine-grained concrete.

For calculating the fine-grained concrete compressive strength at 28 days (f_{cm}) it is possible to use a general Eq. (2) [12]:

$$f_{cm} = AR_c(C/W - b), \tag{2}$$

where A and b – empirical coefficients.

Analysis of experimental data obtained by the authors (Table 1), and by other researchers [10, 11] (Fig. 1) enables to propose average values of coefficient A and b in Eq. (2) (Table 2) taking into account the concrete mixtures workability.

It should be considered at calculating the fine-grained concrete mixture composition that after its compaction a certain volume of air always remains in concrete. The quantity of entrained air is determined by the features of specific air entraining admixtures. A certain volume of air remains due to concrete mixture under-compaction (entrapped air $V_{e.a}$). Approximating the available data [11] it is possible to propose an expression for calculating the volume of air, entrapped in fine-grain concrete mixtures (l):

- for plastic mixtures:

$$V_{e.a} = -6.52\, ln(Sl + 1) + 19.9 \tag{3}$$

- for stiff mixtures:

$$V_{e.a} = 24.95\, ln(Vb + 1) - 8.3 \tag{4}$$

where Sl and Vb are the concrete mixture Slump (cm) and Vebe time (sec).

For semi-dry concrete mixes the air volume significantly depends on the method and parameters of the mixture compacting. Based on the obtained experimental data [13], a model describing of the compacting parameters effect on the residual air amount was compiled and a corresponding nomogram was constructed (Fig. 2).

Table 1. Effect of cement-water ratio and aggregate quality on FGC strength

Aggregate type	Concrete mixture workability	Cement-water ratio						
		1.5	1.75	2	2.25	2.5	2.75	3
		FGC compressive strength at 28 days, MPa						
Quartz sand $M_f^1 = 2.5$; $CSD^2 = 1\%$	Sl = 10 cm	18.8	19.9	26.3	34.6	38.9	–	–
	Vb = 17 s	19.2	27.1	29.4	33.3	43	45.8	–
	__3	30	29.2	35.4	44	51	51.5	58.2
Quartz sand ($M_f = 1.6$; CSD = 3%)	Sl = 10 cm	14	17	25.1	29.8	37	–	–
	Vb = 17 s	18	21.3	27.7	35.1	36.3	44.6	–
	__3	27.7	29.7	32	40.8	45.3	48.9	53.7
Quartz sand ($M_f = 0.8$; CSD = 9%)	Sl = 10 cm	12.1	20.4	20.5	25.5	32.5	–	–
	Vb = 17 s	19.4	22.3	24.5	31.8	32.9	38.6	–
	__3	21.1	25.8	33	35	43.8	47.5	49.2

Note: 1- M_f - fineness modulus; 2- CSD - content of clay, silt and dust; 3 the specimens were produced using semi-dry concrete mixture (humidity 6… 7%) by vibro-pressing (amplitude = 0.5 mm, frequency = 50 Hz, duration = 20 s, pressure = 0.06 MPa).

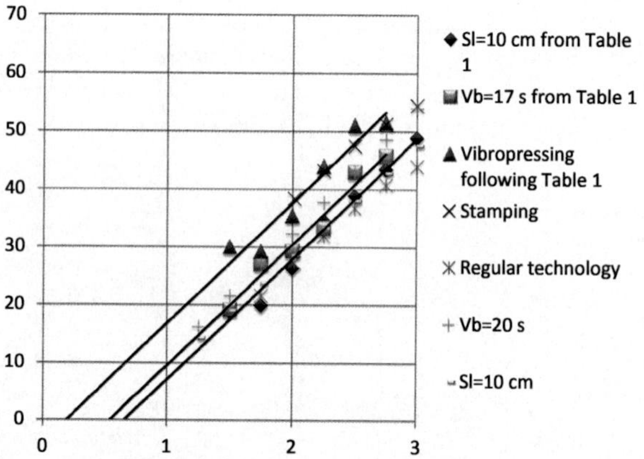

Fig. 1. Dependence of fine-grained concrete strength on C/W: *1* – A = 0.52, b = 0.2; *2* – A = 0.52, b = 0.55; *3* – A = 0.52, b = 0.65.

Table 2. Coefficients A and b in Eq. (2) for calculating fine-grained concrete strength

Aggregates quality	Plastic concrete mixtures	Stiff concrete mixtures	Semi-dry concrete mixtures
High	A = 0.52 b = 0.65	A = 0.52 b = 0.55	A = 0.52 b = 0.2
Average	A = 0.48 b = 0.65	A = 0.48 b = 0.55	A = 0.48 b = 0.2
Low	A = 0.44 b = 0.65	A = 0.44 b = 0.55	A = 0.44 b = 0.2

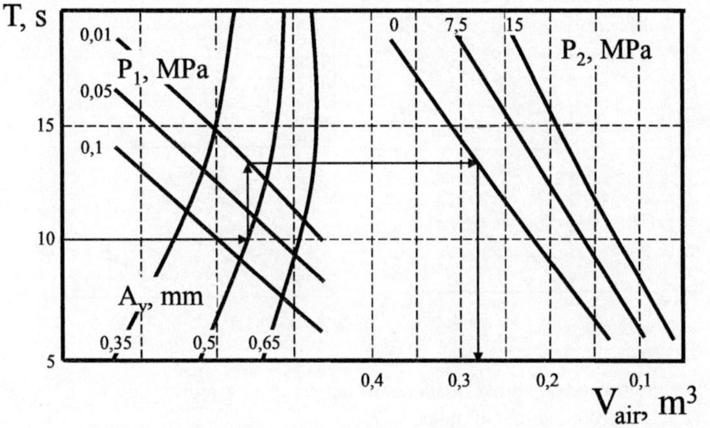

Fig. 2. Nomogram for determining the residual air volume in vibro-pressed concrete: T – compacting duration, sec.; A_v – vibration amplitude, mm; P_1 – load at vibration, MPa; P_2 – compaction pressure, MPa; (vibration frequency is 50 Hz)

Consumptions of all fine-grained concrete mixture components are related by the following expression:

$$V_{cem.p} + V_s + V_{e.a} = 1000l, \quad (5)$$

where $V_{cem.p}$ - cement paste volume, l; V_s - sand volume, l.

The cement paste quantity should be enough to fill all voids between sand grains and form on their surface a film having a certain thickness δ.

It is possible to write the following condition:

$$V_{cem.p.} = Vd_s \frac{S}{\rho_s^b} + \delta S_s \frac{S}{\rho_s^b}, \quad (6)$$

where Vd_s – voidage of aggregate (sand); δ –cement paste film thickness, μm; S_s – specific surface of the aggregate, m^2/m^3; ρ_s^b –aggregate bulk density, kg/m^3; S – aggregate consumption, kg/m^3.

Analysis of known experimental data [12, 14] enables to argue that the film thickness δ depends on concrete mixture workability (*Vb* or *Sl*), *C/W*, specific surface of the aggregate (S_s), aggregates voidage in bulk state (Vd_s) and volume of voids between aggregate grains not filled by cement paste ($V_{n.f}$).

It is possible to find the thickness of the cement film on aggregate grains using nomograms given in Figs. 3, obtained based on experimental data [11].

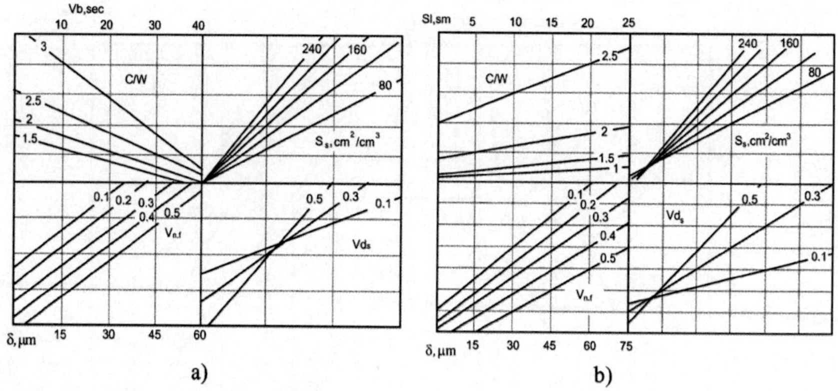

Fig. 3. Nomograms for obtaining thickness of the cement paste film on the aggregate grains in stiff (a) and slamp (b) concrete mixtures.

Specific aggregate surface (S_s, cm^2/cm^3) can be obtained with relatively high accuracy using the following Eq. (7) [12]:

$$S_s = \frac{6.35k}{1000}(0,5a + b + 2c + 4c + 8e + 16f + 32g)\rho_s^b, \tag{7}$$

where k is a coefficient, depending on type of sand; $a...f$ – residues on sieves with holes from 5 to 0.16 mm, %; g – quantity of aggregate passing through sieve with holes 0.16 mm, %.

The value of $V_{n.f}$ characterizes the lack of cement paste for filling voids between grains aggregate and entrapped air presence.

As a first approximation $V_{n.f}$ can be assumed to be equal to the entrapped air volume $V_{e.a}$, however if the obtained value δ will be less than 13 μm, $V_{n.f}$ should be increased, until the condition ($\delta \geq$ 13 μm) will be fulfilled.

Nomograms shown in Fig. 3 are valid at using Portland cement with cement paste normal consistence $N.C$ = 28%, at $N.C$ < 28% δ should be decreased, and at $N.C$ > 28% it should be increased by 5% per 1% of $N.C$ change [11].

Decrease of $V_{n.f}$ value can be achieved by increasing the cement paste quantity using dispersed filler (for example ash or granite dust [7, 13]).

Consumptions of fine-grained concrete components can be found by solving a system of equations:

$$\begin{aligned} V_{cem}V_{cem} + V_w + V_s + V_{e.a.} &= 1000 \\ V_{cem} + V_w + V_{cem.p} &= Vd_s\frac{S}{\rho_s^b} + \delta S_s\frac{S}{\rho_s^b}, \end{aligned} \quad (8)$$

where V_{cem} is volume of cement, m^3; V_w – volume of Water, m^3.

From Eq. (8) consumptions of aggregate and cement are obtained:

$$S = \frac{1000 - 1000 \cdot V_s}{\frac{Vd_s + \frac{\delta S_s}{10^6} - V_{n.f}}{\rho_s^b} + \frac{1}{\rho_s}}, \quad (9)$$

$$C = \frac{1000 - 1000 \cdot V_{e.a} - \frac{S}{\rho_s}}{\frac{1}{\rho_{cem}} + \frac{W}{C}}, \quad (10)$$

where ρ_s and ρ_{cem} are true densities of aggregate and cement.

Water consumption is found from the following conditions:

$$W = C.\ W/C. \quad (11)$$

Numerical example. Design a FGC composition for producing decorative elements. 28-day concrete compressive strength is 25 MPa; Vebe time of concrete mixture Vb = 10 s.

Initial materials: Portland cement (28-day strength R_{cem} = 50 MPa, true density ρ_{cem} = 3.1 g/cm^3, cement paste normal consistency $N.C$ = 28%); quartz sand (specific surface S_s = 210 cm^2/cm^3, fineness modulus M_f= 1.4, true density ρ_s = 2.65 g/cm^3, bulk density ρ_s^b = 1.43 g/cm^3, voidage Vd_s = 0.46).

1. According to the Eq. (2), taking the values of the coefficients A = 0.44, b = 0.55 in the Table 2 (low quality material), we find the necessary C/W.
2. By the Eq. (4) we find the volume of entrapped air.
3. Using the nomogram (Fig. 3,a) we find the conditional film thickness δ taking in a first approximation the value $V_{n.f} = V_{e.a}$. Due to the fact that with this approximation δ less than the conditionally minimum (13 μm), we take δ = 13 μm and from the same nomogram we find the real value of $V_{n.f}$.
4. Using the Eq. (9), (10), (11), we determine the consumption of sand, cement, water, respectively.

The calculated concrete composition:

$$C/W = 1.91;\ V_{e.a} = 0.0518\,\mathrm{m}^3;\ V_{n.f} = 0.373\,\mathrm{m}^3;$$

$$S = 1507\,\text{kg/m}^3;\ C = 438\,\text{kg/m}^3;\ W = 230\,\text{kg/m}^3.$$

The calculated composition is subject to laboratory clarification.

5 Scientific Novelty and Practical Significance

The presented methodology allows for the first time to calculate the composition of fine-grained concrete taking into account the influence of the most common methods of manufacture on the strength, structural features of concrete and the ratio between the components. The developed methodology and algorithm for designing the composition of fine-grained concrete can be effectively used in the development of software for managing enterprises that manufacture decorative, small-piece and thin-walled concrete products

6 Conclusions

The developed methodology for dispensing fine-grained concrete has a number of features compared to the existing ones:

- at selecting the required cement-water ratio the concrete mixture workability, determining the way of concrete products molding, is considered;
- additionally to aggregate grains' fineness their form is taken into account through their specific surface value;
- a physical concept of forming dense concrete mixture structure is used (cement paste fills the voids between the aggregate grains and creates on the grains a film with a certain thickness δ, which affects the concrete mixture workability).

It can be used in the design of concrete compositions and mortars, which are used in the manufacture of thin-walled concrete products, densely reinforced concrete, as well as in structures produced by 3D printing.

References

1. Belhadj, B., Bederina, M., Benguettache, K., Queneudec, M.: Effect of the type of sand on the fracture and mechanical properties of sand concrete. Adv. Concr. Constr. **2**(1), 13–27 (2014)
2. Bouziani, T., Bederina, M., Hadjoudja, M.: Effect of dune sand on the properties of flowing sandconcrete (FSC). Int. J. Concr. Struct. Mater. **6**(1), 59–64 (2012)
3. Kilic, A.: The influence of aggregate type on the strength and abrasion resistance of high strength concrete. Cement Concr. Compos. **30**(4), 290–296 (2008)
4. Abdullahi, M.: Effect of aggregate type on compressive strength of concrete. Int. J. Civ. Struct. Eng. **3**(2), 782–791 (2012)
5. Chandar, K.R.: Experimental investigation for partial replacement of fine aggregates in concrete with sandstone. Adv. Concrete Constr. **4**(4), 243–261 (2016)

6. Varsei, M.: Bending properties of finegrained concrete composite beams reinforced with single-layer carbon/polypropylene woven fabrics with different weave designs and thread densities. J. Text. Inst. **104**(11), 1213–1220 (2013)
7. Le, H.T., Nguyen, S.T., Ludwig, H.A.: Study on high performance fine-grained concrete containing rice husk. Ash. Int. J. Concr. Struct. Mater. **8**, 301–307 (2014)
8. Klyuev, S.V., Klyuev, A.V., Abakarov, A.D., Shorstova, E.S., Gafarova, N.G.: The effect of particulate reinforcement on strength and deformation characteristics of fine-grained concrete. Mag. Civ. Eng. **75**(7), 66–75 (2017)
9. Bazhenov, YuM: Concrete Technology. Matest, Moscow (2007). (in Russian)
10. Lvovich, K.: Sandy concrete and its using in construction. Stroi-Beton, S-Peterburg (2007). (in Russian)
11. Sizov, V.P.: Design of Normal-Weight Concrete Compositions. Stroyizdat, Moscow (1980). (in Russian)
12. Dvorkin, L., Dvorkin, O.: Basics of Concrete. St.-Petersburg, Stroybeton (2006). (in Russian)
13. Dvorkin, L., Dvorkin, O., Ribakov, Y., et al.: Using granite siftings for producing vibro-pressed fine-grained concrete. KSCE J. Civ. Eng. **21**, 2252–2258 (2017)
14. Neville, A.M.: Properties of Concrete, 5th edn. BookVistas, New Delhi, India (2012)

Maximum Daily Stormwater Runoff Flow Rates at the Inlet of the Lviv WWTP Based on the Results of Systematic Hydrologic Observations of the Catchment

Volodymyr Zhuk(✉), Lesya Vovk, Ivan Matlai, and Ihor Popadiuk

Lviv Polytechnic National University, Lviv 79013, Ukraine
volodymyr.m.zhuk@lpnu.ua

Abstract. Method of the estimation of maximum daily stormwater flow rates from the urban catchments of cities with combined sewerage system is presented in the paper. Full-scale sample survey of the territory of the Baltic Sea catchment of the Lviv city allowed to obtain the total area of technical watershed of the Baltic Sea catchment – 40.79 km^2. Based on the regression analysis of results of the survey, the power law dependency between the total and effective imperviousness of the catchment is obtained. Weibull model approximation, based on the actual data for the rainfall events in Lviv for the last 20 years, is obtained and can be recommended for the maximum daily rainfall depth estimation depending on the value of the return period. Estimated maximum flow rates of stormwater runoff at the inlet of Lviv WWTP as the function of the return period *P* in the range of 0.1 yr. to 5 yr. is presented.

Keywords: Stormwater runoff · Return period · Effective imperviousness · Initial abstraction · Potential maximum retention · Daily runoff coefficient

1 Introduction

Estimation of maximum daily stormwater flow rates from the urban catchments is important technical problem, especially in cities with combined sewerage system. Maximum daily discharges are widely used in the water management sector, in particular, for the calculation of the capacities of the sewerage regulating structures and wastewater treatment plants (WWTP) [1, 7, 10]. Due to the rational method stormwater volume is a function of the catchment's area, runoff coefficient and the rainfall depth.

Results of observations from about one hundred meteorological stations in Europe in 1946−1999 [9] showed the increasing of the mean annual precipitation depth on 0.76 mm/year and growth of wet days number on 0.04 1/year.

Cortesi et al. [4] found the statistical structure of daily precipitation in 1971–2010 using 530 daily rainfall series in different locations across Europe. Result was that the maximum daily values were located in the West Mediterranean and the minimum values were located in the Scandinavia.

Z. Blikharskyy (Ed.): EcoComfort 2020, LNCE 100, pp. 514–521, 2021.
https://doi.org/10.1007/978-3-030-57340-9_63

Banaszkiewicz et al. [3], based on data of 4 meteorological stations in the region of North-East Poland in the period of 1951−2000, found that annual rainfall height were decreased for only one station, but increased for other three stations. New results of more detailed study of the upper Vistula basin, Poland, were obtained by Młyński et al. [11], based on the data of 51 rainfall stations in this region. Considerable long-term increasing of the maximum daily rainfall depths is found for 7 rainfall stations and no significant trends for the other 44 stations. Thus, trends of extreme rainfall events are multi-vector across the Europe and they differ even in different locations of the same region. So, long-term trends in the daily maximum precipitation depths cannot be predicted using only general regional model, but they should be evaluated only empirically, based on the local observation data.

One of the important stages of modeling the stormwater drainage systems is a detailed analysis of the type of surface coverage of the urban runoff catchment. Total urban territory, as usual, is divided on to main types: water impervious covers and pervious areas. Analysis shows that total impervious areas (TIA) of the catchment should be divided into two sub-types: directly connected impervious area (DCIA) or effective impervious area (EIA) and non-connected impervious area (NCIA) [2].

Based on the analysis of the results of the study of 19 urban watersheds in the Denver metropolitan area (USA) a generalized formula was obtained in [2]:

$$p_{ef} = p_{tot}^{1.41}, \tag{1}$$

where p_{ef}, p_{tot} – the share of effective and total impervious areas.

In the paper [13] the territory of Marion County (USA), consisting of 100 identical plots with a total area of 900 km^2, was analyzed and other similar power law dependence was obtained:

$$p_{ef} = p_{tot}^{1.259}. \tag{2}$$

In 2017–2018, similar relationships were obtained between the total and effective imperviousness for a number of quarters of the city of Lviv [14, 15].

Many methods are used for determining the runoff coefficients for the design rainfalls. For example, Belov's empirical formula [6] is used in Ukraine for estimation the maximum discharge of the stormwater runoff from urban areas. Green-Ampt method and Horton method are most widely used in computer modeling of stormwater hydrographs [8], that allow modelling the change of the infiltration intensity of permeable covers over time.

CN-method, developed by SCS USDA [12], is the index-type method of estimating the daily runoff coefficients for small urbanized watersheds. CN-method takes into account not only the type of soil, but also depth of initial abstraction h_{in}, the saturation depth h_s and rainfall depth $h_{d.max}$ as function of the return period P. The ratio of initial abstraction h_{in} as usual is a function of saturation depth h_s, for example, in TR-55 [12] h_{in} = 0.2 h_s, resulting the equation:

$$\psi = \frac{h_Q}{h_p} = \frac{(1 - 0.2h_s/h_p)^2}{(1 + 0.8h_s/h_p)}, \tag{3}$$

where h_Q – a runoff depth; h_p – precipitation depth. Saturation depth h_s is determined as a function of the dimensionless curve number parameter *CN*.

2 Materials and Methods

The city of Lviv is located on the Main European watershed line (Fig. 1). Therefore, the Poltva River, originating at the territory of the city, is the main sewer collecting all types of wastewater from the Baltic Sea catchment of the city to the Lviv WWTP.

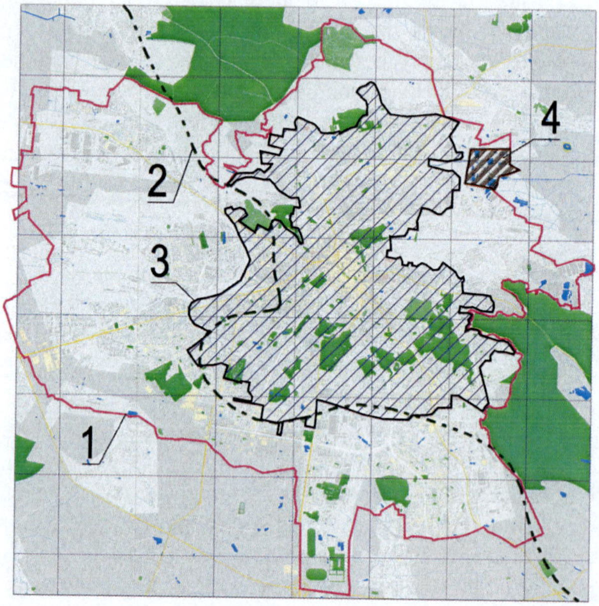

Fig. 1. The layout of the object of research: 1 – official borders of the Lviv city; 2 – Main European watershed (geographical watershed); 3 – technical watershed of the Baltic Sea catchment, 4 – territory of the Lviv WWTP

End to end ranking of the heaviest rainfalls in the Lviv city in the period of 1945–2018 was used for the statistical processing of maximum daily rainfalls. Dependencies of the maximum daily rainfall depths $h_{d.max}$ as functions of the return period *P* are analyzed using four most suitable statistical models: Weibull, Logistic Power, Hoerl and Rational Model.

Parameters *CN* for different cover types of Baltic Sea stormwater catchment of the Lviv city were found according [12] and presented in Table 1. A detailed analysis of the composition and infiltration properties of soils in the city of Lviv [5] makes it possible to classify all soils by 4 hydrological groups according to the SCS classification. The proportions of soils of different hydrological groups are as follows: group A − 13.43%, B − 82.75% and C − 3.82%. Parameter *CN* for all types of water impervious areas (roofs, roads etc.) was taken as equal 98 as usual.

Table 1. Values of the *CN*-parameter for pervious and impervious urbanized areas for soils of different hydrological groups in the city of Lviv

Type of covers	Hydrology group of soil by SCS	Hydrologic condition		
		Poor	Fair	Good
Pervious areas	A	68	49	39
	B	79	69	61
	C	86	79	74
Impervious areas	A, B, C	98	98	98

3 Results and Discussion

3.1 Results of Statistical Processing of the Maximum Rainfalls

As a result of statistical processing, the Weibull model approximation, based on the actual data for the last 20 years was obtained can be recommended:

$$h_{d.\max} = 72.1 - 67.5e^{-0.765 \cdot P^{0.667}} \quad (4)$$

Since the reconstruction of any WWTP have as a rule the long-term horizon of about 30–50 years, it is rational to extrapolate the $h_{d.max} = f(P)$ dependency for last 20 years taking into account current trends of future growth.

3.2 Results of Full-Scale Sample Survey of the Territory of the Baltic Sea Catchment of the Lviv City

Full-scale sample survey of the territory of the Baltic Sea catchment of the Lviv city allowed to obtain the total area of technical watershed of the Baltic Sea catchment – 40.79 km^2 (Fig. 1). By processing the high-resolution satellite image of the catchment and using the simple sampling method, after the analysis of 20000 elementary cell 2 × 2 m, the average value of the total imperviousness was found to be equal $p_{tot} = 0.508$.

To identify the relationship between the total and effective imperviousness within the Baltic Sea runoff catchment of Lviv, a systematic full-scale survey was conducted in 2019. Data on the distribution of types and sub-types of the territory of 75 quarters in 6 administrative districts of the Lviv city with a total area of about 1000 ha were obtained (Table 2).

Table 2. Results of full-scale sample survey of the territory of the Baltic Sea catchment of the Lviv city

City district	Total area, ha	Total impervious area, ha	Impervious areas, ha		Pervious area, ha	p_{tot}	p_{ef}
			Directly connected	Non-connected			
Frankivskyi	370.19	213.49	182.27	31.22	156.71	0.577	0.492
Halytskyi	304.04	193.10	180.37	12.73	110.94	0.635	0.593
Zaliznychnyi	71.34	54.48	51.67	2.82	16.86	0.764	0.724
Lychakivskyi	102.26	76.56	72.55	4.01	25.70	0.749	0.710
Sykhivskyi	106.99	46.43	33.03	13.41	60.55	0.434	0.309
Shevchenkivskyi	46.24	34.40	31.93	2.47	11.84	0.744	0.690
Total	**1001.05**	**618.46**	**551.81**	**66.65**	**382.59**	**0.618**	**0.551**

A graphical dependence between the total and effective imperviousness is presented in Fig. 2. Based on the regression analysis of results the strong power law dependency between the total and effective imperviousness was obtained:

$$p_{ef} = p_{tot}^{1.318} \tag{5}$$

Applying the regression Eq. (5) to the numerical value of the total imperviousness $p_{tot} = 0.508$, the estimated value of the effective imperviousness was found $p_{ef} = 0.410$.

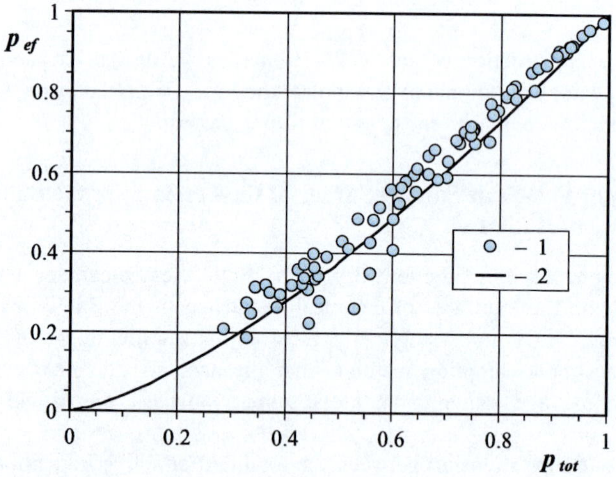

Fig. 2. Dependency between the total and effective imperviousness of the Baltic Sea catchment of the Lviv city: 1 – full-scale survey results of 75 subcatchments; 2 – power law approximation (5)

3.3 Estimated Daily Runoff Coefficients

The numerical simulation of the weighted average values of the daily runoff coefficients ψ of the territory of the Baltic Sea catchment of the Lviv city is performed for different values of the return period P in the range of $0.1 \leq P \leq 5$ yr. The simulation was done by Eq. (3) and takes into account the maximum rainfall depth of different return periods (4), total and effective imperviousness of the catchment, as well as the distribution of pervious covers by hydrological groups.

The calculated values of daily runoff coefficients for climatic and geological conditions of the city of Lviv for maximum daily rainfall events as functions of the return period P for different areas are presented in Fig. 3.

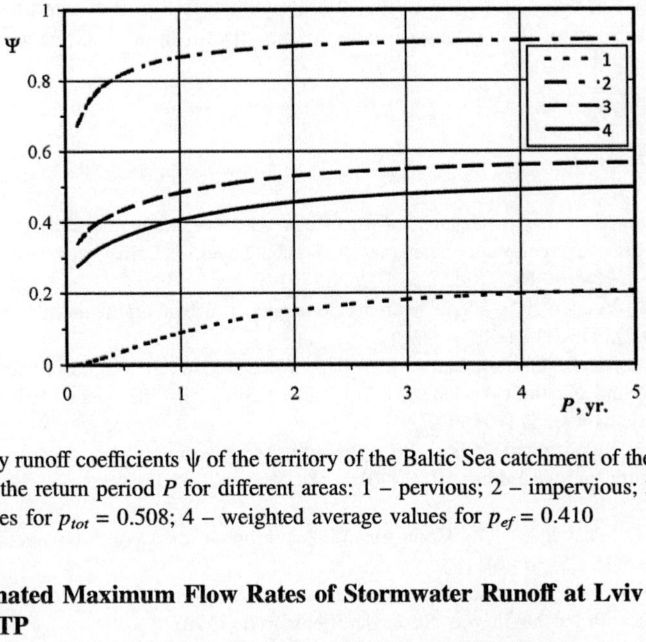

Fig. 3. Daily runoff coefficients ψ of the territory of the Baltic Sea catchment of the Lviv city as functions of the return period P for different areas: 1 – pervious; 2 – impervious; 3 – weighted average values for $p_{tot} = 0.508$; 4 – weighted average values for $p_{ef} = 0.410$

3.4 Estimated Maximum Flow Rates of Stormwater Runoff at Lviv WWTP

The application of the rational formula method [8] to the initial data presented above allowed to obtain the dependence between the maximum flow rates of stormwater runoff at the inlet of Lviv WWTP versus the return period P (Table 3).

Table 3. Estimated maximum flow rates of stormwater runoff at Lviv WWTP

P, yr.	0.1	0.2	0.33	0.5	1	2	3	4	5
$h_{d.max}$, mm/day	14.66	19.33	24.44	30.05	42.24	56.23	63.46	67.43	69.69
ψ	0.277	0.307	0.334	0.359	0.408	0.457	0.479	0.491	0.497
$W_{d.max}$, th.m^3/day	165.7	242.0	332.9	440.1	703.0	1048.2	1239.9	1350.5	1412.8

4 Summary

Full-scale sample survey of the territory of the Baltic Sea catchment of the Lviv city allowed to obtain the total area of technical watershed of the Baltic Sea catchment – 40.79 km^2. Based on the regression analysis of results of the survey, the power law dependency (5) between the total and effective imperviousness of the catchment is obtained.

Weibull model approximation, based on the actual data for the rainfall events in Lviv for the last 20 years (4), is obtained and can be recommended for the maximum daily rainfall depth estimation depending on the value of the return period.

Estimated maximum flow rates of stormwater runoff at the inlet of Lviv WWTP as the function of the return period P in the range of 0.1 yr. to 5 yr. is obtained (Table 3). These results can be used for the Lviv WWTP reconstruction. Presented method of the maximum stormwater flow rates estimation can be useful in other cities with combined sewerage system.

References

1. Adugna, D., Lemma, B., Jensen, M.B., Gebrie, G.: Evaluating the hydraulic capacity of existing drain systems and the management challenges of stormwater in Addis Ababa, Ethiopia. J. Hydrol. Reg. Stud. **25**, 100626 (2019)
2. Alley, W., Veenhuis, J.: Effective impervious area in urban runoff modeling. J. Hydraul. Eng. **109**(2), 313–319 (1983)
3. Banaszkiewicz, B., Grabowska, K., Panfil, M.: Characterization of atmospheric precipitation of Ilawa and Chelminsko-Dobrzynskie lake districts in the years 1951-2000. Acta Agrophysica **13**(3), 575–585 (2009)
4. Cortesi, N., Gonzalez-Hidalgo, J.C., Brunetti, M., Martin-Vide, J.: Daily precipitation concentration across Europe 1971–2010. Nat. Hazards Earth Syst. Sci. **12**, 2799–2810 (2012)
5. Chorna, D., Yavorska, G.: Comparison of microflora of Lviv urboland soils. Studia Biologica **5**(1), 25–36 (2011)
6. Dykarevskyi, V., Kurhanov, A., Nechaev, A., Alekseev, M.: Otvedenie i ochistka poverhnostnyih stochnyih vod. Strojizdat, Leningrad (1990)
7. Epps, T., Hathaway, J.: Establishing a framework for the spatial identification of effective impervious areas in gauged basins: review and case study. J. Sustain. Water. Built Environ. **4** (2), 05018001–1−11 (2018)
8. James, W., Rossman, L.: Water systems models User's guide to SWMM 5, 13 edn. CHI Press Publication, Ontario, p. 905 (2010)
9. Klein Tank, A.M.G., et al.: Daily dataset of 20th-century surface air temperature and precipitation series for the European climate assessment. Int. J. Climatol. **22**(12), 1441–1453 (2002)
10. Liebl, D.: Multi-stakeholder planning for stormwater management in the lake Wingra Watershed. In: Watershed Management Conference 2010, 721–732 (2010)
11. Młyński, D., Wałęga, A., Petroselli, A., Tauro, F., Cebulska, M.: Estimating maximum daily precipitation in the upper Vistula basin Poland. Atmosphere **10**(43), 1–17 (2019)
12. Cronshey, R.: Urban hydrology for small watersheds, TR-55. United States Department of Agriculture. Natural Resources Conservation Service (1986)

13. Yang, G., Bowlinga, L., Cherkauer, K., Pijanowski, B.: The impact of urban development on hydrologic regime from catchment to basin scales. Landscape Urban Plann. **103**, 237–247 (2011)
14. Zhuk, V., Vovk, L., Matlai, I., Popadiuk, I., Mysak, I., Fasuliak, V.: Dependency between the total and effective imperviousness for residential quarters of the Lviv city. J. Ecol. Eng. **21**(5), 56–62 (2020)
15. Zhuk, V., Vovk, L., Matlai, I., Popadiuk, I.: Correlation between the total and effective imperviousness in stormwater modelling. Sci. Bull. UNFU **28**(10), 92–95 (2018)

Author Index

Z. Blikharskyy (Ed.): EcoComfort 2020, LNCE 100, pp. 523–525, 2021.
https://doi.org/10.1007/978-3-030-57340-9